EQUALIZERS FOR DIGITAL MODEMS

To my wife, Pam

EQUALIZERS FOR DIGITAL MODEMS

A.P. Clark, *MA, PhD, DIC, MIERE, CEng*
Professor of Telecommunications,
Loughborough University of Technology

A HALSTED PRESS BOOK

JOHN WILEY & SONS
New York

Published in the U.S.A.
by Halsted Press, a Division of
John Wiley & Sons, Inc., New York

© A.P. Clark, 1985

SMU LIBRARY

Library of Congress Cataloging-in-Publication Data

Clark, A. P.
  Equalizers for digital modems.

  1. Equalizers (Electronics) 2. Modems. I. Title.
TK7872.E7C57   1985   621'398'14   86-14652
ISBN 0-470-20227-0

Printed in Great Britain

Preface

This book is concerned primarily with *equalizers* which are employed in digital data-transmission systems for correcting or removing the *distortion* introduced into the transmitted data-signal by a *linear* channel. A wide range of different *linear* and *decision-feedback* (nonlinear) equalizers are studied, including both conventional equalizers and some novel developments of these. The aim of the book is to give the reader a conceptual understanding of the *basic mechanisms* involved in both linear and decision-feedback equalization processes, thereby enabling him to identify the applications where equalizers can be used with advantage and to determine the particular types of equalizers suited to any given application. Since the *type* of signal distortion introduced by a channel critically affects the choice of equalizer for that channel, a detailed account is presented of the theory of signal distortion. To clarify this analysis, the more important properties of the discrete Fourier transform (DFT) are also derived.

The book has been written as a text for part of a course on digital communication systems, presented either in the final year of an undergraduate course or else in the first year of a postgraduate course. Much of the material in the book has been taught by the author in such courses over the past few years, and the experience gained in teaching this material has significantly influenced its presentation here. The book has also been written for private study by practising and student engineers who are interested in the design and development of digital data-transmission systems. With this in view the treatment of the subject is maintained, for the most part, at an elementary level, the more advanced topics being presented in separate sections and in such a way that they may be omitted without an undue loss in continuity. Care has also been taken in the mathematical analysis to include only those derivations that help to illustrate and elucidate the principles involved. Where no simple or helpful derivation is known to the author, the result is presented without proof. To assist further in the explanation of the more important principles and techniques, these are illustrated by numerous worked examples, by means of which the reader can test and enlarge his understanding of the topics covered. In order to make the worked examples as helpful as possible to the reader who finds the subject matter difficult to grasp, these are quite detailed and involve significant repetition of some of the material. Experience has shown this to be a most effective method of clarifying the basic principles. Considerable use is also made of the worked examples to develop and extend the theory presented in the main text. Problems or examples without worked solutions are not included, since in the author's view these are not usually very helpful in the study of advanced topics of the type considered here.

To avoid needless conceptual difficulties in the explanation of the techniques described, all derivations assume *real-valued* signals, the corresponding results for *complex-valued* signals being presented without proof. Although only a relatively small step is involved in moving from real-valued to complex-valued signals, the corresponding equalizers being also closely related, the precise operation performed by any given equalizer for real-valued signals is conceptually more straightforward than is that of the corresponding equalizer for complex-valued signals. Furthermore, the derivation of some of the optimum equalizers is significantly easier to grasp when using real-valued signals than when using complex-valued signals. Again, and for similar reasons, the equalization of bandpass signals is not studied here. With a clear understanding of the topics covered in this book, the reader should not experience much difficulty in extending his understanding to the more general cases of complex-valued and bandpass signals. To give some idea of the more detailed contents of the book, the material covered in the individual chapters will now be outlined.

Chapter 1 contains some of the more important background theory needed for the proper understanding of the following chapters. Consideration is given here to the possible practical applications of equalizers and to the properties of the actual channels that are likely to be used. The model of the channel assumed throughout the book is studied in some detail, as are the optimum detection process for a received signal message and a matched-filter detector. The relevant properties of vectors and matrices are briefly outlined.

Chapter 2 introduces z-transforms and convolution matrices. It then analyses the discrete Fourier transform (DFT) and derives its more important properties. Particular emphasis is given here to the nature of the DFT matrix.

Chapter 3 uses the results derived in Chapter 2 to develop a theory of signal distortion, which is required for the study of both linear and decision-feedback equalizers. This theory provides an analysis of the composition (or structure) of a received and sampled signal element that has been linearly distorted in transmission, and the theory thereby itself suggests some of the basic functions that must be performed by a good equalizer.

Chapters 4 and 5 form the heart of the book and are concerned, respectively, with linear and decision-feedback (nonlinear) equalizers. Various different arrangements of each type of equalizer are analysed and compared. Numerous worked examples are included here to clarify and illuminate the material presented.

Chapter 6 is concerned with more advanced topics in the theory of linear and decision-feedback equalizers. It considers the conventional methods for adaptively adjusting linear and decision-feedback equalizers. It then studies double-sampling equalizers and various techniques of sharing the equalization process between the transmitter and receiver. Finally, it compares the performances of the more important of the different equalizers studied.

Chapter 7 is concerned with developments of the conventional decision-feedback equalizer, that enable an improved tolerance to noise to be achieved. The systems studied here include those using more sophisticated detectors and also those where a simple threshold-level detector is used, as in a conventional equalizer, but where the design of the equalizer itself is modified. In addition, a description is presented of the Viterbi-algorithm detector, that achieves maximum-likelihood detection of the received signal message, and also of a near-maximum-likelihood detector, that can often achieve a performance close to that of the Viterbi-algorithm detector but with far less complex equipment. The performances of the systems are compared over several different channels, to illustrate the advantages that are likely to be gained by such techniques.

Chapters 1, 4 and 5 are written at a significantly more elementary level than the remaining chapters, and, with the omission of some of the more advanced topics covered, are well suited for use as part of a final-year undergraduate course on digital communication systems. Chapters 2 and 3 are probably the most demanding of the different chapters and are therefore probably better suited for presentation as part of a first-year postgraduate course.

The various available techniques of synchronizing the receiver on to the received signal, for both element timing and carrier phase, are not studied in this book, since an adequate description of synchronization methods requires a somewhat wider treatment than could conveniently be accommodated here. For similar reasons, no consideration is given to error correcting or detecting codes or to the various combinations of modulation and coding that can be used to improve the tolerance of a modem to noise. Finally, the book is concerned essentially with systems and techniques and not with the practical implementation of these. No details are therefore given of the practical hardware, software, circuits or devices that can be used. The emphasis throughout is on basic principles and on the operations or algorithms performed by the modem, without however going into the practical details of precisely how these are best implemented. All systems studied in the book involve the application of computer-like techniques to a sampled digital signal. Following the sampling of the received signal, the receiver operates on a sequence of numbers and therefore it becomes a special purpose digital computer. Such systems can now be implemented remarkably cheaply, leading to highly cost-effective designs.

This book can be considered as a companion volume to 'Principles of Digital Data Transmission' written by the same author. The latter covers a much wider range of topics and is concerned primarily with simple modems, operating over channels that introduce only limited signal distortion, such that equalizers or more sophisticated detectors are not required. There is very little common material between the two books. On the other hand, this book contains part of the material in Chapters 1, 2 and 4 of 'Advanced Data-Transmission Systems' by the same author. However, the latter material has been substantially modified and much new material added.

The original work reported in this book was carried out in the Department of Electronic and Electrical Engineering of Loughborough University of Technology, the extensive computing facilities required being provided by the University Computer Centre. The author gratefully acknowledges the facilities provided by these organisations. The author would also like to acknowledge the valuable contributions by U.S. Tint, M.N. Serinken, M.N.Y. Shum, L.H.C. Lee, R.S. Marshall, B.A. Hussein, K. Parama Raj and M. Slater in the study of the different systems by computer simulation and in the detailed assessment of these systems. The results of the various computer-simulation tests presented in the book, together with some particular items of the theoretical analysis, have been taken from six published papers, details of which are given at the appropriate points in the text. Three of these papers were published by the IEE and the other three by the IERE. The author is indebted to these institutions for their kind permission to include in the book the required material from the published papers. Finally, especial thanks are due to Eileen Barnes for the excellent diagrams and to Mrs. P.G. Bentley and Mrs. A. Hammond for their flawless typing of the manuscript.

A.P. Clark

Contents

1 Background theory — 1
1.1 Introduction — 1
1.2 Model of channel — 11
1.3 Vectors and matrices — 18
1.4 Optimum detection process — 27
1.5 Matched-filter detector — 34
1.6 Example — 42

2 The discrete Fourier transform — 50
2.1 Introduction — 50
2.2 z-transforms — 54
2.3 Convolution matrix — 59
2.4 Definition of the DFT — 61
2.5 Basic properties of the DFT — 65
2.6 Convolution of two sequences — 70
2.7 Linearity of the DFT — 73
2.8 Shift theorem — 73
2.9 Real and imaginary components of a DFT — 79
2.10 Reversal theorem — 84
2.11 Aperiodic autocorrelation function and discrete energy density spectrum — 87
2.12 Multiplication theorem — 91
2.13 Signal energy — 92

3 Theory of signal distortion — 94
3.1 Introduction — 94
3.2 Phase distortion — 98
3.3 Representation of phase distortion by the zeros and poles of the z-transform — 106
3.4 Amplitude distortion — 115
3.5 Representation of amplitude distortion by the zeros and poles of the z-transform — 119
3.6 Assessment of the distortion introduced by a linear baseband channel — 125

	3.7	Complex-valued signals	138
	3.8	Examples	148
4	**Linear equalizers**	**156**
	4.1	Introduction	156
	4.2	Basic assumptions	157
	4.3	Feedforward transversal filter	163
	4.4	Zero forcing	167
	4.5	Separate equalization of the two factors of the channel response	168
	4.6	Examples	183
	4.7	Feedback transversal filter	188
	4.8	Examples	193
	4.9	Tolerance to additive Gaussian noise	195
	4.10	Examples	199
	4.11	Equalizer that minimizes the mean-square error in the equalized signal	205
	4.12	Linear equalization of phase distortion	217
5	**Decision-feedback equalizers**	**230**
	5.1	Decision-directed cancellation of intersymbol interference	230
	5.2	Example	236
	5.3	Tolerance to additive Gaussian noise	239
	5.4	Examples	246
	5.5	Zero forcing equalizer	253
	5.6	Linear and nonlinear equalization of the two factors of the channel response	257
	5.7	Equalizer that minimizes the mean-square error in the equalized signal	262
	5.8	Equalizer that maximizes the signal/noise ratio in the equalized signal, subject to the exact equalization of the channel	275
	5.9	Examples	299
	5.10	Equalizer for severe amplitude distortion	307
6	**Various topics on linear and decision-feedback equalizers**	**324**
	6.1	Adaptive equalizers	324
	6.2	Double-sampling equalizers	334
	6.3	Equalization process shared between transmitter and receiver	356
	6.4	Comparison of different equalizers	382

7 Developments of the conventional decision-feedback equalizer 387
7.1 Introduction 387
7.2 Conventional equalizer 389
7.3 Modified equalizer 393
7.4 Computer-simulation tests 400
7.5 Simplified equalizer 409
7.6 Assessment of systems 419
7.7 Example 433
7.8 Near-maximum-likelihood detectors 435

Chapter 1

Background theory

1.1 INTRODUCTION

With the rapidly increasing use of digital computers and the rapid development of techniques for the digital coding of speech and video (picture) signals, there is at the present time considerable interest in techniques for transmitting digital signals at the highest convenient rates over a wide range of channels and at transmission rates varying from a few hundred bit/s (bits-per-second) to several hundred Mbit/s (10^6 bits-per-second)[1-14]. Perhaps the simplest and most widely used techniques for maximizing the transmission rate over any given channel are those of linear and decision-feedback (nonlinear) equalization[7,12]. These can often enable the transmission rate over the channel to be increased by as much as some five or ten times, leading to a much more effective use of the available transmission paths.

An important piece of equipment involved in the transmission of signals is a *modem* (modulator–demodulator), one of which is placed at each end of the channel carrying the signals[6,7,13]. A *digital modem* is the combination of a *transmitter* and *receiver* that are used to convey information in the form of *digital signals* from one location to another over the appropriate channels, as, for instance, shown in Fig. 1.1. Information is here transmitted in both directions between the two modems over two separate channels. The source of the information fed to the transmitter of a modem and the destination of the information fed from the receiver may involve signal processing in the form of encoding and decoding for error protection and perhaps also, or alternatively, the appropriate analogue-to-digital and digital-to-analogue conversion, if the modems are used for the transmission of digitally coded speech or video signals. The equipment associated with a modem often contains a store which holds the information transmitted or received. The transmitter of a modem converts the input information into the corresponding signal waveform that is suitable for transmission over the given channel. The receiver

2 BACKGROUND THEORY

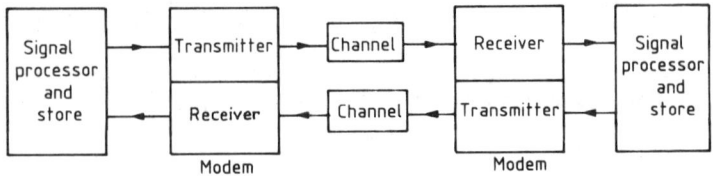

Fig. 1.1 *Transmission of digital data*

converts the received signal waveform into the corresponding information, which is then fed to the associated equipment.

The aim in the design of a modem for the transmission of data over a given channel is to maximize the rate of transmission of information over the channel, subject to an acceptably low error rate in the information extracted from the received signal by the modem at the other end of the channel. However, before considering this problem further, it is necessary first to clarify the meanings of some of the terms used.

The information transmitted by a modem is carried by the *data-symbols* $\{s_i\}$, where $i = 0, 1, 2, \ldots, l-1$, or where i ranges over any other appropriate set of integer values. The small curly brackets $\{.\}$ indicate a sequence or set. Each data-symbol s_i may have any one of m possible values, such that

$$s_i = -(m-1)k, \ -(m-3)k, \ -(m-5)k, \ \ldots, (m-1)k \qquad (1.1)$$

where k is a positive constant. Except where otherwise stated, it is however assumed here that the $\{s_i\}$ are binary (2-level) data symbols, so that

$$s_i = \pm k \qquad (1.2)$$

It is furthermore assumed throughout that the data-symbols $\{s_i\}$ are *statistically independent* and equally likely to have any of their possible values. It follows that the $\{s_i\}$ are *uncorrelated* and have *zero mean*, which implies that they are also *statistically orthogonal*[15-17]. In other words, the expected (or average) value of the product of any two data symbols is zero, as is the expected value of any individual data symbol. A data symbol is, of course, being treated here as a *random variable*, whose statistical properties are as just described, rather than the actual transmitted value of the data symbol, which is a *sample value* of the given random variable. A data symbol is carried by a suitably shaped pulse known as a *signal element*. Thus, for example, the data-symbol s_i (Equation 1.2) could be carried by the signal-element $s_i y(t-iT)$, where $y(t)$ varies in a given manner with time t. The $\{s_i\}$ are here transmitted in sequence and at regular intervals of T

seconds. The resulting signal is said to be an *m-level* signal. The term 'level' does not now refer to the signal amplitude, which may of course be constant, as, for instance, when $s_i = \pm k$ and $y(t)$ is suitably shaped. The simplest digital signal is the 2-level (binary) signal of Equation 1.2. A careful distinction must be made between the received *waveform* $s_i y(t - iT)$ of a signal element, which is considered here as the signal element itself, and the data-symbol s_i that is represented or carried by the signal element. An important reason for this distinction is that the *shape* (waveform) of any given signal-element may be considerably altered as a result of its transmission over the channel and, for certain channels, the resultant shape may also vary with time. Thus the shape of a received signal-element is not necessarily related directly or even in any very simple manner to the symbol represented or carried by that element.

The data carried by the entire group of transmitted signal-elements form a *message*. In most practical applications the data symbols are themselves arranged in separate groups, each group forming a *character* or *word*. Where the message is 'alpha-numeric' it is composed of a sequence of characters, each of which is a letter or numeral. Some of these characters may additionally be punctuation marks and instructions such as 'space' and 'new-line'. In the case of a digitally-coded speech signal, each character or word is a sample of the speech waveform. However, in order to keep the treatment here as general as possible, a transmitted message will be considered as being composed of a sequence of l data-symbols, which are not taken to be grouped into separate characters or words, at least as far as the modem is concerned.

The signal elements, each of which carries the corresponding data-symbol, are normally transmitted in sequence, one following another, to give a *serial* system. Sometimes, however, the signal elements are transmitted in separate groups, the elements in each group being sent simultaneously, to give a *parallel* system. Most often in a serial system the signal elements are transmitted at a steady rate of a given number of elements per second (bauds), the receiver being held in time synchronism with the received signal. Such a system is known as a *synchronous serial system* and, in applications where a relatively high transmission rate is required over a given channel, it is the most commonly used of the different systems.[13] This is the arrangement that is assumed throughout the book.

Consider now a sequence of m-level data-symbols, where $m = 2^n$ and n is a positive integer. A data symbol can here be represented by a sequence of n binary digits (bits), each of which has a possible value of 0 or 1, there being a unique one-to-one relationship between the symbol and n binary digits. Thus, for example, if $m = 8$ and the possible values of a data symbol are $-7k, -5k, -3k, \ldots, 7k$, then $n = 3$ and the corresponding sequences of three binary digits can be taken to be 000, 001, 010, ..., 111. The m-level symbol is here coded into the corresponding sequence of n binary digits, where $n = \log_2 m$. If the data symbols are statistically independent and equally likely to have any of their possible values, as is assumed throughout this book, then so also are the binary digits into which these are coded. Each binary digit now carries *one bit of information*. A 'bit' here is a unit measure of information and is not to be confused with the binary digit itself. However, under the assumed conditions, the number of binary digits into which a data symbol can be coded becomes the *same* as the number of bits of information carried by the data symbol. The term 'bit' can therefore be taken to have *either* meaning. Clearly, the information content per data symbol is n bits, where $n = \log_2 m$. If now the data-symbol rate is b symbols per second, the transmission rate (information rate) of the signal is $b \log_2 m$ bit/s (bits-per-second). Thus the transmission rate can be increased by raising the data-symbol rate b, the number of levels m, or both of these together. However, it can be seen that to double the transmission rate for a given symbol rate, the number of levels m must be replaced by m^2, which means that when m is already large, say 64, no very useful increase in transmission rate can be achieved by any further increase in m, without an undue increase in equipment complexity and an excessive reduction in tolerance to noise.[6,13] On the other hand, to double the transmission rate for a given value of m, the data-symbol rate b, and hence the signal-element rate, must be doubled. But the signal bandwidth is *proportional* to the signal element rate, so that the doubling of the signal-element rate also doubles the *signal bandwidth*. Furthermore, any bandlimiting of the transmitted signal by the channel tends to increase the duration (length in time) of each individual signal-element in an effect known as *time dispersion*. The latter, in turn, adversely affects the detection of any individual data symbol from the corresponding received signal element, when a simple conventional detector is used, because the time dispersion introduces, at the detector input, interference components that originate from the neighbouring signal elements and that are therefore dependent on the neighbouring data symbols. This effect is known as *intersymbol interference*[6,13]. If the signal-element rate is already close to the maximum that can be carried by the given channel, without significant intersymbol interference in the detection of a data-symbol, then the doubling of the signal-element rate greatly increases this interference. One of the aims of the techniques to be described in this book is to remove the intersymbol interference, or at least to reduce it to a negligible level, and hence to enable a higher transmission rate to be achieved.

Most of the data-transmission systems studied in the book can be used not only over baseband channels, such as coaxial cables and twisted pairs, but also over bandpass channels, such as telephone circuits and HF radio links. Voiceband telephone circuits designed for speech are of particular importance because of the highly developed network of the channels that already covers much of the world. The bandwidth of a voiceband channel is typically no wider than 200–3400 Hz, the effective or useful bandwidth being often much less. Furthermore, even though a voiceband channel normally has a bandwidth appreciably wider than the centre frequency, it is not well suited to the transmission of baseband signals. The other bandpass channels of interest here are generally narrowband channels and therefore quite unsuited to the transmission of baseband signals. Thus, over all bandpass channels considered here, it is necessary to use *modulated-carrier* signals. For reasons to be explained shortly, some form of *amplitude modulation* (AM) must in fact be used, the carrier frequency being selected to place the transmitted signal spectrum in the centre of the available frequency band.

In order to achieve the *most efficient use of bandwidth*, together with simple and effective detection processes, it is necessary to use a modulation method in which *both* the modulation and demodulation of the data signal are *linear* operations. There are two alternative linear systems that are generally preferred. The first of these uses a vestigial sideband suppressed carrier AM signal, and the second uses two double-sideband suppressed carrier AM signals whose carriers are in phase quadrature (at the same frequency and at a phase angle of 90°) and form a quadrature amplitude-modulated (QAM) signal. Coherent demodulation is used in each case.[13] A vestigial-sideband signal is a modulated-carrier signal in which a large part of one sideband is removed, so that only one sideband and a vestige of the other are transmitted. In this signal, the carrier cannot itself be used to assist in the detection of the received data signal, so that for the best tolerance to additive noise, at a given transmitted signal power level, the signal carrier is removed. Special arrangements must however now be made to ensure the correct carrier-phase synchronization of the coherent demodulator on to the received signal. Usually these involve the addition of a low-level pilot carrier to the transmitted signal.[13]

When a vestigial-sideband suppressed carrier AM signal is transmitted, the transmission path together with the modulator (at the transmitter) and the demodulator (at the receiver) appear as a *linear baseband channel*[18]. The theoretical analysis of the resultant system therefore reduces to that of a simple linear baseband system, just so long as the correct synchronization of the coherent demodulator on to the received signal is assumed. When two double-sideband suppressed carrier AM signals are transmitted in phase quadrature to give a QAM signal, the transmission path, together with the two modulators at the transmitter and the two coherent demodulators at the receiver, appear as two linear baseband channels, which are coupled (interconnected) in an appropriate manner. Furthermore, by considering one of these baseband channels as carrying a real-valued signal and the other as carrying an imaginary-valued signal, the two baseband channels can be considered as a single baseband channel carrying a *complex-valued* signal (a signal having both real and imaginary components) and the channel having a complex-valued impulse response[18]. The theoretical analysis of the resultant system therefore again reduces to that of a linear baseband channel. Since the emphasis throughout the book is on *real-valued* signals that make an efficient use of bandwidth, it is assumed that whenever the transmission path is a bandpass channel, a suppressed-carrier AM signal is used and, except where otherwise stated, this is assumed to be a vestigial-sideband signal. The resultant channel is therefore in every case a baseband channel with a real-valued impulse-response, and only real-valued baseband signals are used. Modulated-carrier signals will not therefore be considered further. They are studied in detail elsewhere[5-10].

This book is concerned primarily with *equalizers* whose function is to correct the *linear distortion* introduced into a digital signal when the latter is transmitted at a relatively high rate over a linear baseband channel. The digital signals could carry data or could be digitally encoded speech or video signals. The channel could be a coaxial cable or twisted pair, or perhaps a fibre optic link. Alternatively, it could include a linear modulator at the transmitter, a bandpass transmission path and a linear coherent demodulator at the receiver. The bandpass transmission path could be a voiceband telephone circuit or HF radio link. A telephone circuit itself is normally made up of two or more links connected in tandem (end-to-end)[13]. These links may be various different types, such as unloaded audio, loaded audio, carrier and PCM links, where the latter two involve wideband channels. A wideband channel may be a microwave, satellite, coaxial cable, twisted pair or fibre-optic link and often carries a large number of voiceband channels. These channels can be either time-division multiplexed (TDM), as when PCM links are used, or else frequency-division multiplexed (FDM), as when carrier links are used. Again, the bandpass transmission path could be a radio or satellite link designed for either 'mobile' or 'business' systems, provided only that a linear modulation method is used at the transmitter and that no significant nonlinear distortion is applied to the transmitted modulated-carrier signal constraining this to have a constant or near-constant envelope. Unfortunately, in the case of many mobile and business systems, severe nonlinear distortion is introduced into the transmitted signal by the high power amplifier at the transmitter output, thus preventing the use of the techniques described in this book. The transmission rates used over voiceband channels lie typically in the range 600–19,200 bit/s. A mobile system may use a bandwidth wider than that of a voiceband channel, permitting transmission rates as high as 512 kbit/s to be achieved, if required, and a business system could operate in the range 1–8 Mbit/s. Over wideband channels involving transmission paths such as coaxial cables, fibre-optic links, microwave links and satellite links, much higher transmission rates can be achieved, typically up to several hundred Mbit/s.

The attenuation and group-delay distortions in the frequency characteristics of the channels of interest here may, in some cases, be such as to reduce the effective bandwidth of the channel to less than half its nominal bandwidth, thus seriously degrading the performance of any data-transmission system operating at a relatively high transmission rate but not taking due account of the resulting signal distortion. In principle, the distortion in the channel characteristics may be corrected by means of a linear filter whose attenuation and group-delay characteristics are the inverse of those of the channel over the frequency band of the signal, this filter being known as a *linear equalizer*. The problem here is that the distortion may vary considerably from one channel to another, thus preventing the use of a *fixed* equalizer in a modem operating, in turn, over the different channels. In some cases as, for instance, HF radio links, the distortion may even vary considerably *with time*. The particular feature of the equalizers considered here is that they may be adjusted *automatically* or *adaptively*, such that the receiver requires no prior knowledge of the channel and can track variations with time in the channel characteristics. Furthermore, decision-feedback (nonlinear) equalizers can now be used and these are capable of handling considerably more severe distortion than linear equalizers, so that a greatly increased transmission rate can be achieved.

In view of the wide range of channels considered here, there is a correspondingly wide range of different types of *noise* that may be introduced by the channel into the transmitted data signal. The noise may include both *additive* and *multiplicative* components, the latter involving both amplitude and frequency modulation effects[13]. In the case of additive noise, the noise waveform is *added* to the signal waveform, and, in the case of *multiplicative* noise, the noise waveform *amplitude modulates* or else *frequency modulates* the signal waveform. Where the transmission path introduces a *Doppler shift* into the data signal, the latter is modulated *in time*, but, for practical purposes, this can normally be considered as the corresponding *frequency modulation* of the data signal, which here appears as a *frequency offset* (shift in frequency) of the signal carrier.

Perhaps the most important type of additive noise is *impulsive noise*, and this can often predominate over the other types of noise present[13]. Unfortunately, the shape of a typical noise pulse varies widely from one channel to another and its duration may well extend over many adjacent signal-elements, making the noise difficult to simulate reliably on a computer. Furthermore, a burst of impulsive noise of sufficient magnitude and duration, added to the data signal, will cause errors in the detected data-symbols regardless of the modulation method and detection process used. The most important types of multiplicative noise (particularly over telephone circuits) are probably the following transient effects: transient interruptions, sudden signal-level changes and sudden carrier-phase changes[13]. When a transient interruption (which is a temporary loss of signal) exceeds a certain duration or when a sudden level or phase change exceeds a certain magnitude, errors are very likely to occur in the detected data-symbols of the data-transmission systems of the type assumed here. Any advantage gained in making the data-transmission system immune to either sudden signal-level changes or sudden carrier-phase changes, through the appropriate choice of modulation and demodulation processes, is usually (but not always) much more than offset by the corresponding reduction in tolerance to other types of noise[12,13]. Indeed, rather than attempt to make the receiver immune to any given transient effect, it is best to design the receiver so that it recovers as quickly as possible from any transient effect, thus minimizing the length of the ensuing error burst. Extensive tests have been carried out by J.D. Harvey on various techniques for enabling a modem of the type studied here to recover rapidly from transient effects, and some very encouraging results have been obtained.[19] Prolonged error bursts resulting from sudden phase changes in the received signal carrier can be avoided by the appropriate differential coding of the transmitted data signal[13], with only a negligible reduction in tolerance to additive noise, at high signal/noise ratios. Again, a correctly designed receiver should not give an unduly prolonged burst of errors in the detected data-symbols, in response to a transient interruption or sudden signal-level change, except when these become exceptionally severe.

A further type of multiplicative noise is the *frequency offset* (which is a shift of typically up to 1 to 2 Hz in the signal carrier frequency) that occurs in a data signal after transmission over a telephone circuit containing one or more carrier links or over any radio link. This can usually be corrected to the required degree of accuracy in an appropriately designed coherent demodulator, so that frequency offset should not normally affect the performance of a data-transmission system[13].

In view of the preceding considerations and bearing in mind that our main concern here is with the equalization process at the receiver, for general use rather than for any particular application, the various types of additive and multiplicative noise just mentioned will be ignored, and it will be assumed that the only noise introduced in transmission is stationary white Gaussian noise which is added to the data signal at the output of the transmission path. This noise has a constant power spectral density (average power per unit bandwidth) over all frequencies, and is easily analysed theoretically and simulated on a computer. It also has the important property that it is the least predictable of the different types of stationary noise and is therefore, in this sense, the most random[2-4]. Furthermore, the available evidence suggests that, although the tolerance of a data-transmission system to additive white Gaussian noise is not necessarily a good measure of its actual tolerance to the additive noise over any particular type of channel, the relative tolerances to white Gaussian noise of different data-transmission systems are a reasonably good measure of their relative *overall* tolerances to the additive noise over the different types of channel. Finally, at low error rates, additive white Gaussian noise tends to search out the weak points in any detection process, since most errors now tend to occur with the transmitted sequences of data-symbol values for which the detection process is weakest, and hence with the appropriately shaped noise waveforms of minimum energy that can cause errors in the detection of the received signal[12]. It follows that a detection process that has a good tolerance to additive white Gaussian noise is unlikely to have an unacceptably poor tolerance to the additive noise experienced over any practical channel.

This book is not concerned with the design and detailed characteristics of the various transmission paths themselves not is it concerned with the practical implementation of modems. The emphasis throughout is on the basic principles and techniques involved in the equalization of a linear channel. Furthermore, the problems of synchronizing the receiver on to the received signal, for both signal-element timing and signal-carrier phase, are not considered here. These are major problems in their own right, whose investigation is by no means simple and is not closely related to the main topics studied in the book. Considerable work has been done in this area[19,20]. It is therefore assumed throughout that the receiver is correctly synchronized on to the received signal, for both element timing and carrier phase.

Suppose now that a received signal-element is given by $sy(t)$, where $y(t)$ determines the shape of the element waveform and the multiplying parameter s is the data symbol carried by the signal element. It is assumed that $y(t)$ is nonzero only over the time interval $0-T$ (measured in seconds). If the signal is received in the presence of additive white Gaussian noise and in the absence of other signals, the detection process that minimizes the probability of error in the detection of s, feeds the received signal to a linear filter with impulse response $cy(T-t)$, where c is a constant, and samples the output signal from the linear filter at time T. The sample value is compared with the appropriate threshold (or thresholds) to give the detected value of s[5,8-10,13]. The filter maximizes the signal/noise ratio at its output at time T, and is known as a *matched filter*. The whole detection process is one of *matched-filter detection*, and no other detection process can give a lower probability of error in the detection of s[5,8-10,13].

In the detection of a received serial stream of signal elements, the $(i+1)$th of which is $s_i y(t-iT)$, where the $\{s_i\}$ are statistically independent and equally likely to have any of their possible values, and where the signal-elements do not overlap in time, the arrangement of matched-filter detection can be used for each individual received signal element, in turn, and gives the best available tolerance to additive white Gaussian noise[13]. Practical detectors are often designed to approximate to the arrangement just described, even though the additive noise may not be Gaussian, and they usually give a good performance.

The effect of distortion in the attenuation-frequency and group-delay-frequency characteristics of the transmission path is to spread out the individual transmitted signal-elements in time, so that the individual signal-elements at the receiver input overlap each other. Thus, in the detection of a received data-symbol by a matched-filter detector, the output signal from the matched filter contains, in addition to the wanted signal and the noise, intersymbol-interference components that originate from the neighbouring data-symbols. These interfere with the detection of the wanted data-symbol and reduce the tolerance of the system to noise. They may even prevent the correct detection of the received signal in the complete absence of noise.

It is well known that[1-14], over a channel that passes frequency components from 0 to B Hz at a significant (usable) level but no frequency components outside this band, the maximum element rate of a digital signal, for no intersymbol interference in a linear receiver, is $2B$ bauds (elements per second). The linear receiver here is an ideal lowpass filter followed by a sampler. The filter has a constant low attenuation over the frequency band $0-B$ Hz and a very high (ideally infinite) attenuation over all higher frequencies, together with a linear phase characteristic over its passband. The sampler samples the output signal from the lowpass filter at regular intervals of $1/2B$ seconds. With m-level data-symbols that are statistically independent and equally likely to have any of their m possible values, the maximum transmission rate for no intersymbol interference in the samples at the output of the linear receiver is now $2B \log_2 m$ bit/s. If the received signal is sampled over the whole of its duration, it can be accurately reconstructed from the samples and without the use of any further knowledge of the signal. This necessarily implies that all the information in the received signal waveform is contained also in the samples[1-10]. In theory, a strictly bandlimited signal has infinite duration, but the signal at the output of a linear channel with finite bandwidth, resulting from a signal of finite duration at its input, has for practical purposes also a finite duration, and so need only be sampled over this period in order to be able to reconstruct the received waveform and therefore in order to extract all the information from this waveform. Hence, if the detector operates in an appropriate manner on the *samples* of the received waveform rather than on the received waveform itself, there need be no noticeable loss in tolerance to noise and distortion, provided only that a sufficient number of the samples is used. The recent developments in microprocessors and microcomputers mean that it is now often much simpler to handle signals as the corresponding sequences of samples, rather than as continuous waveforms, so that powerful detection processes can be implemented in this way without the use of excessively costly equipment. Thus, following the sampling of the received signal, the receiver operates on a sequence of numbers and therefore it becomes a special purpose computer. This book is concerned entirely with such techniques.

1.2 MODEL OF CHANNEL

To analyse theoretically the performance of a data-transmission system the idealized model of the system in Fig. 1.2 is used. The information to be transmitted is carried by the data-symbols $\{s_i\}$, which are fed to the transmitter in the form of the corresponding

12 BACKGROUND THEORY

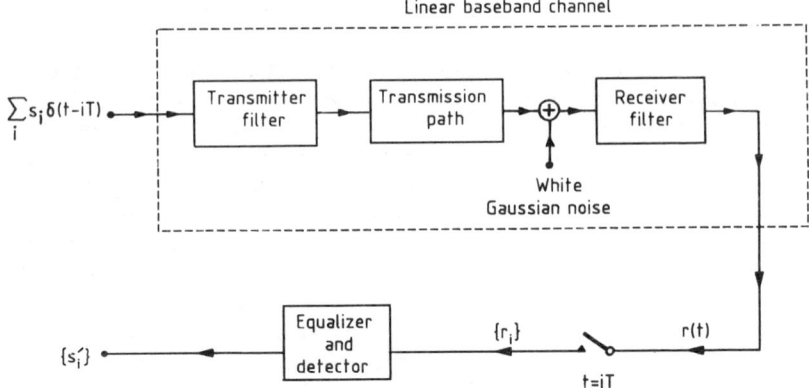

Fig. 1.2 Model of data-transmission system

impulses $\{s_i\delta(t-iT)\}$. The $(i+1)$th signal-element at the input to the transmitter filter is the impulse $s_i\delta(t-iT)$ which has area s_i and occurs at the time instant $t=iT$. Except where specifically stated to the contrary, it is assumed throughout this book that the *first* data symbol of any message is transmitted at time $t=0$, the symbol t being taken to represent *time* measured in *seconds*. The transmitter filter limits the frequency band of the data signal at its output to the available frequency band of the transmission path. The average power (mean-square value) of the data signal at this point is limited to some given value that is determined by the transmission path. All source and load impedances are, for convenience, taken to be purely resistive and equal to one ohm, so that the average power (in watts) of any continuous waveform is equal to its *mean-square value* (in volts or amps squared). The transmission path in Fig. 1.2 is a linear baseband channel that could be a twisted pair or coaxial cable, or could comprise a linear modulator (at the transmitter), a bandpass channel such as a telephone circuit or HF radio link, and a linear coherent demodulator (at the receiver). The representation of a bandpass channel, together with the associated linear modulator and demodulator, as a baseband channel is considered in some detail elsewhere[18], for the general case of a baseband channel with a *complex-valued* impulse response. The baseband channel here has a *real-valued* impulse response and is a special case of the other. The transmitter filter, transmission path and receiver filter together form a linear baseband channel whose impulse response is the real-valued waveform $y(t)$ and has, for practical purposes, a finite duration. It is assumed that $y(t)$ is time-invariant over any one transmission, in the sense that $y(t-iT)$ is a time-shifted version of $y(t-hT)$, for any integers $i \neq h$. Alternatively, and in the particular case of HF radio links, the shape of $y(t-iT)$ may vary slowly with i. It is assumed that the only noise introduced in transmission is stationary white Gaussian noise with zero mean and a flat (frequency independent) two-sided power spectral density of $\frac{1}{2}N_0$, which is added to the data signal at the output of the transmission path, to give the band-limited Gaussian noise waveform $w(t)$ at the output of the receiver filter. The filter is designed to remove as much noise as possible from the received data signal without unduly distorting the latter, to give the noisy and often distorted baseband data signal

$$r(t) = \sum_i s_i y(t-iT) + w(t) \qquad (1.3)$$

where $r(t)$, s_i, $y(t-iT)$ and $w(t)$ are all *real-valued*. The waveform $r(t)$ is sampled once per data symbol, at the time instants $\{iT\}$, to give the received samples $\{r_i\}$, where

$$r_i = \sum_{j=0}^{g} s_{i-j} y_j + w_i \qquad (1.4)$$

and $r_i = r(iT)$, $y_j = y(jT)$, and $w_i = w(iT)$. It is assumed here that, for practical purposes, the duration of the impulse response $y(t)$ of the linear baseband channel in Fig. 1.2 is less than $(g+1)T$ seconds, where g is an appropriate positive integer. Thus the sampled impulse response of the linear baseband channel is given by the $(g+1)$-component row vector

$$Y = [y_0 \quad y_1 \quad \cdots \quad y_g] \qquad (1.5)$$

The transmitter and receiver filters are normally designed to have a bandwidth (measured over positive frequencies) that is close to $1/2T$ Hz, so that the sampling rate of $1/T$ samples/second is close to the Nyquist rate for $r(t)$[1,13]. This ensures that most of the information carried by the received waveform $r(t)$ is contained also in the samples $\{r_i\}$, so that no serious loss in performance is incurred through operating on the $\{r_i\}$ in place of $r(t)$. The samples $\{r_i\}$ are fed to the equalizer and detector (Fig. 1.2).

Except where otherwise stated, it is assumed that the transmitter and receiver filters in Fig. 1.2 have the same transfer function $H^{1/2}(f)$, such that the resultant transfer function of the transmitter and

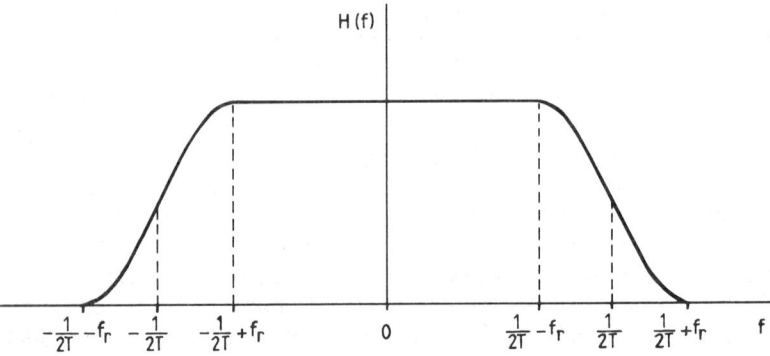

Fig. 1.3 *Resultant transfer function of transmitter and receiver filters in cascade*

receiver filters in cascade is[6,13]

$$H(f) = \begin{cases} T, & -\frac{1}{2T}+f_r < f < \frac{1}{2T}-f_r \\ \frac{1}{2}T\left[1-\sin\left(\frac{\pi\left(|f|-\frac{1}{2T}\right)}{2f_r}\right)\right], & \frac{1}{2T}-f_r \leqslant |f| \leqslant \frac{1}{2T}+f_r \\ 0, & \text{elsewhere} \end{cases}$$
(1.6)

where $f_r \leqslant 1/2T$ and the total bandwidth over positive frequencies is $1/2T+f_r$. $H(f)$ is shown in Fig. 1.3. When the transmission path in Fig. 1.2 introduces no distortion, attenuation or delay, $H(f)$ becomes the transfer function of the linear baseband channel. $H(f)$ satisfies Nyquist's vestigial-symmetry theorem which states that, if $H(f)$ is a real-valued even function of f (symmetrical about $f=0$) and if it has odd symmetry about the nominal cut-off frequencies $\pm 1/2T$ Hz, then the corresponding impulse-response $h(t)$ is a real-valued even time function (symmetrical about $t=0$), and $h(iT)=0$ for all nonzero integer values of i.[6] Furthermore, it can be shown that $h(0)=1$[13]. Thus, under the conditions just assumed and bearing in mind that now $y(t)=h(t)$,

$$y_0 = y(0) = h(0) = 1 \tag{1.7}$$

and
$$y_i = y(iT) = h(iT) = 0 \tag{1.8}$$

for all nonzero integer values of i. Hence, from Equation 1.4,

$$r_i = s_i + w_i \tag{1.9}$$

which means that the data symbols are transmitted at $1/T$ symbols per second with no intersymbol interference in r_i.

Since $h(t) \neq 0$ for $t < 0$, $h(t)$ is not physically realizable. This is for the obvious reason that it is not possible to obtain an output from a filter in response to an impulse at its input, *before* the occurrence of the impulse. However, if a large delay of τ seconds is introduced into the impulse-response $h(t)$, without otherwise changing it, to give the impulse response $h(t-\tau)$ which is such that $h(t-\tau) \simeq 0$ for $t < 0$, the filter then becomes physically realizable, at least for practical purposes. The error caused in $h(t-\tau)$ by setting this accurately to zero for $t < 0$ is negligible so long as $\tau \gg T$, and under these conditions a practical filter can be made to approximate closely to the theoretical ideal, the approximation getting better as τ increases. The fixed delay of τ seconds introduced into the resultant impulse response of the transmitter and receiver filters in Fig. 1.2, to make these physically realizable, will however be ignored throughout this book. Had it not been ignored, the delay could have been compensated for by introducing a delay of τ seconds into the sampling instants $\{iT\}$, without otherwise changing the received samples $\{r_i\}$, but this would have added an unnecessary complication to the description of the system. Thus the equipment filters now to be considered are not physically realizable, but may be made so by introducing a sufficiently large delay into the impulse response.

Since the transmitter and receiver filters in Fig. 1.2 have the *same* transfer function $H^{1/2}(f)$, where $H(f)$ is given by Equation 1.6, the receiver filter is *matched* to the signal at the output of the transmitter filter, so that when the transmission path introduces no signal distortion and no delay, the signal/noise ratio at the sampling instants $\{iT\}$ at the output of the receiver filter, is *maximized*, enabling the optimum available tolerance to noise, for the given transmitted signal, to be achieved by means of the appropriate threshold-level detector[5,8-10,13]

When $f_r = 0$ in Equation 1.6,

$$H(f) = \begin{cases} T, & -\frac{1}{2T} < f < \frac{1}{2T} \\ 0, & \text{elsewhere} \end{cases} \quad (1.10)$$

and the two lowpass filters in Fig. 1.2 become ideal lowpass filters. The received waveform $r(t)$ is now sampled at the Nyquist rate, so that all the information in $r(t)$ is contained also in the samples $\{r_i\}$, and no degradation in tolerance to noise results from operating on the $\{r_i\}$ in place of $r(t)$[1-10]. Equation 1.9, of course, still holds for a perfect transmission path.

It can be seen from Equation 1.6, that, in the general case, $H(f)$ is real-valued and non-negative, and such that

$$\int_{-\infty}^{\infty} H(f)\,df = 1 \qquad (1.11)$$

If now the transmitter and receiver filters in Fig. 1.2 each have the transfer function $H^{1/2}(f)$ and if the impulse response of the transmitter filter is $a(t)$, the $(i+1)$th transmitted signal-element at the input to the transmission path has the waveform $s_i a(t - iT)$, and the Fourier transform (frequency spectrum) of the signal element is

$$s_i \exp(-j2\pi f iT) H^{1/2}(f)$$

Thus the two-sided energy spectral density of an individual transmitted signal-element, at the input to the transmission path, is

$$|s_i \exp(-j2\pi f iT) H^{1/2}(f)|^2 = s_i^2 H(f) \qquad (1.12)$$

and its energy is

$$E_i = s_i^2 \int_{-\infty}^{\infty} H(f)\,df = s_i^2 \qquad (1.13)$$

from Equation 1.11. Since the data-symbols $\{s_i\}$ are statistically orthogonal (uncorrelated and with zero mean), the average transmitted energy per signal element, at the input to the transmission path, is the average or expected value of E_i, $\mathscr{E}[E_i]$, and so is

$$E = \mathscr{E}[s_i^2] \qquad (1.14)$$

which is the mean-square value of s_i.[15-17]

The energy of the $(i+1)$th transmitted signal-element $s_i a(t - iT)$ is equal to s_i^2, as can be seen from Equation 1.13, so that

$$\int_{-\infty}^{\infty} s_i^2 a^2(t - iT)\,dt = s_i^2 \qquad (1.15)$$

The autocorrelation function of this element waveform is[10,15-17]

$$R_s(\tau) = \int_{-\infty}^{\infty} s_i a(t - iT) \cdot s_i a(t - iT + \tau)\,dt \qquad (1.16)$$

and $R_s(\tau)$ is the inverse Fourier transform of the energy spectral density of the element waveform (Equation 1.12), so that

$$R_s(\tau) = \int_{-\infty}^{\infty} s_i^2 H(f) \exp(j2\pi f\tau)\,df \qquad (1.17)$$

But $h(t)$ is the inverse Fourier transform of $H(f)$, such that

$$h(t) = \int_{-\infty}^{\infty} H(f) \exp(j2\pi ft) \, df \tag{1.18}$$

Thus
$$R_s(\tau) = s_i^2 h(\tau) \tag{1.19}$$

Now, from Equation 1.8, $h(iT) = 0$ for any nonzero integer value of i, so that also $R_s(\tau) = 0$ for any value of τ that is a nonzero integral multiple of T.

It follows that, for any two signal-elements $s_i a(t - iT)$ and $s_l a(t - lT)$ at the input to the transmission path (i and l being any two nonnegative integers that are *not* equal),

$$\int_{-\infty}^{\infty} s_i a(t - iT) \cdot s_l a(t - lT) \, dt$$

$$= \int_{-\infty}^{\infty} s_i a(t - iT) \cdot s_l a(t - iT + (i - l)T) \, dt$$

$$= \frac{s_l}{s_i} R_s[(i - l)T] = 0 \tag{1.20}$$

which means that the two element *waveforms* are *orthogonal*.

The energy of any sequence of n signal elements

$$\sum_{i=0}^{n-1} s_i a(t - iT)$$

at the input to the transmission path is

$$\int_{-\infty}^{\infty} \left[\sum_{i=0}^{n-1} s_i a(t - iT) \right]^2 dt = \int_{-\infty}^{\infty} \left[\sum_{i=0}^{n-1} s_i^2 a^2(t - iT) \right] dt$$

$$= \sum_{i=0}^{n-1} s_i^2 \tag{1.21}$$

as can be seen from Equations 1.15 and 1.20. Thus the average transmitted energy per signal element, and hence, with binary signals (Equation 1.2), the average transmitted energy per bit of information, is

$$\mathscr{E}[s_i^2] = s_i^2 = k^2 \tag{1.22}$$

regardless of whether or not the $\{s_i\}$ are uncorrelated or have zero mean.

Since the noise input to the receiver filter is white Gaussian noise with zero mean and two-sided power spectral density of $\frac{1}{2}N_0$, the noise waveform $w(t)$ at the output of the receiver filter is a Gaussian random process with zero mean and a power spectral density of

$$\tfrac{1}{2}N_0|H^{1/2}(f)|^2 = \tfrac{1}{2}N_0 H(f) \tag{1.23}$$

where, of course, $H^{1/2}(f)$ is the transfer function of the receiver filter[15–17]. Thus the average power or mean-square value of the noise waveform $w(t)$ is

$$\sigma^2 = \int_{-\infty}^{\infty} \tfrac{1}{2}N_0 H(f)\,\mathrm{d}f = \tfrac{1}{2}N_0 \tag{1.24}$$

from Equation 1.11. Since $w(t)$ has zero mean, σ^2 is also the *variance* of the noise waveform $w(t)$ and therefore the variance of the noise samples $\{w_i\}$.

The important results of this analysis are now as follows:

(1) The average transmitted energy per signal element is equal to the mean-square value of a data-symbol s_i.
(2) The variance of the noise samples $\{w_i\}$ is equal (numerically) to the value of the two-sided power spectral density of the white Gaussian noise at the input to the receiver filter.

From the Wiener–Kinchine theorem, the autocorrelation function of $w(t)$ is the inverse Fourier transform of its power spectral density $\tfrac{1}{2}N_0 H(f)$ and so is

$$R_n(\tau) = \int_{-\infty}^{\infty} \tfrac{1}{2}N_0 H(f) \exp(\mathrm{j}2\pi f\tau)\,\mathrm{d}f$$
$$= \tfrac{1}{2}N_0 h(\tau) \tag{1.25}$$

from Equation 1.18. Since $h(iT)=0$ for any nonzero integer i, it now follows that

$$R_n(iT) = 0 \tag{1.26}$$

for any nonzero integer i. But the sampling instants of any two noise samples w_i and w_j are separated by a multiple of T seconds, so that, from Equation 1.26, the expected value of the product of w_i and w_j is equal to zero. Furthermore, the noise samples have zero mean, so that the expected value of the product of w_i and w_j is equal to the product of their respective mean values, and hence any two noise samples w_i and w_j are *uncorrelated* Gaussian random variables. It follows that the noise samples $\{w_i\}$ are *statistically independent* Gaussian random variables with zero mean and variance $\sigma^2 = \tfrac{1}{2}N_0$[15–17].

1.3 VECTORS AND MATRICES

Before proceeding further it is necessary to summarize briefly some of the basic properties of vectors and matrices that are used in this and the following chapters. A more detailed study of these can be found elsewhere[21-23].

For our purposes, a vector is an *ordered sequence of numbers* (scalar quantities), which may be real or else complex. These numbers are the *components* of the vector. A *real* vector is one whose components are all *real numbers*.

Consider the n-component row vector

$$X = [x_0 \quad x_1 \quad \ldots \quad x_{n-1}] \quad (1.27)$$

whose components are the real numbers (scalar quantities) $x_0, x_1, \ldots, x_{n-1}$. The row vector X can be considered as the $1 \times n$ matrix X whose component in the $(i+1)$th column is x_i. The corresponding *column* vector is the *transpose* of the row vector X. It can be considered as the transpose of the $1 \times n$ matrix X and therefore as the $n \times 1$ matrix X^T whose component in the $(i+1)$th row is x_i. In the mathematical analysis involving vectors, they are treated as the corresponding matrices, using the normal rules of matrix algebra.

An n-component real vector may be represented as a *point* in an n-dimensional Euclidean vector space. Consider now the n-component real vectors X and Y, where X is given by Equation 1.27 and

$$Y = [y_0 \quad y_1 \quad \ldots \quad y_{n-1}] \quad (1.28)$$

The simple case where $n=2$ is illustrated in Fig. 1.4. Strictly speaking, the orthogonal projection of X on to the first axis (which has the *value* x_0 on this axis) is the two-component vector $x_0 [1 \ 0]$ or $[x_0 \ 0]$, and so on for Y and the other axis, the first and second axes being determined by the vectors $[1 \ 0]$ and $[0 \ 1]$, respectively.

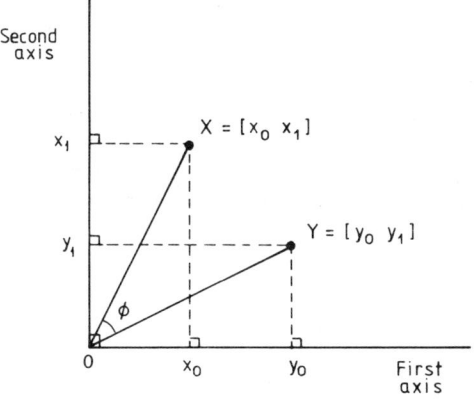

Fig. 1.4 Vectors X and Y in a two-dimensional Euclidean vector space

20 BACKGROUND THEORY

The n axes of the n-dimensional Euclidean vector space are orthogonal (at right angles) and such that the value of the orthogonal projection of the *point* X on to the $(i+1)$th axis is the component x_i of the corresponding *vector* X. The n axes meet at the *origin* O of the vector space, and this point corresponds to the vector whose components are all zero. The latter is known as the *zero vector* and is denoted by O.

The vectors X and Y in Fig. 1.4 can alternatively be considered as the *lines* OX and OY, so that each vector has a *direction* in the vector space. The angle ϕ between the vectors X and Y is simply the angle between the lines OX and OY.

The *sum* of the two vectors X and Y is the n-component row vector

$$X + Y = [x_0 + y_0 \quad x_1 + y_1 \quad \ldots \quad x_{n-1} + y_{n-1}] \tag{1.29}$$

The *inner product* of the real vectors X and Y is the component of the 1×1 matrix

$$X Y^T = x_0 y_0 + x_1 y_1 + \cdots + x_{n-1} y_{n-1} \tag{1.30}$$

and it is evident that

$$X Y^T = Y X^T \tag{1.31}$$

When

$$X Y^T = 0 \tag{1.32}$$

the vectors X and Y are said to be *orthogonal*.

The *length* (or more precisely, the *Euclidean length*) of the real vector X is the Euclidean *distance* from the origin to the corresponding point in the n-dimensional Euclidean vector space, and is defined to be

$$|X| = (X X^T)^{1/2} = \left(\sum_{i=0}^{n-1} x_i^2 \right)^{1/2} \tag{1.33}$$

Clearly

$$|X|^2 = X X^T \tag{1.34}$$

which means that the *squared length* of a vector is the *inner product* of the vector with itself, and is also the sum of the squares of its components.

The *distance* between the real vectors X and Y is the Euclidean distance between the corresponding points in the vector space and is

$$|X - Y| = [(X - Y)(X - Y)^T]^{1/2} = \left(\sum_{i=0}^{n-1} (x_i - y_i)^2 \right)^{1/2} \tag{1.35}$$

which is also the *length* of the vector $X - Y$ or $Y - X$.

The *angle* ϕ between the real vectors X and Y is given by

$$\cos \phi = \frac{1}{|X||Y|} X Y^T \tag{1.36}$$

It can be seen that, when the vectors are *orthogonal* so that $XY^T = 0$, then $\phi = 90°$ or $270°$ and the vectors are at right angles.

It follows from Equation 1.36 that

$$|XY^T| \leqslant |X||Y| \qquad (1.37)$$

where $|XY^T|$ is the absolute value (modulus) of the inner product of X and Y. This is a particular case of the Schwarz inequality[8-10]. Now

$$|X+Y|^2 = (X+Y)(X+Y)^T = XX^T + YY^T + XY^T + YX^T$$
$$= |X|^2 + |Y|^2 + 2XY^T \qquad (1.38)$$

so that

$$XY^T = \tfrac{1}{2}(|X+Y|^2 - |X|^2 - |Y|^2) \qquad (1.39)$$

and, from Equations 1.37 and 1.38,

$$|X+Y| \leqslant |X| + |Y| \qquad (1.40)$$

Equation 1.39 illustrates the close relationship that exists between lengths and inner products.

Strictly speaking, the inner product of two vectors is a *scalar* quantity, so that the inner product of X and Y is the *component* of the 1×1 *matrix* XY^T and *not* the matrix itself. It is with this understanding that the inner product of two vectors is written in terms of the corresponding matrix. Again, in Equation 1.34, $|X|^2$ is a *scalar* quantity whereas XX^T is a 1×1 matrix. $|X|^2$ is of course equal to the *component* of the matrix XX^T and not to the matrix itself. In any equation relating 1×1 matrices and scalars, the 1×1 matrices are taken as the corresponding component values.

Consider next the n-component *complex* vectors X and Y, as given by Equations 1.27 and 1.28, where the components of X and Y are now *complex numbers*. In this case the *inner product* of X and Y is

$$XY^* = x_0 y_0^* + x_1 y_1^* + \cdots + x_{n-1} y_{n-1}^* \qquad (1.41)$$

where Y^* is the conjugate transpose of Y and y_i^* is the complex conjugate of y_i. It is evident that

$$YX^* = (XY^*)^* \qquad (1.42)$$

Now, when

$$XY^* = 0 \qquad (1.43)$$

which means that also $YX^* = 0$, then the vectors X and Y are said to be *orthogonal*.

The *length* (or more precisely the *unitary length*) of the complex vector X is

$$|X| = (XX^*)^{1/2} = \left(\sum_{i=0}^{n-1} |x_i|^2\right)^{1/2} \quad (1.44)$$

where $|x_i|$ is the absolute value (modulus) of x_i. Clearly,

$$|X|^2 = XX^* \quad (1.45)$$

which means that the *squared length* of a vector is the *inner product* of the vector with itself, and is also the sum of the squares of the *absolute values* of its components.

The unitary *distance* between the complex vectors X and Y is

$$|X - Y| = [(X - Y)(X - Y)^*]^{1/2} = \left(\sum_{i=0}^{n-1} |x_i - y_i|^2\right)^{1/2} \quad (1.46)$$

which is also the unitary *length* of the vector $X - Y$ or $Y - X$.

Equation 1.40 holds also for complex vectors, as do Equations 1.37–1.39, when the latter are modified by replacing the *transpose* of a vector (matrix) by its *conjugate transpose*.

Suppose now that the n-component row vectors X and Y, as given by Equations 1.27 and 1.28, are such that

$$Y = XA \quad (1.47)$$

where A is an $n \times n$ matrix whose component in the $(i+1)$th row and $(j+1)$th column is a_{ij}, for $i = 0, 1, \ldots, n-1$ and $j = 0, 1, \ldots, n-1$. Let the $(i+1)$th row of A be given by the n-component row vector

$$A_i = [a_{i,0} \quad a_{i,1} \quad \ldots \quad a_{i,n-1}] \quad (1.48)$$

for $i = 0, 1, \ldots, n-1$. Then

$$y_i = \sum_{j=0}^{n-1} x_j a_{ji} \quad (1.49)$$

for $i = 0, 1, \ldots, n-1$, and

$$Y = \sum_{i=0}^{n-1} x_i A_i \quad (1.50)$$

Equations 1.49 and 1.50 hold regardless of whether the components of X, Y and A are real or complex numbers. $Y = XA$ represents a *linear transformation* that carries the vector X in the n-dimensional vector space into another vector Y of the same space. If two different vectors X in Equation 1.47 always lead to two different vectors Y, the matrix A is *nonsingular*, which means that it has an *inverse* A^{-1} such that

$$AA^{-1} = A^{-1}A = I \quad (1.51)$$

where I is the $n \times n$ *identity* matrix, all of whose components along the main diagonal are of value unity, the remaining components all being zero.

Suppose now that
$$Y_1 = X_1 A \qquad (1.52)$$
and
$$Y_2 = X_2 A \qquad (1.53)$$
where the components of A and of all vectors are *real* numbers, and where X_1 and X_2 are different possible values of X, Y_1 and Y_2 being the corresponding values of Y. Under these conditions let A be an *orthogonal* matrix. Now
$$A^T = A^{-1} \qquad (1.54)$$
and it may readily be shown that
$$|Y_1| = |X_1| \qquad (1.55)$$
$$|Y_2| = |X_2| \qquad (1.56)$$
and
$$Y_1 Y_2^T = X_1 X_2^T \qquad (1.57)$$

This means that the orthogonal transformation performed by the matrix A does not change lengths, inner products, distances or angles. Thus, as a result of the transformation A, there is *no change* in the *relative* positions of the vectors in the n-dimensional Euclidean vector space, the property holding for any number of vectors. The inverse and transpose of an orthogonal matrix are both orthogonal matrices, and the product of two or more $n \times n$ orthogonal matrices is another orthogonal matrix.

Suppose next that the components of A and of all vectors in Equations 1.52 and 1.53 are *complex* numbers, and let A be a *unitary* matrix, whose *conjugate transpose* is A^*. Now
$$A^* = A^{-1} \qquad (1.58)$$
and it may readily be shown that
$$Y_1 Y_2^* = X_1 X_2^* \qquad (1.59)$$
Equations 1.55 and 1.56 holding as before. Thus, as a result of the unitary transformation A, there is *no change* in the *relative* positions of the vectors in the n-dimensional unitary vector space. The inverse and conjugate transpose of a unitary matrix are both unitary matrices, and the product of two or more $n \times n$ unitary matrices is another unitary matrix.

A set of m n-component vectors are said to be *linearly independent* if no one of the vectors can be expressed as a linear combination of one or more of the others. In other words, the m vectors $\{X_i\}$, for $i = 0, 1,$

..., $m-1$, are linearly independent so long as there is *no* set of m scalar quantities $\{c_i\}$ for $i = 0, 1, \ldots, m-1$, not all zero, such that

$$c_0 X_0 + c_1 X_1 + \cdots + c_{m-1} X_{m-1} = 0 \tag{1.60}$$

The set of *all* n-component real vectors that are linear combinations of m given linearly independent real vectors, where $m \leqslant n$, form an m-dimensional *subspace* of the n-dimensional Euclidean vector space. The latter comprises or contains *all* n-component real vectors. The m-dimensional subspace is said to be *spanned* by the given m linearly independent vectors. Any vector lying in the subspace *must* be a linear combination of these vectors, and conversely, any vector not in the subspace *cannot* be a linear combination of the vectors. The zero vector, or origin of the n-dimensional vector space, lies in the subspace. Any vector that is orthogonal to the m linearly independent vectors must be orthogonal to every other vector lying the subspace spanned by these and is therefore orthogonal to the subspace itself. The orthogonal projection of any vector Y in the n-dimensional Euclidean vector space on to the m-dimensional subspace is the vector X in the subspace, such that the vector $Y - X$ is orthogonal to the subspace. X is uniquely determined by Y, and, if Y lies in the subspace, $X = Y$. Orthogonal vectors are necessarily also linearly independent, so that if Y does not lie in the subspace, the vector $Y - X$ cannot be expressed as a linear combination of any set of vectors in the subspace. The *distance* of the vector Y from the m-dimensional subspace is the *length* of the vector $Y - X$.

Two other important *square* matrices are the *real symmetric* matrix A, which is such that

$$A^T = A \tag{1.61}$$

and the *Hermitian* matrix A, which is such that

$$A^* = A \tag{1.62}$$

The components of an Hermitian matrix are complex valued, the components of a real symmetric matrix being, of course, real valued. Just as an *orthogonal* matrix is a special case of a unitary matrix, with all components real-valued, so also a *real symmetric* matrix is a special case of an Hermitian matrix. The inverse and conjugate transpose of an Hermitian matrix are both Hermitian matrices, but the product of two Hermitian matrices A and B is only Hermitian if $AB = BA$, that is, if A and B *commute*.

If an $n \times n$ matrix A and n-component row vector X are such that

$$XA = \lambda X \tag{1.63}$$

where λ is a scalar quantity, then λ is said to be an *eigenvalue* (characteristic root) of A and X is said to be the corresponding *eigenvector* (invariant vector). It is clear that if Equation 1.63 is satisfied by any given vector X, then it is also satisfied by cX where c is any *complex-valued* scalar. If now A is any $n \times n$ *unitary* or *Hermitian* matrix (and therefore also any $n \times n$ orthogonal or real symmetric matrix), then there is always a set of n *mutually orthogonal n-component* vectors $\{X_i\}$ such that

$$X_i A = \lambda_i X_i \tag{1.64}$$

for $i = 0, 1, \ldots, n-1$.

Let U be the $n \times n$ matrix whose $(i+1)$th row is X_i in Equation 1.64, where the lengths of the n $\{X_i\}$ have been set to unity so that U becomes a *unitary* matrix. Also, let D be the $n \times n$ *diagonal* matrix whose $(i+1)$th component along the main diagonal is λ_i. Then, from Equation 1.64,

$$UA = DU \tag{1.65}$$

and

$$UAU^{-1} = D \tag{1.66}$$

By letting $V = U^{-1}$, Equation 1.66 becomes

$$V^{-1}AV = D \tag{1.67}$$

where V is again a unitary matrix. Clearly, the components along the main diagonal of D and the rows of U (or V^{-1}) are the *eigenvalues* and *eigenvectors*, respectively, of A. Equations 1.66 and 1.67 represent a *similarity transformation*. Every $n \times n$ matrix with real- or complex-valued components (regardless of whether it is unitary or Hermitian) has n eigenvalues, which may be real- or complex-values but need not be distinct. The eigenvalues of a matrix A are all nonzero if and only if the matrix is *nonsingular*.

The eigenvalues of a *unitary* matrix (and therefore also of an orthogonal matrix) are of *absolute value* (modulus) *unity*, whereas the eigenvalues of an *Hermitian* matrix (and therefore also of a real symmetric matrix) are *real valued*. An Hermitian matrix all of whose eigenvalues are positive is said to be *positive definite*, and an Hermitian matrix all of whose eigenvalues are either positive or zero is said to be *positive semi-definite*.

A further important matrix is a *circulant*, an example of which is the $n \times n$ matrix

$$A = \begin{bmatrix} a_0 & a_1 & a_2 & \cdots & a_{n-1} \\ a_{n-1} & a_0 & a_1 & \cdots & a_{n-2} \\ a_{n-2} & a_{n-1} & a_0 & \cdots & a_{n-3} \\ \vdots & \vdots & \vdots & \vdots & \vdots \\ a_1 & a_2 & a_3 & \cdots & a_0 \end{bmatrix} \tag{1.68}$$

A circulant matrix A satisfies Equation 1.66, with the appropriate unitary matrix U, which is now *independent* of the component values of A, contrary to the case of a typical unitary or Hermitian matrix. Indeed, as is shown in Section 2.5, the rows of the matrix U, which are the n orthogonal eigenvectors of the $n \times n$ circulant matrix A, are given by

$$U_i = n^{1/2} [1 \quad \omega^i \quad \omega^{2i} \quad \ldots \quad \omega^{(n-1)i}] \tag{1.69}$$

for $i = 0, 1, \ldots, n-1$, where

$$\omega = \exp(j2\pi/n) \tag{1.70}$$

$|\omega^i| = 1$ and the $\{\omega^i\}$, for $i = 0, 1, \ldots, n-1$, are the n distinct nth roots of unity. A matrix may be both a circulant and Hermitian or both a circulant and unitary.

If now A_1 and A_2 are two $n \times n$ circulant matrices, then

$$A_1 A_2 = A_2 A_1 = A \tag{1.71}$$

where A is another *circulant* matrix. If, in addition, A_1 and A_2 are both Hermitian matrices or else both unitary matrices, then A is also an Hermitian or unitary matrix, respectively, in addition to being a circulant. Furthermore, it follows from Equation 1.69 that, for *any* two $n \times n$ circulant matrices A_1 and A_2,

$$U A_1 U^{-1} = D_1 \tag{1.72}$$

and

$$U A_2 U^{-1} = D_2 \tag{1.73}$$

where D_1 and D_2 are $n \times n$ diagonal matrices, so that

$$U A_1 U^{-1} U A_2 U^{-1} = U A_1 A_2 U^{-1} = U A U^{-1} = D \tag{1.74}$$

from Equations 1.66 and 1.71, D being, of course, a diagonal matrix. It follows from Equations 1.72–1.74 that

$$U A_1 A_2 U^{-1} = D_1 D_2 \tag{1.75}$$

where

$$D_1 D_2 = D \tag{1.76}$$

Consider now the n-component row vector

$$X = [x_0 \quad x_1 \quad \ldots \quad x_{n-1}] \tag{1.77}$$

and the $n \times n$ circulant matrix A in Equation 1.68, whose first row is given by the n-component row vector

$$A_0 = [a_0 \quad a_1 \quad \ldots \quad a_{n-1}] \tag{1.78}$$

Under these conditions, the n-component row vector

$$Y = XA \tag{1.79}$$

which is evaluated according to Equations 1.49 and 1.50, is said to be the *circular (periodic) convolution* of the vectors X and A_0. Suppose next that the last $n-m$ components of X are set to zero, such that

$$X = [x_0 \quad x_1 \quad \ldots \quad x_{m-1} \quad 0 \quad 0 \quad \ldots \quad 0] \quad (1.80)$$

and that the components $a_{g+1}, a_{g+2}, \ldots, a_{n-1}$ of the circulant matrix A in Equation 1.68 are set to zero, where

$$n \geqslant m + g \quad (1.81)$$

The first row of A is now given by the n-component row vector

$$A_0 = [a_0 \quad a_1 \quad \ldots \quad a_g \quad 0 \quad 0 \quad \ldots \quad 0] \quad (1.82)$$

and the n-component row vector Y in Equation 1.79 becomes the *linear (aperiodic)* convolution of the vectors X and A_0. Clearly, if

$$X = [1 \quad 0 \quad 0 \quad \ldots \quad 0] \quad (1.83)$$

the circular or linear convolution of X and A_0 is just A_0 itself. If

$$X = [\overbrace{0 \quad 0 \quad \ldots \quad 0}^{h} \quad x_h \quad 0 \quad 0 \quad \ldots \quad 0] \quad (1.84)$$

where x_h is some scalar quantity, the circular convolution of X and A_0 becomes x_h times the $(h+1)$th row of A, which is x_h times A_0, with the components shifted *cyclically to the right* by h places. The *circular* convolution of X and A_0 is here the *linear* convolution with the components shifted *cyclically* to the right by h places. The latter result holds also in the more general case, where the m nonzero components of X in Equation 1.80 are shifted cyclically to the right by h places, and the resulting vector is postmultiplied by the circulant matrix A, to give the corresponding *cyclically shifted* linear convolution of X and A_0. The matrix A, of course, here satisfies Equations 1.81 and 1.82.

1.4 OPTIMUM DETECTION PROCESS

Consider the case where a sequence of n real-valued samples $\{r_j\}$ is obtained at the output of the sampler, in the receiver of the data-transmission system of Fig. 1.2. The sequence of samples can be represented by the n-component row vector

$$R = [r_0 \quad r_1 \quad \ldots \quad r_{n-1}] \quad (1.85)$$

Suppose also that

$$R = S_i + W \quad (1.86)$$

where

$$S_i = [s_{i,0} \quad s_{i,1} \quad \ldots \quad s_{i,n-1}] \quad (1.87)$$

$$W = [w_0 \quad w_1 \quad \ldots \quad w_{n-1}] \tag{1.88}$$

and $i = 0, 1, \ldots, m-1$. In order to simplify the terminology, the *same* symbol S_i will be used for the *actual* received signal vector and for a *possible* received signal vector, the distinction between the two being made clear in the context. The vector S_i may be an individual m-level signal element or alternatively a group of signal elements having m different *combinations* of their corresponding data-symbol values. No assumptions are made about the m possible vectors $\{S_i\}$ other than that these are constant (not subject to change), all different and of finite Euclidean length. The possible vectors $\{S_i\}$ are assumed to be *known* at the receiver, which implies that they are held in store here. The $\{w_j\}$, for $j = 0, 1, \ldots, n-1$, are statistically independent Gaussian random variables with zero mean and variance σ^2. Furthermore, the $\{w_j\}$ are statistically independent of S_i. For convenience, the symbols g, h, i, j, l, m and n are always taken to be *integers* unless specifically stated to the contrary, as for instance in the case of $h(t)$ and $j = \sqrt{-1}$.

The information transmitted is the integer i, and to determine this integer the receiver must deduce from R which of the m possible vectors $\{S_i\}$ has in fact been transmitted. We are thus concerned here with the optimum detection process for i from R. A detection process is the *selection* of one of a finite set of symbols or messages in response to the received signal. Thus it is a *decision* as to which of the possible symbols or messages has been transmitted, and it is essentially a *nonlinear* process. The decision is based on the received signal together with the available *prior knowledge* of this signal. In general, if all the available prior knowledge of the received signal (that is relevant to the detection process) is *not* used, an inferior (suboptimum) detection process results and there is an increased probability of incorrect detection. Thus, in the case considered here, the receiver must have prior knowledge of the m possible vectors $\{S_i\}$ and of their *a priori* probabilities, together with the statistics of the noise vector W. The *a priori* probability $P(S_i)$ of a vector S_i, for some given value of i, is the probability that this vector is received, *before* the receipt of R. $P(S_i)$ is given numerically by the relative frequency of S_i over the receipt of a very large number of vectors $\{R\}$, assuming here that every transmission is independent of every other. The actual received vector S_i is normally unlikely to have an *a priori* probability very different from that of any other possible vector S_h, so that the *a priori* probability $P(S_i)$ of a vector does not of itself enable satisfactory detection of the transmitted symbol i to be achieved. Following the receipt of R, the optimum detector usually has a very much better knowledge of the transmitted signal as a result of the information imparted to it by R, and it can then form the *a posteriori* probability $P(S_i|R)$ for each vector S_i, given the received vector R. $P(S_i|R)$ is simply the *conditional* probability of S_i, given R. Under good conditions, when the signal/noise ratio is very high, $P(S_i|R)$ is usually much greater than $P(S_h|R)$ for any $h \neq i$, where S_i is the *actual* received signal vector.

From Equation 1.86,

$$r_j = s_{i,j} + w_j, \qquad j = 0, 1, \ldots, n-1 \tag{1.89}$$

In Equation 1.86, each of R, S_i and W forms an *ordered set* of n *random variables* and is therefore the corresponding *random vector*. The *value* of a vector is the *ordered set* of values of its n components, and the *sample value* of a random vector is the actual value of the vector obtained in any one occurrence (transmission or reception) of the vector. A random vector itself is the ensemble (collection) of all the different possible sample values of that vector, each value being associated with the corresponding probability or probability density. A random vector with just one component ($n = 1$) becomes a random variable.

A *discrete* random vector is one having a *finite* number of different possible sample values, whereas a *continuous* random vector is one having a continuous range (and hence infinite number) of different possible sample values. The complete *set* of different possible sample values of a discrete random vector X is designated $\{X\}$.

In order to reduce the complexity of the notation, which would otherwise become very cumbersome in certain parts of the book, the *same* symbol will be used for a random vector and its sample value, the correct meaning being in every case made clear in the context.

When the vector R in Equation 1.86 is taken to represent the vector actually received, in some particular transmission, the vectors R, S_i and W in Equation 1.86 become the *sample values* of the corresponding random vectors. Thus their values are now fixed. On the other hand, when considering the mean or mean-square value of any of the vectors R, S_i and W, or of any function of these vectors, or alternatively, when considering the probability of any of these vectors having a value within some given range, the vectors must be taken as *random vectors*, since we are now concerned with *all possible* sample values of the vectors under consideration, and not just with the particular sample values that happen to be received in any given transmission. Equation 1.86 is satisfied, regardless of whether R, S_i and W are random vectors or sample values of random vectors.

S_i often has different values in different transmissions, so that the detected value of S_i is itself a random vector, often differing from one transmission to another. In any *one* detection process, however, S_i becomes a *sample value* of the corresponding random vector as does the detected value of S_i. Of course, the *detected value* of S_i should here, if possible, be S_i.

Before proceeding further with the derivation of the optimum detection process, it is necessary first to clarify an important mathematical relationship that is used in the analysis.

If $P(A|B)$ is the conditional probability of the event A, given the event B, and $P(B|A)$ is the conditional probability of the event B, given the event A, then the two conditional probabilities are related by the following equation

$$P(A|B) = \frac{P(B|A)P(A)}{P(B)} \qquad (1.90)$$

where $P(A)$ is the probability of the event A and $P(B)$ is the probability of the event B. Equation 1.90 is known as *Bayes' theorem*[15-17]. A and B are two *events* and the equation involves only *probabilities*.

It may be shown[5] that this relationship holds also for the case where each of the symbols A and B represents a *set* of sample values of the corresponding set of continuous random variables, so that A and B become sample values of the corresponding continuous random vectors. Equation 1.90 now involves probability *densities* and becomes

$$p(A|B) = \frac{p(B|A)p(A)}{p(B)} \qquad (1.91)$$

where $A = [a_0 \; a_1 \; \ldots \; a_{n-1}]$ and $B = [b_0 \; b_1 \; \ldots \; b_{n-1}]$. $p(A)$ is the value of the probability *density* function of the random vector with sample value A, when the random vector has this sample value. $p(A|B)$ is the value of the corresponding *conditional* probability density function, given B. Similarly for $p(B)$ and $p(B|A)$.

Again, Bayes' theorem also holds where the n random variables, whose sample values are given by A, are *discrete*, so that A may have any one of a finite set of possible values and is a sample value of a *discrete* random vector. B is a sample value of a continuous random vector, as before. The equation now contains the appropriate *mixture* of probabilities and probability densities[5], and becomes

$$P(A|B) = \frac{p(B|A)P(A)}{p(B)} \qquad (1.92)$$

$P(A)$ is here the *probability* that the random vector with sample value A has the sample value. $P(A|B)$ is the corresponding *conditional* probability, given B.

Throughout the following discussion, $P(.)$ will be taken to signify a *probability* and $P(.|.)$ a *conditional probability*, whereas $p(.)$ will be taken to signify a *probability density* and $p(.|.)$ a *conditional probability density*. The quantities inside the brackets here are *sample values* of the corresponding random vectors.

The detector is assumed to have prior knowledge of the m possible vectors $\{S_i\}$, and it also has prior knowledge of their *a priori* probabilities $\{P(S_i)\}$. The whole of this prior knowledge is used to optimize the detection process, that is, to *minimize the probability of error* in the detection of the transmitted data-symbol i.

From the value of the received vector R the detector selects one of the m possible vectors $\{S_i\}$ as the detected signal, and the corresponding value of i gives the detected data-symbol.

In order to minimize the probability of error in a detection process[5] it is necessary to *maximize* the probability of a *correct* decision, $P(C)$. Now

$$P(C) = \int_{-\infty}^{\infty} \cdots \int_{-\infty}^{\infty} P(C|R) p(r_0, r_1, \ldots, r_{n-1}) \, dr_0 \, dr_1 \ldots dr_{n-1} \tag{1.93}$$

where $P(C|R)$ is the conditional probability of a correct decision *given* the received vector R, and $p(r_0, r_1, \ldots, r_{n-1})$ is the value of the joint probability density function of the n random variables, corresponding to the n components of R, at the given value of R. The vector R may here have *any* of its possible values and is not confined to the *particular* value received. Equation 1.93 may be written more simply as

$$P(C) = \int_{-\infty}^{\infty} P(C|R) p(R) \, dR \tag{1.94}$$

Since $p(R)$ is non-negative, it can be seen that $P(C)$ is maximized by maximizing $P(C|R)$ for *every possible* value of the vector R. For any *given* received vector R, $P(C|R)$ is maximized by selecting as the detected value of S_i, the vector S_h for which

$$P(S_h|R) > P(S_i|R) \tag{1.95}$$

for $i = 0, 1, \ldots, m-1$ and $i \neq h$. $P(S_i|R)$ is the conditional probability of S_i, given R, and so is the *a posteriori* probability of S_i.

Thus the detection process that *minimizes the probability of error*, selects from the m possible vectors $\{S_i\}$ the vector which *maximizes* $P(S_i|R)$ and which therefore has the greatest *a posteriori* probability of being correct. R is here the *given* received vector. Clearly, the detected data-symbol is h, where h is the value of i that maximizes $P(S_i|R)$. If the maximum *a posteriori* probability is shared by two or more of the m

possible vectors $\{S_i\}$, the detector may select *any one* of these vectors as the detected vector S_h.

From Equation 1.92,

$$P(S_i|R) = \frac{p(R|S_i)P(S_i)}{p(R)} \quad (1.96)$$

where the terms are defined as follows. $P(S_i)$ is the *a priori* probability of S_i.

$$p(R) = p(r_0, r_1, \ldots, r_{n-1}) \quad (1.97)$$

is the value of the joint probability density function of the random variables with sample values $r_0, r_1, \ldots, r_{n-1}$, when the random variables have the given values $r_0, r_1, \ldots, r_{n-1}$.

$$p(R|S_i) = p(r_0, r_1, \ldots, r_{n-1}|s_{i,0}, s_{i,1}, \ldots, s_{i,n-1}) \quad (1.98)$$

is the value of the conditional joint probability density function of the random variables with sample values $r_0, r_1, \ldots, r_{n-1}$, when the random variables have the given values $r_0, r_1, \ldots, r_{n-1}$ and given the values $s_{i,0}, s_{i,1}, \ldots, s_{i,n-1}$.

From Equations 1.95 and 1.96, the detection process that minimizes the probability of error, selects from the m possible vectors $\{S_i\}$, the vector S_h for which

$$P(S_h)p(R|S_h) > P(S_i)p(R|S_i) \quad (1.99)$$

for $i = 0, 1, \ldots, m-1$, $i \neq h$.

The detection process given by Equation 1.99 minimizes the probability of error in the detection of the data-symbol i, for *any* joint probability density function of the n noise samples $\{w_j\}$ and not just for the joint Gaussian probability density assumed here.

In the important case now to be assumed, where the m possible vectors $\{S_i\}$ are *equally likely* to be received,

$$P(S_i) = \frac{1}{m} \quad (1.100)$$

and the detection process, that minimizes the probability of error, selects from the m possible vectors $\{S_i\}$ the vector S_h for which

$$p(R|S_h) > p(R|S_i) \quad (1.101)$$

for $i = 0, 1, \ldots, m-1$ and $i \neq h$. Thus the detected data symbol is h, where h is the value of i that *maximizes* $p(R|S_i)$. Now, $p(R|S_i)$ is the conditional probability density function of R, given S_i, and is known as the *likelihood function* of S_i. Since the detector selects the value of i that maximizes $p(R|S_i)$, it is known as a *maximum-likelihood detector*[5,12,16].

Since the noise components $\{w_j\}$ in the given received vector R are sample values of statistically independent Gaussian random variables with zero mean and variance σ^2, which are also statistically independent of S_i, it follows from Equation 1.89 that, for a given i, r_j is a sample value of a Gaussian random variable with a mean value of $s_{i,j}$ and a variance σ^2. Furthermore, the different $\{r_j\}$ are sample values of statistically independent Gaussian random variables. Thus the conditional probability density function of the random variable with sample value r_j, given S_i, is

$$p(r_j|S_i) = \frac{1}{\sqrt{(2\pi\sigma^2)}} \exp\left(-\frac{(r_j - s_{i,j})^2}{2\sigma^2}\right) \quad (1.102)$$

where $j = 0, 1, \ldots, n-1$, and so

$$p(R|S_i) = p(r_0, r_1, \ldots, r_{n-1}|S_i) = p(r_0|S_i)p(r_1|S_i) \ldots p(r_{n-1}|S_i)$$

$$= \sum_{j=0}^{n-1} \frac{1}{\sqrt{(2\pi\sigma^2)}} \exp\left(-\frac{(r_j - s_{i,j})^2}{2\sigma^2}\right)$$

$$= \frac{1}{(2\pi\sigma^2)^{n/2}} \exp\left(-\frac{1}{2\sigma^2} \sum_{j=0}^{n-1} (r_j - s_{i,j})^2\right)$$

$$= \frac{1}{(2\pi\sigma^2)^{n/2}} \exp\left(-\frac{1}{2\sigma^2} |R - S_i|^2\right) \quad (1.103)$$

where $|R - S_i|$ is the *Euclidean distance* between the vectors R and S_i. Equation 1.103 uses the fact that the joint probability density function $p(u, v)$ of two statistically independent random variables u and v is the *product* of the individual probability densities $p_1(u)$ and $p_2(v)$, such that

$$p(u, v) = p_1(u)p_2(v) \quad (1.104)$$

The symbols u and v in Equation 1.104 are, of course, the *sample values* of the corresponding random variables.

It can be seen from Equation 1.103 that when $p(R|S_i)$ is *maximum*, $|R - S_i|$ is *minimum*, so that the detected data-symbol h satisfies

$$|R - S_h| < |R - S_i| \quad (1.105)$$

for $i = 0, 1, \ldots, m-1$ and $i \neq h$, which means that the distance between R and S_h is *less* than the distance between R and any of the other $\{S_i\}$. Thus the detection process, that *minimizes the probability of error*, in the detection of i from R under the assumed conditions, is a *maximum-likelihood detector* that selects the data-symbol h for which S_h is *closest* to R. Provided (as is assumed here) that the received samples $\{r_j\}$, for $j < 0$ and $j > n-1$, do not contain samples of the received data signal, so that these are *independent* of both S_i and R (bearing in mind that the noise components $\{w_j\}$ are statistically independent), no reduction in the error probability is achieved through increasing the number of components of R by including in the vector any $\{r_j\}$ for $j<0$ or $j>n-1$. The detection process here is *optimum* in the sense that it minimizes the probability of error in the detection of the complete received *signal message* S_i.

In the most general case, where a *multilevel QAM* or *PM* (phase modulated) signal is transmitted and *linear* modulation and demodulation processes are used, such that the transmitted signal and channel sampled-impulse-response in Fig. 1.2 are *complex valued*, the vectors R, S_i and W in Equation 1.86 have complex-valued components and lie in an n-dimensional *unitary* vector space. By means of a derivation along the same lines as that in Equations 1.93–1.105, it can now be shown that the detection process, minimizing the probability of error in the detection of i from R, is a *maximum-likelihood detector* that selects the data-symbol h for which S_h is at the minimum unitary distance from R. It is assumed here that the data-symbols $\{i\}$ are equally likely and that the *real* and *imaginary parts* of the noise components $\{w_j\}$ are statistically independent Gaussian random variables with zero mean and variance σ^2.

1.5 MATCHED-FILTER DETECTOR

Consider now the important case where the n-component vector S_i in Equation 1.86 is such that

$$S_i = s_i Y \tag{1.106}$$

where Y is the n-component vector

$$Y = [y_0 \quad y_1 \quad \ldots \quad y_{n-1}] \tag{1.107}$$

and, of course, $y_j = 0$ for $j < 0$ and $j > n-1$. All signals here are *real valued*. It is assumed that the transmitted signal-element is $s_i \delta(t)$, which is an impulse of area s_i and occurs at time $t = 0$. The impulse is fed to a linear baseband channel with impulse response $y(t)$, where $y(t) = 0$ for $t < 0$ and $t \geq nT$, T being the sampling interval. The output signal from the channel is sampled at the time instants $\{jT\}$ to give the sequence of samples $\{s_i y_j\}$ of the received signal-element, where $y_j = y(jT)$. Thus, from Equations 1.86 and 1.106, the received sequence of n samples is given by the n-component vector

$$R = s_i Y + W \tag{1.108}$$

It is evident from the previous analysis that the optimum detection

BACKGROUND THEORY 35

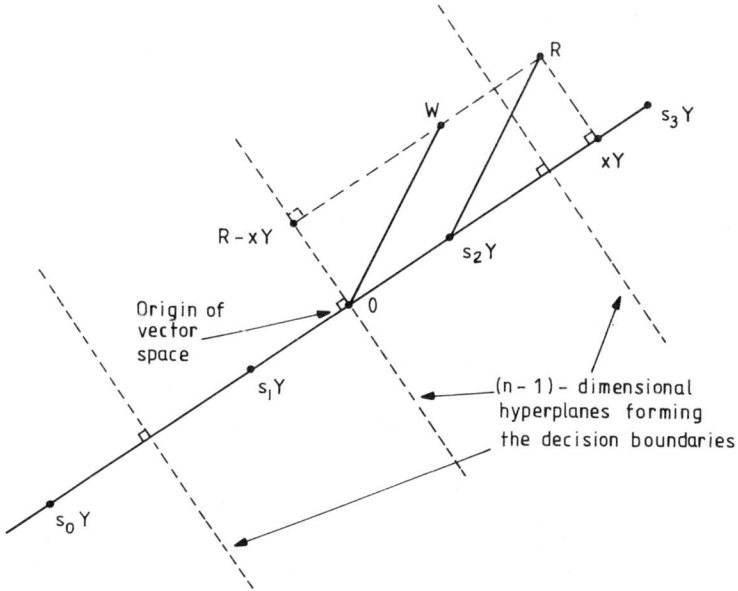

Fig. 1.5 *Decision boundaries in the matched-filter detection of* s_i

process for the data-symbol i from the received vector R, is to take its possible value h such that $s_h Y$ is at the minimum Euclidean distance from R, in other words, such that $|R - s_h Y|$ is minimum. But the m possible vectors $\{s_i Y\}$, for $i = 0, 1, \ldots, m-1$, all lie in the one-dimensional subspace spanned by Y, so that they lie on the straight line that passes through the origin and the point given by the vector Y in the n-dimensional Euclidean vector space containing Y. This is illustrated in Fig. 1.5, where it is assumed that the possible values of i are 0, 1, 2 and 3 and the actual transmitted value of i is 2. Thus the received signal-element is $s_2 Y$. Suppose that the noise-vector W is as shown in Fig. 1.5, to give the received vector

$$R = s_2 Y + W \qquad (1.109)$$

It is clear that if the data-symbol s_i is detected as its possible value s_h that minimizes $|R - s_h Y|$, as in a maximum-likelihood detector, then, in the particular case considered here, s_i is detected as s_3, since $s_3 Y$ is closest to R. The symbol i is now detected as 3. In general, if R is closest to $s_h Y$, then s_i is detected as s_h, so that i is detected as h.

The detection process just described for s_i is equivalent to that in which the n-dimensional vector-space containing R is divided into m decision regions, such that, if R is in the fourth of these as shown in

Fig. 1.5, then the data-symbol s_3 is detected. The decision regions are here separated by decision boundaries each of which is an $(n-1)$-dimensional hyperplane that perpendicularly bisects the line joining the two adjacent possible signal-vectors $s_i Y$ and $s_{i+1} Y$, for $i=0, 1$, and 2, as shown in Fig. 1.5. It can be seen that if R lies at any point *on* the decision boundary between $s_i Y$ and $s_{i+1} Y$, then R is equidistant from $s_i Y$ and $s_{i+1} Y$, this decision boundary being in fact the locus of *all* points equidistant between $s_i Y$ and $s_{i+1} Y$. Clearly, the hyperplane separates the two regions for which R is closer to $s_i Y$ and R is closer to $s_{i+1} Y$.

Figure 1.5 shows the orthogonal projection of R on to the one-dimensional subspace spanned by Y. Since this projection must lie in the given one-dimensional subspace, it must be a real scalar multiple of Y, say xY, where x is the appropriate scalar quantity. It can now be seen that if R is closer to $s_j Y$ than it is to $s_i Y$ (for $i=0, 1, 2, 3, j=0, 1, 2, 3$ and $i \neq j$) then xY must also be closer to $s_j Y$ than it is to $s_i Y$. This follows from the fact that the decision boundary between $s_i Y$ and $s_j Y$ (with which R can be considered to be compared in the detection process) is, in fact, the locus of all possible vectors $\{R\}$ whose *orthogonal projection* on to the one-dimensional subspace spanned by Y is the point (vector) $\frac{1}{2}(s_i + s_j) Y$.

Now the vector $R - xY$ must be *orthogonal* to Y, so that

$$(R - xY)Y^T = 0 \tag{1.110}$$

or

$$xYY^T = RY^T \tag{1.111}$$

or

$$x = RY^T(YY^T)^{-1} = |Y|^{-2} RY^T \tag{1.112}$$

where $|Y|$ is the length of the vector Y. The symbol x is said to be the *maximum likelihood estimate* of s_i, since it is the *unconstrained* possible value of s_i having the maximum likelihood function $p(R|x)$, such that the value of x minimizes $|R - xY|$. Thus x is the *unconstrained* possible value of s_i for which xY is at the minimum Euclidean distance from R (Fig. 1.5). The symbol x may here take on any *real* value and is not limited to the actual possible values of s_i, so that x is an *estimate* of s_i. Now, instead of measuring the distances between the vectors $\{s_i Y\}$ and R and taking as the detected data-symbol its possible value s_h for which $s_h Y$ is closest to R, the following detection process can be used, giving the same detected value of s_h. The orthogonal projection xY of R on to the one-dimensional subspace spanned by Y is first determined to give the value of x. This means, of course, that x is determined according to Equation 1.112. The detected data-symbol s_h is then determined as its possible value closest to x. The detection process can be implemented as follows. The received vector R is fed into a linear feedforward transversal filter with tap gains given by the

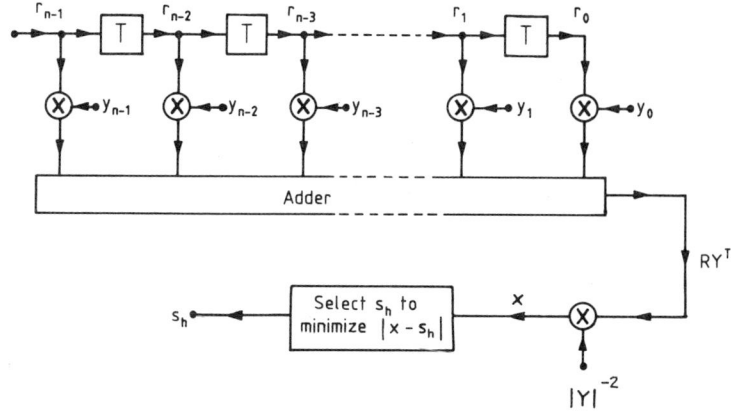

Fig. 1.6 Matched-filter detection of s_i

components of the vector Y but in the *reverse* order, as shown in Fig. 1.6. A square marked T here is a store that effectively introduces a delay of T seconds. The output signal from the transversal filter is RY^T, and the output signal from the following multiplier is the required signal x in Equation 1.112. The detected data-symbol s_h is then determined as its possible value closest to x, and this is achieved by means of a simple threshold-level detector. The latter compares x with thresholds half-way between adjacent possible values of s_i, and takes as s_h the possible value of s_i that lies between the same thresholds as x. As before, the detector requires prior knowledge of Y and of the possible values of s_i.

The detection process just described is an arrangement of *matched-filter detection* and is the *optimum* detection process for the received data-symbol s_i from the vector R, since it gives the *same* detected value as does the corresponding maximum-likelihood detector.

In the most general case where a *multilevel QAM* or *PM* signal is transmitted, but the other conditions are the same as or equivalent to those assumed before, the signals and channel sampled-impulse-response in Fig. 1.2 are *complex-valued*. The matched-filter detector is now as that in Fig. 1.6 except that each *tap gain* of the filter becomes the *complex conjugate* of the value shown. Thus RY^T is replaced by RY^*, and the maximum-likelihood estimate of s_i, given by the matched filter, becomes $x = RY^*(YY^*)^{-1}$, which minimizes the *unitary* distance $|R - xY|$. However, since our main concern is with *real-valued* signals, the above case will not be considered further here.

Consider now the particular case where $s_i = \pm k$ and the received vector R is given by Equation 1.108. Regardless of the possible values of s_i, the output signal from the matched filter in Fig. 1.6 is

38 BACKGROUND THEORY

$$x = |Y|^{-2} R Y^T = |Y|^{-2}(s_i Y Y^T + W Y^T) = s_i + |Y|^{-2} W Y^T \quad (1.113)$$

since, of course, $Y Y^T = |Y|^2$. The noise component in x is

$$u = |Y|^{-2} W Y^T = |Y|^{-2}(w_0 y_0 + w_1 y_1 + \cdots + w_{n-1} y_{n-1}) \quad (1.114)$$

which is a linear combination of the noise components $\{w_j\}$. But the $\{w_j\}$ are statistically independent Gaussian random variables with zero mean and variance σ^2, so that the $\{w_j y_j\}$ are statistically independent Gaussian random variables and are therefore also uncorrelated[15-17]. It follows that the *variance* of the sum of the $\{w_j y_j\}$ is the *sum* of their individual variances. Since u is a *linear combination* of the $\{w_j\}$ it is also a Gaussian random variable with zero mean[15-17]. Its variance is therefore

$$\eta^2 = |Y|^{-4} \sigma^2 (y_0^2 + y_1^2 + \cdots + y_{n-1}^2)$$

$$= |Y|^{-4} \sigma^2 |Y|^2 = |Y|^{-2} \sigma^2 = \frac{\sigma^2}{E} \quad (1.115)$$

where
$$E = |Y|^2 \quad (1.116)$$

and E is the *energy* (or squared length) of the vector (sequence) Y. The probability density function of u is now

$$p(u) = \frac{1}{\sqrt{2\pi\eta^2}} \exp\left(-\frac{u^2}{2\eta^2}\right) \quad (1.117)$$

But
$$x = s_i + u \quad (1.118)$$

where $s_i = \pm k$, and s_i is detected as its possible value s_h closest to x. Thus, when $x > 0$, $s_h = k$, and when $x < 0$, $s_h = -k$.

The $\{w_j\}$ are statistically independent of s_i, so that u and s_i are statistically independent, the value of u being in no way influenced by the value of s_i.

If $s_i = -k$, an error occurs in s_h when $x > 0$, that is, when $u > k$, and the probability of this occurring is

$$\int_k^\infty p(u)\, du = \int_k^\infty \frac{1}{\sqrt{2\pi\eta^2}} \exp\left(-\frac{u^2}{2\eta^2}\right) du$$

$$= \int_{k/\eta}^\infty \frac{1}{\sqrt{2\pi}} \exp\left(-\tfrac{1}{2} u^2\right) du = Q\left(\frac{k}{\eta}\right) \quad (1.119)$$

where
$$Q(v) = \int_v^\infty \frac{1}{\sqrt{2\pi}} \exp\left(-\tfrac{1}{2} u^2\right) du \quad (1.120)$$

If $s_i = k$, an error occurs in s_h when $x < 0$, that is, when $u < -k$, and the probability of this occurring is

$$\int_{-\infty}^{-k} p(u)\,du = \int_{k}^{\infty} p(u)\,du = Q\left(\frac{k}{\eta}\right) \quad (1.121)$$

Thus, *regardless* of the binary value of s_i, the probability of error in s_h is

$$Q\left(\frac{k}{\eta}\right) = Q\left(k\sqrt{\frac{E}{\sigma^2}}\right) \quad (1.122)$$

which is therefore also the *average error rate* in the repetitive reception of R and detection of s_i. An important feature of this result is that the error probability is a function only of the *energy* of the vector (sequence) Y and not of the particular values of its components.

Suppose now that the tap gains of the transversal filter in Fig. 1.6 are changed from their given values $y_{n-1}, y_{n-2}, \ldots, y_0$ to the corresponding values $z_{n-1}, z_{n-2}, \ldots, z_0$, and let Z be the n-component vector

$$Z = [z_0 \quad z_1 \quad \ldots \quad z_{n-1}] \quad (1.123]$$

Assume also that the vector Z is allowed to have *any direction* in the n-dimensional vector space containing Z but that Z is constrained to have a *fixed length* equal to that of Y. Thus

$$|Z|^2 = E \quad (1.124)$$

Then, if the vector R (Equation 1.108) is fed into the transversal filter to the position shown in Fig. 1.6, the output signal from the transversal filter is

$$RZ^T = s_i Y Z^T + W Z^T \quad (1.125)$$

from Equation 1.108, and the signal/noise ratio here is

$$\mathscr{E}[(s_i Y Z^T)^2] / \mathscr{E}[(W Z^T)^2] \quad (1.126)$$

$s_i Y Z^T$ is the *signal* component in RZ^T and WZ^T is the *noise* component.. But

$$WZ^T = w_0 z_0 + w_1 z_1 + \cdots + w_{n-1} z_{n-1} \quad (1.127)$$

which means that WZ^T is a Gaussian random variable with zero mean and variance

$$\sigma^2(z_0^2 + z_1^2 + \cdots + z_{n-1}^2) = \sigma^2 |Z|^2 = \sigma^2 E \quad (1.128)$$

from Equation 1.124. Thus the mean-square value of WZ^T is also

$$\mathscr{E}[(WZ^T)^2] = \sigma^2 E \quad (1.129)$$

which is constant and therefore *independent* of the direction of the vector Z in the vector space.

Now, let ϕ be the angle between the vectors Y and Z, such that

$$|Y||Z|\cos\phi = YZ^T \qquad (1.130)$$

as can be seen from Equation 1.36. But

$$|Y||Z| = |Y|^2 = E \qquad (1.131)$$

so that

$$YZ^T = E\cos\phi \qquad (1.132)$$

and the signal component in RZ^T (Equation 1.125) is

$$s_i YZ^T = s_i E \cos\phi \qquad (1.133)$$

The mean-square value of this signal component is

$$\mathscr{E}[(s_i E \cos\phi)^2] = \mathscr{E}[s_i^2] E^2 \cos^2\phi \qquad (1.134)$$

which, for given possible values of s_i, is maximum when

$$\cos\phi = \pm 1 \qquad (1.135)$$

or
$$\phi = 0 \quad \text{or} \quad \pi \qquad (1.136)$$

or
$$Z = \pm Y \qquad (1.137)$$

It follows that the signal/noise ratio at the output of the filter is *maximized* when Equation 1.137 holds, in other words, when the filter is *matched* to the received signal (the vector $s_i Y$) as in Fig. 1.6. The signal/noise ratio at the output of the filter is not affected if all tap gains are multiplied by some real scalar *constant*, since this changes the output levels of the signal and noise by exactly the same amount, so that the tap gains may in fact be in any constant positive or negative ratio to the values shown in Fig. 1.6, in order to maximize the output signal/noise ratio.

It follows from the above analysis that the matched filter of Fig. 1.6 not only enables the best available detection process for s_i to be achieved from its output signal, but it also *maximizes* the signal/noise ratio here. The latter is in fact a *necessary* condition for the former to be satisfied.

The signal x in Fig. 1.6 is the *unbiased* estimate of s_i having the minimum mean-square error. The estimate is *unbiased* in the sense that

$$\mathscr{E}[x] = \mathscr{E}[s_i] \qquad (1.138)$$

as can be seen from Equation 1.113, and it has the *minimum mean-*

square error

$$\mathcal{E}[(x-s_i)^2] = \mathcal{E}[(|Y|^{-2}WY^T)^2]$$
$$= |Y|^{-4}\sigma^2|Y|^2 = \frac{\sigma^2}{E} \qquad (1.139)$$

from Equations 1.114 and 1.115. The mean-square error must be *minimum* for an unbiased estimate because the signal/noise ratio in this estimate has been *maximized*.

Returning now to the probability of error in the detection of s_i from R, as given by Equation 1.122, the following important result follows. For *given* values of k, E and σ^2, the same error probability is obtained with *any* vector Y, for example, whether

$$Y = [1 \quad 0 \quad 0 \quad \cdots \quad 0] \qquad (1.140)$$

or $\quad Y = [n^{-1/2} \quad n^{-1/2} \quad n^{-1/2} \quad \cdots \quad n^{-1/2}] \qquad (1.141)$

The former value of Y is attractive, because the matched filter detector degenerates into a simple threshold-level detector, which here becomes the optimum detection process. Furthermore, n data-symbols may now be transmitted in the vector R, such that

$$R = s_0 Y_0 + s_1 Y_1 + \cdots + s_{n-1} Y_{n-1} + W \qquad (1.142)$$

where now $s_i = \pm k$, for $i = 0, 1, \ldots, n-1$, and

$$Y_i = [\overbrace{0 \quad 0 \quad \cdots \quad 0}^{i} \quad 1 \quad 0 \quad 0 \quad \cdots \quad 0] \qquad (1.143)$$

so that Y_i is obtained from Y_0 by shifting its nonzero component i places to the right. The subscript of s here identifies *one* of n transmitted *data-symbols* rather than the *possible value* of a *given* data symbol, and is the terminology to be adopted in the subsequent chapters of the book. Clearly, $Y_i Y_h^T = 0$ for every $i \neq h$, so that the $\{Y_i\}$ are *mutually orthogonal*, and the n data-symbols $\{s_i\}$ may here be detected optimally and without the interference of any one data symbol in the detection process for another. This is, of course, the ideal *serial* transmission system often used. The basic conditions that are satisfied here are illustrated in more detail in the worked example of Section 1.6. Unfortunately, over practical channels the sampled impulse-response Y may often have several nonzero components such that, if more than one data symbol is transmitted, for example, if

$$R = s_0 Y_0 + s_1 Y_1 + W \qquad (1.144)$$

where Y_1 is obtained from Y_0 by shifting its nonzero components one place to the right, then it is possible that $Y_0 Y_1^T \neq 0$. Under these

42 BACKGROUND THEORY

conditions, the matched filter detection of either s_0 or s_1 (as in Fig. 1.6) is *no longer optimum*, essentially because of the *intersymbol interference* at the output of each matched filter caused by the unwanted data-symbol. For instance, the output signal from the filter matched to Y_0 together with the following multiplier (Fig. 1.6) is

$$x_0 = |Y_0|^{-2} R Y_0^T = s_0 + |Y_0|^{-2} s_1 Y_1 Y_0^T + u_0 \qquad (1.145)$$

where $|Y_0|^{-2} s_1 Y_1 Y_0^T$ is the *intersymbol interference* and u_0 is the noise. The optimum detection process for s_0 and s_1 now becomes more complex, since due account must be taken of the intersymbol interference if the optimum detection of s_0 and s_1 is to be achieved. The various topics under consideration in this book are, in fact, all aimed at reducing the effects of *intersymbol interference* in a received sampled signal.

1.6 EXAMPLE

Problem

Two received signals are

$$r_0 = s_0 + s_1 + w_0 \qquad (1.146)$$

$$r_1 = s_0 - s_1 + w_1 \qquad (1.147)$$

where s_0 and s_1 are statistically independent binary polar data-symbols such that $s_0 = \pm 2$ and $s_1 = \pm 1$, and where w_0 and w_1 are statistically independent Gaussian random variables with zero mean and variance σ^2. The data symbols are equally likely to have either binary value.

Give the appropriate linear combination of r_0 and r_1, (a) whereby s_0 may be detected without intersymbol interference from s_1, and (b) whereby s_1 may be detected without intersymbol interference from s_0.

Describe a detection process for s_0 and s_1 for the case where they are detected (a) from r_0 only, and (b) from both r_0 and r_1.

Derive, for each case, an expression for the probability of error in the detection of s_0 and s_1.

Compare the tolerances to noise of the two detection processes for each of s_0 and s_1.

Analyse the detection processes in terms of the appropriate two-component vectors in a two-dimensional Euclidean vector space, by deriving the maximum-likelihood estimates of s_0 and s_1, and show the relevant signals as points in a diagram of this space.

Solution

(a) The linear combination of r_0 and r_1, whereby s_0 may be detected without intersymbol interference from s_1, is

$$r_0 + r_1 = 2s_0 + w_0 + w_1 \qquad (1.148)$$

(b) The linear combination of r_0 and r_1, whereby s_1 may be detected without intersymbol interference from s_0, is

$$r_0 - r_1 = 2s_1 + w_0 - w_1 \qquad (1.149)$$

When s_0 and s_1 are detected from r_0 only, use is made of the fact that $s_0 + s_1$ uniquely determines both s_0 and s_1, as shown in Table 1.1. Thus $s_0 + s_1$ is detected as its possible value closest to r_0 and the corresponding possible value then gives the detected values of s_0 and s_1. This is implemented by comparing r_0 with thresholds at 0 and ± 2, as follows.

If $r_0 > 2$, $\quad s'_0 = 2 \quad$ and $\quad s'_1 = 1$

If $2 > r_0 > 0$, $\quad s'_0 = 2 \quad$ and $\quad s'_1 = -1$

If $0 > r_0 > -2$, $\quad s'_0 = -2 \quad$ and $\quad s'_1 = 1$

If $-2 > r_0$, $\quad s'_0 = -2 \quad$ and $\quad s'_1 = -1$

s'_0 and s'_1 are here the *detected* values of s_0 and s_1, respectively.

When s_0 and s_1 are detected from both r_0 and r_1, s_0 is detected from

$$r_0 + r_1 = 2s_0 + w_0 + w_1 \qquad (1.150)$$

and s_1 is detected from

$$r_0 - r_1 = 2s_1 + w_0 - w_1 \qquad (1.151)$$

Thus s_0 is detected as its possible value for which $2s_0$ is closest to $r_0 + r_1$, and s_1 is detected as its possible value for which $2s_1$ is closest to $r_0 - r_1$. These detection processes are implemented by comparing each of $r_0 + r_1$ and $r_0 - r_1$ with a threshold of zero, as follows.

Table 1.1 RELATIONSHIP BETWEEN s_0, s_1 AND $s_0 + s_1$

s_0	s_1	$s_0 + s_1$
2	1	3
2	−1	1
−2	1	−1
−2	−1	−3

44 BACKGROUND THEORY

If $r_0 + r_1 > 0$, $\quad s_0' = 2$

If $r_0 + r_1 < 0$, $\quad s_0' = -2$

If $r_0 - r_1 > 0$, $\quad s_1' = 1$

If $r_0 - r_1 < 0$, $\quad s_1' = -1$

In the detection of s_0 and s_1 from

$$r_0 = s_0 + s_1 + w_0 \tag{1.152}$$

the noise component w_0 is a Gaussian random variable with zero mean and variance σ^2. Thus the probability density function of w_0, at a sample value w, is

$$p_0(w) = \frac{1}{\sqrt{2\pi\sigma^2}} \exp\left(-\frac{w^2}{2\sigma^2}\right) \tag{1.153}$$

Now, if $s_0 + s_1 = \pm 3$, an error occurs in the detection of s_1 when w_0 has a magnitude greater than unity and the opposite sign to $s_0 + s_1$. An error occurs in the detection of s_0 when w_0 has a magnitude greater than 3 and the opposite sign to $s_0 + s_1$. If $s_0 + s_1 = \pm 1$, an error occurs in the detection of s_1 when w_0 has a magnitude greater than unity and the same sign as $s_0 + s_1$, and an error occurs in the detection of both s_0 and s_1 when w_0 has a magnitude between 1 and 3 and the opposite sign to $s_0 + s_1$. When the magnitude of w_0 here exceeds 3, there is an error only in the detection of s_0.

In the repeated detection of s_0 and s_1 from r_0, at high signal/noise ratios, effectively all errors are caused by the magnitude of w_0 exceeding *unity* and w_0 having the appropriate sign. Thus, when $s_0 + s_1 = \pm 3$, the probability of error in the detection of s_0 is negligible, and the probability of error in the detection of s_1 is effectively

$$\int_1^\infty p_0(w)\,dw = \int_1^\infty \frac{1}{\sqrt{2\pi\sigma^2}} \exp\left(-\frac{w^2}{2\sigma^2}\right) dw$$

$$= \int_{1/\sigma}^\infty \frac{1}{\sqrt{2\pi}} \exp\left(-\tfrac{1}{2}w^2\right) dw = Q\left(\frac{1}{\sigma}\right) \tag{1.154}$$

When $s_0 + s_1 = \pm 1$, the probability of error in the detection of s_0 is

$$\int_1^\infty p_0(w)\,dw = Q\left(\frac{1}{\sigma}\right) \tag{1.155}$$

and the probability of error in the detection of s_1 is effectively

$$2 \int_1^\infty p_0(w)\,dw = 2Q\left(\frac{1}{\sigma}\right) \tag{1.156}$$

Thus on average, over the four possible values of s_0+s_1, the probability of error in the detection of s_0 is $\tfrac{1}{2}Q(1/\sigma)$, and the probability of error in the detection of s_1 is $\tfrac{3}{2}Q(1/\sigma)$. At low error rates, each of these can be taken to be $Q(1/\sigma)$, with only a small error in the signal/noise ratio for a given error rate.

When s_0 is detected from

$$r_0 + r_1 = 2s_0 + w_0 + w_1 \tag{1.157}$$

$w_0 + w_1$ is a Gaussian random variable with zero mean and variance $2\sigma^2$. This follows because w_0 and w_1 are statistically independent and therefore uncorrelated Gaussian random variables, with zero mean and variance σ^2. Thus the probability density function of w_0+w_1, at a sample value w, is

$$p_1(w) = \frac{1}{\sqrt{4\pi\sigma^2}} \exp\left(-\frac{w^2}{4\sigma^2}\right) \tag{1.158}$$

An error occurs in the detection of s_0 from r_0+r_1, when w_0+w_1 has a magnitude greater than 4 and the opposite sign to s_0. Thus the probability of error in the detection of s_0 from r_0+r_1 is

$$\int_4^\infty p_1(w)\,dw = \int_4^\infty \frac{1}{\sqrt{4\pi\sigma^2}} \exp\left(-\frac{w^2}{4\sigma^2}\right) dw$$

$$= \int_{4/\sqrt{2}\sigma}^\infty \frac{1}{\sqrt{2\pi}} \exp(-\tfrac{1}{2}w^2)\,dw = Q\left(\frac{4}{\sqrt{2}\sigma}\right)$$

$$= Q\left(\frac{2\sqrt{2}}{\sigma}\right) \tag{1.159}$$

When s_1 is detected from

$$r_0 - r_1 = 2s_1 + w_0 - w_1 \tag{1.160}$$

$w_0 - w_1$ is a Gaussian random variable with zero mean and variance $2\sigma^2$, for the same reasons as for $w_0 + w_1$. Thus the probability density function of $w_0 - w_1$, at a sample value w, is $p_1(w)$ (Equation 1.158) as for $w_0 + w_1$. An error occurs in the detection of s_1 from $r_0 - r_1$, when $w_0 - w_1$ has a magnitude greater than 2 and the opposite sign to s_1.

46 BACKGROUND THEORY

Thus the probability of error in the detection of s_1 from $r_0 - r_1$ is

$$\int_2^\infty p_1(w)\,dw = \int_2^\infty \frac{1}{\sqrt{4\pi\sigma^2}} \exp\left(-\frac{w^2}{4\sigma^2}\right) dw$$

$$= \int_{2/\sqrt{2\sigma}}^\infty \frac{1}{\sqrt{2\pi}} \exp(-\tfrac{1}{2}w^2)\,dw = Q\left(\frac{2}{\sqrt{2\sigma}}\right)$$

$$= Q\left(\frac{\sqrt{2}}{\sigma}\right) \tag{1.161}$$

At a given low error probability, the value of σ in $Q(2\sqrt{2}/\sigma)$, the probability of error in the detection of s_0 from $r_0 + r_1$, is $2\sqrt{2}$ times that in $Q(1/\sigma)$, the effective probability of error in the detection of s_0 from r_0. Thus the detection of s_0 from $r_0 + r_1$ gains an advantage of approximately

$$20 \log_{10}(2\sqrt{2}) = 9 \text{ dB} \tag{1.162}$$

in tolerance to noise over the detection of s_0 from r_0.

Again, at a given low error probability, the value of σ in $Q(\sqrt{2}/\sigma)$, the probability of error in the detection of s_1 from $r_0 - r_1$, is $\sqrt{2}$ times that in $Q(1/\sigma)$, the effective probability of error in the detection of s_1 from r_0. Thus the detection of s_1 from $r_0 - r_1$ gains an advantage of approximately

$$20 \log_{10}(\sqrt{2}) = 3 \text{ dB} \tag{1.163}$$

in tolerance to noise over the detection of s_1 from r_0. There is, for practical purposes, the same tolerance to noise in the detection of s_0 or s_1 from r_0.

Consider now the two-component vectors

$$R = [r_0 \quad r_1] \tag{1.164}$$

$$Y_0 = [1 \quad 1] \tag{1.165}$$

$$Y_1 = [1 \quad -1] \tag{1.166}$$

and

$$W = [w_0 \quad w_1] \tag{1.167}$$

Then, from Equations 1.144 and 1.145,

$$R = s_0 Y_0 + s_1 Y_1 + W \tag{1.168}$$

where

$$|Y_0| = |Y_1| = \sqrt{2} \tag{1.169}$$

From Equation 1.112, the maximum likelihood *estimate* of s_0 is

$$x_0 = \tfrac{1}{2} R Y_0^T = \tfrac{1}{2} s_0 Y_0 Y_0^T + \tfrac{1}{2} s_1 Y_1 Y_0^T + \tfrac{1}{2} W Y_0^T$$

$$= s_0 + \tfrac{1}{2}(w_0 + w_1) \tag{1.170}$$

BACKGROUND THEORY 47

which is *unbiased*, and the maximum likelihood *estimate* of s_1 is

$$x_1 = \tfrac{1}{2} R Y_1^T = \tfrac{1}{2} s_0 Y_0 Y_1^T + \tfrac{1}{2} s_1 Y_1 Y_1^T + \tfrac{1}{2} W Y_1^T$$
$$= s_1 + \tfrac{1}{2}(w_0 - w_1) \tag{1.171}$$

which is also *unbiased*. The above equations use the fact that

$$Y_0 Y_1^T = Y_1 Y_0^T = 0 \tag{1.172}$$

Clearly, Y_0 and Y_1 are orthogonal, and there is no intersymbol interference in the estimate of either s_0 or s_1. If Equation 1.172 did not hold here, Equations 1.170 and 1.171 would *not* give the maximum-likelihood estimates of s_0 and s_1, respectively, bearing in mind that Equation 1.112 assumes the *absence* of intersymbol interference.

It is evident now that the detection of s_0 from $r_0 + r_1$ and the detection of s_1 from $r_0 - r_1$ are both processes of *matched-filter detection*. Furthermore, since there is no intersymbol interference in either case, each is an *optimum* detection process. Some further light is thrown on these detection processes by letting

$$S = s_0 Y_0 + s_1 Y_1 \tag{1.173}$$

so that

$$R = S + W \tag{1.174}$$

from Equation 1.168. S is here a two-component row vector whose possible values are

$$S_0 = [3 \quad 1] \tag{1.175}$$
$$S_1 = [1 \quad 3] \tag{1.176}$$
$$S_2 = [-1 \quad -3] \tag{1.177}$$
$$S_3 = [-3 \quad -1] \tag{1.178}$$

as can be seen from Equations 1.165 and 1.166, bearing in mind that $s_0 = \pm 2$ and $s_1 = \pm 1$. The possible values of S can now be plotted in the two-dimensional Euclidean vector space containing all possible vectors $\{R\}$, as shown in Fig. 1.7. This shows the relationships between $S_0, S_1, S_2, S_3, s_0 Y_0$ and $s_1 Y_1$ in the vector space, and it is assumed here that

$$R = 2Y_0 - Y_1 + W \tag{1.179}$$

so that $s_0 = 2$, $s_1 = -1$, and $S = S_1$ (Equation 1.176).

If the received vector R is fed to a filter matched to Y_0 and the output signal from the filter is suitably scaled to give the unbiased maximum-likelihood estimate x_0 of s_0 (as in Equation 1.170 and similarly to Fig. 1.6) then $x_0 Y_0$ is the *orthogonal projection* of R on to

48 BACKGROUND THEORY

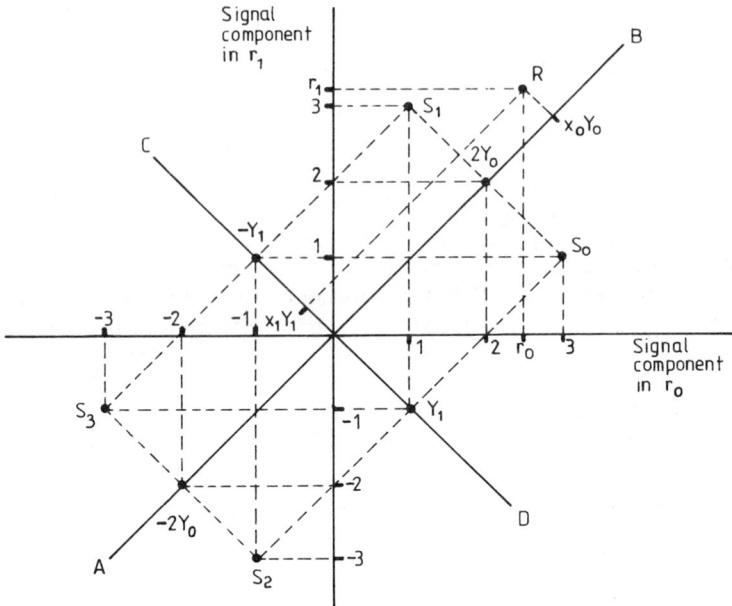

Fig. 1.7 Two-dimensional Euclidean vector space containing R

the line AB in Fig. 1.7. Since Y_1 is orthogonal to Y_0, $s_1 Y_1$ is eliminated from $x_0 Y_0$, as a result of the orthogonal projection, as is the component of W that is orthogonal to Y_0. Similarly, if R is fed to a filter matched to Y_1 and the output signal from the filter is suitably scaled to give the unbiased maximum-likelihood estimate x_1 of s_1 (as in Equation 1.171 and similarly to Fig. 1.6) then $x_1 Y_1$ is the *orthogonal projection* of R on to the line CD in Fig. 1.7. Since Y_0 is orthogonal to Y_1, $s_0 Y_0$ is eliminated from $x_1 Y_1$, as a result of the orthogonal projection, as is the component of W that is orthogonal to Y_1.

Since r_0 is the orthogonal projection of R on to the horizontal axis in Fig. 1.7, not only is the magnitude of the signal component $2Y_0$ in R reduced here relative to that in the output signal x_0 from the filter matched to Y_0, but also a significant intersymbol-interference component dependent on s_1 is introduced. These two effects together greatly reduce the tolerance to noise in the detection of s_0 from r_0, relative to that in the detection of s_0 from x_0. A similar relationship holds between the detection of s_1 from r_0 and the detection of s_1 from x_1. Indeed, in Fig. 1.7, s_0 and s_1 are correctly detected from x_0 and x_1, respectively, as is s_0 from r_0, whereas s_1 is incorrectly detected from r_0.

REFERENCES

1. Nyquist, H. 'Certain topics in telegraph transmission theory', *AIEE Trans.*, **47**, 617–644 (1928)
2. Shannon, C.E. 'A mathematical theory of communication', *Bell Syst. Tech. J.*, **27**, 379–423 and 623–656 (1948)
3. Shannon, C.E. and Weaver, W. *The Mathematical Theory of Communication*, University of Illinois Press, Urbana (1949)
4. Woodward, P.M. *Probability and Information Theory, with Applications to Radar*, Pergamon Press, Oxford (1953)
5. Wozencraft, J.M. and Jacobs, I.M. *Principles of Communication Engineering*, Wiley, New York (1965)
6. Bennett, W.R. and Davey, J.R. *Data Transmission*, McGraw-Hill, New York (1965)
7. Lucky, R.W., Salz, J. and Weldon, E.J. *Principles of Data Communication*, McGraw-Hill, New York (1968)
8. Lathi, B.P. *An Introduction to Random Signals and Communication Theory*, Intertext Books, London (1968)
9. Schwartz, M. *Information, Transmission, Modulation and Noise*, McGraw-Hill, Kogakusha, Tokyo (1970)
10. Taub, H. and Schilling, D.L. *Principles of Communication Systems*, McGraw-Hill, New York (1971)
11. Lucky, R.W. 'A survey of the communication theory literature: 1968–1973', *IEEE Trans. Inform. Theory*, **IT-19**, 725–739 (1973)
12. Clark, A.P. *Advanced Data-Transmission Systems*, Pentech Press, London (1977)
13. Clark, A.P. *Principles of Digital Data Transmission*, Second Edition, Pentech Press, London (1983)
14. Proakis, J.G. *Digital Communications*, McGraw-Hill, New York (1983)
15. Papoulis, A. *Probability, Random Variables and Stochastic Processes*, McGraw-Hill, New York (1965)
16. Thomas, J.B. *An Introduction to Statistical Communication Theory*, Wiley, New York (1968)
17. Davenport, W.B. *Probability and Random Processes*, McGraw-Hill, New York (1970)
18. Kurzweil, J. and Bradley, S.D. 'Modeling passband distortion in QAM transmission systems', *Midwest Symposium on Circuits and Systems*, University of Toledo, 114–118 (1980)
19. Harvey, J.D. 'Synchronisation of a Synchronous Modem', SERC Report GR/A/1200.7, SERC, Swindon, England (1980)
20. Stiffler, J.J. *Theory of Synchronous Communications*, Prentice Hall, Englewood Cliffs, NJ (1971)
21. Browne, E.T. *Introduction to the Theory of Determinants and Matrices*, Chapel Hill, University of North Carolina Press (1958)
22. Paige, L.J. and Swift, J.D. *Elements of Linear Algebra*, Blaisdell Publishing Co., New York (1961)
23. Ayres, F. *Matrices*, McGraw-Hill, New York (1962)

Chapter 2

The discrete Fourier transform

2.1 INTRODUCTION

The model of the data-transmission system is assumed to be as shown in Fig. 2.1, where all signals are *real valued*. This is a synchronous serial digital system in which the receiver is held synchronized in time with the received stream of signal elements. The properties of the transmission path and the properties of the transmitter and receiver filters are assumed to be as described in Section 1.2. Each received signal-element at the detector input here appears as a sequence of samples, and the signal distortion introduced by the transmission path is the *change* from the ideal form in this sequence of samples. An essential tool for the classification and study of signal distortion is the discrete Fourier transform (DFT). The DFT has, of course, many other important applications and is of great interest in its own right.

It is assumed that the transmission path in Fig. 2.1 is a linear baseband channel or else that it is a linear bandpass channel and the modulation and demodulation processes used in the data-transmission system are both linear. In the latter case, the modulator (at the transmitter) and the demodulator (at the receiver) are both considered to be part of the transmission path, which is therefore always a *baseband* channel. Furthermore, the filter at the output of the transmitter, that limits the transmitted signal spectrum to the available frequency band of the transmission path, and the filter at the input to the receiver, that removes the noise frequency components outside the signal frequency band, are now always lowpass filters that operate on a *baseband* signal. The transmitter filter, transmission path and receiver filter together form a *linear baseband channel*, as shown in Fig. 2.1. It is assumed that this channel is time invariant, so that its impulse response does not vary with time. The baseband channel can now introduce only two types of signal distortion, known as *amplitude* distortion and *phase* distortion.

In the following discussion the DFT is derived from first principles in order to bring out the true significance of the transform. In Chapter

THE DISCRETE FOURIER TRANSFORM 51

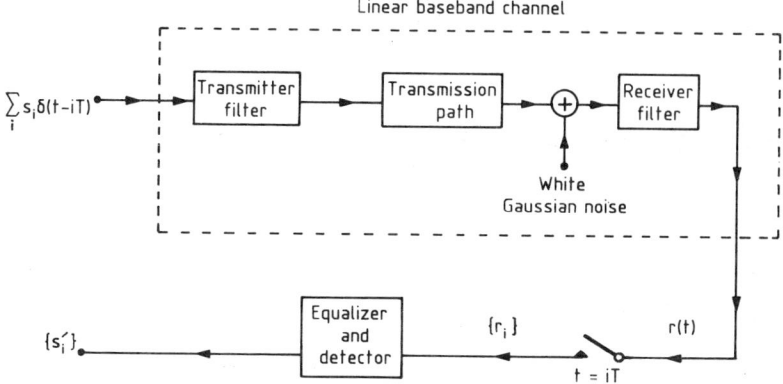

Fig. 2.1 Data-transmission system

3 the DFT is used to define both amplitude distortion and phase distortion, in the sampled impulse-response of a linear baseband channel. By means of this analysis it is possible not only to estimate the nature and severity of the signal distortion represented by any given sampled impulse-response, but also to determine which general type of detection process is likely to be the most cost-effective for a data signal transmitted over the baseband channel. The different possible detection processes are not, however, studied in this chapter.

The input signal to the baseband channel is a sequence of impulses

$$\sum_i s_i \delta(t - iT)$$

regularly spaced at intervals of T seconds and starting with $s_0 \delta(t)$. Each impulse is a signal element carrying the corresponding data-symbol s_i. The data symbols may be binary or multilevel, and each s_i may take on any one of the specified set of possible values. The transmitted signal may be of infinite or finite duration, so that the integer i may take on all nonnegative integer values or just those over a given finite range.

In practice, a rectangular or rounded waveform would be used in place of the sequence of impulses at the input to the baseband channel, and the appropriate change would be made to the transmitter filter to give the same signal at the input to the transmission path as that obtained in Fig. 2.1. A sequence of *impulses* is assumed because this greatly simplifies the theoretical analysis of the system. The data-symbol s_i is, of course, carried by the *area* of the corresponding impulse.

The signal at the output of the baseband channel in Fig. 2.1 is the

52 THE DISCRETE FOURIER TRANSFORM

real-valued waveform

$$r(t) = \sum_i s_i y(t - iT) + w(t) \tag{2.1}$$

where $y(t)$ is the impulse response of the baseband channel, such that the impulse $s_i \delta(t - iT)$ at the input to the baseband channel becomes the waveform $s_i y(t - iT)$ at its output. $w(t)$ is the bandlimited Gaussian waveform at the output of the receiver filter, originating from the additive white Gaussian noise at its input. Since we are not, for the present, concerned with the effects of the noise on the detection of the received data signal, but solely with the distortion introduced by the linear baseband channel of Fig. 2.1 into the transmitted data signal, we shall now assume that the noise is set to zero, so that

$$r(t) = \sum_i s_i y(t - iT) \tag{2.2}$$

The distortion introduced into the data signal is clearly determined by $y(t)$.

It is assumed that the waveform $r(t)$ at the output of the baseband channel is sampled once per signal element, at the time instants $\{iT\}$, to give the corresponding samples $\{r_i\}$, where $r_i = r(iT)$. These are fed to the detector which operates on the $\{r_i\}$ to give the detected values of the $\{s_i\}$.

Suppose now that $\delta(t)$, a unit impulse at the time instant $t = 0$, is fed to the baseband channel in the absence of any other signals. The output signal from the baseband channel is $y(t)$, the impulse response of the channel. This waveform, which is real-valued, is sampled at the time instants $\{iT\}$, for all integer values of i, to give the sequence of impulses

$$y_0 \delta(t) + y_1 \delta(t - T) + y_2 \delta(t - 2T) + \cdots + y_g \delta(t - gT) \tag{2.3}$$

The representation of the sampled signal as a sequence of *impulses* is a mathematical technique for identifying each sample (sample value) with its precise time-instant[1-4]. The impulses may be considered more simply as the corresponding sequence of real-valued samples

$$[y_0 \; y_1 \; y_2 \; \cdots \; y_g] \tag{2.4}$$

where $y_i = y(iT)$. It is assumed here that $y_i = 0$ when $i < 0$ and $i > g$, so that the $\{y_i\}$ for which $i < 0$ and $i > g$ can be ignored. The sequence given by Expression 2.4 is said to be the *sampled impulse-response* of the baseband channel.

In practice, the sampler in Fig. 2.1 determines the sample of the received signal $r(t)$ at the time instant $t = iT$, for each integer value of i, and feeds each sample to the detector where it is stored for use in the detection process. The sampled impulse-response of the channel is therefore in fact just a sequence of numbers or values, which are represented more realistically by Expression 2.4 than by Expression 2.3. However, it is theoretically correct and often mathematically convenient (as is shown in Section 2.2) to consider a sequence of samples in terms of the corresponding sequence of *impulses*. It must therefore be borne in mind that this is an idealized representation of the signal at the output of the sampler and, except in the *values* of the samples, it does not correspond even approximately to the signal waveform normally obtained in practice.

The sampled impulse-response is physically realizable and of finite duration. Furthermore, if the delay in transmission is neglected, $y_0 \neq 0$. In the sampled impulse-response of any practical channel, the sample values become negligible after a certain time delay, so that, with the appropriate value of g, Expression 2.4 can be considered, within any required degree of accuracy, to be the sampled impulse-response of *any* given practical baseband channel. In all cases to be considered here, the sum of the squares of the $\{y_i\}$ is strictly *finite*.

If only the data-symbol s_i is transmitted, so that a single impulse $s_i \delta(t - iT)$ of value (area) s_i is fed to the baseband channel in Fig. 2.1, at the time instant $t = iT$, then the corresponding received samples $\{r_j\}$, starting with r_i, are

$$s_i y_0, s_i y_1, \ldots, s_i y_g$$

at the time instants $iT, (i+1)T, \ldots, (i+g)T$, respectively. All preceding and following $\{r_j\}$ are zero. Table 2.1 shows how the $\{r_j\}$ are formed from the $\{s_j\}$ and $\{y_j\}$, when a continuous (uninterrupted) stream of data-symbols $\{s_j\}$ is transmitted. It can be seen that the received

Table 2.1 RELATIONSHIP BETWEEN r_i AND THE $\{s_j\}$ AND $\{y_j\}$

Time	$(i-g)T$	$(i-g+1)T$	$(i-g+2)T$	...	$(i-2)T$	$(i-1)T$	iT
$(i-g+1)$th element	$s_{i-g} y_0$	$s_{i-g} y_1$	$s_{i-g} y_2$	...	$s_{i-g} y_{g-2}$	$s_{i-g} y_{g-1}$	$s_{i-g} y_g$
$(i-g+2)$th element	0	$s_{i-g+1} y_0$	$s_{i-g+1} y_1$	...	$s_{i-g+1} y_{g-3}$	$s_{i-g+1} y_{g-2}$	$s_{i-g+1} y_{g-1}$
ith element	0	0	0	... 0		$s_{i-1} y_0$	$s_{i-1} y_1$
$(i+1)$th element	0	0	0	... 0		0	$s_i y_0$
Received samples	r_{i-g}	r_{i-g+1}	r_{i-g+2}	...	r_{i-2}	r_{i-1}	r_i

sample at time $t = iT$ is now

$$r_i = \sum_{h=0}^{g} s_{i-h} y_h \qquad (2.5)$$

for any given integer value of i. Thus it may well not be possible to detect any one of the data-symbols $s_i, s_{i-1}, \ldots, s_{i-g}$ from r_i, due to the interference from the other data symbols, even in the complete absence of noise as assumed here. If it is required to detect s_i from r_i then, in Equation 2.5, $s_i y_0$ is the *wanted* signal component and the terms $s_{i-1} y_1$ to $s_{i-g} y_g$ are the *intersymbol-interference* components. Similarly, if it is required to detect s_{i-j} from r_i, $s_{i-j} y_j$ is the wanted signal component and the terms $\{s_{i-h} y_h\}$, for $0 \leqslant h \leqslant g$ and $h \neq j$, are the intersymbol-interference components. Unfortunately, the ease with which the distortion may be eliminated and the reduction in tolerance to additive noise resulting from the signal distortion, are not necessarily determined by just the number and magnitudes of the $\{y_i\}$ in Expression 2.4. In order to determine how a detector can best handle the linear signal distortion given by Equation 2.5, it is necessary first to express this distortion in terms of *amplitude distortion* and *phase distortion*. The magnitudes of these two parameters give a much better measure of the effects of the signal distortion on the complexity and performance of the detector for the received digital signal, than do the number and magnitudes of the $\{y_i\}$ in Expression 2.4. Since the amplitude distortion and phase distortion are defined and evaluated by means of the DFT, it is necessary first to study the DFT. Another aim of this study is to explain some of the useful properties of the DFT and to show some of the ways in which it may be used in the analysis of digital signals.

2.2 Z-TRANSFORMS

Assume that the signal at the input to the baseband channel in Fig. 2.1 is the sequence of impulses

$$\sum_i s_i \delta(t - iT) \qquad (2.6)$$

where i now takes on all integer values. Now the Fourier transform of $\delta(t - iT)$ is $\exp(-j2\pi f iT)$, so that the Fourier transform of Expression 2.6 is

$$\sum_i s_i \exp(-j2\pi f iT) = \sum_i s_i z^{-i} \qquad (2.7)$$

THE DISCRETE FOURIER TRANSFORM

where
$$z = \exp(j2\pi f T) \quad (2.8)$$
and
$$j = \sqrt{-1} \quad (2.9)$$

The z-transform of the signal in Expression 2.6 is defined to be

$$S(z) = \sum_i s_i z^{-i} \quad (2.10)$$

The sampled impulse-response of the baseband channel, expressed as the corresponding sequence of impulses, is

$$\sum_{i=0}^{g} y_i \delta(t - iT) \quad (2.11)$$

so that the z-transform of the sampled impulse-response of the channel is

$$Y(z) = \sum_{i=0}^{g} y_i z^{-i} \quad (2.12)$$

$Y(z)$ is, of course, the z-transform of the *sequence* of samples given by the $(g+1)$-component row vector

$$Y = [y_0 \quad y_1 \quad \cdots \quad y_g] \quad (2.13)$$

The vector (sequence) Y itself is taken to be the *sampled impulse-response* of the channel and $Y(z)$ is said to be the *z-transform of Y*. For convenience, $Y(z)$ is often referred to simply as the *z-transform of the channel*. For the purposes of Chapters 2 and 3, it is assumed that Y is time invariant and is known at the receiver. This implies, of course, that Y is held in store at the receiver.

The relationship between $y(t)$, Y and $Y(z)$ for a typical example is illustrated in Fig. 2.2. In deriving $Y(z)$ from Y here, the sampled impulse-response of the channel is considered mathematically as

$$y_0 \delta(t) + y_1 \delta(t-T) + \cdots + y_5 \delta(t-5T)$$

and the delay in the transmission of the first sample, y_0, is for convenience set to zero. It is, in fact, normal practice to ignore the delay in transmission of the first component of the sampled impulse-response of the channel, because in a synchronous serial system the receiver is synchronized in time to the received signal, with usually an *unknown* delay or displacement between the timing waveform at the receiver and that at the transmitter. Thus the *time scale* used at the receiver is shifted so as to remove the time delay.

Again, the z-transform of the samples $\{r_i\}$ of the received signal in Fig. 2.1 is

$$R(z) = \sum_i r_i z^{-i} \quad (2.14)$$

56 THE DISCRETE FOURIER TRANSFORM

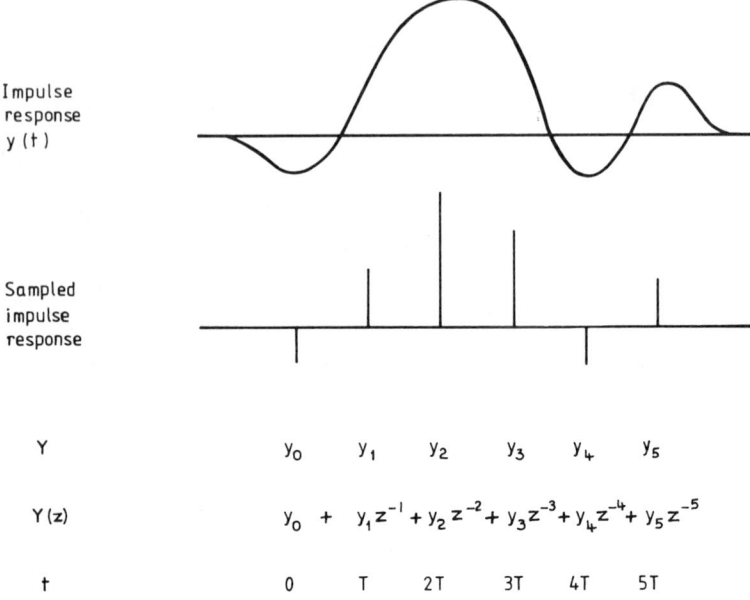

Fig. 2.2 *Relationship between $y(t)$, Y and $Y(z)$ for a typical baseband channel*

In taking the z-transform of a sequence of samples, these are treated as the corresponding sequence of *impulses*, as for instance in Expression 2.11. This may be done so long as the samples are taken at regular intervals of T seconds, at the time instants $\{iT\}$, regardless of whether or not the sampled signal itself, as actually generated, approximates to the corresponding sequence of impulses.

The value of the *transfer function* of the baseband channel and sampler in Fig. 2.1, at a frequency f Hz, is the value of the Fourier transform of the sequence of impulses in Expression 2.11, at f Hz. It can be seen that this is simply the value of $Y(z)$ when $z = \exp(j2\pi fT)$. Clearly, z must here lie *on the unit circle* (the circle of unit radius with its centre at the origin) in the complex number plane. This plane contains all possible values of z (not just those confined to the unit circle) and is often referred to as the z *plane*. Thus, at f Hz, the transfer function has the value $Y(z)$, where z lies on the unit circle and at an angle of $2\pi fT$ radians with the positive real axis, as shown in Fig. 2.3. Furthermore, the value of z is the same at all frequencies $f + (i/T)$, where i takes on all integer values, and z moves *once* round the unit circle in the z plane as f increases from i/T to $(i+1)/T$, for any integer i. This means that the transfer function of the baseband channel and

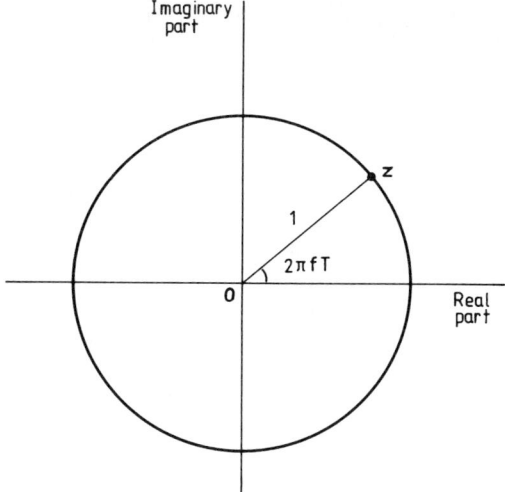

Fig. 2.3 Relationship between z and f

sampler repeats itself cyclically over successive frequency intervals of $1/T$ Hz.

It can be seen from Equations 2.10, 2.12 and 2.14 that the z-transform of any sample x_i, at the time instant $t = iT$, is $x_i z^{-i}$. Thus z^{-i} corresponds to the time instant $t = iT$, and its *coefficient* x_i, in the z-transform, is the *sample value* at this time instant. Furthermore, if a set of samples is advanced in time by iT seconds, where i is an integer, the corresponding z-transform is multiplied by z^i. Similarly, if the samples are delayed by iT seconds, the z-transform is multiplied by z^{-i}.

Suppose next that $z = \exp[(\sigma + j2\pi f)T]$ where σ may have any positive or negative real value. z can now have *any* complex value and so can lie anywhere in the z plane. It can be seen that, just as any value of z on the unit circle in the z plane may be obtained via the Fourier transform of a sampled signal, so any value of z not on the unit circle may be obtained via the Laplace transform of the sampled signal. The transformation $z = \exp[(\sigma + j2\pi f)T]$ corresponds to the mapping of the s plane on to the z plane. All points to the left of the imaginary axis in the s plane are transformed (mapped) into points inside the unit circle in the z plane, and all points to the right of the imaginary axis are transformed into points outside the unit circle. This gives some idea of the significance of any given value of z. However, unless specifically considering values of z that do *not* lie on the unit circle in the z plane, such as the location of the *zeros* (*roots*) of the z-transform, it will now be assumed that $z = \exp(j2\pi fT)$.

58 THE DISCRETE FOURIER TRANSFORM

One reason for constraining the possible values of z such that $|z|=1$ (except when considering the zeros (roots) of the z-transform as just mentioned) is that this identifies the z-transform closely with the Fourier transform, and so avoids the problems of defining the z-transform over negative time instants, that is, for z^i where $i>0$. The simple one-sided or single-sided Laplace transform of a time waveform, of course, assumes the latter to be zero over negative time. A second reason is that it avoids the serious problems of convergence associated with an infinite sequence, at values of $|z|$ well away from unity, the z-transform of any such sequence being usually only valid over some range of values of $|z|$[2,3]. A further, and perhaps the most important single reason, is that, for the most part in Chapters 4–7, when not considering the zeros or poles of a z-transform, the latter is used simply as a convenient means of identifying the *times of occurrence* of the samples of a sequence, this being achieved by associating the coefficient of z^{-i} with the time instant $t=iT$. The powers of z in a z-transform do not here have any values or magnitudes as such, being essentially *operators* that determine the time locations or delays of the corresponding coefficients.

Now, from Equations 2.10 and 2.12,

$$S(z)Y(z)=\sum_i s_i z^{-i} \sum_{h=0}^{g} y_h z^{-h} \qquad (2.15)$$

so that the coefficient of z^{-i} in $S(z)Y(z)$ is

$$\sum_{h=0}^{g} s_{i-h} y_h = r_i \qquad (2.16)$$

from Equation 2.5. Thus

$$S(z)Y(z)=R(z) \qquad (2.17)$$

Consider next a finite sequence of m signal elements, at the input to the baseband channel in Fig. 2.1, starting with the element $s_0 \delta(t)$ at the time instant $t=0$. The corresponding sequence of data symbols is

$$[s_0 \quad s_1 \quad s_2 \quad \cdots \quad s_{m-1}] \qquad (2.18)$$

and the resulting sequence of samples at the detector input is

$$[r_0 \quad r_1 \quad r_2 \quad \cdots \quad r_{m+g-1}] \qquad (2.19)$$

The z-transform of the sequence of the $\{s_i\}$ in Expression 2.18 is

$$S(z)=s_0+s_1 z^{-1}+\cdots+s_{m-1} z^{-m+1} \qquad (2.20)$$

and the z-transform of the sequence of the $\{r_i\}$ in Expression 2.19 is

$$R(z)=r_0+r_1 z^{-1}+\cdots+r_{m+g-1} z^{-m-g+1} \qquad (2.21)$$

It can be seen from Equation 2.5 that, since $s_i = 0$ for $i < 0$ and $i > m-1$, Equation 2.17 is again satisfied. Thus it is clear that, whether the sequence of the $\{s_i\}$ is finite or infinite, the z-transform of the corresponding sequence of the $\{r_i\}$ is the *product* of the z-transforms of the sequences of the $\{s_i\}$ and $\{y_i\}$. On the other hand, the *sequence* of the $\{r_i\}$ itself is the *convolution* of the sequences of the $\{s_i\}$ and $\{y_i\}$, whether the sequence of the $\{s_i\}$ is finite or infinite. The great value of z-transforms is that the *convolution* of two sequences becomes the *product* of their respective z-transforms, which is often easier to handle and evaluate mathematically. Furthermore, as has previously been mentioned, the z-transform of a sequence of samples can be considered both as a convenient method of representing the *values* and *times of occurrence* of the samples, in terms of a polynomial in z, and also as a means of evaluating the Fourier transform of the sequence, for any value of frequency, using the fact that $z = \exp(j2\pi fT)$.

Equation 2.17 obviously holds also in the case where $R(z)$ is the z-transform of the sequence of samples obtained at the output of a *digital filter*, whose sampled impulse-response has the z-transform $Y(z)$, and whose input signal is the sequence of samples with z-transform $S(z)$. Indeed the whole of this and the following analysis applies also in the case where the baseband channel in Fig. 2.1 is replaced by the corresponding digital filter.

The values of the zeros (*roots*) of the z-transform of a finite sequence are those values of z, *not* necessarily satisfying $z = \exp(j2\pi fT)$, for which the corresponding polynomial in z goes to zero.

2.3 CONVOLUTION MATRIX

With the sequence of m signal-elements at the input to the baseband channel, the convolution of the sequence of the m $\{s_i\}$ with the sequence of the $g+1$ $\{y_i\}$, that gives the corresponding sequence of the $m+g$ $\{r_i\}$, can be expressed in terms of matrices, as follows.

Let S and R be the n-component row vectors

$$S = [s_0 \quad s_1 \quad \ldots \quad s_{m-1} \quad 0 \quad \ldots \quad 0] \qquad (2.22)$$

and
$$R = [r_0 \quad r_1 \quad \ldots \quad r_{n-1}] \qquad (2.23)$$

where
$$n = m + g \qquad (2.24)$$

The vectors S and R are, of course, both matrices having one row and n columns, and each represents the corresponding *sequence* of samples. Let Y_c be the $n \times n$ matrix

$$Y_c = \begin{bmatrix} y_0 & y_1 & \cdots & y_{g-2} & y_{g-1} & y_g & 0 & 0 & \cdots & 0 \\ 0 & y_0 & \cdots & y_{g-3} & y_{g-2} & y_{g-1} & y_g & 0 & \cdots & 0 \\ \vdots & \vdots & \vdots\vdots\vdots & \vdots & \vdots & \vdots & \vdots\vdots\vdots & \vdots\vdots\vdots & \vdots\vdots\vdots & \vdots \\ y_2 & y_3 & \cdots & y_g & 0 & 0 & \cdots & \cdots & y_0 & y_1 \\ y_1 & y_2 & \cdots & y_{g-1} & y_g & 0 & \cdots & \cdots & 0 & y_0 \end{bmatrix} \quad (2.25)$$

whose $(i+1)$th row, for $i = 0, 1, \ldots, m-1$, is given by the n-component row vector

$$Y_i = [\overbrace{0 \ \cdots \ 0}^{i} \ y_0 \ y_1 \ \cdots \ y_g \ 0 \ \cdots \ 0] \quad (2.26)$$

The matrix Y_c is not to be confused with $Y(z)$ which is the z-transform of the sequence (vector) Y (Equation 2.13). Both of these, however, perform the same linear transformation on the signal elements involved. Since Y_0 (the first row of Y_c) is obtained from Y by adding $n - g - 1$ zeros to the latter, Y_0 may also be taken as the sampled impulse-response of the channel.

Y_c is a circulant matrix, which is a square matrix such that the $(i+1)$th row is obtained by a cyclic shift of the components of the first row by i places to the right. A cyclic shift of one place to the right of a component at the right-hand end of a row, transfers this component to the left-hand end of the row. Clearly, each row of Y_c contains the same components and in the same cyclic order, and all the components along the main diagonal of Y_c are y_0. Some important properties of a circulant matrix are given by Equations 1.68–1.84 in Section 1.3.

It can now be seen from Equation 2.5 that

$$R = S Y_c \quad (2.27)$$

bearing in mind that $s_i = 0$ for $i < 0$ and $i > m - 1$. Again, R is obtained by the *convolution* of the sequences of the $\{s_i\}$ and $\{y_i\}$, and the sequences are here represented by the n-component row vectors S and Y_0. Y_c may therefore be considered as the *convolution matrix* that transforms S to R. The sequence (vector) R is the *linear (aperiodic)* convolution of the sequence of data symbols given by S and the sampled impulse-response of the channel given by Y (Equation 2.13).

Throughout the following analysis the terms *sequence* and *vector* are used interchangeably and each is treated mathematically as a matrix having one row and n columns (assuming, of course, an n-component sequence or row vector).

A further technique for evaluating the convolution of two

sequences or vectors is via the *discrete Fourier Transform* (DFT) which will now be studied in some detail.

2.4 DEFINITION OF THE DFT

Consider a sequence of n samples $\{r_i\}$ of the continuous waveform $r(t)$ at the output of the baseband channel in Fig. 2.1. The samples are regularly spaced at intervals of T seconds and the first sample r_0 is at time $t=0$. The samples may be represented as impulses whose areas are equal to the sample values. The $(i+1)$th impulse occurs at time $t=iT$ and has a value (area) r_i, for $i=0, 1, \ldots, n-1$. Thus the sequence of impulses is

$$\sum_{i=0}^{n-1} r_i \delta(t-iT) \qquad (2.28)$$

The Fourier transform of this sequence is

$$\sum_{i=0}^{n-1} r_i \exp(-j2\pi fiT) \qquad (2.29)$$

where $j=\sqrt{-1}$.

The value of the Fourier transform at the discrete frequency h/nT Hz, where h is any integer, is

$$G\left(\frac{h}{nT}\right) = \sum_{i=0}^{n-1} r_i \exp\left(-j2\pi \frac{h}{nT} iT\right)$$

$$= \sum_{i=0}^{n-1} r_i \exp\left(-j2\pi \frac{hi}{n}\right) = \sum_{i=0}^{n-1} r_i \omega^{-hi} \qquad (2.30)$$

where

$$\omega = \exp\left(j\frac{2\pi}{n}\right) \qquad (2.31)$$

Clearly,

$$\omega^i = \exp\left(j\frac{2\pi i}{n}\right) \qquad (2.32)$$

so that ω^i is a unit vector in the complex number plane, at an angle of $2\pi i/n$ radians with the positive real axis, as shown in Fig. 2.4. ω^i, for $i=0, 1, \ldots, n-1$, gives each of the n distinct nth roots of unity, which means, of course, that

$$(\omega^i)^n = \omega^{in} = 1 \qquad (2.33)$$

Furthermore, for any integer i,

$$\omega^i = \omega^{(i \bmod{-n})} \qquad (2.34)$$

where

$$0 \leqslant i \bmod{-n} < n \qquad (2.35)$$

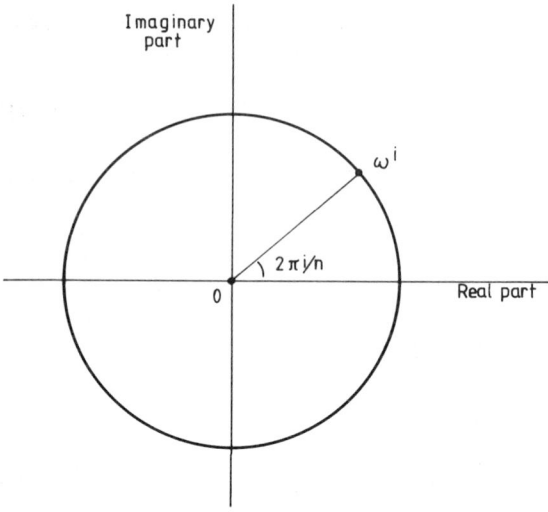

Fig. 2.4 Location of ω^i in the complex-number plane

and
$$i \text{ modulo} - n = i - ln \tag{2.36}$$

l being the most positive integer such that $i - ln$ is nonnegative. In particular, when $i < 0$, l is the least negative integer such that $i - ln \geq 0$, and ω^{-i}, for $i = 0, 1, \ldots, n-1$, again gives each of the n distinct nth roots of unity. In general, l is the appropriate integer such that Equation 2.35 is satisfied.

It is evident that in Equation 2.30

$$G\left(\frac{h}{nT}\right) = G\left(\frac{h+in}{nT}\right) \tag{2.37}$$

for any integer i, so that there are at most n different values of $G(h/nT)$ for all integer values of h.

At zero frequency (d.c.), $h = 0$ and the Fourier transform becomes

$$G(0) = \sum_{i=0}^{n-1} r_i \tag{2.38}$$

Let F_h be the n-component row vector

$$F_h = [1 \quad \omega^{-h} \quad \omega^{-2h} \quad \ldots \quad \omega^{-(n-1)h}] \tag{2.39}$$

where $h = 0, 1, \ldots, n-1$, and ω is given by Equation 2.31. Let R be the n-component row vector

$$R = [r_0 \quad r_1 \quad \ldots \quad r_{n-1}] \tag{2.40}$$

Then, from Equation 2.30, the Fourier transform of the sequence of samples $r_0, r_1, \ldots, r_{n-1}$, at the discrete frequency h/nT Hz, is

$$G\left(\frac{h}{nT}\right) = RF_h^T \tag{2.41}$$

where F_h^T is the column vector ($n \times 1$ matrix) formed by the transpose of the row vector ($1 \times n$ matrix) F_h.

Let F be the $n \times n$ matrix whose $(h+1)$th row is F_h, so that

$$F = \begin{bmatrix} 1 & 1 & 1 & \cdots & 1 \\ 1 & \omega^{-1} & \omega^{-2} & \cdots & \omega^{-(n-1)} \\ 1 & \omega^{-2} & \omega^{-4} & \cdots & \omega^{-2(n-1)} \\ \vdots & \vdots & \vdots & \vdots & \vdots \\ 1 & \omega^{-(n-1)} & \omega^{-2(n-1)} & \cdots & \omega^{-(n-1)^2} \end{bmatrix} \tag{2.42}$$

Clearly, F is a symmetric matrix (but not a real matrix) and the $(h+1)$th column of F is F_h^T.

$G(h/nT)$ is the $(h+1)$th component of the n-component row vector RF, for $h = 0, 1, \ldots, n-1$, so that the n values of $G(h/nT)$ are given by the n components of RF. The vector RF is said to be the *discrete Fourier transform* (DFT) of R.

Whereas the components of R are real-valued, the components of RF are often complex-valued.

From Equations 2.37 and 2.41 it can be seen that the n components of the row vector RF contain the values of $G(h/nT)$ for *all* values of the integer h. The z-transform of the sequence R is $R(z)$ in Equation 2.21, where $n = m + g$. But $G(h/nT)$ is the value of $R(z)$ when $z = \exp(j2\pi h/n)$. Thus the n values of $G(h/nT)$ for $h = 0, 1, \ldots, n-1$, as given by the n components of RF, are the n values of $R(z)$ for values of z equally spaced on the unit circle in the z plane, at angles of $2\pi h/n$ radians with the positive real axis. In the particular case where $n = 8$, the values of z (on the unit circle) corresponding to the eight discrete frequencies $\{h/8T\}$, for $h = 0, 1, \ldots, 7$, are as shown in Fig. 2.5, where, at the given values, $z = \omega^h$ for $h = 0, 1, \ldots, 7$. It is evident from Equation 2.31 that now $\omega^h = \omega^{h+8i}$ for each value of h and any integer i.

The inverse of the matrix F is

$$F^{-1} = n^{-1} \begin{bmatrix} 1 & 1 & 1 & \cdots & 1 \\ 1 & \omega & \omega^2 & \cdots & \omega^{n-1} \\ 1 & \omega^2 & \omega^4 & \cdots & \omega^{2(n-1)} \\ \vdots & \vdots & \vdots & \vdots & \vdots \\ 1 & \omega^{n-1} & \omega^{2(n-1)} & \cdots & \omega^{(n-1)^2} \end{bmatrix} \tag{2.43}$$

64 THE DISCRETE FOURIER TRANSFORM

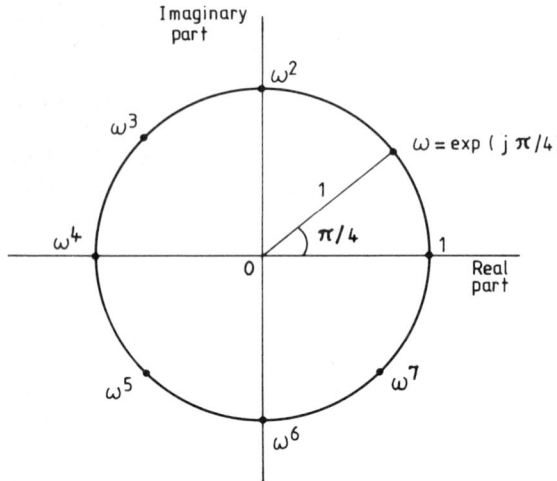

Fig. 2.5 Values of z corresponding to the eight discrete frequencies

whose $(h+1)$th row is the n-component row vector $n^{-1}E_h$, where

$$E_h = [1 \quad \omega^h \quad \omega^{2h} \quad \ldots \quad \omega^{(n-1)h}] \qquad (2.44)$$

The $(h+1)$th column of F^{-1} is $n^{-1}E_h^T$.

The n-component row vector RF^{-1} is said to be the *inverse discrete Fourier transform* (inverse DFT) of R.

The n discrete frequencies of the DFT of a sequence are not chosen quite as arbitrarily as might be suggested by the above analysis. Suppose, for instance, that the sequence of n impulses $\{r_i\delta(t-iT)\}$ is repeated cyclically to give the *periodic* sequence of impulses $\{r_h\delta(t-hT)\}$ over all positive and negative integers $\{h\}$. In the latter sequence, $r_h = r_i$ whenever $i = h - kn$, where k is the appropriate integer such that $0 \leq i < n$. The Fourier transform of the resulting periodic sequence (with period nT seconds) becomes a Fourier series, having frequency components *only* at integral multiples of $1/nT$ Hz. These frequencies clearly include the n discrete frequencies of the DFT of R (Equation 2.40). Furthermore, the Fourier series is itself *periodic*, repeating itself cyclically after every n components, as for the $\{r_h\}$ and in the same way as in Equation 2.37. Again, if the sequence R is determined from its DFT, that is, from its Fourier transform sampled at the discrete frequencies $\{h/nT\}$, for $h = 0, 1, \ldots, n-1$, and using, of course, the *inverse* Fourier transform, the sequence actually evaluated is the *periodic* sequence derived from R in the manner previously described. Thus, in the determination of the DFT of a sequence of *finite* duration, the sequence is normally considered to be *periodic*,

THE DISCRETE FOURIER TRANSFORM   65

which means that the sequence has been modified to have infinite duration. However, since we are here primarily concerned with the evaluation of the DFT of a sequence and *not* with the evaluation of a sequence from its DFT or sampled spectrum, the given sequence being *always* of finite duration, all the required results can, in fact, be derived without extending the sequence to make it periodic. This approach is to be preferred, since the sequence under investigation does not now need to be modified in any way.

2.5 BASIC PROPERTIES OF THE DFT

The $n \times n$ matrices F and nF^{-1} are nonsingular symmetric matrices each of whose components has an absolute value (modulus) of unity. Furthermore, nF^{-1} is both the *complex conjugate* and also the *conjugate transpose* of F.

To show that the matrix F^{-1} given in Equation 2.43 is in fact the inverse of F, let $F^{-1} = E$.

Since F_h is the n-component row vector formed by the $(h+1)$th row of F, and $n^{-1}E_i^T$ is the $(i+1)$th column of E, it follows that the component in the $(h+1)$th row and $(i+1)$th column of the $n \times n$ matrix FE is

$$n^{-1}F_h E_i^T = n^{-1}(1 + \omega^{-h}\omega^i + \omega^{-2h}\omega^{2i} + \cdots + \omega^{-(n-1)h}\omega^{(n-1)i})$$
$$= n^{-1}(1 + \omega^{i-h} + \omega^{2(i-h)} + \cdots + \omega^{(n-1)(i-h)}) \qquad (2.45)$$

Now let

$$\alpha = \omega^{i-h} \qquad (2.46)$$

so that

$$n^{-1}F_h E_i^T = n^{-1}(1 + \alpha + \alpha^2 + \cdots + \alpha^{n-1})$$
$$= n^{-1}\frac{1-\alpha^n}{1-\alpha} \qquad (2.47)$$

from a well-known theorem[5,6]. Furthermore,

$$\frac{1-\alpha^n}{1-\alpha} = 0 \qquad (2.48)$$

when $\alpha^n = 1$ and $\alpha \neq 1$. But

$$\alpha^n = (\omega^{i-h})^n = \omega^{n(i-h)} = 1 \qquad (2.49)$$

for all integers h and i, and

$$\alpha = \omega^{i-h} = 1 \qquad (2.50)$$

only when $i = h$, bearing in mind that $i = 0, 1, \ldots, n-1$ and $h = 0, 1, \ldots, n-1$. Thus

$$n^{-1} F_h E_i^T = 0 \qquad (2.51)$$

whenever $i \neq h$. This means that all components *off* the main diagonal of FE are equal to zero, so that FE is a *diagonal* matrix.

The $(h+1)$th component along the *main diagonal* of FE is

$$n^{-1} F_h E_h^T = n^{-1}(1 + \omega^{-h}\omega^h + \omega^{-2h}\omega^{2h} + \cdots + \omega^{-(n-1)h}\omega^{(n-1)h})$$
$$= n^{-1}(1 + \omega^0 + \omega^0 + \cdots + \omega^0) = 1 \qquad (2.52)$$

Thus
$$FE = I \qquad (2.53)$$

where I is the $n \times n$ identity matrix, which means that

$$E = F^{-1} \qquad (2.54)$$

and Equation 2.43 is correct.

The $n \times n$ matrix obtained by multiplying F by itself is

$$FF = F^2 \qquad (2.55)$$

so that the component in the $(h+1)$th row and $(i+1)$th column of F^2 is

$$F_h F_i^T = 1 + \omega^{-h}\omega^{-i} + \omega^{-2h}\omega^{-2i} + \cdots + \omega^{-(n-1)h}\omega^{-(n-1)i}$$
$$= 1 + \omega^{-(h+i)} + \omega^{-2(h+i)} + \cdots + \omega^{-(n-1)(h+i)} \qquad (2.56)$$

Now let
$$\beta = \omega^{-(h+i)} \qquad (2.57)$$

so that
$$F_h F_i^T = 1 + \beta + \beta^2 + \cdots + \beta^{n-1}$$
$$= \frac{1 - \beta^n}{1 - \beta} \qquad (2.58)$$

from a well-known theorem[5,6]. Clearly,

$$\frac{1 - \beta^n}{1 - \beta} = 0 \qquad (2.59)$$

when $\beta^n = 1$ and $\beta \neq 1$. But

$$\beta^n = (\omega^{-(h+i)})^n = \omega^{-n(h+i)} = 1 \qquad (2.60)$$

for all integers h and i, and

$$\beta = \omega^{-(h+i)} = 1 \qquad (2.61)$$

only when $h + i = 0$ or $h + i = n$, bearing in mind that $h = 0, 1, \ldots, n-1$

and $i = 0, 1, \ldots, n-1$. Thus
$$F_h F_i^T = 0 \tag{2.62}$$
whenever $h+i \neq 0$ or n. This means that, when $h+i \neq 0$ or n, the component of F^2 in the $(h+1)$th row and $(i+1)$th column is zero. When $h+i = 0$ or n,
$$F_h F_i^T = 1 + \omega^{-(h+i)} + \omega^{-2(h+i)} + \cdots + \omega^{-(n-1)(h+i)}$$
$$= 1 + 1 + 1 + \cdots + 1 = n \tag{2.63}$$
from Equation 2.56, so that each nonzero component of F^2 has the value n. Hence

$$F^2 = n \begin{bmatrix} 1 & 0 & 0 & \cdots & 0 & 0 \\ 0 & 0 & 0 & \cdots & 0 & 1 \\ 0 & 0 & 0 & \cdots & 1 & 0 \\ \vdots & \vdots & \vdots & \vdots\vdots\vdots & \vdots & \vdots \\ 0 & 0 & 1 & \cdots & 0 & 0 \\ 0 & 1 & 0 & \cdots & 0 & 0 \end{bmatrix} \tag{2.64}$$

which is, of course, nonsingular.

From Equations 2.40 and 2.64,
$$n^{-1} R F^2 = \begin{bmatrix} r_0 & r_{n-1} & r_{n-2} & \cdots & r_1 \end{bmatrix} \tag{2.65}$$
and the postmultiplication of R by $n^{-1} F^2$ merely rearranges the order of the components of R. The matrix $n^{-1} F^2$ is therefore known as a *permutation* matrix.

It is clear from Equation 2.64 that the matrix $n^{-1} F^2$ is real, orthogonal and symmetric. Furthermore, the $n \times n$ matrix formed by multiplying $n^{-1} F^2$ by itself is
$$n^{-2} F^4 = I \tag{2.66}$$
where I is an $n \times n$ identity matrix. Thus $n^{-1/2} F$ is a fourth root of the identity matrix.

From the previous analysis it can be seen that

$$F^{-2} = n^{-1} \begin{bmatrix} 1 & 0 & 0 & \cdots & 0 & 0 \\ 0 & 0 & 0 & \cdots & 0 & 1 \\ 0 & 0 & 0 & \cdots & 1 & 0 \\ \vdots & \vdots & \vdots & \vdots\vdots\vdots & \vdots & \vdots \\ 0 & 0 & 1 & \cdots & 0 & 0 \\ 0 & 1 & 0 & \cdots & 0 & 0 \end{bmatrix} \tag{2.67}$$

so that nF^{-2} is a permutation matrix and is real, orthogonal and symmetric. The $n \times n$ matrix formed by multiplying nF^{-2} by itself is

$$n^2 F^{-4} = I \tag{2.68}$$

and $n^{1/2} F^{-1}$ is a fourth root of the identity matrix. It is clear from Equations 2.64 and 2.67 that $n^{-1} F^2$ and nF^{-2} are the *same* permutation matrix. It is also interesting to observe that

$$F^{-1} = F F^{-2} \tag{2.69}$$

so that

$$X F^{-1} = (XF) F^{-2} \tag{2.70}$$

Hence, if the *inverse* DFT of a sequence is used in place of the required DFT, the result is to reverse the order of the second to last components of the required sequence and multiply each component by n^{-1}, as can be seen from Equation 2.67. This result is related to the fact that F_h (Equation 2.39) and E_h (Equation 2.44) are such that

$$E_h = F_{n-h} \tag{2.71}$$

for $h = 0, 1, \ldots, n-1$, where $F_n = F_0$.

It is now of interest to consider the relationships that exist between the matrices F and Y_c, where Y_c is the circulant matrix given by

$$Y_c = \begin{bmatrix} y_0 & y_1 & \cdots & y_{g-2} & y_{g-1} & y_g & 0 & 0 & \cdots & 0 \\ 0 & y_0 & \cdots & y_{g-3} & y_{g-2} & y_{g-1} & y_g & 0 & \cdots & 0 \\ \vdots & \vdots & \vdots\vdots\vdots & \vdots & \vdots & \vdots & \vdots & \vdots & \vdots\vdots\vdots & \vdots \\ y_2 & y_3 & \cdots & y_g & 0 & 0 & \cdots & \cdots & y_0 & y_1 \\ y_1 & y_2 & \cdots & y_{g-1} & y_g & 0 & \cdots & \cdots & 0 & y_0 \end{bmatrix} \tag{2.72}$$

as in Equation 2.25. Let C_i be the n-component row vector given by the transpose of the $(i+1)$th column of Y_c. Thus

$$C_i = [y_i \quad y_{i-1} \quad y_{i-2} \quad \cdots \quad y_{i-n+1}] \tag{2.73}$$

where $y_i = 0$ for $i > g$ and $y_{-i} = y_{n-i}$ for $0 < i < n$. Now, from Equation 2.44,

$$\begin{aligned} E_h C_i^T &= y_i + y_{i-1} \omega^h + y_{i-2} \omega^{2h} + \cdots + y_{i-n+1} \omega^{(n-1)h} \\ &= \omega^{ih} (y_i \omega^{-ih} + y_{i-1} \omega^{-(i-1)h} + y_{i-2} \omega^{-(i-2)h} + \cdots \\ &\quad + y_{i-n+1} \omega^{-(i-n+1)h}) \\ &= \omega^{ih} (y_i \omega^{-ih} + y_{i-1} \omega^{-(i-1)h} + \cdots + y_1 \omega^{-h} + y_0 \\ &\quad + y_{n-1} \omega^{-(n-1)h} + \cdots + y_{i+1} \omega^{-(i+1)h} \end{aligned} \tag{2.74}$$

since $y_{-i} = y_{n-i}$ for $0 < i < n$, and $\omega^{-(i-n)h} = \omega^{-ih}$ for any integer i.

Equation 2.74 assumes, for convenience, that $2 < i < n-2$, but it can readily be seen from Equation 2.74 that, in general,

$$E_h C_i^T = \omega^{ih}(y_{n-1}\omega^{-(n-1)h} + y_{n-2}\omega^{-(n-2)h} + \cdots + y_1\omega^{-h} + y_0)$$
$$= \omega^{ih}(y_0 + y_1\omega^{-h} + y_2\omega^{-2h} + \cdots + y_g\omega^{-gh}) \quad (2.75)$$

since $y_i = 0$ for $i > g$.

Clearly, $\quad y_0 + y_1\omega^{-h} + y_2\omega^{-2h} + \cdots + y_g\omega^{-gh} = \lambda_h \quad (2.76)$

where λ_h is a function of h, for $h = 0, 1, \ldots, n-1$, but λ_h is *independent* of i.

It now follows that

$$E_h C_i^T = \lambda_h \omega^{ih} \quad (2.77)$$

where ω^{ih} is the $(i+1)$th component of the vector E_h. But C_i^T is the n-component column vector given by the $(i+1)$th column of the $n \times n$ matrix Y_c, so that

$$E_h Y_c = \lambda_h E_h \quad (2.78)$$

Thus, for $h = 0, 1, \ldots, n-1$, E_h is an *eigenvector* of the matrix Y_c in Equation 2.72 and λ_h is the corresponding *eigenvalue*[7]. The relationship derived in Equation 2.78 for the circulant matrix Y_c holds also for *any* other $n \times n$ circulant matrix used in place of Y_c, since the basic result (excluding the actual value of λ_h) is not affected by the component values of Y_c[8].

From Equation 2.78,

$$n^{-1} E_h Y_c = \lambda_h n^{-1} E_h \quad (2.79)$$

Let D be the $n \times n$ diagonal matrix whose $(h+1)$th component along the main diagonal is λ_h in Equation 2.76. Then, since $n^{-1}E_h$ is the $(h+1)$th row of F^{-1}, the $(h+1)$th row of DF^{-1} is $\lambda_h n^{-1} E_h$. But $n^{-1} E_h Y_c$ is the $(h+1)$th row of $F^{-1} Y_c$, as can be seen from Equations 2.43 and 2.44. Thus, from Equation 2.79,

$$F^{-1} Y_c = DF^{-1} \quad (2.80)$$

and $\quad F^{-1} Y_c F = D \quad (2.81)$

which shows that D is obtained from Y_c by a *similarity transformation*, so that Y_c is *similar* to the diagonal matrix D.

Clearly, Y_c has n linearly independent eigenvectors $\{E_h\}$, given by Equation 2.44, and n eigenvalues $\{\lambda_h\}$, given by Equation 2.76. Furthermore, for *any* circulant matrix Y_c, the eigenvectors of Y_c are *independent* of the component values of Y_c. This is a most important result which means that the *same* similarity transformation converts any $n \times n$ circulant matrix into the corresponding diagonal matrix.

Again, the n eigenvalues of Y_c are completely determined by the n-component row vector

$$Y_0 = [y_0 \quad y_1 \quad \ldots \quad y_g \quad 0 \quad \ldots \quad 0] \tag{2.82}$$

which is the first row of Y_c and, from Equation 2.13, is also the sampled impulse-response of the baseband channel in Fig. 2.1.

It can be seen from Equations 2.42 and 2.76 that the DFT of Y_0 is

$$Y_0 F = [\lambda_0 \quad \lambda_1 \quad \ldots \quad \lambda_{n-1}] \tag{2.83}$$

so that the n components of $Y_0 F$ are the *eigenvalues* of Y_c. In other words, the n components of the DFT of the sampled impulse-response of the channel are the n eigenvalues of the $n \times n$ convolution matrix corresponding to this sampled impulse-response.

The result that has just been derived holds for any sampled impulse-response, regardless of the number of zeros in Y_0, so that the result holds also for the particular case where $g = n-1$ and there are *no* zero-valued components in Y_0. For our applications here, however, Y_0 must always have a sufficient number of zeros, such that the product of the n-component vector giving the input sequence and the $n \times n$ convolution matrix Y_c, leads to the required *linear (aperiodic)* convolution of the input sequence and Y (Equation 2.13). This is illustrated by the analysis in Sections 2.2 and 2.3.

From Equations 2.41 and 2.42 it follows that

$$\lambda_h = Y_0 F_h^T = J\left(\frac{h}{nT}\right) \tag{2.84}$$

for $h = 0, 1, \ldots, n-1$. $J(h/nT)$ is here the value of the transfer function of the baseband channel and sampler in Fig. 2.1, at the frequency h/nT Hz, and F_h^T is the $(h+1)$th column of F. Clearly, the n components of the DFT of the sampled impulse-response of the channel are the values of the transfer function of the baseband channel and sampler, at the frequencies $\{h/nT\}$ Hz, for $h = 0, 1, \ldots, n-1$. $J(h/nT)$ has, of course, the same property as $G(h/nT)$ in Equation 2.37.

2.6 CONVOLUTION OF TWO SEQUENCES

Consider again the finite sequence of m signal-elements

$$\sum_{i=0}^{m-1} s_i \delta(t - iT) \tag{2.85}$$

at the input to the baseband channel in Fig. 2.1, where the values of

THE DISCRETE FOURIER TRANSFORM

the data symbols, given by the areas of the corresponding impulses, are the first m components $\{s_i\}$ of the n-component row vector

$$S = [s_0 \quad s_1 \quad \ldots \quad s_{m-1} \quad 0 \quad \ldots \quad 0] \tag{2.86}$$

The corresponding sequence of n samples at the detector input are the components $\{r_i\}$ of the n-component row vector

$$R = [r_0 \quad r_1 \quad \ldots \quad r_{n-1}] \tag{2.87}$$

where
$$n = m + g \tag{2.88}$$

The sampled impulse-response of the baseband channel in Fig. 2.1 is given by the n-component row vector Y_0 in Equation 2.82 which is the first row of the matrix Y_c in Equation 2.72. Thus, from Equation 2.27,

$$R = SY_c \tag{2.89}$$

and R is the *linear (aperiodic) convolution* of the sequences (vectors) S and Y_0.

Let the n-component row vectors

$$P = [p_0 \quad p_1 \quad \ldots \quad p_{n-1}] \tag{2.90}$$

and
$$Q = [q_0 \quad q_1 \quad \ldots \quad q_{n-1}] \tag{2.91}$$

be the discrete Fourier transforms of R and S, respectively, so that

$$P = RF \tag{2.92}$$

and
$$Q = SF \tag{2.93}$$

Thus
$$R = PF^{-1} \tag{2.94}$$

and
$$S = QF^{-1} \tag{2.95}$$

so that, from Equation 2.89,

$$PF^{-1} = QF^{-1}Y_c \tag{2.96}$$

or
$$P = QF^{-1}Y_c F \tag{2.97}$$

It follows immediately from Equation 2.81 that

$$P = QD \tag{2.98}$$

where D is the $n \times n$ diagonal matrix whose $(h+1)$th component along the main diagonal is λ_h in Equations 2.76 and 2.83. Thus

$$p_i = q_i \lambda_i \tag{2.99}$$

for $i = 0, 1, \ldots, n-1$.

It is clear from Equation 2.99 that each component of the DFT of the *convolution* of two sequences is simply the *product* of the corresponding components of the DFT's of the two sequences. This is a most important property of DFT's. It holds for the convolution of *any* two sequences provided only that the same number of components, n, is used for each DFT and n is large enough ($\geqslant m+g$) to include all the nonzero components in the *convolution* of the two sequences. The result holds, of course, in the general case where a sequence of values is fed to a digital filter, whose output sequence is the convolution of the input sequence and the sampled impulse-response of the filter.

Suppose now that
$$S = [s_0 \quad s_1 \quad \ldots \quad s_{n-1}] \tag{2.100}$$
and
$$Y_0 = [y_0 \quad y_1 \quad \ldots \quad y_{n-1}] \tag{2.101}$$
where all components of S and Y_0 are *nonzero*. Clearly, Y_c is now the $n \times n$ circulant matrix whose $(i+1)$th row is

$$Y_i = [y_{n-i} \quad y_{n-i+1} \quad \ldots \quad y_{n-1} \quad y_0 \quad \ldots \quad y_{n-i-1}] \tag{2.102}$$

It can be seen from the preceding analysis that, if $R = SY_c$ (Equation 2.89), the results given by Equations 2.98 and 2.99 *still hold*, even though R is no longer the linear (aperiodic) convolution of S and Y_0. The essential requirement is that Y_c is a *circulant matrix*. The sequence R is in fact now the *circular (periodic)* convolution of S and Y_0, linear convolution being a *special case* of circular convolution. Unless specifically stated to the contrary, the convolution of two sequences is taken to be a *linear* convolution.

Since each component of P (Equation 2.98) is simply the *product* of the corresponding components of Q and Y_0F, P is very easily determined from Q and Y_0F. On the other hand, since R is the *convolution* of S and Y_0, the evaluation of R from S and Y_0 is considerably more complex than is the evaluation of P from Q and Y_0F. Clearly, R can be evaluated from S and Y_0 by forming Q and Y_0F, multiplying the corresponding pairs of components to give P and then forming $PF^{-1} = R$. By suitably simplifying and speeding up the implementation of the DFT into one of the possible forms known as fast Fourier transforms, it is often simpler to evaluate R by the method just described than by the direct convolution of S and Y_0[1-3]. We are, however, not here concerned with the method of evaluating R but rather with the nature and severity of the signal distortion corresponding to any given vector Y_0. We shall not therefore consider further the actual implementation of a DFT.

The relationships derived in the following sections are of very general application and are not just confined to an arrangement of the type shown in Fig. 2.1. In order to emphasize this fact, the two sequences (vectors) under consideration are renamed W and X.

2.7 LINEARITY OF THE DFT

Let W and X be two n-component row vectors (sequences) and let U and V be their respective DFT's, such that

$$U = WF \qquad (2.103)$$
and
$$V = XF \qquad (2.104)$$

where F is the $n \times n$ matrix given by Equation 2.42.

Since F is a linear transformation matrix,

$$aU + bV = (aW + bX)F \qquad (2.105)$$

where a and b are two arbitrary scalars. In particular, when $a = b = 1$,

$$U + V = (W + X)F \qquad (2.106)$$

so that the DFT of the sum of two vectors is the sum of the individual DFT's. Similarly, the inverse DFT of the sum of two vectors is the sum of the individual inverse DFT's.

2.8 SHIFT THEOREM

Assume that the n components of the sequence given by the row vector

$$X = [x_0 \quad x_1 \quad \ldots \quad x_{n-1}] \qquad (2.107)$$

are samples of a received signal waveform, the samples being spaced at regular intervals of T seconds, with x_0 at time $t = 0$. This is, of course, the same as the assumption made for the vector R in Section 2.4, as can be seen from Expression 2.28. Thus the set of samples $\{x_i\}$ may be represented by the sequence of impulses

$$\sum_{i=0}^{n-1} x_i \delta(t - iT) \qquad (2.108)$$

Suppose now that the samples $\{x_i\}$ are each *delayed* by lT seconds (shifted to the right by l places), where l is an integer $0 < l < n$, so that the sample x_i occurs at the time instant $t = (i + l)T$. The set of samples may be represented by the sequence of impulses

$$\sum_{i=0}^{n-1} x_i \delta(t - (l+i)T) \qquad (2.109)$$

The Fourier transform of these impulses, at a frequency h/nT Hz, is

$$G\left(\frac{h}{nT}\right) = \sum_{i=0}^{n-1} x_i \exp\left(-j2\pi \frac{h}{nT}(l+i)T\right)$$

$$= \sum_{i=0}^{n-1} x_i \omega^{-h(l+i)} = X_l F_h^T \qquad (2.110)$$

where ω is given by Equation 2.31, F_h^T is the n-component column vector formed by the $(h+1)$th column of the matrix F in Equation 2.42, and

$$X_l = [x_{n-l} \quad x_{n-l+1} \quad \ldots \quad x_{n-1} \quad x_0 \quad \ldots \quad x_{n-l-1}] \qquad (2.111)$$

But $G(h/nT)$ is the $(h+1)$th component of the n-component DFT of the given sequence, and $X_l F_h^T$ is the $(h+1)$th component of the n-component vector $X_l F$. Thus the DFT of the sequence of samples delayed by lT seconds is $X_l F$.

Similarly, when the samples $\{x_i\}$ are each *advanced* by lT seconds (shifted to the left by l places) relative to their original positions in time, the DFT of the sequence of samples becomes $X_{-l} F$, where

$$X_{-l} = [x_l \quad x_{l+1} \quad \ldots \quad x_{n-1} \quad x_0 \quad \ldots \quad x_{l-1}] \qquad (2.112)$$

Clearly, a shift in time of the sequence of samples by a whole number of sampling intervals, appears in the DFT of the sequence as though the sequence had in fact been shifted *cyclically* in the same direction and by the same number of places. Thus a *time* shift of a sequence by a whole number of sampling intervals can be considered as the corresponding *cyclic* shift of the samples, without in any way affecting the DFT of the sequence. It follows immediately from the above analysis that an advance or delay of the original sequence X (Equation 2.107) by hn sampling intervals, where h is any integer, leaves the DFT of the sequence *unchanged*, and an advance or delay by $hn + l$ sampling intervals, where $0 < l < n$, gives a sequence whose DFT is the same as that of the original sequence advanced or delayed, respectively, by l sampling intervals. This property will now be studied further by means of a simple example.

Consider the sequence (row vector)

$$X = [x_0 \quad x_1 \quad x_2 \quad x_3 \quad x_4] \qquad (2.113)$$

where the time of occurrence (in seconds) of a component of X is given by the corresponding component of the 5-component vector

$$\Delta = [0 \quad T \quad 2T \quad 3T \quad 4T] \qquad (2.114)$$

Suppose that the sequence X is now *advanced* in time by $2T$ seconds (two sampling intervals). The advance is represented by the corresponding cyclic shift of two places to the left in the components of X, to give the sequence

$$X_{-2}=[x_2 \quad x_3 \quad x_4 \quad x_0 \quad x_1] \qquad (2.115)$$

where the times of occurrence of the components of X_{-2} are given by

$$\Delta_{-2}=[0 \quad T \quad 2T \quad -2T \quad -T] \qquad (2.116)$$

To obtain the correct DFT of the sequence it is, of course, necessary that its first component (as written) occurs at time $t=0$, as is in fact the case for both X and X_{-2}.

Consider next the sequence

$$X=[x_0 \quad x_1 \quad x_2 \quad x_3 \quad x_4] \qquad (2.117)$$

for which Δ has been changed to

$$\Delta_{5h}=[5hT \quad (5h+1)T \quad (5h+2)T \quad (5h+3)T \quad (5h+4)T] \qquad (2.118)$$

where h is any positive or negative integer. X in Equation 2.117 has the *same* DFT as X in Equation 2.113, for which the times of occurrence of the components are given by Δ (Equation 2.114). Thus, shifts in time of integral multiples of $5T$ (nT in the general case of an n-component sequence) can be ignored. It is in fact only necessary to consider the *remainder* of the time shift that is obtained after subtracting from the actual time shift (in seconds) the appropriate integral multiple of $5T$ (nT in the general case) to keep the remainder as an *advance* in time lying in the range 0 to $4T$ (0 to $(n-1)T$ in the general case). Thus, for example, if X in Equation 2.113 is advanced in time by $12T$, then the sequence can be taken to be X_{-2} in Equation 2.115, the times of occurrence of its components being given by Δ_{-2} in Equation 2.116.

Suppose now that the original sequence X is *delayed* in time by $2T$ seconds. The delay is represented by the corresponding cyclic shift of two places to the right in the components of X, to give the sequence

$$X_2=[x_3 \quad x_4 \quad x_0 \quad x_1 \quad x_2] \qquad (2.119)$$

where the delay of $2T$ seconds applied to each component of X gives the times of occurrence of $2T$, $3T$, $4T$, $5T$ and $6T$, respectively, to the components x_0, x_1, x_2, x_3 and x_4 of X_2. To keep the time of occurrence of the *first* component of X_2 (*as written* in Equation 2.119) at time $t=0$, this being an important assumption that is made when forming the DFT of the sequence, an advance of $5T$ is introduced into the time of occurrence of each component of X_2, the advance not changing the DFT of the sequence. The times of occurrence of the components of the sequence X_2 in Equation 2.119 are therefore now

given by the vector
$$\Delta_2 = [0 \quad T \quad -3T \quad -2T \quad -T] \quad (2.120)$$

It can be seen here that
$$X_2 = X_{-3} \quad (2.121)$$
and
$$\Delta_2 = \Delta_{-3} \quad (2.122)$$

In general, for an n-component sequence, a *delay* in time of l places (lT seconds or l sampling intervals) is taken to be an *advance* of $(-l)$ modulo $-n$ places, an advance of l places being considered as an advance of l modulo $-n$ places. The quantity l modulo $-n$ is here an integer that satisfies the following two equations:

$$0 \leqslant l \text{ modulo} - n < n \quad (2.123)$$
and
$$l \text{ modulo} - n = l - hn \quad (2.124)$$

where h is the appropriate integer that causes Equation 2.123 to be satisfied. This arrangement has the advantage that x_0, which is the first component, *in time*, of the original sequence X, is always the first component, *in time*, of the cyclically shifted sequence, and x_{n-1} is the last component, in time, of the latter sequence. The first component, *as written*, of the original or cyclically shifted sequence is always that occurring at time $t = 0$.

Suppose now that a sequence of samples given by the n-component vector
$$X = [x_0 \quad x_1 \quad \ldots \quad x_{m-1} \quad 0 \quad \ldots \quad 0] \quad (2.125)$$

is fed to a digital filter whose sampled impulse-response is given by the n-component vector
$$W = [w_0 \quad w_1 \quad \ldots \quad w_g \quad 0 \quad \ldots \quad 0] \quad (2.126)$$
where
$$n = m + g \quad (2.127)$$

and w_i is the sample at time $t = iT$, for $i = 0, 1, \ldots, g$.

If M is the $n \times n$ circulant matrix whose first row is given by W, so that M is the convolution matrix corresponding to the sequence (vector) W, then the output sequence from the digital filter is given by the n-component row vector XM.

Consider now the $n \times n$ circulant matrix M_{-l} whose first row is the n-component vector
$$W_{-l} = [w_l \quad \ldots \quad w_g \quad 0 \quad \ldots \quad 0 \quad w_0 \quad \ldots \quad w_{l-1}] \quad (2.128)$$

W_{-l} is derived from W by a cyclic shift of l places to the left in its components. Clearly, M_{-l} is obtained from M by a cyclic shift of l

places to the left in the *columns* of M, so that XM_{-l} is obtained from XM by a cyclic shift of l places to the left in the *components* of XM. Similarly, if M_l is the $n \times n$ circulant matrix whose first row is obtained from W by a cyclic shift of l places to the right in the components of W, then XM_l is obtained from XM by a cyclic shift of l places to the right in the components of XM.

W_{-l} in Equation 2.128 can be considered to represent an advance in time of lT seconds in the sampled impulse-response of the digital filter, so that, in Equation 2.126, w_i is now the sample at time $t=(i-l)T$ instead of at time $t=iT$. Hence, in Equation 2.128, a *time* shift is represented by the corresponding *cyclic* shift. This cyclic shift gives the *same* cyclic shift in the output sequence from the digital filter. But the given *time* shift in W must also give the *same* time shift in the output sequence. Thus the output sequence can here be represented by XM_{-l}.

It is clear that an advance (or delay) in *time* of lT seconds in a sequence of samples having a sampling interval of T seconds, normally represented as a shift of l places to the left (or right) in the sequence of samples, can alternatively be represented by a *cyclic* shift of l places to the left (or right) in the components of the sequence, whether the sequence represents a signal or the sampled impulse-response of a filter or channel. This important relationship is known here as the Shift theorem, and it holds whether $-n<l<n$ or whether $l\leqslant -n$ or $l\geqslant n$.

The representation of a time shift as a cyclic shift is, of course, not necessarily unique, since a cyclic shift of l places to the left (or right) corresponds to a shift in time of $l+hn$ places to the left (or right), where h is any integer. However, in all cases of real interest or importance here, the time shifts in either direction are always less than $\frac{1}{2}n$ places, so that the corresponding cyclic shifts are always less than $\frac{1}{2}n$ places to the left or right, and there is now a unique relationship between the time shift and the corresponding cyclic shift. Under these conditions, a time shift is represented unambiguously by the corresponding cyclic shift. The resultant sequence, *as written*, is always such that it *starts* at time $t=0$.

An immediate and sometimes useful consequence of the shift theorem is that the components $\{x_{n-i}\}$ of the vector

$$X = [x_0 \quad x_1 \quad \ldots \quad x_{n-1}] \tag{2.129}$$

where n is odd and $i=1, 2, \ldots, \frac{1}{2}(n-1)$, can be taken as the samples at the time instants $\{-iT\}$, and the components $\{x_i\}$ can be taken as the samples at the time instants $\{iT\}$, so that $x_{(n+1)/2}$ is the sample at time $t=-\frac{1}{2}(n-1)T$ and is the *first* component (in time) of the sequence, and $x_{(n-1)/2}$ is the sample at time $t=\frac{1}{2}(n-1)T$ and is the

78 THE DISCRETE FOURIER TRANSFORM

last component (in time) of the sequence. Furthermore, if the vector X in Equation 2.129 has been shifted cyclically from that in Equation 2.107 (without renaming the components) and if now the last h components of the latter are all zero, where $h \geqslant \frac{1}{2}(n-1)$, then all the important results and relationships to be developed here, involving any shift in *time* of the sequence X, apply also to the corresponding *cyclic* shift, which can therefore be considered to be *identically equivalent* to the time shift. Extensive use is made of this fact in Chapter 3.

It is interesting now to evaluate the relationship between the DFT's of two sequences, where one of the sequences is given by a *cyclic shift* of the other.

Consider the n-component sequence (row vector) X in Equation 2.129, with x_0 at time $t=0$. The $(h+1)$th component of XF (the DFT of X) is XF_h^T, where

$$F_h = [1 \quad \omega^{-h} \quad \omega^{-2h} \quad \ldots \quad \omega^{-(n-1)h}] \tag{2.130}$$

so that

$$XF_h^T = \sum_{i=0}^{n-1} x_i \omega^{-ih} \tag{2.131}$$

Suppose that the sequence X is *delayed* in time by l places (sampling intervals), this being represented by the corresponding cyclic shift of X to give X_l (Equation 2.111). The $(h+1)$th component of $X_l F$ (the DFT of X_l) is

$$X_l F_h^T = \sum_{i=0}^{n-1} x_i \omega^{-h(i+l)}$$

$$= \sum_{i=0}^{n-1} x_i \omega^{-hi} \omega^{-hl} = XF_h^T \omega^{-hl} \tag{2.132}$$

using Equation 2.34. Thus the DFT of the delayed sequence is

$$X_l F = XFD_l \tag{2.133}$$

where D_l is the diagonal matrix whose $(h+1)$th component along the main diagonal is ω^{-hl}.

Suppose next that the sequence X is *advanced* in time by l places, this being represented by the corresponding cyclic shift of X to give X_{-l} (Equation 2.112). The $(h+1)$th component of $X_{-l}F$ is

$$X_{-l} F_h^T = \sum_{i=0}^{n-1} x_i \omega^{-h(i-l)}$$

$$= \sum_{i=0}^{n-1} x_i \omega^{-hi} \omega^{hl} = XF_h^T \omega^{hl} \tag{2.134}$$

using Equation 2.34. Thus the DFT of the advanced sequence is

$$X_{-l}F = XFD_{-l} \qquad (2.135)$$

where D_{-l} is the diagonal matrix whose $(h+1)$th component along the main diagonal is ω^{hl}.

In Equations 2.133 and 2.135, l may be any positive integer. However, from the definitions of D_l and D_{-l}, it is evident that

$$D_l = D_{l+hn} \qquad (2.136)$$

for any integers l and h (positive or negative or zero), so that D_l has only n different values. In particular, when $0 < l < n$, $D_{-l} = D_{n-l}$. Similar relationships of course also hold for X_l and X_{-l}.

Clearly, a cyclic shift of l places to the left (or right) in the n components of the vector X results in the multiplication of the corresponding DFT by D_{-l} (or D_l). Furthermore, it can be seen that

$$D_{-l} = D_l^{-1} \qquad (2.137)$$

It follows from the Shift theorem that if a sequence of samples is advanced or delayed in *time* by l places, the DFT of the sequence is postmultiplied by the diagonal matrix D_{-l} or D_l, respectively. Again, if the DFT of the sequence is postmultiplied by D_{-l} or D_l, where now $0 < l < n$, the sequence is advanced or delayed, respectively, by $l + hn$ places in time, where h is any positive or negative integer or zero.

2.9 REAL AND IMAGINARY COMPONENTS OF A DFT

From Equation 2.42,

$$F = \begin{bmatrix} 1 & 1 & 1 & \cdots & 1 \\ 1 & \omega^{-1} & \omega^{-2} & \cdots & \omega^{-(n-1)} \\ 1 & \omega^{-2} & \omega^{-4} & \cdots & \omega^{-2(n-1)} \\ \vdots & \vdots & \vdots & \vdots\vdots\vdots & \vdots \\ 1 & \omega^{-(n-1)} & \omega^{-2(n-1)} & \cdots & \omega^{-(n-1)^2} \end{bmatrix} \qquad (2.138)$$

where

$$\omega^{-i} = \exp\left(-j\frac{2\pi i}{n}\right)$$

$$= \cos\frac{2\pi i}{n} - j\sin\frac{2\pi i}{n} \qquad (2.139)$$

so that

$$F = F_a - jF_b \qquad (2.140)$$

where

$$F_a = \begin{bmatrix} 1 & 1 & 1 & \cdots & 1 \\ 1 & \cos\dfrac{2\pi}{n} & \cos\dfrac{2\pi 2}{n} & \cdots & \cos\dfrac{2\pi(n-1)}{n} \\ 1 & \cos\dfrac{2\pi 2}{n} & \cos\dfrac{2\pi 4}{n} & \cdots & \cos\dfrac{2\pi 2(n-1)}{n} \\ \vdots & \vdots & \vdots & \vdots\vdots\vdots & \vdots \\ 1 & \cos\dfrac{2\pi(n-1)}{n} & \cos\dfrac{2\pi 2(n-1)}{n} & \cdots & \cos\dfrac{2\pi(n-1)^2}{n} \end{bmatrix}$$

(2.141)

and

$$F_b = \begin{bmatrix} 0 & 0 & 0 & \cdots & 0 \\ 0 & \sin\dfrac{2\pi}{n} & \sin\dfrac{2\pi 2}{n} & \cdots & \sin\dfrac{2\pi(n-1)}{n} \\ 0 & \sin\dfrac{2\pi 2}{n} & \sin\dfrac{2\pi 4}{n} & \cdots & \sin\dfrac{2\pi 2(n-1)}{n} \\ \vdots & \vdots & \vdots & \vdots\vdots\vdots & \vdots \\ 0 & \sin\dfrac{2\pi(n-1)}{n} & \sin\dfrac{2\pi 2(n-1)}{n} & \cdots & \sin\dfrac{2\pi(n-1)^2}{n} \end{bmatrix}$$

(2.142)

Consider the n-component row vector X in Equation 2.130, all of whose components are *real-valued*. Clearly, the DFT of X is

$$XF = XF_a - jXF_b \tag{2.143}$$

Since $$\cos\frac{2\pi hi}{n} = \cos\frac{2\pi h(n-i)}{n} \tag{2.144}$$

for integers h and i having values in the range 0 to $n-1$, it can be seen that the n-component vectors formed by the 2nd and nth columns of F_a are the same. So are the vectors formed by the 3rd and $(n-1)$th columns, the 4th and $(n-2)$th columns, and so on. But, for values of the integer i in the range 0 to $n-1$, the *real part* of the $(i+1)$th component of XF is

$$\sum_{h=0}^{n-1} x_h \cos\frac{2\pi hi}{n}$$

where $\cos(2\pi hi/n)$ is the $(h+1)$th component of the $(i+1)$th column of

THE DISCRETE FOURIER TRANSFORM 81

F_a. It follows that for any n-component real vector X, the real parts of the 2nd and nth components of XF are the same, as are the real parts of the 3rd and $(n-1)$th components, the 4th and $(n-2)$th components, and so on.

Since
$$\sin\frac{2\pi hi}{n} = -\sin\frac{2\pi h(n-i)}{n} \tag{2.145}$$

for integers h and i having values in the range 0 to $n-1$, the n-component vectors formed by the 2nd and nth columns of F_b are the negatives of each other. So are the vectors formed by the 3rd and $(n-1)$th columns of F_b, the 4th and $(n-2)$th columns, and so on. But for values of the integer i in the range 0 to $n-1$, the *imaginary part* of the $(i+1)$th component of XF is

$$-\sum_{h=0}^{n-1} x_h \sin\frac{2\pi hi}{n}$$

where $\sin(2\pi hi/n)$ is the $(h+1)$th component of the $(i+1)$th column of F_b. Thus, for any n-component real vector X, the imaginary parts of the 2nd and nth components of XF are the negatives of each other, as are the imaginary parts of the 3rd and $(n-1)$th components, the 4th and $(n-2)$th components, and so on.

Clearly, for $i=1, 2, \ldots, n-1$, the n-component vectors formed by the $(i+1)$th and $(n-i+1)$th columns of F are *complex conjugates* so that, for any n-component *real* vector X, the $(i+1)$th and $(n-i+1)$th *components* of XF are themselves *complex conjugates*. It can be seen from Equation 2.138 that the *first* component of XF must be *real-valued*. Thus, if

$$XF = [\mu_0 \quad \mu_1 \quad \ldots \quad \mu_{n-1}] \tag{2.146}$$

so that μ_i is the $(i+1)$th component of the DFT of X, then

$$\mu_i = \mu_{n-i}^* \tag{2.147}$$

for $i=0, 1, \ldots, n-1$, where μ_{n-i}^* is the complex conjugate of μ_{n-i}, and $\mu_n = \mu_0$.

Since F is a symmetric (not Hermitian) matrix, the same relationships hold for the rows of F as for the columns. Thus the n-component vectors formed by the $(i+1)$th and $(n-i+1)$th rows of F are complex conjugates. Also, for any n-component column vector X^T with real-valued components, the $(i+1)$th and $(n-i+1)$th components of FX^T are complex conjugates. Again, the *first* component of FX^T must be real-valued. Obviously, $F^T = F$.

It can be seen from Equations 2.43, 2.141 and 2.142 that

$$nF^{-1} = F_a + jF_b \tag{2.148}$$

so that, from Equation 2.140,

$$nF^{-1} = \bar{F} \qquad (2.149)$$

where $\bar{F}$ is the complex conjugate of the matrix F. This can in fact be seen directly from Equations 2.42, 2.43 and 2.32. Thus, for any n-component row vector X with real-valued components

$$nXF^{-1} = X\bar{F} = \overline{(XF)} \qquad (2.150)$$

where $\overline{(XF)}$ is, of course, the complex conjugate of the DFT of X. This is an important result which is used in the derivation of several useful relationships.

Since F is a symmetric matrix, the conjugate transpose of F is

$$F^* = \bar{F} \qquad (2.151)$$

It also follows from Equation 2.149 that

$$F^* = nF^{-1} \qquad (2.152)$$

so that the conjugate transpose of $n^{-1/2}F$ is

$$(n^{-1/2}F)^* = n^{-1/2}F^* = n^{1/2}F^{-1} \qquad (2.153)$$

But $n^{1/2}F^{-1}$ is the *inverse* of $n^{-1/2}F$. Thus,

$$(n^{-1/2}F)^* = (n^{-1/2}F)^{-1} \qquad (2.154)$$

which means that $n^{-1/2}F$ is a *unitary* matrix, as is $n^{1/2}F^{-1}$. As has previously been shown, each of these unitary matrices, when raised to the fourth power, becomes the $n \times n$ identity matrix.

It is of interest now to consider the DFT's of *symmetric* and *skew-symmetric* sequences whose components are all *real-valued*. A symmetric sequence is defined to be a sequence with an odd number of components, such that the central component has any real value and, for each i, the component i places to the left of the central component is the same as the component i places to the right. A *skew-symmetric* sequence is defined to be a sequence with an odd number of components, such that the central component is zero and, for each i, the component i places to the left of the central component is the negative of the component i places to the right.

When considering the DFT of a symmetric or skew-symmetric sequence it is convenient to shift the sequence in time by the appropriate number of places (sampling intervals) so that the central component (sample) occurs at time $t=0$. By the Shift theorem, the components of the sequence are shifted *cyclically* by the appropriate number of places such that the first component, *as written*, of the resultant sequence, which is always the component at time $t=0$, is the *central* component, *in time*, of the original sequence.

Consider, for example, the 5-component sequence

$$X = [x_0 \quad x_1 \quad x_2 \quad x_3 \quad x_4] \tag{2.155}$$

which is symmetrical about x_2, so that $x_1 = x_3$ and $x_0 = x_4$. The times of occurrence (in seconds) of the components of X are given by the corresponding components of

$$\Delta = [0 \quad T \quad 2T \quad 3T \quad 4T] \tag{2.156}$$

The sequence X is now advanced *in time* by $2T$ seconds and the time shift is represented by the corresponding *cyclic shift* in the components of X, to form the sequence

$$X_{-2} = [x_2 \quad x_3 \quad x_4 \quad x_0 \quad x_1] \tag{2.157}$$

The times of occurrence of the components of X_{-2} are given by the corresponding components of

$$\Delta_{-2} = [0 \quad T \quad 2T \quad -2T \quad -T] \tag{2.158}$$

Since it is not convenient to operate with the sequence X_{-2}, the sequence X_{-2} is renamed X, as in Equation 2.155, but the times of occurrence of the components of X are given by Δ_{-2} (Equation 2.158) and *not* by Δ. The first component, *in time*, of the sequence X is now x_3, and the sequence X is symmetrical about x_0, such that, $x_1 = x_4$ and $x_2 = x_3$.

Consider next the general n-component symmetric or skew-symmetric sequence X, which has been shifted in time such that its central component (about which symmetry occurs) is located at time $t = 0$. The components of the resultant sequence, in the order *as written*, are now renamed $x_0, x_1, \ldots, x_{n-1}$, to give the sequence

$$X = [x_0 \quad x_1 \quad x_2 \quad \ldots \quad x_{n-2} \quad x_{n-1}] \tag{2.159}$$

where the times of occurrence of the components of X are given by

$$\Delta = [0 \quad T \quad 2T \quad \ldots \quad -2T \quad -T] \tag{2.160}$$

In the resultant sequence X (Equation 2.159) n is assumed to be odd. Furthermore, for $i = 1, 2, \ldots, \frac{1}{2}(n-1)$, x_i occurs at time iT and x_{n-i} occurs at time $-iT$. Thus x_i and x_{n-i} are the components i places *in time* to the right and left, respectively, of the central component x_0. The cyclic shift in the components of X eliminates the delay terms in its DFT without otherwise changing the DFT, so that the basic properties of the DFT are more easily identified.

Suppose first that X is *real* and *symmetric*, such that

$$x_i = x_{n-i} \tag{2.161}$$

for $i = 1, 2, \ldots, n-1$, which means that the $(i+1)$th and $(n-i+1)$th

84 THE DISCRETE FOURIER TRANSFORM

components of X are the same. The $(i+1)$th component of XF is

$$\sum_{h=0}^{n-1} x_h \omega^{-hi} = \sum_{h=0}^{n-1} x_h \left(\cos \frac{2\pi hi}{n} - j \sin \frac{2\pi hi}{n} \right) \quad (2.162)$$

It can be seen from Equations 2.161 and 2.162 that the imaginary-valued components of XF now cancel out to give a *real* vector XF. Since the $(i+1)$th and $(n-i+1)$th components of XF are complex conjugates of each other, it follows that XF is not only *real* but is also *symmetric*, such that its $(i+1)$th component equals its $(n-i+1)$th component, for $i = 1, 2, \ldots, n-1$. Similarly, it may be shown that if XF is real and symmetric, then so is X.

Suppose next that X is *real* and *skew-symmetric*, such that

$$x_0 = 0 \quad (2.163)$$

and $$x_i = -x_{n-i} \quad (2.164)$$

for $i = 1, 2, \ldots, n-1$, which means that the $(i+1)$th and $(n-i+1)$th components of X are the negatives of each other. It can be seen from Equations 2.162–2.164 that the real-valued components of XF now cancel out to give an *imaginary* vector XF. Since the $(i+1)$th and $(n-i+1)$th components of XF are complex conjugates and its first component is the sum of the $\{x_i\}$, XF is not only *imaginary* but is also *skew-symmetric*, such that its first component is zero and its $(i+1)$th component is the negative of its $(n-i+1)$th component, for $i = 1, 2, \ldots, n-1$. Similarly, it may be shown that if XF is imaginary and skew-symmetric, then X is real and skew-symmetric.

It must be emphasized again, that for both the symmetric and skew-symmetric sequences X, the components of X have been shifted cyclically to the left to make x_0 the *central* component (*in time*) of the original sequence. This corresponds to a shift in time of the sequence to bring the central component to the time instant $t = 0$.

The general relationships in symmetry that exist between a sequence and its DFT have been gathered together in Table 2.2.

2.10 REVERSAL THEOREM

Let X be the n-component sequence (row vector)

$$X = [x_0 \quad x_1 \quad \ldots \quad x_{n-1}] \quad (2.165)$$

whose components are real valued, some of the components possibly also being zero. The time of occurrence of a component of X is given by the corresponding component of the n-component vector

$$\Delta = [0 \quad T \quad 2T \quad \ldots \quad (n-1)T] \quad (2.166)$$

THE DISCRETE FOURIER TRANSFORM 85

Table 2.2 RELATIONSHIPS BETWEEN A SEQUENCE AND ITS DFT

Sequence X	DFT of X
Symmetric	Symmetric
Skew symmetric	Skew symmetric
Symmetric and real	Symmetric and real
Skew symmetric and real	Skew symmetric and imaginary
Symmetric and imaginary	Symmetric and imaginary
Skew symmetric and imaginary	Skew symmetric and real
Real	Real part symmetric, imaginary part skew symmetric
Imaginary	Real part skew symmetric, imaginary part symmetric
Real part symmetric, imaginary part skew symmetric	Real
Real part skew symmetric, imaginary part symmetric	Imaginary

As before, the set of samples $\{x_i\}$ may be represented by the sequence of impulses

$$\sum_{i=0}^{n-1} x_i \delta(t - iT) \tag{2.167}$$

Suppose now that the sequence X is *reversed in time*, the reversal being pivoted about the component x_0, which, of course, occurs at time $t = 0$. The resulting sequence of impulses is

$$\sum_{i=0}^{n-1} x_i \delta(t + iT) \tag{2.168}$$

The Fourier transform of this sequence of impulses, at a frequency h/nT Hz where h is an integer, is

$$G\left(\frac{h}{nT}\right) = \sum_{i=0}^{n-1} x_i \exp\left(j2\pi \frac{h}{nT} iT\right)$$

$$= \sum_{i=0}^{n-1} x_i \exp\left(j2\pi \frac{hi}{n}\right) = \sum_{i=0}^{n-1} x_i \omega^{hi} \tag{2.169}$$

where, from Equation 2.31,

$$\omega = \exp\left(j\frac{2\pi}{n}\right) \tag{2.170}$$

Now, from Equation 2.44,

$$E_h = [1 \quad \omega^h \quad \omega^{2h} \quad \ldots \quad \omega^{(n-1)h}] \tag{2.171}$$

and E_h is the complex conjugate of

$$F_h = [1 \quad \omega^{-h} \quad \omega^{-2h} \quad \ldots \quad \omega^{-(n-1)h}] \tag{2.172}$$

which is the $(h+1)$th row of the DFT matrix F. Thus, from Equation 2.169,

$$G\left(\frac{h}{nT}\right) = XE_h^T = XF_h^* = \overline{(XF_h^T)} \tag{2.173}$$

where F_h^* is the conjugate transpose of F_h and $\overline{(XF_h^T)}$ is the complex conjugate of XF_h^T. But $G(h/nT)$ is the $(h+1)$th component of the n-component DFT of the *time reverse* of X, and F_h^T is the $(h+1)$th column of F, so that the DFT of the *reversed sequence* is $\overline{(XF)}$, which is the *complex conjugate* of the DFT of X.

The same result may be obtained in a slightly different form, as follows. First, let

$$x_n = x_0 \tag{2.174}$$

and bear in mind that

$$\omega^{hi} = \omega^{hi-hn} = \omega^{-h(n-i)} \tag{2.175}$$

Now, it can be seen from Equation 2.169, that the $(h+1)$th component of the n-component DFT of the *time reverse* of X is

$$G\left(\frac{h}{nT}\right) = \sum_{i=0}^{n-1} x_i \omega^{-h(n-i)}$$

$$= \sum_{i=0}^{n-1} x_{n-i} \omega^{-hi} = X_r F_h^T \tag{2.176}$$

where
$$X_r = [x_0 \quad x_{n-1} \quad x_{n-2} \quad \ldots \quad x_1] \tag{2.177}$$

Thus the DFT of the reversed sequence is $X_r F$.

It follows that the reversal of a sequence pivoted about its first component x_0 (that occurs at time $t = 0$) can be represented as X_r in Equation 2.177, where the time of occurrence of a component of X_r is given by the corresponding component of the n-component vector

$$\Delta_r = [0 \quad -(n-1)T \quad -(n-2)T \quad \ldots \quad -T] \tag{2.178}$$

From Equations 2.172, 2.173 and 2.176, the DFT of the reversed sequence is now

$$X_r F = \overline{(XF)} \tag{2.179}$$

The Reversal theorem therefore states that, if a sequence is reversed in time, the reversal being pivoted about the component at time $t = 0$, the DFT of the sequence is replaced by its *complex conjugate*. The converse, of course, also holds. In the case where the components of a sequence are *complex valued*, the Reversal theorem is modified as follows. If the sequence is replaced by its complex conjugate and the latter is reversed in time, the reversal being pivoted about the component at time $t=0$, the DFT of the resulting sequence is the complex conjugate of the DFT of the original sequence.

The Shift and Reversal theorems may be combined, as illustrated in the following example. Let X be the n-component sequence (row vector)

$$X = [x_0 \quad x_1 \quad \ldots \quad x_g \quad 0 \quad \ldots \quad 0] \qquad (2.180)$$

and let W be the n-component sequence

$$W = [x_g \quad x_{g-1} \quad \ldots \quad x_0 \quad 0 \quad \ldots \quad 0] \qquad (2.181)$$

formed from X by reversing the order of its first $g+1$ components, where g is any positive integer less than n. The $\{x_i\}$ here are *real-valued*.

The DFT of the sequence obtained from X by reversing X in time, the reversal being pivoted about the component x_0, is $X_r F = \overline{(XF)}$, where

$$X_r = [x_0 \quad 0 \quad 0 \quad \ldots \quad 0 \quad x_g \quad x_{g-1} \quad \ldots \quad x_1] \qquad (2.182)$$

But W is obtained from X reversed in time by delaying the latter by g places (sampling intervals), so that, from Equations 2.133 and 2.179,

$$WF = \overline{(XF)} D_g \qquad (2.183)$$

where D_g is the $n \times n$ diagonal matrix whose $(h \times 1)$th component along the main diagonal is ω^{-hg}.

2.11 APERIODIC AUTOCORRELATION FUNCTION AND DISCRETE ENERGY DENSITY SPECTRUM

Consider the n-component sequence given by the vector

$$X = [x_0 \quad x_1 \quad \ldots \quad x_g \quad 0 \quad 0 \quad \ldots \quad 0] \qquad (2.184)$$

with z-transform

$$X(z) = x_0 + x_1 z^{-1} + \cdots + x_g z^{-g} \qquad (2.185)$$

and let

$$n \geqslant 2g + 1 \qquad (2.186)$$

Assume that the sequence X is fed to a digital filter whose sampled impulse-response is the n-component sequence

$$W = [x_g \quad x_{g-1} \quad \ldots \quad x_0 \quad 0 \quad 0 \quad \ldots \quad 0] \qquad (2.187)$$

with z-transform
$$W(z) = x_g + x_{g-1}z^{-1} + \cdots + x_0 z^{-g} \tag{2.188}$$
Thus the z-transform of the output sequence from the filter is
$$\begin{aligned}X(z)W(z) &= x_0 x_g + (x_0 x_{g-1} + x_1 x_g)z^{-1} + \cdots \\ &\quad + (x_{g-1}x_0 + x_g x_1)z^{-2g+1} + x_g x_0 z^{-2g}\end{aligned} \tag{2.189}$$
which has $2g+1$ terms. Let the output sequence from the filter be given by the n-component row vector
$$A_g = [a_{-g} \quad a_{-g+1} \quad \cdots \quad a_{-1} \quad a_0 \quad a_1 \quad \cdots \quad a_g \quad 0 \quad 0 \quad \cdots \quad 0] \tag{2.190}$$
with z-transform
$$A_g(z) = a_{-g} + a_{-g+1}z^{-1} + \cdots + a_g z^{-2g} \tag{2.191}$$
Clearly, the output sequence A_g is the *linear convolution* of the sequences X and W. It can be seen from Equations 2.189–2.191 that the $(g+1+i)$th component of the sequence A_g is
$$a_i = \sum_{h=-\infty}^{\infty} x_h x_{h+i} \tag{2.192}$$
where $x_i = 0$ for $i < 0$ and $i > g$. The n components of the output sequence A_g contain the $2g+1$ potentially nonzero components of the discrete *aperiodic autocorrelation function* of the sequence X, given by the coefficients in $A_g(z)$ (Equation 2.191).

The value of the autocorrelation function of X, for a time interval of i places (sampling intervals), is given by a_i in Equation 2.192 and is the time correlation of the sequence X with the sequence X delayed (or advanced) by i places. The value of the autocorrelation function for a time interval of zero is
$$a_0 = \sum_{h=0}^{g} x_h^2 \tag{2.193}$$
which is the *squared length* of the vector X. The components of A_g are *real-valued* and the sequence A_g is symmetrical about the $(g+1)$th component a_0. The component a_0 is positive and has a greater magnitude than any other component of A_g. Also, $a_i = 0$ for $i < g$ and $i > g$.

It can be seen from Equation 2.187 that the sampled impulse-response of the digital filter is
$$\sum_{i=0}^{g} x_{g-i}\delta(t - iT) \tag{2.194}$$
If the time response of the filter is advanced by gT seconds, that is, by

g sampling intervals, the sampled impulse-response of the digital filter becomes

$$\sum_{i=0}^{g} x_{g-i}\delta(t+(g-i)T) \quad (2.195)$$

which is, of course, non-physical, since the output signal from the filter now *precedes* the input signal from which it originates. The study of this hypothetical filter is, nevertheless, of great interest. The sequence representing the resulting sampled impulse-response of the filter is now the *time reverse* of the sequence X (Equation 2.184), the reversal being pivoted about the component x_0 at time $t=0$. Since the sampled impulse-response of the digital filter is advanced by g sampling intervals, so also is the sequence A_g at the output of the filter, assuming that the same sequence X is fed to the filter in each case. The central component a_0 of the output sequence now occurs at time $t=0$, and, from the Shift theorem, the corresponding output sequence is given by the n-component row vector

$$A = [a_0 \quad a_1 \quad \ldots \quad a_g \quad 0 \quad \ldots \quad 0 \quad a_{-g} \quad \ldots \quad a_{-2} \quad a_{-1}] \quad (2.196)$$

bearing in mind that the component a_i occurs at time $t=iT$, for $-g \leqslant i \leqslant g$. The same result is obtained by applying the Shift theorem to the sequence W (Equation 2.187) advanced in time by g sampling intervals. The sampled impulse-response of the filter is now given by the n-component vector

$$W_{-g} = [x_0 \quad 0 \quad 0 \quad \ldots \quad 0 \quad x_g \quad x_{g-1} \quad \ldots \quad x_1] \quad (2.197)$$

which is obtained from W by shifting its components cyclically to the left by g places. But, as is shown in Section 2.8, a cyclic shift of g places to the left in the components of W results also in a cyclic shift of g places to the left in the output sequence from the filter. Thus, again the output sequence A_g becomes the sequence A in Equation 2.196. A is the *circular convolution* of X and W_{-g} and is also the *linear convolution* of X and W shifted cyclically to the left by g places.

The sequence of n terms given by the n-component vector A is defined to be the *aperiodic autocorrelation function* of the sequence given by the vector X in Equation 2.184, where $n \geqslant 2g+1$. As before, a_0 is the squared length of the vector X, and is also the largest component. The components of A are *real-valued* and symmetrical about a_0, such that

$$a_i = a_{-i} \quad (2.198)$$

for $i=1, 2, \ldots, g$, and

$$a_i = 0 \quad (2.199)$$

for $i < -g$ and $i > g$.

So long as $n \geqslant 2g+1$, A contains *all* the coefficients in $A_g(z)$ (Equation 2.191) and is the aperiodic autocorrelation function of X, as just defined. However, when $n < 2g+1$, the correct autocorrelation function is not obtained, because A now has an insufficient number of terms.

It can be seen from Equation 2.192 that the component a_i of A is formed from two identical sequences X, one of which is shifted by i places (sampling intervals) with respect to the other, each $(g+1)$-component sequence being preceded and followed by zeros to make it an infinite sequence. a_i is now given by the sum of the products of all coincident pairs of components (samples), one from each of the two sequences. There is no question here of one of the two sequences being shifted *cyclically* with respect to the other.

Let the n-component DFT's of W_{-g} and X be

$$W_{-g}F = [u_0 \quad u_1 \quad \ldots \quad u_{n-1}] \qquad (2.200)$$

and $\qquad XF = [v_0 \quad v_1 \quad \ldots \quad v_{n-1}] \qquad (2.201)$

where $n \geqslant 2g+1$. From Equation 2.99, which applies to both *linear* and *circular* convolutions, the n-component DFT of A is

$$AF = [u_0 v_0 \quad u_1 v_1 \quad \ldots \quad u_{n-1} v_{n-1}] \qquad (2.202)$$

so that the $(i+1)$th component of AF is $u_i v_i$. This follows because A is the *circular convolution* of X and W_{-g}, or alternatively, because A_g is formed by the *linear convolution* of W and X, as can be seen from Equation 2.189, and because $W_{-g}F$ and AF are the DFT's of the sequences formed from W and A_g through shifting them to the left by g places, so that AF is the DFT of the *linear convolution* of X and the sequence W advanced in time by g places.

But, from Equation 2.179, $W_{-g}F$ is the complex conjugate of XF, since W_{-g} is the true time reverse of X, represented as the corresponding cyclic shift, so that

$$u_i = v_i^* \qquad (2.203)$$

where v_i^* is the *complex conjugate* of v_i, and

$$u_i v_i = |v_i|^2 \qquad (2.204)$$

for $i = 0, 1, \ldots, n-1$. $|v_i|$ is the absolute value (modulus) of v_i. Thus the DFT of A is

$$AF = [|v_0|^2 \quad |v_1|^2 \quad \ldots \quad |v_{n-1}|^2] \qquad (2.205)$$

the $(i+1)$th component of AF being the square of the absolute value of the $(i+1)$th component of XF.

The n-component vector AF gives the *discrete energy-density spectrum* of the sequence X. All components of AF are real-valued and nonnegative. Furthermore, since X is real, v_i is the complex conjugate of v_{n-i}, for $i = 1, 2, \ldots, n-1$. Thus

$$|v_i| = |v_{n-i}| \qquad (2.206)$$

for $i = 1, 2, \ldots, n-1$, and AF is symmetric in the sense that its $(i+1)$th component is equal to its $(n-i+1)$th component, for each i. Alternatively, it can be seen that AF is a symmetric sequence since it is the DFT of the real symmetric sequence A.

To summarize, the discrete energy-density spectrum of the sequence X is the DFT of the aperiodic autocorrelation function of this sequence. The discrete energy-density spectrum and the aperiodic autocorrelation function are both real and symmetric sequences, and the components of the discrete energy-density spectrum are nonnegative.

2.12 MULTIPLICATION THEOREM

Let W and X be two n-component row vectors

$$W = [w_0 \quad w_1 \quad \ldots \quad w_g \quad 0 \quad \ldots \quad 0] \qquad (2.207)$$

and

$$X = [x_0 \quad x_1 \quad \ldots \quad x_g \quad 0 \quad \ldots \quad 0] \qquad (2.208)$$

where g is any nonnegative integer less than n. The $\{w_i\}$ and $\{x_i\}$ are real-valued, and W is *not* now a function of X.

Let the n-component DFT's of W and X be

$$WF = [u_0 \quad u_1 \quad \ldots \quad u_{n-1}] \qquad (2.209)$$

and

$$XF = [v_0 \quad v_1 \quad \ldots \quad v_{n-1}] \qquad (2.210)$$

From Equation 2.150, nXF^{-1} is the complex conjugate of XF, so that

$$nXF^{-1} = [v_0^* \quad v_1^* \quad \ldots \quad v_{n-1}^*] \qquad (2.211)$$

It now follows that

$$\sum_{i=0}^{n-1} u_i v_i^* = WF(nXF^{-1})^T$$

$$= nWFF^{-1}X^T = nWX^T \qquad (2.212)$$

so that

$$\sum_{i=0}^{n-1} u_i v_i^* = n \sum_{i=0}^{g} w_i x_i \qquad (2.213)$$

Taking the complex conjugate of each side of Equation 2.213,

$$\sum_{i=0}^{n-1} u_i^* v_i = n \sum_{i=0}^{g} w_i x_i \tag{2.214}$$

The relationships given by Equations 2.213 and 2.214 hold for *all* real vectors W and X satisfying Equations 2.207 and 2.208.

2.13 SIGNAL ENERGY

The *energy* of the real sequence given by the vector X in Equation 2.208 is defined as the *squared length* of X, and is given by a_0 in Equation 2.193. The *energy* of the complex sequence XF in Equation 2.210 is defined as

$$\sum_{i=0}^{n-1} |v_i|^2 = \sum_{i=0}^{n-1} v_i v_i^* \tag{2.215}$$

and Equation 2.215 may be taken as the general definition of the energy of an n-component sequence whose $(i+1)$th component is v_i, whether the sequence is real or complex.

If the vector W in Equation 2.207 is set equal to X, then, from Equation 2.213,

$$\sum_{i=0}^{n-1} |v_i|^2 = n \sum_{i=0}^{g} x_i^2 \tag{2.216}$$

so that the energy of XF is equal to n times the energy of X.

It is evident from Equations 2.193 and 2.196, that, when $n \geqslant 2g+1$, the energy of X is equal to a_0, which is the *first term* of A, the aperiodic autocorrelation function of X. Also, from Equations 2.205 and 2.215, the energy of XF is equal to the *sum* of the n components of AF, the discrete energy-density spectrum of X. Clearly, if $|v_i|^2 = 1$ for $i = 0, 1, \ldots, n-1$, then XF has n units of energy and X has one unit of energy.

REFERENCES

1. Gold, B. and Rader, C.M. *Digital Processing of Signals*, McGraw-Hill, New York (1969)
2. Rabiner, L.R. and Gold, B. *Theory and Application of Digital Signal Processing*, Prentice-Hall, Englewood Cliffs, New Jersey (1975)
3. Oppenheim, A.V. and Schafer, R.W. *Digital Signal Processing*, Prentice-Hall, Englewood Cliffs, New Jersey (1975)
4. Ziemer, R.E., Tranter, W.H. and Fannin, D.R. *Signals and Systems: Continuous and Discrete*, Macmillan, New York (1983)

5. Dwight, H.B. *Tables of Integrals and other Mathematical Data*, Fourth Edition, Macmillan, New York (1961)
6. Swokowski, E. *Fundamentals of College Algebra*, Prindle, Weber and Schmidt, Boston, Massachusetts (1971)
7. Paige, L.J. and Swift, J.D. *Elements of Linear Algebra*, Blaisdell Publishing Co., New York (1961)
8. Browne, E.T. *Introduction to the Theory of Determinants and Matrices*, Chapel Hill, University of North Carolina Press (1958)

Chapter 3

Theory of signal distortion

3.1 INTRODUCTION

In the synchronous serial digital data transmission system shown in Fig. 3.1, the transmission path together with the transmitter and receiver filters are assumed to be as described in Section 1.2. The received sample at time $t = iT$ is the *real-valued* quantity

$$r_i = \sum_{h=0}^{g} s_{i-h} y_h + w_i \qquad (3.1)$$

and the sampled impulse-response of the linear baseband channel in Fig. 3.1 is given by the $(g+1)$-component sequence (row vector)

$$Y = [y_0 \quad y_1 \quad \ldots \quad y_g] \qquad (3.2)$$

whose z-transform is

$$Y(z) = y_0 + y_1 z^{-1} + \cdots + y_g z^{-g} \qquad (3.3)$$

The *distortion* introduced by the channel is determined by the sequence Y, whose components $\{y_i\}$ are *real valued*. The noise component w_i originates from the additive white Gaussian noise at the input to the receiver filter, the $\{w_i\}$ being statistically independent real-valued Gaussian random variables with zero mean and fixed variance. The noise will be ignored during much of the discussion of signal distortion.

If the data-symbol s_{i-l} is detected directly from r_i, where $l = 0, 1, \ldots, g$, the received sample can be written as

$$r_i = s_{i-l} y_l + \sum_{\substack{h=0 \\ h \neq l}}^{g} s_{i-h} y_h + w_i \qquad (3.4)$$

where $s_{i-l} y_l$ is the *wanted signal*,

$$\sum_{\substack{h=0 \\ h \neq l}}^{g} s_{i-h} y_h \qquad (3.5)$$

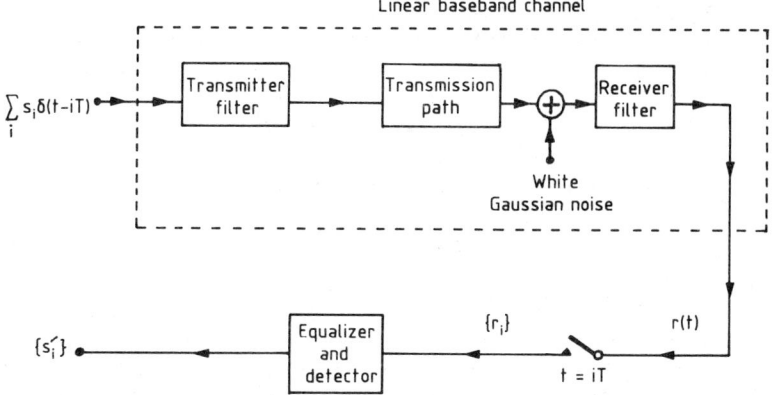

Fig. 3.1 Data-transmission system

is the *intersymbol interference* and w_i is the *noise*. The intersymbol interference is the direct result of the *change* introduced into the signal by the *transmission path*, the transmitter and receiver filters in Fig. 3.1 being such that, with an ideal transmission path, there is no intersymbol interference in r_i. Thus, whenever there are *two or more* nonzero components $\{y_h\}$ in the sampled impulse-response Y, the channel is said to introduce *signal distortion*. If there is only *one* nonzero component of Y, say y_l, then the intersymbol interference component in Equation 3.4 goes to zero and the channel now introduces *no* signal distortion. The fact that y_l may not be equal to unity or may even be negative does not, of itself, represent signal distortion, since in any practical receiver the value of y_l is always appropriately adjusted or suitably taken account of by means of the automatic gain control (AGC) and synchronization circuits. The only adverse effect, that can occur under the assumed conditions, is that $|y_l|$ may become small, with the corresponding increase in the probability of an error being introduced by the noise component w_i in the detection of the data-symbol s_{i-l} from r_i. However, due allowance is always made for the value of y_l in the theoretical analysis of the system, without having to regard the signal as being distorted. It is, of course, assumed here that the receiver has prior knowledge of y_l so that the detection process may be appropriately optimized. Thus signal distortion is here taken to imply the presence of *intersymbol interference*, that is, *more than one* nonzero component of Y.

When $y_0 = 1$ and $y_h = 0$ for $1 \leqslant h \leqslant g$ in Equation 3.1, the baseband channel introduces no attenuation, delay or distortion, and now

$$r_i = s_i + w_i \tag{3.6}$$

so that s_i is readily detected from r_i.

When $y_l = 1$ for an integer l in the range 1 to g, and $y_h = 0$ for $0 \leqslant h \leqslant g$ and $h \neq l$, the baseband channel introduces a delay of lT seconds but no attenuation or distortion. Now

$$r_i = s_{i-l} + w_i \qquad (3.7)$$

and s_i is detected from r_{i+l}.

It can be seen from Equation 3.4 that in general the ratio of the magnitude of the wanted signal component to the number and magnitudes of the intersymbol-interference components, gives some idea of the signal distortion introduced by the transmission path. However, it can be shown that the ease with which the distortion may be eliminated and the reduction in tolerance to additive noise resulting from the signal distortion, are not necessarily determined by just the number and magnitudes of the $\{y_h\}$ in Equation 3.4. The purpose of this analysis is to describe the sampled impulse-response of the channel in terms of both *amplitude distortion* and *phase distortion*, since the magnitudes of these two parameters give a useful guide as to the severity of the signal distortion and the ease with which it may be eliminated.

Suppose now that the sampled impulse-response of the baseband channel in Fig. 3.1 is given by the n-component real row vector

$$Y_0 = [y_0 \quad y_1 \quad \ldots \quad y_g \quad 0 \quad \ldots \quad 0] \qquad (3.8)$$

where
$$n \geqslant 2g + 1 \qquad (3.9)$$

Under these conditions, the aperiodic autocorrelation function of Y_0 involves a sequence of no more than n components, and, as far as the DFT of Y_0 is concerned, a *time* shift in the sequence Y_0, of up to g places (sampling intervals) in either direction, is identically equivalent to the corresponding *cyclic* shift.

The DFT of Y_0 is the n-component row vector

$$Y_0 F = [\lambda_0 \quad \lambda_1 \quad \ldots \quad \lambda_{n-1}] \qquad (3.10)$$

where
$$\lambda_i = J\left(\frac{i}{nT}\right) \qquad (3.11)$$

for $i = 0, 1, \ldots, n-1$, and $J(i/nT)$ is the value of the transfer function of the linear baseband channel and sampler in Fig. 3.1, at the discrete frequency i/nT Hz. Furthermore, the $\{\lambda_i\}$ are the n eigenvalues of the

$n \times n$ circulant matrix

$$Y_c = \begin{bmatrix} y_0 & y_1 & \cdots & y_{g-2} & y_{g-1} & y_g & 0 & \cdots & 0 & 0 \\ 0 & y_0 & \cdots & y_{g-3} & y_{g-2} & y_{g-1} & y_g & \cdots & 0 & 0 \\ \vdots & \vdots & \vdots\vdots\vdots & \vdots & \vdots & \vdots & \vdots & \vdots\vdots\vdots & \vdots & \vdots \\ y_2 & y_3 & \cdots & y_g & 0 & 0 & 0 & \cdots & y_0 & y_1 \\ y_1 & y_2 & \cdots & y_{g-1} & y_g & 0 & 0 & \cdots & 0 & y_0 \end{bmatrix}$$
(3.12)

whose first row is Y_0 in Equation 3.8. The circulant matrix is, of course, the convolution matrix Y_c in Equation 2.25 where n is now given by Equation 3.9 and not Equation 2.24. From the analysis in Section 2.9, λ_0 is real valued and

$$\lambda_i = \lambda^*_{n-i} \tag{3.13}$$

for $i = 1, 2, \ldots, n-1$, where λ^*_{n-i} is the complex conjugate of λ_{n-i}.

The aperiodic autocorrelation function of the sequence (vector) Y_0 is the n-component row vector

$$B = [b_0 \quad b_1 \quad \cdots \quad b_g \quad 0 \quad \cdots \quad 0 \quad b_{-g} \quad \cdots \quad b_{-1}] \tag{3.14}$$

where

$$b_i = \sum_{h=-\infty}^{\infty} y_h y_{h+i} \tag{3.15}$$

and $y_i = 0$ for $i < 0$ and $i > g$. All the $\{b_i\}$ are real valued, b_0 is positive, $b_i = 0$ for $i < -g$ and $i > g$, and the remaining $\{b_i\}$ may be positive or negative (or zero) with magnitudes less than b_0. B is symmetric in the sense that

$$b_i = b_{-i} \tag{3.16}$$

for $i = 1, 2, \ldots, g$.

The discrete energy-density spectrum of the sequence Y_0 is the n-component row vector

$$BF = [|\lambda_0|^2 \quad |\lambda_1|^2 \quad \cdots \quad |\lambda_{n-1}|^2] \tag{3.17}$$

where, of course, the components of BF are all real valued and nonnegative, and BF is symmetric in the sense that

$$|\lambda_i|^2 = |\lambda_{n-i}|^2 \tag{3.18}$$

for $i = 1, 2, \ldots, n-1$.

It is clear from the previous analysis that none of the relationships summarized in Equations 3.8 to 3.18 is dependent upon the particular relationship between n and g, provided only that $n > 2g$. n may, of course, be odd or even.

A linear baseband channel, such as that in Fig. 3.1, introduces four

98 THEORY OF SIGNAL DISTORTION

different effects on the transmitted signal. These are signal attenuation, signal delay, amplitude distortion and phase distortion. *Signal attenuation* is taken to include a possible *increase* or *decrease* in signal level (signal gain or loss) together with a possible *change of sign*. Thus signal attenuation is the multiplication of the signal waveform by a real-valued *scalar constant* that may have either sign and any appropriate magnitude. A change in signal level and/or sign, together with a delay in the signal, are not considered as *distortion* effects, since they do not of themselves either complicate or adversely affect the detection process. It is now convenient to consider signal attenuation together with amplitude-distortion, and signal delay together with phase distortion, because of the obvious relationship between each pair of effects.

Both amplitude distortion and phase distortion will be *defined* in terms of the n components of $Y_0 F$, these being the values of the transfer function of the baseband channel and sampler at the n discrete frequencies $\{i/nT\}$, for $i = 0, 1, \ldots, n-1$. The values of the transfer function at other frequencies will be *ignored*. Y_0 is, of course, uniquely determined by $Y_0 F$, and vice versa, since F is nonsingular. This method of describing signal distortion leads to a simple and powerful technique for analysing its properties.

3.2 PHASE DISTORTION

When the baseband channel and sampler in Fig. 3.1 introduce *amplitude distortion*, this is taken to mean that the n values of their transfer function $J(i/nT)$, for $i = 0, 1, \ldots, n-1$, do *not* all have the same absolute value (modulus). Thus, from Equations 3.10 and 3.11, the n components $\{\lambda_i\}$ of $Y_0 F$ (the DFT of the sampled impulse-response of the channel) do *not* all have the same absolute value.

Pure phase distortion is here defined as signal distortion (more than one nonzero component of Y_0) that includes *no amplitude distortion* and *no signal attenuation*. The absence of amplitude distortion implies that, in Equation 3.17,

$$|\lambda_i| = c \tag{3.19}$$

for $i = 0, 1, \ldots, n-1$, where c is a positive constant. The absence of signal attenuation implies that the energy (or squared length) of Y_0 is equal to unity, so that

$$|Y_0|^2 = \sum_{i=0}^{n-1} y_i^2 = 1 \tag{3.20}$$

But, from Equation 2.216,

$$\sum_{i=0}^{n-1} |\lambda_i|^2 = n \sum_{i=0}^{n-1} y_i^2 \tag{3.21}$$

so that, from Equation 3.20,

$$\sum_{i=0}^{n-1} |\lambda_i|^2 = n \tag{3.22}$$

and, from Equation 3.19,

$$|\lambda_i| = 1 \tag{3.23}$$

for $i = 0, 1, \ldots, n-1$. Also, since λ_0 is real valued, $\lambda_0 = \pm 1$.

Equation 3.23 is the condition that *must* be satisfied when the baseband channel introduces no amplitude distortion and no signal attenuation. It follows that, in the absence of both amplitude distortion and signal attenuation, each component $|\lambda_i|^2$ of the discrete energy-density spectrum in Equation 3.17 has the value unity. It is interesting to observe that a unit impulse has an energy spectral-density of unity over all frequencies (positive and negative), the unit impulse having, of course, infinite energy.

Since the $\{\lambda_i\}$ are the eigenvalues of the real $n \times n$ convolution matrix Y_c, it is clear that all the eigenvalues of Y_c are of absolute value unity. From Equation 2.80,

$$Y_c = FDF^{-1} \tag{3.24}$$

where D is the $n \times n$ diagonal matrix whose $(i+1)$th component along the main diagonal is λ_i. Also

$$Y_c^T = F^{-1}DF \tag{3.25}$$

since the matrices F, F^{-1} and D are all symmetric. It follows from Equation 2.67 that

$$Y_c Y_c^T = FDF^{-2}DF = n^{-1}FDKDF \tag{3.26}$$

where K is the $n \times n$ permutation matrix

$$K = \begin{bmatrix} 1 & 0 & 0 & \ldots & 0 & 0 \\ 0 & 0 & 0 & \ldots & 0 & 1 \\ 0 & 0 & 0 & \ldots & 1 & 0 \\ \vdots & \vdots & \vdots & \vdots\vdots\vdots & \vdots & \vdots \\ 0 & 0 & 1 & \ldots & 0 & 0 \\ 0 & 1 & 0 & \ldots & 0 & 0 \end{bmatrix} \tag{3.27}$$

But, for each i, $|\lambda_i| = 1$ and $\lambda_i = \lambda_{n-i}^*$. Also $\lambda_0 = \pm 1$. Thus

$$DKD = K \tag{3.28}$$

and, from Equations 3.26 and 2.67,

$$Y_c Y_c^T = n^{-1} FKF = FF^{-2}F = I \tag{3.29}$$

where I is an $n \times n$ identity matrix. Thus

$$Y_c^T = Y_c^{-1} \tag{3.30}$$

and Y_c is a *real orthogonal* matrix. The baseband channel and sampler in Fig. 3.1 therefore perform an *orthogonal transformation* on the transmitted data signal.

Conversely, if Y is a real orthogonal matrix, all its eigenvalues have an absolute value of unity, which, of course, means that all components of $Y_0 F$ have an absolute value of unity and the baseband channel and sampler in Fig. 3.1 introduce no amplitude distortion and no attenuation. It follows that a necessary and sufficient condition for no amplitude distortion and no attenuation is that the real convolution matrix Y is an *orthogonal* matrix.

Assume next that, in the synchronous serial data-transmission system of Fig. 3.1, a sequence of m signal-elements, in the form of the m impulses

$$\sum_{i=0}^{m-1} s_i \delta(t - iT) \tag{3.31}$$

is fed to the input of the baseband channel. The sequence of data-symbols $\{s_i\}$ may be represented more simply by the n-component row vector

$$S = [s_0 \quad s_1 \quad \ldots \quad s_{m-1} \quad 0 \quad \ldots \quad 0] \tag{3.32}$$

where now $n = m + 2g$. The sequence could be represented by just the first m components of S, but the reason for choosing the given value of n will become apparent later.

After transmission over the baseband channel and sampling at the channel output, the sequence (vector) S is transformed into the sequence given by the n-component row vector

$$SY_c = \sum_{i=0}^{m-1} s_i Y_i \tag{3.33}$$

where Y_c is the $n \times n$ circulant matrix in Equation 3.12, whose $(i+1)$th row (for $i = 0, 1, \ldots, m + g - 1$) is the n-component row vector

$$Y_i = [\overbrace{0 \quad \ldots \quad 0}^{i} \quad y_0 \quad y_1 \quad \ldots \quad y_g \quad \overbrace{0 \quad \ldots \quad 0}^{m+g-i-1}] \tag{3.34}$$

The last g of the n components of SY_c are all zero.

THEORY OF SIGNAL DISTORTION 101

Since Y_c is an orthogonal matrix, the n $\{Y_i\}$ are orthogonal vectors, which means that

$$Y_h Y_i^T = 0 \qquad (3.35)$$

whenever $h \neq i$. Clearly, the m signal-elements $\{s_i Y_i\}$ at the output of the channel are orthogonal, so that the channel uses an orthogonal transformation to convert the m orthogonal signal-elements $\{s_i \delta(t - iT)\}$ at the input of the channel into the m orthogonal signal-elements $\{s_i Y_i\}$ at its output. The signal-elements $\{s_i \delta(t - iT)\}$ at the input to the linear baseband channel in Fig. 3.1 are orthogonal, of course, because they are *disjoint in time*. For convenience, the $(i+1)$th signal-element $s_i \delta(t - iT)$, for $i = 0, 1, \ldots, m-1$, may be considered instead as the n-component row vector

$$S_i = [\overbrace{0 \quad 0 \quad \ldots \quad 0}^{i} \quad s_i \quad 0 \quad 0 \quad \ldots \quad 0] \qquad (3.36)$$

It is evident that the m vectors $\{S_i\}$ are mutually orthogonal. For the vectors Y_h and Y_i (for any $h \neq i$) to be *exactly* orthogonal, as in Equation 3.35, it is necessary that $g \to \infty$, and therefore also that $n \to \infty$, but a good approximation to this is normally obtained in practice without the need for an unduly large value of g. Since the error in the equation can be made as small as required through the use of a sufficiently large value of g, it is ignored and Equation 3.35 is assumed to hold *exactly*. If Equation 3.35 is not satisfied exactly, then, of course, neither is Equation 3.23.

Suppose now that the n components of the vector (sequence) SY_c in Equation 3.33 are fed sequentially and in the correct order to a digital filter whose sampled impulse-response is given by the n-component row vector

$$W = [y_g \quad y_{g-1} \quad \ldots \quad y_0 \quad 0 \quad \ldots \quad 0] \qquad (3.37)$$

W is derived from Y_0 by reversing the order of its first $g+1$ components. The sampled impulse-response of the filter is clearly the *reverse* of that of the baseband channel in Fig. 3.1, so that the filter is *matched* to the channel and is therefore also *matched* to each individual signal-element $s_i Y_i$.

It can be seen that the sequence at the input to the digital filter has up to $m+g$ nonzero components, whereas the output sequence from the digital filter has up to $m+2g$ non-zero components. It is for this reason that n has been given the value $m+2g$.

Let M be the $n \times n$ circulant matrix whose first row is the vector W, so that M is the convolution matrix corresponding to the sampled impulse-response W. But

$$M = Y_c^T L \qquad (3.38)$$

where L is the $n \times n$ permutation matrix obtained from the identity matrix by a cyclic shift of g places to the right of its columns, so that $Y_c^T L$ is the $n \times n$ matrix obtained from Y_c^T by shifting its columns cyclically by g places to the right.

The received signal at the input to the digital filter is the n-component sequence given by the vector SY_c, and the output signal from the digital filter is therefore the n-component sequence given by the vector

$$SY_c M = SY_c Y_c^T L = SL \tag{3.39}$$

since, of course, Y_c is an orthogonal matrix. This means that the output signal from the digital filter is the sequence S shifted to the right by g places and so delayed in time by gT seconds.

From Equations 3.30 and 3.38,

$$M = Y_c^{-1} L \tag{3.40}$$

where Y_c^{-1} is an orthogonal matrix. Thus the digital filter performs an orthogonal transformation on the received signal which is the *inverse* of that introduced by the channel, and the filter also delays the signal by gT seconds. Thus the data signal is restored to its original form, but delayed by g places. Clearly, the digital filter is not only *matched* to the channel but is also the *inverse* of the channel.

It is now of interest to consider in more detail the aperiodic autocorrelation function B and the discrete energy-density spectrum BF of the sampled impulse-response of the baseband channel, where the channel introduces no amplitude distortion and no signal attenuation.

It can be seen from Equation 3.23 that, in the presence of pure phase distortion,

$$|\lambda_i|^2 = 1 \tag{3.41}$$

for $i = 0, 1, \ldots, n-1$. Thus it follows from Equation 2.84 that the sequence Y_0 (the sampled impulse-response of the baseband channel in Fig. 3.1) now has a Fourier transform of absolute value unity at each of the n discrete frequencies $\{i/nT\}$ Hz and therefore unit energy spectral density at each of these frequencies. This implies that the baseband channel has unit gain (or no attenuation) at each frequency, which is to be expected in the absence of both amplitude distortion and signal attenuation. Now, from Equations 3.17 and 3.41,

$$BF = [1 \quad 1 \quad \ldots \quad 1] \tag{3.42}$$

so that

$$B = [1 \quad 1 \quad \ldots \quad 1] F^{-1} \tag{3.43}$$

where

$$B = [b_0 \quad b_1 \quad b_2 \quad \ldots \quad b_{n-1}] \tag{3.44}$$

and $n \geqslant 2g+1$. For consistency with the terminology in Equation 3.14,

$$b_{n-i} = b_{-i} \tag{3.45}$$

for $i = 1, 2, \ldots, g$. From Equation 3.43,

$$b_i = n^{-1}(1 + \omega^i + \omega^{2i} + \cdots + \omega^{(n-1)i}) \tag{3.46}$$

for $i = 0, 1, \ldots, n-1$. When $0 < i < n$, it can be seen from Equation 2.47 that

$$1 + \omega^i + \omega^{2i} + \cdots + \omega^{(n-1)i} = 0 \tag{3.47}$$

since ω^i is here one of the nth roots of unity not including unity itself Thus, when $0 < i < n$,

$$b_i = 0 \tag{3.48}$$

When $i = 0$, b_i is the first component of B and has the value

$$b_0 = 1 \tag{3.49}$$

as can be seen from Equation 2.52. Hence the aperiodic autocorrelation function of the sequence Y_0 becomes

$$B = [1 \quad 0 \quad 0 \quad \ldots \quad 0] \tag{3.50}$$

Since the $(i+1)$th component of B, for $i = 0, 1, \ldots, g$, is the value of the aperiodic autocorrelation function of the sequence Y_0, for a time interval (relative time delay) of i places, it follows from Equations 3.14, 3.15 and 3.50 that

$$\sum_{h=-\infty}^{\infty} y_h y_{i+h} = 0 \tag{3.51}$$

when $i \neq 0$, assuming that $y_i = 0$ for $i < 0$ and $i > g$. This means that the sequence Y_0 is *orthogonal* to the sequence obtained from Y_0 by shifting its components one or more places to the left or right. Y_0 is here assumed to be shifted in *time* by a multiple of T seconds, where T is the sampling interval used for the sampled impulse-response of the baseband channel in Fig. 3.1. Finally, Equation 3.49 shows that Equation 3.20 must hold. The result given by Equation 3.51 is entirely consistent with that given by Equation 3.35, and confirms the fact that the baseband channel performs an orthogonal transformation on the transmitted data signal. For exact equality in Equation 3.51, $g \to \infty$.

It is interesting now to compare the previous analysis by matrix algebra, of the transmission of a finite sequence of m signal elements, with an analysis of the transmission of an *infinite* sequence of signal elements, in terms of the z-transforms of the transmitted signals. Suppose therefore that, in the synchronous serial data-transmission

104 THEORY OF SIGNAL DISTORTION

system of Fig. 3.1, a *continuous* sequence of signal elements is transmitted, such that the z-transform of the sequence of impulses at the input to the baseband channel is

$$S(z) = \sum_i s_i z^{-i} \qquad (3.52)$$

The z-transform of the sampled impulse-response of the baseband channel is

$$Y(z) = \sum_{i=0}^{g} y_i z^{-i} \qquad (3.53)$$

and the z-transform of the sequence obtained after sampling the output signal from the baseband channel, at the time instants $\{iT\}$, is

$$S(z)Y(z) = \sum_i s_i z^{-i} Y(z) \qquad (3.54)$$

assuming here that there is no noise. It is evident from Equation 3.51 that each individual signal-element in this sampled signal is orthogonal with respect to every other signal-element.

Suppose now that the sequence given by Equation 3.54 is fed to a digital filter whose sampled impulse-response is given by W in Equation 3.37 and whose z-transform is

$$W(z) = \sum_{i=0}^{g} y_{g-i} z^{-i} \qquad (3.55)$$

The filter is *matched* to the channel, and, from Equation 3.53, $W(z)$ is obtained from $Y(z)$ by *reversing* the order of its coefficients.

It can be seen from Equations 2.189 and 3.14 that the coefficient of z^{-i} in $Y(z)W(z)$ is b_{i-g}, so that, from Equation 3.50,

$$Y(z)W(z) = z^{-g} \qquad (3.56)$$

Thus the z-transform of the output sequence from the filter is

$$\sum_i s_i z^{-i} Y(z) W(z) = \sum_i s_i z^{-i-g} \qquad (3.57)$$

This means that the signal at the filter output, at time $t = (h+g)T$, is s_h, so that s_h can be detected from this signal without interference from any of the other $\{s_i\}$.

From Equation 3.56,

$$W(z) = z^{-g} Y^{-1}(z) \qquad (3.58)$$

where $Y(z)Y^{-1}(z) = 1$. As before, the matched filter introduces a transformation into the received signal that is the *inverse* of that introduced by the channel, and the filter also *delays* the signal by gT seconds (g places). The filter gives a sequence of orthogonal signal-elements at its output that are a delayed copy of the signal elements at the input to the channel, as can be seen from Equation 3.57. With a finite value value of g, Equations 3.56 to 3.58 hold only approximately, $Y^{-1}(z)$ being an infinite series.[1]

It is clear from the preceding discussion that $W(z)$ represents an *orthogonal* transformation and is both the *reverse* and *inverse* of $Y(z)$, neglecting the delay introduced. Similarly, the *orthogonal* transformation $Y(z)$ is both the *reverse* and *inverse* of $W(z)$, again neglecting the signal delay. Thus it seems that just as an orthogonal matrix is one whose inverse is also its transpose, so the z-transform corresponding to an orthogonal transformation is one whose inverse is also its reverse.

The important points emerging from the preceding discussion can now be summarized as follows. If the baseband channel and sampler in Fig. 3.1 introduce no amplitude distortion and no signal attenuation, then $|\lambda_i| = 1$, for $i = 0, 1, \ldots, n-1$ where $n \geq 2g + 1$, and every component $|\lambda_i|^2$ of the discrete energy-density spectrum of the sampled impulse-response of the channel has the value *unity*. The channel (including the sampler) may also introduce no phase distortion and no signal delay, in which case the sampled impulse-response of the channel is

$$Y_0 = [1 \quad 0 \quad \ldots \quad 0] \qquad (3.59)$$

which represents perfect transmission. Alternatively, the channel may introduce a signal delay of iT seconds (i places) in which case

$$Y_0 = [\overbrace{0 \quad \ldots \quad 0}^{i} \quad 1 \quad 0 \quad \ldots \quad 0] \qquad (3.60)$$

On the other hand, the channel may introduce *pure phase distortion* in which case the sequence Y_0 has *more* than one nonzero component (ideally an infinite number) but its discrete energy-density spectrum again has all its components equal to unity. The combination of these two conditions may be taken as a definition of pure phase distortion. The Euclidean length of the vector Y_0 must here be equal to unity, to satisfy the condition of no signal attenuation.

Pure phase distortion and/or delay are always an orthogonal transformation of the transmitted signal, and vice versa. The corresponding z-transform is such that the reversal of the order of its coefficients gives the inverse orthogonal transformation together with the appropriate delay. A linear equalizer for a baseband channel introducing pure phase distortion is a filter *matched* to the channel. Its z-transform is both the reverse and inverse of that of the channel (neglecting the delay involved) and it introduces itself both the reverse and inverse of the phase distortion introduced by the channel.

3.3 REPRESENTATION OF PHASE DISTORTION BY THE ZEROS AND POLES OF THE Z-TRANSFORM

Suppose first that the baseband channel in Fig. 3.1 has the sampled impulse-response

$$Y = [y_0 \quad y_1 \quad \ldots \quad y_g] \quad (3.61)$$

with z-transform

$$Y(z) = y_0 + y_1 z^{-1} + \cdots + y_g z^{-g}$$
$$= y_0 (1 - \alpha_1 z^{-1})(1 - \alpha_2 z^{-1}) \ldots (1 - \alpha_g z^{-1}) \quad (3.62)$$

where the $\{y_i\}$ are real valued and the $\{\alpha_i\}$ are real or complex valued. Also $y_0 \neq 0$. The values of z such that $Y(z) = 0$ are said to be the *zeros* of $Y(z)$ in the z plane, and the values of z such that $Y(z) \to \infty$ are said to be the *poles* of $Y(z)$. Clearly a *root* of $Y(z)$ is also a zero of $Y(z)$, and the z plane is simply the complex number plane containing all possible values of z. As has previously been mentioned, z is normally constrained to lie on the unit circle, so that $|z| = 1$. However, when considering zeros and poles of $Y(z)$, z must be allowed to take on *any* complex value. Furthermore, except where otherwise stated, it is assumed throughout that there are *no* zeros or poles of $Y(z)$ that lie *on* the unit circle. If now, for example, $1 - \alpha z^{-1}$ is a factor of $Y(z)$, $Y(z)$ has a zero at $z = \alpha$ and a pole at $z = 0$, and if $(1 - \alpha z^{-1})^{-1}$ is a factor of $Y(z)$, $Y(z)$ has a pole at $z = \alpha$ and a zero at $z = 0$. Thus $Y(z)$ in Equation 3.62 has a zero at each of $z = \alpha_1, \alpha_2, \ldots, \alpha_g$ and g poles at $z = 0$. Again, if $1 - \alpha z$ is a factor of $Y(z)$, $Y(z)$ has a zero at $z = 1/\alpha$ and a pole at $z \to \infty$, and if $(1 - \alpha z)^{-1}$ is a factor of $Y(z)$, $Y(z)$ has a pole at $z = 1/\alpha$ and a zero at $z \to \infty$. Furthermore, if $(1 - \alpha z^{-1})^m$ is a factor of $Y(z)$, $Y(z)$ has m zeros at $z = \alpha$ and m poles at $z = 0$, and if $(1 - \alpha z^{-1})^{-m}$ is a factor of $Y(z)$, $Y(z)$ has m poles at $z = \alpha$ and m zeros at $z = 0$. Hence there may be more than one zero or pole at any one value of z. When $Y(z)$ has one or more *poles* that are *not* at the origin or infinity, it has ideally an infinite number of terms,[1] but a good approximation to the ideal can be obtained with a finite number of terms, except where there are one or more poles *on* the unit circle in the z plane. If z^{-1} is a factor if $Y(z)$, it corresponds to a *delay* of one sampling interval of T seconds and introduces a pole at the origin together with a zero at infinity. If z is a factor of $Y(z)$, it corresponds to an *advance* of one sampling interval and introduces a zero at the origin together with a pole at infinity. Because the zeros and poles of $Y(z)$ that lie at the origin and infinity do not represent or imply the presence of signal distortion, being associated with *time shifts* rather than with signal distortion, these zeros and poles are *ignored* here. Conversely, any zeros or poles *not* at the origin or infinity imply the presence of *some* signal distortion.

It has been shown in Section 3.2 that, in order for the sampled impulse-response Y of the linear baseband channel in Fig. 3.1 (given by Equation 3.61) to introduce an *orthogonal transformation* into the received signal and so to introduce *pure phase distortion*, it is necessary that the z-transform, formed by reversing the order of the coefficients in $Y(z)$, is also the *inverse* z-transform suitably delayed. Suppose now that

$$Y(z) = \pm z^{-h} X(z) W^{-1}(z) \tag{3.63}$$

where
$$X(z) = x_0 + x_1 z^{-1} + \cdots + x_m z^{-m} \tag{3.64}$$

and
$$W(z) = x_0 + x_1 z + \cdots + x_m z^{m} \tag{3.65}$$

The $\{x_i\}$, for $i = 0, 1, \ldots, m$, are *real valued*, h and m being appropriate positive integers. Clearly, $X(z)$ and $W(z)$ have the same number of terms, and $W(z)$ is obtained from $X(z)$ by changing the sign of each power of z, that is, by replacing each component z^{-i} by z^i, without however changing the coefficients. Clearly, the *sequence* with z-transform $X(z)$ is

$$[x_0 \quad x_1 \quad \ldots \quad x_m] \tag{3.66}$$

the components occurring at the time instants $0, T, \ldots, mT$, respectively, and the *sequence* with z-transform $W(z)$ is

$$[x_m \quad x_{m-1} \quad \ldots \quad x_0] \tag{3.67}$$

the components occurring at the time instants $-mT, -(m-1)T, \ldots, 0$, respectively. By definition, $W(z)W^{-1}(z) = 1$.

Although $W^{-1}(z)$ has an infinite number of terms, it can be approximated to as accurately as required by expressing it to a sufficiently large but strictly finite number of terms, which will now be assumed. The sampled impulse-response Y of the channel, with z-transform $Y(z)$ in Equation 3.63, can thus be considered as the *convolution* of two finite-length sequences, one with the z-transform $X(z)$ and the other with the z-transform $W^{-1}(z)$. Also involved in Y is a delay of h sampling intervals, whose z-transform is z^{-h}, and perhaps a negative sign. Clearly, $Y(z)$ need not be physically realizable since it may contain positive powers of z, indicating that an output signal is obtained from the channel and sampler in Fig. 3.1 ahead of the corresponding input signal. $Y(z)$ can, of course, be made physically realizable (for practical purposes) by making the positive integer h in z^{-h} sufficiently large, such that the coefficients of any resulting positive powers of z in $Y(z)$ become small. The delay of h sampling intervals now introduced into the sampled impulse-response Y by z^{-h} is sufficiently large to ensure that all components of Y that occur at negative time instants can be ignored (set to zero) without noticeably affecting the signal.

It is evident that the sequence with z-transform $W(z)$ is obtained from the sequence with z-transform $X(z)$, by reversing the order of the latter sequence, the reversal being pivoted about the component x_0 that occurs at time $t=0$. If now the sequence with z-transform $z^{-h}X(z)W^{-1}(z)$ is reversed in order, the reversal being pivoted about the component at time $t=0$, then so also are the two constituent sequences with z-transforms $X(z)$ and $W^{-1}(z)$, the delay with z-transform z^{-h} being replaced by the corresponding advance with z-transform z^h. This is because the *reversal* in time of the *convolution* of two sequences is the same as the *convolution* of the two sequences *reversed* in time. The reversal of the first of the two sequences causes its z-transform to be changed from $X(z)$ to $W(z)$, and the reversal of the second sequence causes its z-transform to be changed from $W^{-1}(z)$ to $X^{-1}(z)$. The latter is evident from the fact that the reversal of the sequence with z-transform $W^{-1}(z)$, pivoted about the component at time $t=0$, is achieved by changing the sign of each power of z in $W^{-1}(z)$, and this, in turn, converts $W^{-1}(z)$ into $X^{-1}(z)$. Thus the z-transform of the reversed sequence is $z^h W(z) X^{-1}(z)$ and this is the inverse of $z^{-h}X(z)W^{-1}(z)$ which is the z-transform of the original sequence. Clearly, the inverse of $-z^{-h}X(z)W^{-1}(z)$ is $-z^h W(z) X^{-1}(z)$, which is again the z-transform of the reversed sequence. It follows from the above that the *reversal* of the sequence with z-transform $\pm z^{-h}X(z)W^{-1}(z)$ gives the *inverse* sequence, so that both the original and reversed sequences represent *pure phase distortion*.

It is well known that if the coefficients of a polynomial in z are *reversed* in order, without otherwise changing the polynomial, then the zeros (roots) of the polynomial are transferred to the corresponding *reciprocal* values of z. This can in fact be seen from Equation 3.62 where the reversal of the order of the coefficients in $Y(z)$ converts $y_0 + y_1 z^{-1} + \cdots + y_g z^{-g}$ to $y_g + y_{g-1} z^{-1} + \cdots + y_0 z^{-g}$ and at the same time converts $y_0(1 - \alpha_1 z^{-1})(1 - \alpha_2 z^{-1}) \ldots (1 - \alpha_g z^{-1})$ to $y_0(-\alpha_1 + z^{-1})(-\alpha_2 + z^{-1}) \ldots (-\alpha_g + z^{-1})$, as can be seen by multiplying out each of the last two expressions. Now, the ith zero of $Y(z)$ is α_i, since $1 - \alpha_i z^{-1} = 0$ when $z = \alpha_i$, and the ith zero of the corresponding z-transform with the coefficients in reverse order is α_i^{-1}, since $-\alpha_i + z^{-1}$

$=0$ when $z=\alpha_i^{-1}$. Thus the ith zero of the latter polynomial is the *reciprocal* of the ith zero of $Y(z)$. But, as has been mentioned before, the sequence with z-transform $W(z)$ is obtained from the sequence with z-transform $X(z)$ by reversing the order of the latter sequence, the reversal being pivoted about the component at time $t=0$. Thus the zeros of $W(z)$ are the *reciprocals* of those of $X(z)$, ignoring any at the origin or infinity in the z plane. Since both $X(z)$ and $W(z)$ have a finite number of terms, they have *no poles*, other than any at the origin or infinity, which are of course neglected here. If x_0 and x_m are nonzero, both $X(z)$ and $W(z)$ have m zeros and no poles. Clearly, the zeros of $W(z)$ are given by transferring the zeros of $X(z)$ to the respective *reciprocal* values of z.

The *inverse* of $X(z)$ is $X^{-1}(z)$, which is such that $X(z)X^{-1}(z)=1$. $X^{-1}(z)$ has as many poles as $X(z)$ has zeros, the poles of $X^{-1}(z)$ occurring at the same values of z as the zeros of $X(z)$. $X^{-1}(z)$ and $W^{-1}(z)$ have no zeros, other than any at the origin or infinity. It follows that the transfer of the poles of $X^{-1}(z)$ to the respective reciprocal values of z gives the poles of $W^{-1}(z)$. Furthermore, if there is a reversal in the order of the coefficients in $X(z)$, then, because $X(z)X^{-1}(z)=1$, there must also be a reversal in the order of the coefficients in $X^{-1}(z)$. This means that $W^{-1}(z)$ is obtained from $X^{-1}(z)$ by reversing the order of its coefficients, as has, in fact, previously been shown by means of an alternative argument. Thus the reversal of the order of the coefficients in a polynomial containing only poles transfers these poles to the corresponding reciprocal values of z. Since this applies also to zeros, the result may be extended to the more general case where a polynomial contains both zeros and poles. Hence, the *reversal* of the order of the coefficients in a z-transform, transfers its zeros and poles to the corresponding reciprocal values of z. For comparison, the *inversion* of the z-transform replaces its zeros and poles by poles and zeros, respectively. It can now be seen that each *zero* of $Y(z)$ in Equation 3.63 is accompanied by a *pole* at the reciprocal value of z and similarly each *pole* is accompanied by a *zero* at the reciprocal value of z. Thus the zeros and poles occur in reciprocal pairs, each pair involving one zero and one pole.

Consider now the n-component sequences

$$X = [x_0 \quad x_1 \quad \ldots \quad x_m \quad \overbrace{0 \quad 0 \quad \ldots \quad 0}^{n-m-1}] \quad (3.68)$$

and

$$W = [x_0 \quad \overbrace{0 \quad 0 \quad \ldots \quad 0}^{n-m-1} \quad x_m \quad x_{m-1} \quad \ldots \quad x_1] \quad (3.69)$$

where the components $x_0, x_1, \ldots, x_m$ occur at the time instants $0, T, \ldots, mT$, respectively, in X and at the time instants $0, -T, \ldots, -mT$,

110 THEORY OF SIGNAL DISTORTION

respectively, in W. Since W is the time reverse of X, the reversal being pivoted about the component x_0 at time $t=0$, it follows from Equation 2.179 that the DFT of W is

$$WF = (\overline{XF}) \qquad (3.70)$$

where F is the $n \times n$ DFT matrix (Equation 2.42) and $(\overline{XF})$ is the complex conjugate of XF. Clearly, the ith component of the DFT of W is the *complex conjugate* of the ith component of the DFT of X.

As has previously been stated, $W^{-1}(z)$ is taken to have a finite number of terms, the number being made sufficiently large to reduce the error in the z-transform to negligible proportions. In fact, $W^{-1}(z)$ is represented by the most significant adjacent set of $n-m$ of its terms, where n is the number of components of X and W (Equations 3.68 and 3.69) and n is the number of components of X and W (Equations 3.68 and 3.69) and n is made so large that the $n-m$ terms of $W^{-1}(z)$ taken to represent this function include all terms of significant magnitude, the remaining terms of $W^{-1}(z)$ being set to zero. Now let

$$V = [v_0 \quad v_1 \quad \ldots \quad v_h \quad v_{h+1} \quad \ldots \quad v_{h+m} \quad v_{h+m+1} \quad \ldots \quad v_{n-1}] \qquad (3.71)$$

be the n-component sequence whose components are given by the coefficients in $W^{-1}(z)$, the components of V being located according to their times of occurrence as given by $W^{-1}(z)$. Thus, for example, v_0, v_1 and v_{n-1} occur at the time instants $t=0$, $t=T$ and $t=-T$, respectively. Furthermore, the appropriate portion of V, involving the m components v_{h+1} to v_{h+m}, comprises m successive *zero-valued* components, such that the convolution of V and W, determined in the same way as the convolution of S and Y_0 in Equations 2.22–2.27, gives the required linear (aperiodic) convolution of the two sequences. The convolution is, of course, cyclically shifted, such that its first component, *as written*, is that at time $t=0$. Since, by definition,

$$W(z)W^{-1}(z) = 1 \qquad (3.72)$$

the n-component sequence with z-transform $W(z)W^{-1}(z)$ is

$$[1 \quad 0 \quad 0 \quad \ldots \quad 0] \qquad (3.73)$$

and the n-component DFT of this sequence is

$$[1 \quad 1 \quad 1 \quad \ldots \quad 1] \qquad (3.74)$$

But the sequence with z-transform $W(z)W^{-1}(z)$ is the *convolution* of the sequences W and V, so that the ith component of the DFT of the sequence with z-transform $W(z)W^{-1}(z)$ is the *product* of the ith components of the DFT's of W and V. Since *each* of these products is unity, it is evident that the ith component of the DFT of V is the *reciprocal* (inverse) of the ith component of the DFT of W. It now follows from Equation 3.70 that the ith component of the DFT of V is the *complex conjugate* of the *reciprocal* of the ith component of the DFT of X.

Suppose next that the z-transform of the sampled impulse-response of the linear baseband channel in Fig. 3.1 is

$$Y(z) = X(z)W^{-1}(z) \tag{3.75}$$

where the n-component sampled impulse-response of the channel is (for practical purposes)

$$Y_0 = [y_0 \quad y_1 \quad \ldots \quad y_g \quad 0 \quad 0 \quad \ldots \quad 0] \tag{3.76}$$

g being made so large that a negligible error is introduced into Equation 3.75 by assuming a *finite* number of nonzero components of Y_0 and hence by assuming the corresponding finite number of terms in $W^{-1}(z)$. Now Y_0 is the *convolution* of X and V, which means that the ith component of the DFT of Y_0 is the *product* of the ith components of the DFT's of X and V, one of which is the *complex conjugate* of the *reciprocal* of the other. Thus, the absolute value (modulus) of the ith component of the DFT of Y_0 must be equal to unity for each i.

In the more general case where

$$Y(z) = \pm z^{-h} X(z) W^{-1}(z) \tag{3.77}$$

as in Equation 3.63, the delay of h sampling intervals together with a possible change of sign, represented by $\pm z^{-h}$, corresponds to the *convolution* of the n-component sequence Y_0 (having the z-transform $X(z)W^{-1}(z)$) with an n-component sequence

$$[\overbrace{0 \quad 0 \quad \ldots \quad 0}^{h} \quad \pm 1 \quad 0 \quad 0 \quad \ldots \quad 0] \tag{3.78}$$

whose DFT has all components of absolute value unity. Hence, in the case of the sampled impulse-response Y_0 of the channel with a z-transform given by Equation 3.77, the absolute value (modulus) of the ith component of the DFT of Y_0 is again equal to unity for each i. Thus, from the definition of phase distortion as given by Equation 3.23, it follows that Equation 3.77 is a *sufficient* condition to ensure that $Y(z)$ represents pure phase distortion. It remains now to consider whether or not the equation represents a *necessary* condition.

It is assumed that $Y(z)$ has a strictly *finite* number of zeros and poles. Now, in $Y^{-1}(z)$, which is the inverse of $Y(z)$, each zero or pole of $Y(z)$ is replaced by a pole or zero, respectively. But, if $Y(z)$ represents pure phase distortion, it must be such that the z-transform formed from $Y(z)$ by reversing the order of its coefficients, the reversal being pivoted about the coefficient of z^0, is also $Y^{-1}(z)$. This is demonstrated by the analysis in Section 3.2, but can be shown more directly via the Reversal theorem, as follows. If the n-component sampled impulse-response Y_0, of the channel with z-transform $Y(z)$, represents pure phase distortion, each component of its DFT must have an absolute value of *unity*. Now, if Y_0 is reversed in time, the reversal being pivoted about its component at time $t=0$, then, by the Reversal theorem, each component of its DFT must be replaced by its *complex conjugate*. But the complex conjugate of a complex number with an absolute value of unity is also the *reciprocal* (*inverse*) of this number, so that the DFT of the reversed sequence is the *inverse* of the DFT of Y_0. Thus the z-transform of the reversed sequence must be $Y^{-1}(z)$, provided only that n is sufficiently large. Furthermore, the reversal of the order of the coefficients in $Y(z)$ transfers each zero and pole to the corresponding reciprocal value of z. For these conditions to be satisfied, each zero and pole of $Y(z)$ must be accompanied by a pole or zero, respectively, at the reciprocal value of z, and for this to apply, $Y(z)$ must satisfy Equation 3.77, modified by the inclusion of an arbitrary real-valued multiplying constant c on the right hand side. However, since for pure phase distortion all components of the DFT of Y_0 must have absolute values of *unity* (Equation 3.23) and since this condition is satisfied by the sequence Y_0 with z-transform $Y(z)$ given by Equation 3.77, it follows that $c=1$. Thus, under the conditions assumed here where $Y(z)$ has a strictly *finite* number of zeros and poles, Equation 3.77 is a condition that must *necessarily* be satisfied if the channel introduces pure phase distortion. If now $Y(z)$ is permitted to have an *infinite* number of zeros (that cannot be represented exactly by a finite number of poles) or alternatively if $Y(z)$ need only *approximate* to the ideal as $g \to \infty$, then pure phase distortion can be obtained under conditions *other* than those given by Equation 3.77. An important class of sequences $\{Y_0\}$ that approximate to pure phase distortion, having a finite number of zeros and no poles, are known as Huffman sequences[2-4]. For our purposes, however, Equation 3.77 includes all cases of real interest.

Since pure phase distortion is represented by the z-transform $Y(z)$ in Equation 3.77, it is such that each zero and pole of the z-transform is accompanied by a pole or zero, respectively, at the reciprocal value of z. Since the coefficients in the z-transform are all *real valued*, all zeros and all poles at complex values of z occur in *complex conjugate pairs*, so that each zero or pole is also accompanied by a pole or zero, respectively, at the *complex conjugate* of the reciprocal value of z. This latter condition applies to $Y(z)$, when it represents pure phase distortion and when its coefficients may be real or complex valued. It therefore applies to both *real*- and *complex-valued* signals and so is of more general application than the condition previously given, which of course applies only to real-valued signals. Complex-valued signals are considered further in Section 3.7.

It has been assumed so far that $Y(z)$ and therefore also $X(z)$ and $W^{-1}(z)$ have *no* zeros or poles *on* the unit circle in the z plane. If now $X(z)$ (Equation 3.64) has one or more zeros on the unit circle in the z plane, the poles of $W^{-1}(z)$ on the unit circle always coincide with the corresponding zeros of $X(z)$, so that the zeros and poles *on* the unit circle cancel out and therefore do not appear in $Y(z)$ (Equation 3.77). Pure phase distortion itself has *no* zeros or poles *on* the unit circle.

The value of the transfer function $J(f)$ of the baseband channel and sampler in Fig. 3.1, at a frequency f Hz, is the value of $Y(z)$ when $z = \exp(j2\pi fT)$, and so is the value of $Y(z)$ at the point on the unit circle in the z plane making an angle of $2\pi fT$ radians with the positive real axis. Thus a zero or pole *near* the unit circle results in a decrease or increase, respectively, in the magnitudes of $J(f)$, at values of f that correspond to points on the unit circle close to the zero or pole. Furthermore, the closer the zero or pole is to the unit circle in the z plane, the greater is the decrease or increase in the magnitudes of $J(f)$ at these values of f. When the zero or pole lies *on* the unit circle, at $z = e$ where of course $|e| = 1$, then, at the corresponding values of f, $J(f)$ is zero or infinite, respectively. The corresponding values of f are given by $\exp(j2\pi fT) = e$. A pole on the unit circle is not physically realizable, since it implies infinite gain at the corresponding frequencies.

It can be seen that $Y(z)$ in Equation 3.77 must have at least one *pole* not at the origin or infinity (except in the degenerate case where $Y(z)$ represents no distortion), so that $Y(z)$ is not a polynomial of finite degree. Thus pure phase distortion can only be represented exactly by an infinite sequence Y_0 and is only approximately realized by the finite sequence assumed. The approximation to the ideal infinite sequence is achieved through the use of the following simple relationships.

A pole may be represented approximately by a set of zeros, as follows. Consider a pole at $z = \alpha$, represented by the z-transform $(1 - \alpha z^{-1})^{-1}$. If $|\alpha| < 1$,

$$(1 - \alpha z^{-1})^{-1} \simeq 1 + \alpha z^{-1} + \alpha^2 z^{-2} + \cdots + \alpha^{n-1} z^{-(n-1)} \quad (3.79)$$

which has zeros (roots) at

$$z = \alpha \omega^i \quad (3.80)$$

for $i = 1, 2, \ldots, n-1$, where

$$\omega^i = \exp\left(j\frac{2\pi i}{n}\right) \quad (3.81)$$

114 THEORY OF SIGNAL DISTORTION

If $|\alpha| > 1$,

$$(1 - \alpha z^{-1})^{-1} = -\alpha^{-1} z (1 - \alpha^{-1} z)^{-1}$$
$$\simeq -\alpha^{-1} z (1 + \alpha^{-1} z + \alpha^{-2} z^2 + \cdots + \alpha^{-(n-1)} z^{n-1}) \quad (3.82)$$

The fact that the expansion in Equation 3.79 or 3.82 holds for $z = \pm 1$ and for any real value of α satisfying the given inequality,[1] implies that the expansion holds also for any value of z on the unit circle in the z plane and for any complex value of α satisfying the given inequality.

Since z^i is the z-transform of an advance in time of iT seconds, and since we are here considering the z-transform of the sampled impulse-response of the channel, the z-transform given by Equation 3.82 is not physically realizable. However, if a delay of nT seconds is included in the response given by Equation 3.82, this has the z-transform

$$z^{-n}(1 - \alpha z^{-1})^{-1}$$
$$\simeq -\alpha^{-1} z^{-(n-1)} (1 + \alpha^{-1} z + \alpha^{-2} z^2 + \cdots + \alpha^{-(n-1)} z^{n-1})$$
$$= -\alpha^{-n} (1 + \alpha z^{-1} + \alpha^2 z^{-2} + \cdots + \alpha^{n-1} z^{-(n-1)}) \quad (3.83)$$

which is physically realizable. Furthermore, the zeros of this z-transform are given by Equation 3.80, as before.

When $|\alpha| = 1$, the z-transform $(1 - \alpha z^{-1})^{-1}$ represents a pole on the unit circle in the z plane, and this is, of course, not physically realizable.

Evidently a pole at $z = \alpha$ (where $|\alpha| \neq 1$), together with a delay of nT seconds when $|\alpha| > 1$, may be approximately represented by $n - 1$ zeros, equally spaced on the circle of radius $|\alpha|$ in the z plane at the $n - 1$ of the n possible values of $\alpha \omega^i$ that exclude α (where $\omega = \exp(j2\pi/n)$).

A z-transform having one or more poles *inside* the unit circle in the z plane but no poles outside, can be realized exactly in practice by means of the appropriate linear feedback transversal filter, which is quite stable (Section 4.7). A z-transform having one or more poles *outside* the unit circle, however, can only be realized *exactly* by the appropriate linear feedforward transversal filter with an infinite number of taps and then only when accompanied by infinite signal delay (see Section 4.5). Nevertheless, as can be seen from Equations 3.79 and 3.83, in a z-transform having poles both inside and outside the unit circle in the z plane, each factor representing a pole can be expressed approximately by the appropriate polynomial in z^{-1}, with a finite number of terms and of finite degree, so long as the necessary signal delay is introduced into the original (ideal) z-transform, for each pole outside the unit circle. The approximation to the ideal z-

transform represents a system that is both stable and physically realizable, and the approximation may be made as close to the ideal as required by taking a sufficient number of terms in each polynomial that approximately represents a pole.

Although the z-transform of any *finite* sequence has no poles but only zeros, the basic structure of the z-transform of a sequence exhibiting phase distortion is more simply represented in terms of both zeros and poles, assuming therefore that the sequence is infinite. Where the z-transform of the ideal infinite sequence has one or more poles, these may appear in the z-transform of the finite sequence that approximates to it, as the corresponding sets of zeros given by Equation 3.80.

In conclusion, when a filter introducing pure phase distortion is connected in cascade with a channel, and when the poles of the z-transform of the filter coincide with zeros of the z-transform of the channel, then the action of the filter is to replace these zeros by the corresponding set of zeros at the reciprocal values of z. Thus the replacement of one or more zeros in a z-transform, by zeros at the reciprocal values of z, is pure phase distortion and is therefore an orthogonal transformation, as has already been shown.

When the filter is matched to the baseband channel introducing pure phase distortion, the zeros and poles of the filter z-transform coincide with the poles and zeros, respectively, of the channel z-transform, so that the channel and filter in cascade have a z-transform with no zeros or poles (other than any at the origin or infinity in the z plane). Thus the matched filter is also a *linear equalizer* for the channel, since it equalizes or corrects the channel response. The equalizer has a z-transform with an infinite number of terms and usually it can only be realized approximately in practice. Furthermore, the equalizer always introduces some signal delay.

3.4 AMPLITUDE DISTORTION

Pure amplitude distortion is here defined as signal distortion (more than one nonzero component in the sampled impulse-response of the channel) that includes no phase distortion and no signal delay.

When the transfer function $J(i/nT)$ of the baseband channel and sampler has a *complex value* at one or more of the n discrete frequencies $\{i/nT\}$, this necessarily implies phase shifts at the corresponding frequencies and therefore *some* combination of phase distortion and signal delay in the received signal. Indeed, the absence of phase distortion and signal delay in the received signal could, if so required, be taken to imply that the n values of $J(i/nT)$, for $i = 0, 1, \ldots, n-1$, must all be *real* and *positive*. Now, from Equations 3.10 and 3.11,

$$J(i/nT) = \lambda_i \qquad (3.84)$$

and $\qquad Y_0 F = [\lambda_0 \quad \lambda_1 \quad \ldots \quad \lambda_{n-1}] \qquad (3.85)$

where $\qquad Y_0 = [y_0 \quad y_1 \quad \ldots \quad y_{n-1}] \qquad (3.86)$

so that the n values of $J(i/nT)$, for $i = 0, 1, \ldots, n-1$, are the n components of the DFT of the n-component sampled impulse-response Y_0 of the linear baseband channel in Fig. 3.1. Thus, in the absence of phase distortion and signal delay, the $\{\lambda_i\}$ could be taken to be all *real-valued* and *positive*. Bearing in mind that, in the *absence* of amplitude distortion and attenuation, the $\{\lambda_i\}$ all have an absolute value (modulus) of unity, the DFT of any sampled impulse-response Y_0 can now be split *uniquely* into two factors, one of which is a measure of the amplitude distortion introduced by the channel and sampler, and the other of which is a measure of the phase distortion. More precisely, the absolute value of λ_i is here the $(i+1)$th component of the DFT of the factor of Y_0 representing the *amplitude distortion* and the exponential of j times the phase angle of λ_i is the $(i+1)$th component of the DFT of the factor of Y_0 representing the *phase distortion*. Amplitude and phase distortions are now equated with the amplitude and phase *responses* of the linear baseband channel and sampler in Fig. 3.1. Unfortunately, this very simple definition of amplitude distortion is rather too narrow for reasons that will be explained later. A slightly wider definition of pure amplitude distortion and one that moves some way towards a *linear-phase* signal is obtained by constraining the $\{\lambda_i\}$ to be *real valued* but allowing some of them to be positive and the remainder negative. A linear-phase sampled impulse-response is one in which the phase angle of the Fourier transform of the signal is a constant multiple of the frequency[5]. This corresponds to a constant group delay and a constant phase delay, where the group delay is the derivative of the phase with respect to angular frequency, and the phase delay is the phase divided by the angular frequency. The definition of pure amplitude distortion to be assumed here is in fact a *linear-phase* sampled impulse-response in which both the group delay and phase delay are *zero*. Any further relaxation in the definition of pure amplitude distortion to include an even wider range of signals, although bringing this closer into line with linear-phase and piecewise-linear-phase signals[5], can lead to unacceptable ambiguity in the definitions of amplitude and phase distortions, and so is not used here. A piecewise-linear-phase sampled impulse-response is one having a constant group delay but no longer a constant phase delay over the frequency band of the signal[5].

Now, for no phase distortion and no signal delay, the n values of $J(i/nT)$, for $i = 0, 1, \ldots, n-1$, and hence the n components $\{\lambda_i\}$ of $Y_0 F$ (Equation 3.85), must all be *real valued*. But, from Equation 3.13, λ_i and λ_{n-i} are complex conjugates, so that

$$\lambda_i = \lambda_{n-i} \tag{3.87}$$

for $i = 1, 2, \ldots, n-1$. This means that $Y_0 F$ is *real* and *symmetric*. It is evident from the analysis in Section 2.9 that, under these conditions, if the sampled impulse-response of the baseband channel in Fig. 3.1 is given by the n-component row vector Y_0 in Equation 3.86, then

$$y_i = y_{n-i} \tag{3.88}$$

for $i = 1, 2, \ldots, n-1$, and the vector Y_0 is *real* and *symmetric*. Conversely, as is shown in Section 2.9, if Y_0 is real and symmetric, then so also is $Y_0 F$.

It is assumed here that in the symmetric sequence Y_0 the components have been shifted cyclically to the left to make y_0 the *central* component (in time) of the original sequence. This corresponds to an advance in time of the sampled impulse-response of the channel, to bring the central component of the symmetric sequence to the time instant $t = 0$. Clearly, this is not physically realizable, and, in practice, amplitude distortion must always be accompanied by the corresponding signal delay, to make the distortion physically realizable. It is however simpler to study amplitude distortion in the absence of signal delay, since the delay terms are now eliminated from $Y_0 F$. The signal delay will therefore be ignored, bearing in mind that the system can always be made physically realizable by reintroducing the appropriate delay.

It is clear from Equations 3.8 and 3.86 that the sampled impulse-response of the baseband channel in Fig. 3.1 can now be rewritten as the n-component vector

$$Y_0 = [y_0 \quad y_1 \quad \ldots \quad y_{g/2} \quad 0 \quad \ldots \quad 0 \quad y_{-g/2} \quad \ldots \quad y_{-1}] \tag{3.89}$$

where $n \geq 2g + 1$, g is even, and

$$y_i = y_{-i} \tag{3.90}$$

for $i = 1, 2, \ldots, \frac{1}{2}g$. The $n \times n$ convolution matrix Y_c corresponding to the sampled impulse-response Y_0 in Equation 3.89 is the $n \times n$ circulant matrix whose first row is the vector Y_0. It is evident from Equation 3.90 that Y_c is now a *real symmetric* matrix, and this may be shown more rigorously as follows.

In the presence of no phase distortion and no delay, the $\{\lambda_i\}$, which are the eigenvalues of Y_c, must all be real valued. Also, from Equations 2.80 and 2.149,

$$Y_c = FDF^{-1} = n^{-1} FD\bar{F} \tag{3.91}$$

where D is the real diagonal matrix whose $(i+1)$th component along the main diagonal is λ_i, and $\bar{F}$ is the complex conjugate of F. Since F, D and $\bar{F}$ are symmetric matrices, it follows that

$$Y_c^T = n^{-1} \bar{F}DF \tag{3.92}$$

But, since D is real,

$$\bar{F}DF = \overline{(FD\bar{F})} \tag{3.93}$$

so that, from Equations 3.91–3.93,

$$Y_c^T = n^{-1} \overline{(FD\bar{F})} = \overline{Y_c} = Y_c \tag{3.94}$$

since Y_c is real. Thus Y_c is a symmetric matrix.

Conversely, if Y_c is a real symmetric matrix, all its eigenvalues are real valued and the baseband channel and sampler in Fig. 3.1 introduce no phase distortion and no delay. It follows that a necessary and sufficient condition for no phase distortion and no delay is that the real convolution matrix Y_c is a *symmetric* matrix. This means that the sequence Y_0 is symmetrical about its component at time $t=0$, so that a reversal in time of the sequence, pivoted about this component, *does not change* the sequence.

For comparison, a necessary and sufficient condition for no amplitude distortion and no attenuation is that the real convolution matrix Y_c is an *orthogonal* matrix, which means that a reversal in time of the sequence Y_0, pivoted about its component at time $t=0$, gives the *inverse* sequence.

Another important distinction between amplitude and phase effects is that amplitude distortion and signal attenuation are always accompanied by a change both in the discrete energy-density spectrum and in the aperiodic autocorrelation function of the sampled impulse-response of the channel, from their ideal values in Equations 3.42 and 3.50, respectively, whereas phase distortion and signal delay are not accompanied by a change either in the discrete energy-density spectrum or in the aperiodic autocorrelation function.

As before, it is instructive to consider the transmission of a continuous sequence of signal elements. Assume therefore that, in the synchronous serial data-transmission system of Fig. 3.1, a continuous sequence of signal elements is transmitted, such that the z-transform of the sequence of impulses at the input to the baseband channel is $S(z)$ in Equation 3.52. The z-transform of the sequence obtained after

THEORY OF SIGNAL DISTORTION   119

sampling the output signal from the baseband channel, is $S(z)Y(z)$ in Equation 3.54, where $Y(z)$ in Equation 3.53 is the z-transform of the sampled impulse-response of the channel. This is assumed to include sufficient delay to be physically realizable.

Suppose now that the received sequence with z-transform $S(z)Y(z)$ is fed to a digital filter with z-transform $z^{-h}Y^{-1}(z)$, where z^{-h} is the z-transform of the delay of hT seconds needed to make the filter response physically realizable, for practical purposes. The output sequence from the digital filter has the z-transform

$$S(z)Y(z)z^{-h}Y^{-1}(z) = z^{-h}S(z) = \sum_i s_i z^{-i-h} \quad (3.95)$$

so that this is simply a delayed version of the transmitted sequence of values, and s_l can be detected from the output sample at time $t = (l+h)T$, without interference from the other $\{s_i\}$.

The digital filter with z-transform $z^{-h}Y^{-1}(z)$ is a *linear equalizer* for the channel, since it equalizes or corrects the channel response. Contrary to the case where the channel introduces pure phase distortion, the sampled impulse-response of the linear equalizer here is not very simply related to the sampled impulse-response of the channel. The equalizer is not matched to the channel.

3.5 REPRESENTATION OF AMPLITUDE DISTORTION BY THE ZEROS AND POLES OF THE Z-TRANSFORM

Assume now that the linear baseband channel in Fig. 3.1 introduces *pure amplitude distortion* (with no delay), so that it is, of course, not physically realizable. The n-component sampled impulse-response of the channel

$$Y_0 = [y_0 \quad y_1 \quad \ldots \quad y_m \quad 0 \quad \ldots \quad 0 \quad y_{-m} \quad \ldots \quad y_{-1}] \quad (3.96)$$

is now *symmetrical* about the component y_0 that occurs at time $t = 0$, its first nonzero component, in time, being y_{-m}. Clearly, Y_0 has $2m+1$ potentially nonzero components, so that

$$g = 2m \quad \text{and} \quad n \geq 2g+1 \quad (3.97)$$

Also, the component y_i occurs at time $t = iT$, and

$$y_i = y_{-i} \quad (3.98)$$

for $i = 1, 2, \ldots, m$. As before, all $\{y_i\}$ are *real-valued*. The z-transform of Y_0 is

$$Y(z) = y_{-m}z^m + \cdots + y_{-1}z + y_0 + y_1 z^{-1} + \cdots + y_m z^{-m} \quad (3.99)$$

which contains an equal number of positive and negative powers of z. It is evident that if the sequence Y_0 is *reversed in time*, the reversal being pivoted about the component y_0, at time $t=0$, then Y_0 remains *unchanged* and so does $Y(z)$.

Consider next a channel whose sampled impulse-response has the z-transform

$$Y(z) = \pm X(z)W(z) \qquad (3.100)$$

where
$$X(z) = x_0 + x_1 z^{-1} + \cdots + x_m z^{-m} \qquad (3.101)$$

and
$$W(z) = x_0 + x_1 z + \cdots + x_m z^m \qquad (3.102)$$

and where all $\{x_i\}$ are real valued. It is assumed, as before, that $Y(z)$ has no zeros on the unit circle, and $Y(z)$ now has no poles (other than any at the origin or infinity, which are ignored here).

It can be seen from Equations 3.101 and 3.102 that $W(z)$ is obtained from $X(z)$ by reversing the order of its coefficients, the reversal being pivoted about the coefficient of z^0. Clearly, the sequence with z-transform $X(z)W(z)$ is the *convolution* of the sequences with z-transforms $X(z)$ and $W(z)$. If now the sequence with z-transform $X(z)W(z)$ is reversed in order, the reversal being pivoted about the component at time $t=0$, then so also are the two constituent sequences with z-transforms $X(z)$ and $W(z)$. The reversal of the first of the two sequences causes its z-transform to be changed from $X(z)$ to $W(z)$, and the reversal of the second of the two sequences causes its z-transform to be changed from $W(z)$ to $X(z)$. Thus the reversal of the resultant sequence leaves it *unchanged*, which means that it is *symmetrical* about the component at time $t=0$, and so represents *pure amplitude distortion*. The sampled impulse-response of this channel clearly satisfies Equations 3.96–3.98.

The values of z at the zeros of $W(z)$ are the *reciprocals* of those at the zeros of $X(z)$. Thus all zeros of $Y(z)$ occur in *reciprocal pairs*, and $X(z)$ can always be selected to have all its zeros inside the unit circle in the z plane and $W(z)$ to have all its zeros outside the unit circle. If $W(z)$ and $X(z)$ have an infinite number of terms, then they may have poles as well as zeros. In this case the values of z at the poles of $W(z)$ are the reciprocals of those at the poles of $X(z)$, and again $X(z)$ can always be selected to have all its poles inside the unit circle and $W(z)$ to have all its poles outside. Since all coefficients in $Y(z)$ are *real valued*, all its zeros and all its poles at complex values of z occur not only in reciprocal pairs but also in complex-conjugate pairs. Clearly, each zero is accompanied by another zero at the *complex conjugate* of the *reciprocal* value of z, and similarly for each pole. This condition is in fact satisfied by both *real-* and *complex-valued* signals, and is therefore of more general application than the simpler condition previously described, which applies only to real-valued signals. Complex-valued signals are considered further in Section 3.7.

Now, let

$$X = [x_0 \quad x_1 \quad \ldots \quad x_m \quad 0 \quad 0 \quad \ldots \quad 0] \qquad (3.103)$$

be the n-component sequence with z-transform $X(z)$ (Equation 3.101) and let

$$W = [x_0 \quad 0 \quad 0 \quad \ldots \quad 0 \quad x_m \quad x_{m-1} \quad \ldots \quad x_1] \qquad (3.104)$$

be the n-component sequence with z-transform $W(z)$ (Equation 3.102). As before, $n \geq 2g + 1$ where $g = 2m$ (Equation 3.97). In the sequence X, the component x_i occurs at time $t = iT$, whereas in the sequence W, x_i occurs at time $t = -iT$. By the Reversal theorem (Section 2.10), the n-component DFT's of X and W, given by XF and WF, must satisfy the equation

$$WF = \overline{(XF)} \qquad (3.105)$$

where $\overline{(XF)}$ is the *complex conjugate* of XF. This means that the ith component of the DFT of W is the complex conjugate of the ith component of the DFT of X. But the ith component of the DFT of the n-component sequence given by the *convolution* of X and W, with z-transform $X(z)W(z)$, must be the *product* of the two ith components just mentioned (see Equation 2.99) and so must be the absolute value squared of the given complex number. Hence each component of the resultant DFT must be not only *real valued* but *positive*, so that, from Equation 3.100, the components of the DFT of Y_0 must be real valued and either all positive or else all negative. Clearly, $Y(z)$ in Equation 3.100 must represent pure amplitude distortion. Of course, all sequences here are of *finite* length and $Y(z)$ has *no poles* (other than any at the origin or infinity, which are ignored here).

Equation 3.100 represents a *sufficient* condition that must be satisfied by $Y(z)$ to ensure that it represents pure amplitude distortion. A *necessary* condition for pure amplitude distortion, that must be satisfied by the real sequence Y_0, is that it is symmetrical about its central component which occurs at time $t = 0$. Under this condition, the reversal of the order of the coefficients in $Y(z)$, without changing the position of the coefficient of z^0, leaves $Y(z)$ *unchanged*. However, this reversal must transfer each zero of $Y(z)$ to the corresponding reciprocal value of z. For this not to change $Y(z)$, each zero of $Y(z)$ must be accompanied by a zero at the reciprocal value of z, and, for this to be true, $Y(z)$ must satisfy Equation 3.100. The z-transforms $W(z)$ and $X(z)$ can now always be found to satisfy both Equation 3.100 and the other assumed conditions. The requirement that the sequence W be obtained from the reversal of the sequence X, by pivoting this about its component at time $t=0$, follows from the fact that the *central* component of Y_0 is at time $t=0$. Clearly Equation 3.100 is a condition that must *necessarily* be satisfied by $Y(z)$, when the latter has a finite number of zeros, no poles and no zeros on the unit circle in the z plane. However, the location of the zeros just described is not itself a *sufficient* condition to ensure pure amplitude distortion, since the sampled impulse-response of the channel may be advanced or delayed by an arbitrary number of sampling intervals, without affecting the location of the zeros (other than of any at the origin or infinity, which are ignored here), so that the *incorrect* position in *time* of the sampled impulse-response is not recognized by the location of the zeros.

In the general case where $Y(z)$ has both poles and zeros it has an infinite number of terms. Each pole of $X(z)$ is now accompanied by a pole of $W(z)$ at the reciprocal value of z, so that the poles of $Y(z)$ occur in reciprocal pairs, as do its zeros.

Any channel (and sampler) together with the appropriate matched filter have a z-transform that is the product of the z-transforms of the channel and matched filter. Furthermore, since the sampled impulse-response of the matched filter is the reverse of that of the channel, the values of z at the zeros and poles of the z-transform of the matched filter are the reciprocals of those at the zeros and poles, respectively, of the z-transform of the channel. Thus all zeros and all poles of the z-transform of the channel and matched filter occur in reciprocal pairs, so that the channel and matched filter together introduce pure amplitude distortion (together with some delay). The resultant sampled impulse-response is symmetric and is, of course, the aperiodic autocorrelation function of the channel itself. Clearly, a matched filter eliminates all phase distortion. It is not, however, a pure phase equalizer except in the particular case where the channel introduces pure phase distortion. Whenever the channel introduces some amplitude distortion the matched filter always *increases* this distortion.

It is clear that the z-transform $z^{-h}Y^{-1}(z)$ of the linear equalizer for a channel, introducing pure amplitude distortion together with an appropriate delay to make it physically realizable, may be obtained from $Y(z)$ by replacing all zeros by poles and all poles by zeros, and, of course, introducing a delay of hT seconds. Thus, in $z^{-h}Y^{-1}(z)$, all zeros and all poles occur in reciprocal pairs and the sampled impulse-response of the linear equalizer is symmetric. Also, the channel and linear equalizer connected in cascade have a z-transform with no zeros or poles (other than zeros at infinity and poles at the origin, introduced by the delay in transmission). A linear equalizer for a practical channel introducing pure amplitude distortion has a z-

transform with an infinite number of terms and this can only be realized approximately in practice. Furthermore, the equalizer always introduces some signal delay.

Suppose next that $Y(z)$ (Equation 3.3) has the factor

$$U(z) = u_0 + u_1 z^{-1} + \cdots + u_m z^{-m} \qquad (3.106)$$

whose coefficients are real valued and whose m zeros all lie *on* the unit circle. The integer m here is not related to that in Equations 3.96–3.104. Clearly, $m \leqslant g$, and the g zeros of $Y(z)$ include the m zeros of $U(z)$. The zeros of $U(z)$ are now ± 1 or complex-conjugate pairs of absolute value (modulus) unity. If $U(z)$ represents pure amplitude distortion together with the appropriate delay to make it physically realizable, and if u_0 and u_m are nonzero, it follows that m is even and $z^{m/2} U(z)$ represents pure amplitude distortion with no delay. Now,

$$u_{(m/2)+i} = u_{(m/2)-i} \qquad (3.107)$$

for $i = 1, 2, \ldots, \tfrac{1}{2}m$. It can be seen that for these conditions to hold, the zeros of $U(z)$ must all be in reciprocal pairs which are now also complex-conjugate pairs and may, of course, include one or more *pairs* of zeros at 1 or -1. The zeros of $U(z)$ could, for example, be as shown in Fig. 3.2. If $U(z)$ contains an *odd* number of zeros at 1 or -1, the $\{u_i\}$ do not form a symmetric sequence (as do the $\{y_i\}$ in Equation 3.96 and as defined in Section 2.9) and $z^h U(z)$ *cannot* represent pure amplitude distortion, regardless of the value of the integer h.

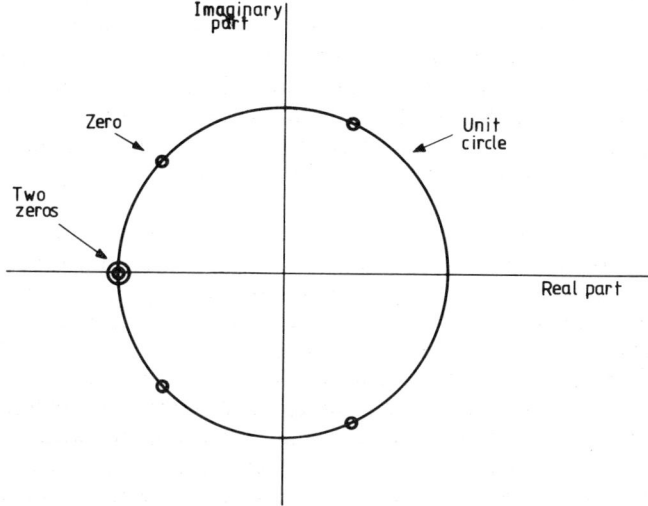

Fig. 3.2 Zeros of $U(z)$ in the complex number plane

124 THEORY OF SIGNAL DISTORTION

It is interesting to observe that when $z^{m/2}U(z)$ represents pure amplitude distortion, then

$$z^{m/2}U(z) = \pm U_1(z)U_2(z) \qquad (3.108)$$

where the n-component sequence U_2 with z-transform $U_2(z)$ is the time reverse of the n-component sequence U_1 with z-transform $U_1(z)$, the reversal being pivoted about the component at time $t=0$. However, the components of these sequences (and hence the coefficients in $U_1(z)$ and $U_2(z)$) are *not* necessarily *real-valued*, even though the coefficients in $U(z)$ are real-valued. Let U be the n-component sequence with z-transform $z^{m/2}U(z)$ and formed therefore by the *convolution* of U_1 and U_2 (see Equation 3.108). As before $n \geqslant 2g+1$ so that $n \geqslant 2m+1$. Bearing in mind that Equation 3.105 holds only when the components of X and W are *real-valued*, it can be seen from the discussion on the convolution of the sequences X and W (Equations 3.103 and 3.104), that when some of the components of the sequence U_1 (and therefore also of the sequence U_2) are *complex-valued*, then the n-components of the DFT of U are no longer necessarily all positive or else all negative. They are still *real-valued*, since U is a real-valued symmetric sequence, but some may be positive and the remainder negative. The latter follows since U is *not* now the *aperiodic autocorrelation function* of U_1 or U_2, contrary to the case where *all* components of the latter are real valued (see Section 2.11).

When $Y(z)$ represents pure amplitude distortion and has *no* zeros on the unit circle, then, from Equation 3.100,

$$Y(z) = \pm X(z)W(z) \qquad (3.109)$$

where $X(z)$ and $W(z)$ are given by Equations 3.101 and 3.102 and *all* coefficients are real valued. This means that when $Y(z)$ has *no* zeros on the unit circle and represents pure amplitude distortion, the components of the DFT of Y_0 (Equation 3.85) must be *real valued* and either *all positive* or else *all negative*. Whenever Y_0 represents pure amplitude distortion and has *both* positive and negative components in its DFT, $Y(z)$ must have an even number of zeros on the unit circle that are not repeated zeros (not occurring in pairs or multiple pairs at the different values of z occupied by these zeros). So long as *all* zeros of $Y(z)$ that lie on the unit circle are repeated zeros (the zeros occurring in pairs or multiple pairs at each different value of z on the unit circle), then, when Y_0 represents pure amplitude distortion, the components of its DFT must be real-valued and either all positive or else all negative. A sufficient condition to ensure that the latter holds is, in fact, that *all* zeros of $Y(z)$, whether on or off the unit circle, occur in *complex-conjugate reciprocal* pairs.

Clearly, whenever $Y(z)$ in Equation 3.3 is physically realizable and represents pure amplitude distortion and delay, its zeros, whether off or on the unit circle, occur in *reciprocal pairs*, all complex zeros being at the same time in *complex-conjugate pairs*. The corresponding sampled impulse-response of the baseband channel is always symmetrical about the central sample at time $t = \frac{1}{2}gT$, g of course being even.

3.6 ASSESSMENT OF THE DISTORTION INTRODUCED BY A LINEAR BASEBAND CHANNEL

Consider the z-transform of the sampled impulse-response of the baseband channel in Fig. 3.1, where the z-transform may have an infinite number of terms, with all coefficients real-valued, and is represented by means of its zeros and poles in the z plane.

In the case of pure amplitude distortion, all zeros and all poles occur in reciprocal pairs, whereas with pure phase distortion, each zero is accompanied by a pole at the reciprocal value of z, and each pole by a zero. All zeros and all poles at complex values of z occur in complex-conjugate pairs.

Any linear baseband channel whose output signal is sampled at regular intervals of time, as in Fig. 3.1, can be considered to be composed of a number of separate segments, each of which may be represented by the appropriate digital filter. Since the z-transform of the channel is the *product* of the z-transforms of its segments, the zeros and poles of the z-transform of the channel are given by the *sum* of the zeros and poles of its individual segments.

Suppose now that the linear baseband channel in Fig. 3.1 is composed of two separate segments connected in cascade, with a sampler sampling at the time instants $\{iT\}$ between the two segments. The z-transform of the channel is the product of the z-transforms of the two individual segments. For convenience, any attenuation or delay introduced by either segment will now be ignored. Suppose first that the z-transform of the first segment (channel A) has just two zeros, which occur at reciprocal real values of z, as shown in Fig. 3.3. This segment clearly introduces pure amplitude distortion. The z-transform of the second segment (channel B) has a zero coincident with one of the two zeros of the first segment and a pole coincident with the other zero, as shown in Fig. 3.4. This segment clearly introduces pure phase distortion. The resultant z-transform of the channel has just two coincident zeros, as shown in Fig. 3.5, since the coincident zero and pole cancel out. Nevertheless, the channel clearly introduces the amplitude distortion of Fig. 3.3 and the phase

126 THEORY OF SIGNAL DISTORTION

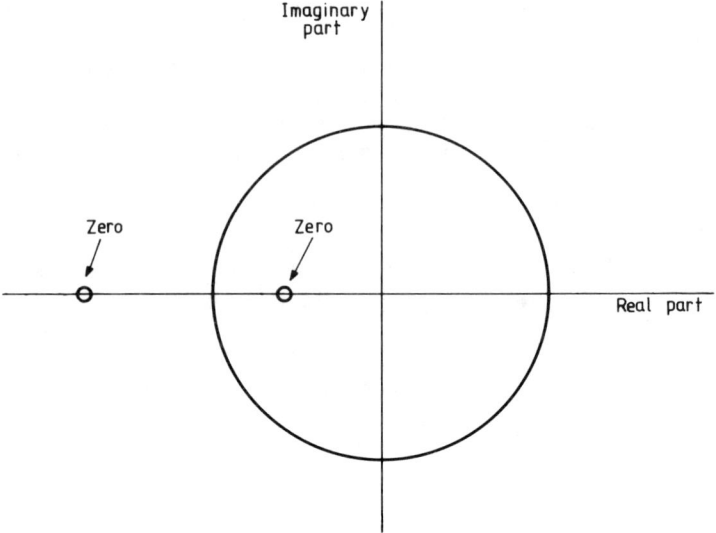

Fig. 3.3 Zeros of channel A

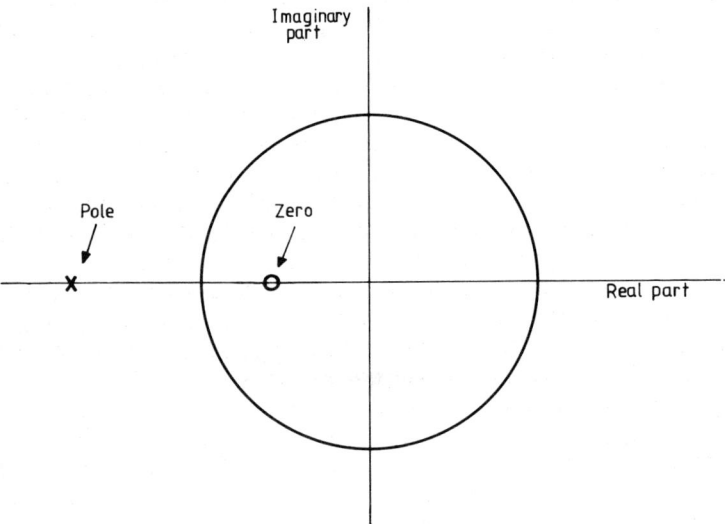

Fig. 3.4 Zero and pole of channel B

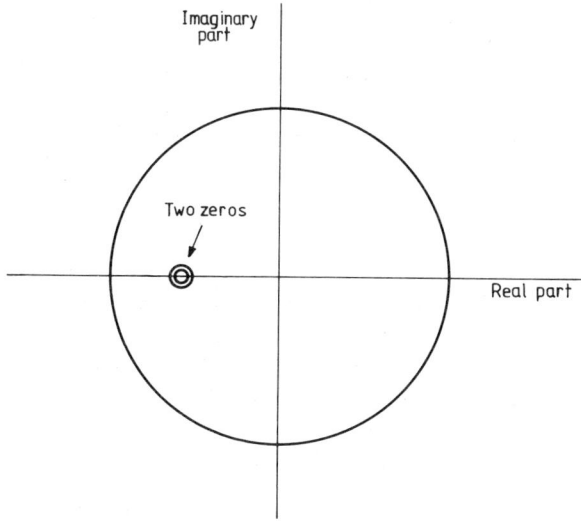

Fig. 3.5 Zeros of resultant channel formed by connecting channels A and B in cascade

distortion of Fig. 3.4. It follows that any two coincident zeros at a real value of z, with no pole or zero at the reciprocal value of z, are formed by the combination of the corresponding amplitude and phase distortions. Thus any single zero or pole at a real value of z, with no zero or pole at the reciprocal value of z, represents a z-transform which when multiplied by itself is a combination of amplitude and phase distortions, so that any such zero or pole itself represents a combination of amplitude and phase distortions. Although the resultant levels of the amplitude and phase distortions of the channel are determined by the distribution of *all* zeros and poles in the z plane, any single zero or pole at a real value of z, with no zero or pole at the reciprocal value of z, implies the presence of at least *some* amplitude distortion and *some* phase distortion. A strictly *finite* number of zeros and poles is assumed here, with no zeros or poles *on* the unit circle.

Suppose now that the z-transform of the first segment of the channel has four zeros at two pairs of complex-conjugate values of z which also form two pairs of reciprocal values of z, as shown in Fig. 3.6. The z-transform of the second segment has two zeros and two poles at the same four values of z, both zeros and poles being at complex-conjugate values of z, and the values of z at the poles being the reciprocals of those at the zeros, as shown in Fig. 3.7. The first segment clearly introduces pure amplitude distortion and the second introduces pure phase distortion. The resultant z-transform of the

128 THEORY OF SIGNAL DISTORTION

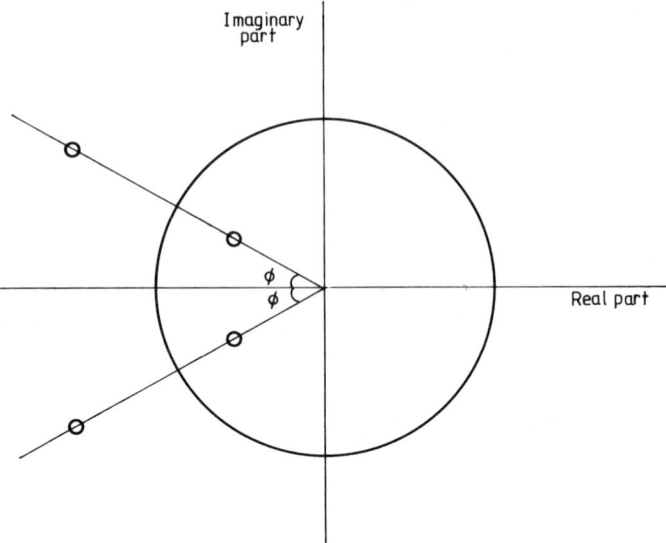

Fig. 3.6   Zeros of channel A

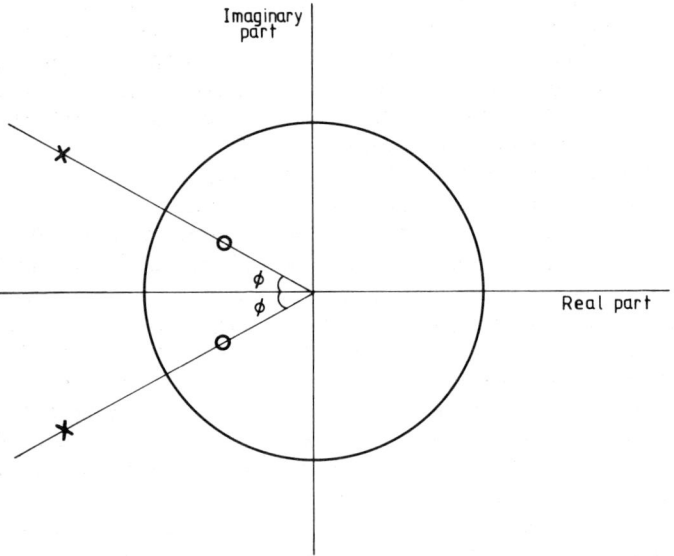

Fig. 3.7   Zeros and poles of channel B

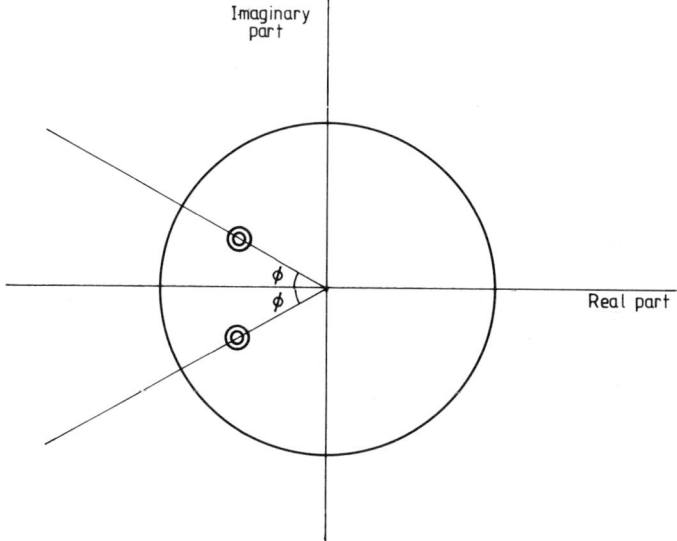

Fig. 3.8 Zeros of resultant channel formed by connecting channels A and B in cascade

channel has four zeros and no poles, as shown in Fig. 3.8. This channel introduces the amplitude distortion of Fig. 3.6 and the phase distortion of Fig. 3.7. By an exactly similar line of reasoning to that used before, it can be seen that any pair of zeros at complex-conjugate values of z, not accompanied by a pair of zeros or poles at the reciprocal values of z, imply the presence or at least *some* amplitude distortion and *some* phase distortion.

So long as the z-transform of the sampled impulse-response of the channel does not have too many zeros and poles, a study of the distribution of these can indicate not only the predominant type of distortion but also its severity. Some idea of the latter can be obtained by noting the following two points. First, two zeros or else two poles, at values of z one of which is the *complex-conjugate* of the *reciprocal* of the other, each represent the *same* level of *amplitude* distortion. This follows from the fact that when a zero or pole is transferred from a value of z to the complex conjugate of its reciprocal, as, for example, occurs when channel B (Fig. 3.4) is connected in cascade with channel A (Fig. 3.3) to give the resultant channel (Fig. 3.5), the transformation is achieved by means of a linear channel that introduces pure *phase* distortion (channel B) and therefore does not change the resultant *amplitude* distortion. Secondly, a zero and a pole at *complex-conjugate reciprocal* values of z each represent the *same* level of *phase* distortion. This follows from the fact that when a zero or pole is replaced by a pole or zero, respectively, at the complex-conjugate reciprocal value of z, as, for example, occurs when channel A (Fig. 3.3) is connected in cascade with channel B (Fig. 3.4) to give the resultant channel (Fig. 3.5), the transformation is achieved by means of a linear channel that introduces pure *amplitude* distortion (channel A) and therefore does not change the resultant *phase* distortion. Of course, no zeros or poles here lie *on* the unit circle in the z plane. When all zeros and poles of the z-transform of a channel line *inside* the unit circle, as, for example, in Figs. 3.5 and 3.8, the channel is said to be *minimum phase*. When all zeros and all poles occur in complex-conjugate reciprocal pairs, as, for example, in Figs. 3.3 and 3.6, the channel is *linear phase*. Thus, when channel B is connected in cascade with channel A, for either of the two examples in Figs. 3.3–3.8, it converts a *linear phase* channel into the corresponding *minimum phase* channel introducing the *same amplitude distortion*.

Finally, the nearer a zero or pole is to the unit circle in the z plane, the greater is the corresponding signal distortion. In particular, a zero or pole at a value of z such that $\frac{1}{2} < |z| < 2$ represents appreciable signal distortion, the distortion often increasing rapidly with the *number* of zeros or poles lying within this area. Zeros in the same neighbourhood tend to reinforce each other, as do poles, but zeros and poles in the same neighbourhood tend to cancel. Zeros or poles, regularly spaced over the circumference of a circle of any radius and with its centre at the origin, tend to cancel, the degree of cancellation increasing with the number of zeros or poles.

The signal distortion introduced by a linear baseband channel is perhaps best studied by considering the channel to be composed of two constituent channels, A and B, connected in cascade and with a sampler in between, where channel A introduces pure amplitude distortion and channel B introduces pure phase distortion. Figures 3.3–3.8 are examples of such channels.

The n-component DFT of the sampled impulse-response of channel A is

$$[\theta_0 \quad \theta_1 \quad \ldots \quad \theta_{n-1}] \qquad (3.110)$$

where each θ_i is *real-valued*, and the n-component DFT of the sampled impulse-response of channel B is

$$[\phi_0 \quad \phi_1 \quad \ldots \quad \phi_{n-1}] \qquad (3.111)$$

where $|\phi_i| = 1$ for each i. Now

$$\theta_i \phi_i = \lambda_i \qquad (3.112)$$

for $i = 0, 1, \ldots, n-1$, where, of course, the n-component DFT of the whole baseband channel is

$$Y_0 F = [\lambda_0 \quad \lambda_1 \quad \ldots \quad \lambda_{n-1}] \tag{3.113}$$

Furthermore,

$$\theta_i = \theta_{n-i} = \pm |\lambda_i| \tag{3.114}$$

and

$$\phi_i = \phi_{n-i}^* = \frac{\lambda_i}{\theta_i} \tag{3.115}$$

for each i, $|\phi_i|$ being held at unity when $\lambda_i = 0$. Ideally $n \to \infty$ here, since n must be large enough to accommodate the convolution of the sampled impulse-responses of channels A and B, channel B having a sampled impulse-response with theoretically an infinite number of components, as also often does channel A. Thus, in practice, $n \gg 2g + 1$. The sampled impulse-response of each of the channels A and B can be made real-valued, but for the purposes of the present analysis it is not necessary that either channel be physically realizable.

It will now be shown that when the baseband channel and sampler in Fig. 3.1 have a sampled impulse-response with a finite number of nonzero values (all of which are real), the baseband channel and sampler can always be split into the two separate channels A and B, as just described. The analysis demonstrates some interesting properties of these channels[6,7].

The sampled impulse-response of the baseband channel in Fig. 3.1 is the sequence of real values $y_0, y_1, \ldots, y_g$, whose z-transform is

$$\begin{aligned} Y(z) &= y_0 + y_1 z^{-1} + y_2 z^{-2} + \cdots + y_g z^{-g} \\ &= y_0 (1 - \alpha_1 z^{-1})(1 - \alpha_2 z^{-1}) \ldots (1 - \alpha_g z^{-1}) \\ &= Y_1(z) Y_2(z) \end{aligned} \tag{3.116}$$

where the $\{\alpha_i\}$ are the g zeros (roots) of $Y(z)$. $Y_1(z)$ and $Y_2(z)$ are the two factors of $Y(z)$ whose zeros lie, respectively, inside and outside the unit circle in the z plane. Thus, for each α_i in $Y_1(z)$, $|\alpha_i| < 1$ where $|\alpha_i|$ is the absolute value (modulus) of α_i, and for each α_i in $Y_2(z)$, $|\alpha_i| > 1$. The coefficients in $Y_1(z)$ and $Y_2(z)$ are real valued, which means that all complex-valued zeros occur in complex-conjugate pairs. It is assumed that $y_0 \neq 0$ and *no* zeros of $Y(z)$ lie *on* the unit circle.

The polynomial in z, which when multiplied by itself gives $Y(z)$, is

$$Y^{1/2}(z) = Y_1^{1/2}(z) Y_2^{1/2}(z) \tag{3.117}$$

where

$$\begin{aligned} Y_1(z) &= (1 - \alpha_1 z^{-1})(1 - \alpha_2 z^{-1}) \ldots (1 - \alpha_h z^{-1}) \\ &= 1 + p_1 z^{-1} + p_2 z^{-2} + \cdots + p_h z^{-h} \end{aligned} \tag{3.118}$$

132 THEORY OF SIGNAL DISTORTION

and $\alpha_1, \alpha_2, \ldots, \alpha_h$ are the h zeros of $Y(z)$ that lie inside the unit circle in the z plane. The $\{p_i\}$ are all real valued. $Y_1^{1/2}(z)$ may be evaluated as a convergent series directly from $Y_1(z)$, using the second of the two expressions for $Y_1(z)$ in Equation 3.118[1], or alternatively by using the fact that

$$Y_1^{1/2}(z) = (1-\alpha_1 z^{-1})^{1/2}(1-\alpha_2 z^{-1})^{1/2} \ldots (1-\alpha_h z^{-1})^{1/2} \quad (3.119)$$

The binomial expansion may here be applied to each of the h factors of $Y_1^{1/2}(z)$, giving a convergent series in every case.[1] $Y_2^{1/2}(z)$ is determined from $Y_2(z)$ by using the fact that

$$\begin{aligned}Y_2^{1/2}(z) &= [y_0(1-\alpha_{h+1}z^{-1})(1-\alpha_{h+2}z^{-1})\ldots(1-\alpha_g z^{-1})]^{1/2} \\ &= y_0^{1/2}(1-\alpha_{h+1}z^{-1})^{1/2}(1-\alpha_{h+2}z^{-1})^{1/2}\ldots(1-\alpha_g z^{-1})^{1/2} \\ &= f(1-\alpha_{h+1}^{-1}z)^{1/2}(1-\alpha_{h+2}^{-1}z)^{1/2}\ldots(1-\alpha_g^{-1}z)^{1/2} \\ &= f(1+q_1 z + q_2 z^2 + \cdots + q_{g-h}z^{g-h})^{1/2} \end{aligned} \quad (3.120)$$

where $f = (y_0(-\alpha_{h+1})(-\alpha_{h+2})\ldots(-\alpha_g)z^{-g+h})^{1/2}$ (3.121)

The $\{\alpha_i\}$ for $i = h+1, h+2, \ldots, g$ are the $g-h$ zeros of $Y_2(z)$, so that the corresponding $\{\alpha_i^{-1}\}$ all lie *inside* the unit circle in the z plane. The $\{q_i\}$ are all real valued. It can be seen from Equation 3.120 that $Y_2^{1/2}(z)$ may be evaluated from $Y_2(z)$ by either of the two methods mentioned for $Y_1^{1/2}(z)$[1]. Finally, $Y^{1/2}(z)$ is determined from Equation 3.117. In general, $Y^{1/2}(z)$ is the sum of an infinite number of terms $\{e_i z^{i+k}\}$ where i takes on all integer values (positive, negative and zero). The $\{e_i\}$ are either all real valued or all imaginary valued, and $k = 0$ or $\frac{1}{2}$. Thus the sampled impulse-response with z-transform $Y^{1/2}(z)$ need not be physically realizable.

Consider now the polynomial $V(z)$ which is such that[7]

$$cz^k V(z) = Y^{1/2}(z) \quad (3.122)$$

where $c = 1$ or $\sqrt{-1}$ and $k = 0$ or $\frac{1}{2}$, c being chosen so that the coefficients in $V(z)$ are real valued and k being chosen to give integral powers of z in $V(z)$. When $Y(z)$ has one or more zeros outside the unit circle in the z plane, $V(z)$ includes positive powers of z, as can be seen from Equation 3.120, so that the sampled impulse-response corresponding to $V(z)$ is not now physically realizable. Let $U(z)$ be the z-transform of the sequence obtained by reversing the order of the sequence with z-transform $V(z)$, the reversal being pivoted about the component at time $t = 0$. Clearly, the coefficients in $U(z)$ are obtained from those in $V(z)$ by reversing their order, the reversal being pivoted about the coefficient of z^0. Let $X(z)$ be the z-transform of the sequence obtained by reversing the order of the sequence with z-transform $Y(z)$, the reversal being pivoted about the component at time $t = 0$. Thus

$$X(z) = y_0 + y_1 z + y_2 z^2 + \cdots + y_g z^g \qquad (3.123)$$

But the reversal of the sequence with z-transform $Y(z)$, pivoted about the component at time $t=0$, to give the sequence with z-transform $X(z)$, similarly reverses the sequence whose z-transform is the square root of that of the given sequence. Thus $X^{1/2}(z)$ is the z-transform of the sequence obtained by reversing the order of the sequence with z-transform $Y^{1/2}(z)$, the reversal being pivoted about the time instant $t=0$. It follows that

$$cz^{-k} U(z) = X^{1/2}(z) \qquad (3.124)$$

But clearly,

$$Y(z) = Y^{1/2}(z) X^{1/2}(z) Y^{1/2}(z) X^{-1/2}(z) \qquad (3.125)$$

since $X^{1/2}(z) X^{-1/2}(z) = 1$. $X^{-1/2}(z)$ may be evaluated from either $X^{1/2}(z)$ or $X(z)$, in each case using either of the two basic techniques involved in the determination of $Y^{1/2}(z)$ from $Y(z)^1$. Now, from Equation 3.124,

$$c^{-1} z^k U^{-1}(z) = X^{-1/2}(z) \qquad (3.126)$$

$V(z)$, $U(z)$ and $U^{-1}(z)$ may readily be evaluated from Equations 3.122, 3.124 and 3.126, and, from Equation 3.125[7],

$$Y(z) = cz^k V(z) cz^{-k} U(z) cz^k V(z) c^{-1} z^k U^{-1}(z)$$
$$= c^2 V(z) U(z) z^{2k} V(z) U^{-1}(z) \qquad (3.127)$$

where $c^2 = \pm 1$ and $2k = 0$ or 1. The polynomials, formed by the expansion of $V(z)$, $U(z)$ and $U^{-1}(z)$ into the corresponding convergent infinite series, contain only integral powers of z and real-valued coefficients. By expressing each of these polynomials to a sufficiently large number of terms, Equation 3.127 may be satisfied to any required degree of accuracy. It will therefore be assumed that $V(z)$, $U(z)$ and $U^{-1}(z)$ are all expressed to a finite number of terms, with the same number for $V(z)$ and $U(z)$. Clearly, the sampled impulse-response of the channel can be considered as the convolution of two finite sequences, one with the z-transform $c^2 V(z) U(z)$ and the other with z-transform $z^{2k} V(z) U^{-1}(z)$. Similarly, the sequence with z-transform $V(z) U(z)$ is the convolution of the sequences with z-transforms $V(z)$ and $U(z)$, and the sequence with z-transform $V(z) U^{-1}(z)$ is the convolution of the sequences with z-transforms $V(z)$ and $U^{-1}(z)$.

It can be seen from Equation 3.100 that the sequence with z-transform $c^2 V(z) U(z)$ represents *pure amplitude distortion*, and it can be seen from Equation 3.63 that the sequence with z-transform $z^{2k} V(z) U^{-1}(z)$ represents *pure phase distortion*. Thus the reversal of the sequence with z-transform $c^2 V(z)U(z)$ about its component at time $t=0$ leaves it *unchanged*, whereas the corresponding reversal of the sequence with z-transform $z^{2k}V(z)U^{-1}(z)$ gives the *inverse* sequence. All components of the DFT of the former sequence are real-valued, whereas all components of the DFT of the latter sequence have an absolute value (modulus) of unity. Finally, $c^2 V(z)U(z)$ has no poles and an even number of zeros which occur in complex-conjugate reciprocal pairs, whereas $z^{2k}V(z)U^{-1}(z)$ has as many poles as zeros, each zero being accompanied by a pole at the complex-conjugate reciprocal value of z. Each zero in $U(z)$ is replaced by a pole in $U^{-1}(z)$, and all complex-valued zeros or poles, of course, occur in complex-conjugate pairs. The reversal in time of the sequence with z-transform $c^2 V(z)U(z)$, pivoted about the component at time $t=0$, transfers each zero to the complex-conjugate reciprocal value of z and so leaves the zeros of the z-transform *unchanged*. The corresponding reversal in time of the sequence with z-transform $z^{2k}V(z)U^{-1}(z)$ transfers each zero and pole of the z-transform to the complex-conjugate reciprocal value of z, so that it replaces each zero by a pole and each pole by a zero, to give the *inverse* z-transform and hence the inverse sequence. Strictly speaking, since the zeros and poles of any z-transform are unaffected if the polynomial is multiplied by any constant quantity, the fact that the zeros and poles of the z-transform are interchanged upon the reversal of the sequence, is only sufficient to ensure that the reversal of the sequence gives the inverse sequence, so long as the sum of the squares of its components is equal to unity. The latter condition is, however, satisfied here.

It is clear from Equation 3.127 that the baseband channel in Fig. 3.1 can be replaced by two baseband channels in cascade, the channels being separated by a sampler that samples its input signal at the time instants $\{iT\}$ and feeds the corresponding impulses to the following channel. The first channel (channel A) has the z-transform $c^2 V(z)U(z)$ and introduces pure amplitude distortion. The second channel (channel B) has the z-transform $z^{2k}V(z)U^{-1}(z)$ and introduces pure phase distortion.

Amplitude or phase distortion is defined in terms of the components of the DFT of the sampled impulse-response of the corresponding channel. In theory, of course, the sampled impulse-response may have an infinite number of components and so therefore must the DFT. However, in practice, a good approximation to a sampled impulse-response representing pure amplitude or phase distortion can be achieved with a finite sequence of m components, where usually $m \gg 2g+1$. The DFT must now have at least $2m-1$ components, the sampled impulse-response being appropriately lengthened by adding the required number of zeros to the end of the sequence. The DFT is uniquely determined by the corresponding sampled impulse-response and vice versa. The sampled impulse-response can here be made physically realizable by introducing sufficient delay.

If $Y(z)$ represents pure amplitude distortion, then so does $V(z)$ (Equation 3.122), so that $V(z)U^{-1}(z)=1$. If $Y(z)$ represents pure phase distortion, then so does $V(z)$, so that $V(z)U(z)=1$. If all the zeros of $Y(z)$ lie inside the unit circle in the z plane, then so do all the zeros of $V(z)$. The zeros of both $U(z)$ and $U^{-1}(z)$ must now lie outside the unit circle, so that both $V(z)U(z)$ and $V(z)U^{-1}(z)$ have more than one term with nonzero coefficients, which means that $Y(z)$ represents both amplitude and phase distortion. The same applies if all the zeros of $Y(z)$ lie outside the unit circle. It can be seen that if $Y(z)$ has any factor representing pure amplitude distortion or pure phase distortion, this must appear as a factor of $c^2 V(z)U(z)$ or $z^{2k}V(z)U^{-1}(z)$, respectively, and need not therefore be involved further in the evaluation of the polynomial.

It is assumed throughout the above analysis that $Y(z)$ has real-valued coefficients, g zeros and no poles. Furthermore, $Y(z)$ has *no* zeros *on* the unit circle. Suppose now that $Y(z)$ has one or more zeros *on* the unit circle in the z plane. Under these conditions, $Y^{1/2}(z)$ and $X^{1/2}(z)$ may be determined as before, but $X^{-1/2}(z)$ is no longer properly defined, essentially because the expansion of $X^{-1/2}(z)$ is not now convergent. Thus $c^2 V(z)U(z)$ may be evaluated as previously described but not $z^{2k}V(z)U^{-1}(z)$. Any factor of $Y(z)$ containing zeros *on* the unit circle and representing pure amplitude distortion, such as, for example,

$$z(1-\alpha z^{-1})(1-\alpha^* z^{-1}) = z - (\alpha + \alpha^*) + z^{-1} \qquad (3.128)$$

where $|\alpha|=1$, is removed from the analysis by writing

$$Y(z) = (z - (\alpha + \alpha^*) + z^{-1})Y'(z) \qquad (3.129)$$

and analysing $Y'(z)$ in place of $Y(z)$. The factor $(z-(\alpha+\alpha^*)+z^{-1})$ is subsequently multiplied by $c^2 V(z)U(z)$ for $Y'(z)$ to give the pure amplitude distortion factor of $Y(z)$. Any zeros, that cannot be accommodated into factors representing pure amplitude distortion, must represent some combination of amplitude distortion and either or both delay and phase distortion (according to the definitions of these terms adopted here). However, since the delay and phase distortion components of the latter factors cannot be isolated from the amplitude distortion, they are best combined with the pure amplitude distortion factor, bearing in mind that this now involves either or both delay and phase distortion. Since neither the delay nor the phase distortion here in any way degrade the operation of a detector that takes due account of the effects, they are quite harmless and may therefore be ignored in assessing the severity of the distortion. The alternative approach, previously discussed, of widening the definition of pure amplitude distortion to accommodate the effects, has not been adopted because of the unacceptable ambiguity that results in the definitions of pure amplitude distortion and pure phase distortion. An even more powerful reason, however, will now be considered.

Our interest in the theory of signal distortion, is concerned solely with its application to the design of a *linear filter* that is employed ahead of the detector in the receiver. If now the channel and sampler in Fig. 3.1 have a z-transform with one or more zeros *on* the unit circle, this represents *infinite* attenuation at the corresponding frequencies, with the consequent total loss of the corresponding signal-frequency components. In order to restore these frequency components, the linear filter would need to introduce infinite gain at the appropriate frequencies. In other words, its z-transform would need to have poles at the corresponding points on the unit circle, and this is clearly not possible. Thus the linear filter cannot remove any zeros of the z-transform of the channel and sampler that lie *on* the unit circle, and it can only operate on zeros that lie *off* the unit circle. We are therefore not concerned with the distortion represented by any factor of $Y(z)$ containing only zeros on the unit circle, any such factor being ignored in the design of the linear filter.

It follows that, for our purposes, $Y(z)$ has no zeros *on* the unit circle, which not only avoids the problems introduced by such zeros but leads to the following useful results. All coefficients in both $V(z)$ and $U(z)$ of the factor $c^2 V(z) U(z)$ of $Y(z)$ in Equation 3.127 are now *real-valued*. This is, of course, not necessarily the case if $Y(z)$ has any zeros *on* the unit circle, and it follows from the Reversal theorem (Section 2.10) that all components of the DFT of the sequence with z-transform $V(z)U(z)$ are not only real-valued but *positive*. Thus, when $Y(z)$, with no zeros on the unit circle, satisfies Equation 3.100, the components of the DFT of the sampled impulse-response of the channel are *real-valued* and are either *all positive* or else *all negative*. It is the presence of zeros of $Y(z)$ *on* the unit circle that can cause some of the components of the DFT of the sampled impulse-response to the positive and the remainder negative. Finally, with *no* zeros of $Y(z)$ on the unit circle, the coefficients in $V(z)U(z)$ give the *aperiodic autocorrelation function* of the sequence with z-transform $V(z)$, the function being sensitive only to amplitude distortion and attenuation. The coefficients in $V(z)U^{-1}(z)$ give the corresponding function that is sensitive only to phase distortion and delay.

It is shown in Section 4.12 that phase distortion can be equalized linearly with no reduction in the tolerance of the data-transmission system to additive white Gaussian noise, relative to the case where there is no signal distortion of any kind. In other words, pure phase distortion does not itself introduce any irreparable degradation into the system performance and can usually be equalized quite easily. Furthermore, it can be shown that, in the presence of pure phase distortion and additive white Gaussian noise, the *best* detection process is achieved by a linear equalizer, followed by a threshold-level detector using the appropriate decision thresholds, no advantage being gained here through the use of any more sophisticated detection process[6]. These statements are intuitively reasonable in view of the fact that pure phase distortion is an orthogonal transformation and a phase equalizer is a matched filter (Section 1.5). On the other hand, pure amplitude distortion often results in a reduction in tolerance to noise that cannot be avoided, no matter what detection process is used. Again, in the presence of amplitude distortion an advantage in tolerance to additive white Gaussian noise can always be gained by replacing a linear equalizer by the appropriate decision-feedback (nonlinear) equalizer or more sophisticated detection process[6]. This is considered further in Chapters 5 and 7.

It follows that, with the model of the baseband channel just assumed, the channel B can be equalized linearly at the receiver, and the combination of the whole channel and linear equalizer can now be considered to be exactly equivalent to just the channel A on its own. This is, of course, only possible because the linear equalization of channel B gives no reduction in tolerance to additive white Gaussian noise relative to the case where the channel A is present on its own.

Thus, in assessing the severity of the signal distortion introduced by a channel, it is clearly the *amplitude distortion* that is the most important factor to be considered. The amplitude distortion can be determined by taking the DFT of the sampled impulse-response of the channel to give $Y_0 F$ in Equation 3.10. In the absence of amplitude distortion, $|\lambda_i|$ has the same value for each i, this value being a measure of the signal attenuation. The degree to which the $\{|\lambda_i|\}$ differ in value is a measure of the amplitude distortion.

Alternatively, it can be seen from Equation 3.17 that the $(i+1)$th component of the discrete energy-density spectrum is $|\lambda_i|^2$, and the inverse DFT of the discrete energy-density spectrum is the aperiodic autocorrelation function B in Equation 3.14. This means that B is completely determined by the amplitude distortion and signal attenuation, so that B depends *only* on these two parameters. In the absence of amplitude distortion and signal attenuation, B is given by Equation 3.50, regardless of the phase distortion or signal delay, so that B is obviously unaffected by phase distortion and signal delay.

The advantage of using B as a measure of amplitude distortion is that it is very simply evaluated from the sampled impulse-response of the channel and it is more closely related to the sampled impulse-response (at least conceptually) than is the absolute value (modulus) of the DFT of this response, or the discrete energy-density spectrum.

Signal attenuation or gain affects the magnitudes of all components of B equally. Furthermore, $b_0 > 0$, $b_i = b_{-i}$ for $i = 1, 2, \ldots, g$, and the remaining $\{b_i\}$ are all zero. In the absence of amplitude distortion, $b_i = 0$ for each $i \neq 0$. Thus the quantity

$$d = \frac{1}{b_0^2} \sum_{i=1}^{g} b_i^2 \tag{3.130}$$

gives a measure of the severity of the amplitude distortion and therefore of the irreducible degradation in performance likely to be experienced in the data-transmission system of Fig. 3.1. The quantity d is, of course, independent of signal attenuation, signal delay and phase distortion.

When $d > 0.1$ there is significant amplitude distortion and when $d > 0.5$ the distortion is severe. Although this is necessarily a crude measure of the amplitude distortion, since it takes no account of the detailed structure or shape of the distortion, it does give a general guide as to the type of detection process that is likely to be the most cost-effective. If $d < 0.1$, a simple linear equalizer is probably the most suitable arrangement at the receiver. If $0.1 < d < 0.5$, a decision-feedback (nonlinear) equalizer would be more cost-effective than a linear equalizer (see Chapter 5), and, if $d > 0.5$, a maximum-likelihood (optimum) detector should preferably be used (see Sections 1.4 and 7.8)[6,7]. In every case, any phase distortion introduced by the channel is removed by means of a linear filter, with, of course, no degradation in tolerance to additive white Gaussian noise.

G.W. Irwin has shown that the parameter d can be used to *predict* the performance of a maximum-likelihood detector over quite a wide range of channels[8].

3.7 COMPLEX-VALUED SIGNALS

Where the transmission path in Fig. 3.1 is a *bandpass* channel, together with a linear modulator at the transmitter and a linear demodulator at the receiver, and where a QAM signal is transmitted, then the data-symbols $\{s_i\}$ and usually also the impulse-response of the transmission path are *complex-valued*. Now

$$s_i = s_{0,i} + j s_{1,i} \tag{3.131}$$

where $j = \sqrt{-1}$, and both $s_{0,i}$ and $s_{1,i}$ can take on independently any one of m different values

$$-(m-1)k, \ -(m-3)k, \ \ldots, \ (m-1)k \qquad (3.132)$$

just like s_i in Equation 1.1. An element of the QAM signal carrying the data-symbol s_i is itself the sum of two double sideband suppressed carrier AM signal elements in phase quadrature, carrying the data-symbols $s_{0,i}$ and $s_{1,i}$. The two signal-elements have the same carrier frequency and there is a phase angle of $\pi/2$ radians (90°) between the two signal carriers. Thus the QAM signal comprises two AM signals in phase quadrature, one of which carries the data-symbols $\{s_{0,i}\}$ and the other carries the data-symbols $\{s_{1,i}\}$. Each AM signal is associated with the corresponding modulator and demodulator, to give a baseband channel. By allocating *real* values to the baseband signals at both the input and output of the baseband channel fed by the $\{s_{0,i}\}$ and imaginary values to the corresponding signals associated with the baseband channel fed by the $\{s_{1,i}\}$, a *complex-valued* baseband signal is obtained at both the input and output of the resultant channel in Fig. 3.1. The detailed properties of this channel are analysed elsewhere[9,10].

The sampled impulse-response of the linear baseband channel in Fig. 3.1 is, of course, given by the sequence of samples at the output of the sampler, obtained in response to a unit impulse, that is *real valued* and occurs at time $t = 0$, at the input of the baseband channel. The real and imaginary parts of any component of this sequence (vector) are the samples of the respective baseband signals at the outputs of the two constituent parallel channels that here together make up the linear baseband channel in Fig. 3.1. It can now be seen that, if the sampled impulse-response of the linear baseband channel is *real valued*, there is no coupling between the two parallel channels, the signal distortion in each of these being the *same* and being determined exactly as previously described. However, when the sampled impulse-response of the resultant channel is *complex valued*, coupling is introduced between the two constituent parallel channels by the *imaginary parts* of the components of the sampled impulse-response of the resultant channel. Nevertheless, the transmitted sequence (vector) of data-symbols $\{s_i\}$ and the sampled impulse-response of the linear baseband channel are similar to the corresponding real-valued sequences S and Y previously considered, and Equation 2.27 still holds, except that the components of R, S and Y_c are now complex-valued. Before proceeding to a detailed study of distortion in complex-valued signals, some of its basic properties will first be summarized.

When the baseband channel introduces *pure phase distortion*, it introduces a *unitary transformation* into the transmitted signal, such that the received signal-elements corresponding to the resultant (complex-valued) transmitted signal-elements are orthogonal, their inner products being zero. An *orthogonal* transformation is the special case of a *unitary* transformation, where all signals are *real valued*. The sampled impulse-response of the baseband channel (Equation 3.8) is now such that the sequence, formed from Y_0 by reversing the order of its first $g+1$ components and then replacing each of these by its complex conjugate, is the *inverse* of Y_0 delayed by g sampling intervals. The sequence Y_0 and its inverse are both such that the sum of the squares of the absolute values (moduli) of the n components is equal to unity. The sequence obtained by *convolving* the inverse (and delayed) sequence with Y_0 has all its components equal to zero except for the $(g+1)$th component which has the value unity. The matrix Y_c in Equation 3.30 now becomes a *unitary* matrix, whose inverse is therefore equal to its conjugate transpose. The eigenvalues of this matrix all have an absolute value of unity, so that all components of the DFT of the sampled impulse-response of the baseband channel have an absolute value of unity, which is the condition that must be satisfied in the presence of pure phase distortion. It is, of course, assumed here that the channel introduces no signal attenuation. With the appropriate detection process, pure phase distortion again introduces no reduction in tolerance to additive white Gaussian noise relative to the case where there is no distortion.

When the baseband channel introduces *pure amplitude distortion* and no delay, the sampled impulse-response of the baseband channel is given by Equation 3.89, where now y_i is equal to the complex conjugate of y_{-i}, for $i=1, 2, \ldots, \frac{1}{2}g$, and y_0 is real valued. If the first $g+1$ components (in time) of this sequence are reversed in order and if each component is then replaced by its complex conjugate, the sequence is left *unchanged*. The convolution matrix Y_c corresponding to the sampled impulse-response in Equation 3.89 is an *Hermitian* matrix and is therefore equal to its conjugate transpose. The eigenvalues of this matrix are all *real valued*, so that all components of the DFT of the sampled impulse-response of the baseband channel are real valued, which is the condition that must be satisfied in the presence of pure amplitude distortion.

Consider now the case where the sampled impulse-response of the linear baseband channel and sampler in Fig. 3.1 is given by the n-component sequence

$$Y_0 = [\, y_0 \quad y_1 \quad \ldots \quad y_g \quad 0 \quad 0 \quad \ldots \quad 0\,] \qquad (3.133)$$

with z-transform

$$Y(z) = y_0 + y_1 z^{-1} + \cdots + y_g z^{-g} \qquad (3.134)$$

where the $\{y_i\}$ are *complex valued*, and $n \geq 2g+1$. As before, it is assumed throughout that $Y(z)$ has *no* zeros *on* the unit circle in the z plane.

We must first examine a particular complication that occurs in the theoretical analysis of complex-valued signals. Suppose that

$$Y_0 = [\, y_0 \quad 0 \quad 0 \quad \ldots \quad 0\,] \tag{3.135}$$

so that the received sample, at time $t = iT$ at the detector input in Fig. 3.1, is

$$r_i = s_i y_0 + w_i \tag{3.136}$$

where w_i is the noise component in r_i. It is clear that there is no interference in the detection of s_i from r_i due to any of the other data-symbols $\{s_h\}$. Indeed, if s_i is *real valued*, there is no intersymbol interference and therefore *no signal distortion*, regardless of the value of y_0, which, of course, may take on any complex value. However, when

$$s_i = s_{0,i} + j s_{1,i} \tag{3.137}$$

as in Equation 3.131, so that

$$r_i = s_{0,i} y_0 + j s_{1,i} y_0 + w_i \tag{3.138}$$

it is evident that, whenever y_0 is *complex valued*, $s_{0,i}$ can no longer be detected directly from the *real part* of r_i in an optimum detection process, nor can $s_{1,i}$ be detected directly from the *imaginary part* of r_i. In this sense there is intersymbol interference and therefore *signal distortion*, bearing in mind that, in practice, $s_{0,i}$ and $s_{1,i}$ are handled as *separate* data-symbols. Furthermore, when

$$|y_0| = 1 \tag{3.139}$$

there is no attenuation, since now, in the absence of noise,

$$|r_i|^2 = s_{0,i}^2 + s_{1,i}^2 \tag{3.140}$$

However, if in addition $y_0 \neq \pm 1$, the channel introduces *pure phase distortion*, as can be seen from the fact that each of the real and imaginary parts of r_i is here dependent on *both* $s_{0,i}$ and $s_{1,i}$, and because, from Equation 3.139, all components of the DFT of Y_0 (Equation 3.135) have absolute values of unity. If now Y_0 in Equation 3.133 is *convolved* with Y_0 in Equation 3.135, each component of the former is multiplied by the first component of the latter, thus rotating its phase by a fixed amount. It follows that the introduction of a given phase rotation (that is not an integral multiple of π radians) into each component y_i of the sampled impulse-response of the channel (Y_0 in

Equation 3.133) is the corresponding pure phase distortion included into the channel distortion.

Suppose next that, for practical purposes,
$$Y(z) = \pm z^{-h} X(z) W^{-1}(z) \tag{3.141}$$
where
$$X(z) = x_0 + x_1 z^{-1} + \cdots + x_m z^{-m} \tag{3.142}$$
$$W(z) = x_0^* + x_1^* z + \cdots + x_m^* z^m \tag{3.143}$$
and the integer h in Equation 3.141 is sufficiently large to make $Y(z)$ physically realizable, as assumed in Equation 3.134. $Y(z)$ has *no* zeros on the unit circle. The $\{x_i\}$, for $i = 0, 1, \ldots, m$, are *complex valued* and x_i^* is the complex conjugate of x_i. Consider now the n-component sequences
$$X = [x_0 \quad x_1 \quad \ldots \quad x_m \quad 0 \quad 0 \quad \ldots \quad 0] \tag{3.144}$$
and
$$W = [x_0^* \quad 0 \quad 0 \quad \ldots \quad 0 \quad x_m^* \quad x_{m-1}^* \quad \ldots \quad x_1^*] \tag{3.145}$$
whose z-transforms are $X(z)$ and $W(z)$, respectively. Also let
$$n \geq 2g + 1 \gg 4m \tag{3.146}$$
where $g + 1$ is the number of potentially nonzero components in Y_0 (Equation 3.133). It is evident that the sequence W is obtained from X by reversing the order of the *complex conjugate* of the latter sequence, the reversal being pivoted about the component x_0 that occurs at time $t = 0$. From the Reversal theorem, the ith component of the DFT of W is the *complex conjugate* of the ith component of the DFT of X. Since $W(z)W^{-1}(z) = 1$, the ith component of the DFT of the n-component sequence with z-transform $W^{-1}(z)$ is the *reciprocal* (inverse) of the ith component of the DFT of W, and so is the *complex conjugate* of the *reciprocal* of the ith component of the DFT of X. The sequence with z-transform $X(z)W^{-1}(z)$ is here the *convolution* of the sequence X and the sequence with z-transform $W^{-1}(z)$. Hence, the ith component of the DFT of the n-component sequence with z-transform $X(z)W^{-1}(z)$ is the *product* of the ith components of the DFT's of X and the sequence with z-transform $W^{-1}(z)$, so that it has an *absolute value of unity*. A sufficiently large value of n is assumed here. Since each component of the DFT of the sequence with z-transform $X(z)W^{-1}(z)$ has an absolute value of unity, so also does each component of the DFT of the n-component sequence with z-transform $\pm z^{-h} X(z) W^{-1}(z)$. Thus $Y(z)$ in Equation 3.141 represents *pure phase distortion*.

If the sequence Y_0 (Equation 3.133) is reversed in time, the reversal being pivoted about its component y_0 at time $t = 0$, and if each component of the reversed sequence is replaced by its complex

THEORY OF SIGNAL DISTORTION   143

conjugate, then the z-transform of the resulting sequence is

$$\pm z^h W(z) X^{-1}(z) = Y^{-1}(z) \tag{3.147}$$

This follows because the complex conjugate of the reversal (conjugate reverse) of the sequence Y_0 is the *convolution* of the conjugate reverse of X, which is W, and the conjugate reverse of the sequence with z-transform $W^{-1}(z)$, which is the sequence with z-transform $X^{-1}(z)$, together with an *advance* of h sampling intervals and the appropriate sign. Thus the *reversal* of the *complex conjugate* of the sequence Y_0, pivoted about its first component (that at time $t=0$) gives the *inverse* sequence.

It is evident from Equations 3.142 and 3.143 that $X(z)$ and $W(z)$ have the same number of zeros and no poles, and furthermore each zero of $X(z)$ is accompanied by a zero of $W(z)$ that is at the *complex-conjugate reciprocal* value of z. The z-transform $W^{-1}(z)$ has no zeros but as many poles as $W(z)$ has zeros, the poles of $W^{-1}(z)$ occurring at the same values of z as the zeros of $W(z)$. Thus each *zero* of $X(z)$ is accompanied by a *pole* of $W^{-1}(z)$ that is at the *complex-conjugate reciprocal* value of z. It follows that $Y(z)$ in Equation 3.141 has as many poles as zeros, each zero being accompanied by a pole at the *complex-conjugate reciprocal* value of z, and similarly each pole being accompanied by a zero at the *complex-conjugate reciprocal* value of z. In other words, the zeros and poles occur in *complex-conjugate reciprocal* pairs, each pair involving one zero and one pole. The location of the zeros and poles just described is not itself a *sufficient* condition to ensure pure phase distortion, since $Y(z)$ may be multiplied by any scalar quantity without affecting the zeros and poles, but resulting in the fact that the $\{y_i\}$ no longer satisfy the equation

$$\sum_{i=0}^{g} |y_i|^2 = 1 \tag{3.148}$$

which is a *necessary* condition for pure phase distortion.

Finally, a linear equalizer for a baseband channel introducing pure phase distortion is, as before, a filter *matched* to the channel. Its z-transform is both the *complex conjugate* of the *reverse* (the *conjugate reverse*) of that of the channel and also the *inverse* of that of the channel (in each case neglecting the delay involved). The equalizer introduces itself the *conjugate reverse* and *inverse* of the phase distortion introduced by the channel.

Suppose next that

$$Y(z) = \pm X(z) W(z) \tag{3.149}$$

where $X(z)$ and $W(z)$ are given by Equations 3.142 and 3.143, and the

144 THEORY OF SIGNAL DISTORTION

corresponding n-component sequences X and W are given by Equations 3.144 and 3.145. As before, $Y(z)$ has *no* zeros *on* the unit circle. The sequence Y_0 with z-transform $Y(z)$ is now

$$Y_0 = [\, y_0 \quad y_1 \quad \ldots \quad y_m \quad 0 \quad \ldots \quad 0 \quad y_{-m} \quad \ldots \quad y_{-1} \,] \quad (3.150)$$

where the component y_i occurs at time $t = iT$, for $-m \leqslant i \leqslant m$. Clearly, the sampled impulse-response of the channel is not physically realizable. The sequence Y_0 is here plus or minus times the *convolution* of the sequences X and W. As has previously been stated, the ith component of the DFT of W is the *complex conjugate* of the ith component of the DFT of X. But the ith component of the DFT of Y_0 must be plus or minus times the *product* of the ith components of the DFT's of X and W, and one of these is the complex conjugate of the other. Thus the ith component of the DFT of Y_0 must be plus or minus times the absolute value squared of the ith component of the DFT of X or W, and so must be *real valued*. Furthermore, the components of the DFT of Y_0 must be either all positive or else all negative. Thus $Y(z)$ in Equation 3.149 represents *pure amplitude distortion*.

If the sequence Y_0 (Equation 3.150) is reversed in time, the reversal being pivoted about its component y_0 at time $t = 0$, and if each component of the reversed sequence is replaced by its complex conjugate, then the z-transform of the resulting sequence is

$$\pm W(z)X(z) = Y(z) \quad (3.151)$$

This follows because the conjugate reverse of Y_0 is the convolution of the conjugate reverse of X, which is W, and the conjugate reverse of W, which is X, together with the appropriate sign. Thus the *reversal* of the *complex conjugate* of the sequence Y_0, pivoted about its component at time $t = 0$, leaves the sequence *unchanged*. This means, of course, that, in the sequence Y_0,

$$y_i = y_{-i}^* \quad (3.152)$$

for $i = 1, 2, \ldots, m$.

As before, each zero of $X(z)$ is accompanied by a zero of $W(z)$ that is at the *complex-conjugate reciprocal* value of z. Thus $Y(z)$ in Equation 3.149 has an even number of zeros, each zero being accompanied by another zero at the *complex-conjugate reciprocal* value of z. In other words, the zeros occur in complex-conjugate reciprocal pairs. $Y(z)$, of course, has no poles. The location of the zeros just described is not itself a *sufficient* condition to ensure pure amplitude distortion, since, if $Y(z)$ is multiplied by any complex-valued quantity, the DFT of Y_0 is multiplied by the same quantity but the zeros of $Y(z)$ are *unaffected*. Thus all components of the DFT of Y_0 may here by complex valued, with the *same* given arbitrary phase angle, which means that the channel does not introduce pure amplitude distortion. Of course, where the sampled impulse-response of the channel is *real valued*, this cannot happen. Again, if the sampled impulse-response of the channel is advanced or delayed by an arbitrary number of sampling intervals, this does not affect the location of the zeros (other than of any at the origin or infinity, which are ignored here), so that the *incorrect* position in time of the sampled impulse-response is not recognized by the location of the zeros.

In conclusion, since a real-valued sequence is a special case of a complex-valued sequence, it can be seen that the following results hold for both real- and complex-valued sequences, where these sequences are the *sampled impulse-responses* of the linear-baseband channel in Fig. 3.1. When the *reversal* of the *complex-conjugate* of a sequence, pivoted about its component at time $t=0$, leaves the sequence *unchanged*, the sequence represents *pure amplitude distortion*. When the *reversal* of the *complex-conjugate* of a sequence, pivoted about its component at time $t=0$, gives the *inverse sequence*, the sequence represents *pure phase distortion*. The inverse of a sequence is, of course, that sequence which, when convolved with the original sequence, gives a sequence all of whose components are zero except for the component at time $t=0$ which is unity.

Consider next the case where the linear baseband channel in Fig. 3.1 is replaced by two baseband channels in cascade, the channels being separated by a sampler that samples its input signal at the time instants $\{iT\}$ and feeds the corresponding complex-valued impulses to the following channel. As before, the first channel (channel A) introduces pure amplitude distortion, and the second channel (channel B) introduces pure phase distortion. When the sampled impulse-response of the channel is *complex-valued*, the zeros or poles of its z-transform do not occur in complex-conjugate pairs. Suppose now that the z-transform of the channel A has two zeros at complex-conjugate reciprocal values of z, as in Fig. 3.9, and that the z-transform of the channel B has a zero and a pole at the given complex-conjugate reciprocal values of z, as in Fig. 3.10. The resultant channel has just two zeros, as in Fig. 3.11, and is therefore a *minimum-phase* channel. It introduces the amplitude distortion of Fig. 3.9 and the phase distortion of Fig. 3.10. Thus, when there are only a *finite* number of zeros and poles, a z-transform having *any* zeros or poles *not* accompanied by zeros or poles at the complex conjugates of the reciprocal values of z, represents a *combination* of amplitude and phase distortions. Furthermore, in the z-transform of the sampled impulse-response of any linear baseband channel, any two zeros or poles at values of z, one of which is the complex conjugate of the

146 THEORY OF SIGNAL DISTORTION

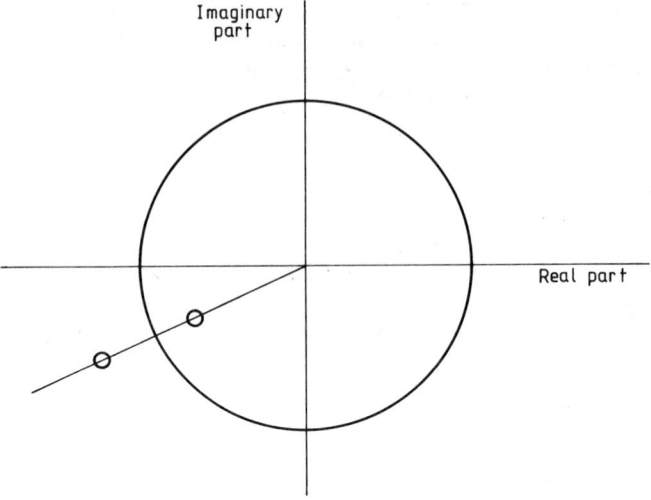

Fig. 3.9 Zeros of channel A

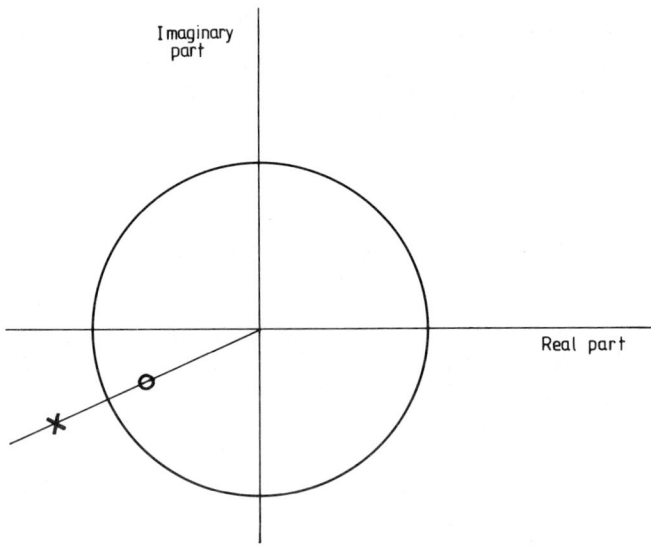

Fig. 3.10 Zero and pole of channel B

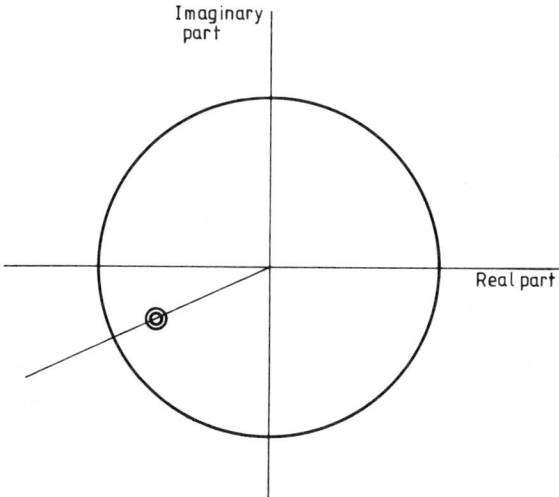

Fig. 3.11   Zeros of resultant channel formed by connecting channels A and B in cascade

reciprocal of the other, imply the presence of some amplitude distortion, and any zero or pole having a pole or zero, respectively, at the complex conjugate of the reciprocal value of z implies the presence of some phase distortion. If there are no zeros or poles, except at the origin and infinity, then there is no signal distortion, other than possibly a constant phase rotation in each component of the sampled impulse-response of the channel, which represents the corresponding phase distortion, and there may of course also be signal attenuation and delay.

The simple relationships of the type just considered, which exist between signals that are real valued and those that are complex valued, carry over, as might be expected, to the other properties of these signals, as well as to the various available techniques for detecting and estimating them. For instance, the different techniques studied in Chapters 4–7 are not greatly affected by whether real-valued or complex-valued signals are used. Furthermore, the detailed study of real-valued signals brings out very clearly the important properties of *both* real-valued and complex-valued signals and at the same time avoids the more serious conceptual difficulties involved with the latter. The remaining chapters of the book are therefore concerned almost entirely with *real-valued* signals. Once the principles and techniques have been mastered for these, it is a relatively simple matter to extend them to complex-valued signals,

148 THEORY OF SIGNAL DISTORTION

particularly in the light of the relationships that have just been developed.

3.8 EXAMPLES

Problem 1
Two channels, M and N, are available for the transmission of digital data. The z-transform of the sampled impulse-response of channel M is

$$M(z) = 0.6963(1 + 0.5z)(1 + 0.5z^{-1}) \qquad (3.153)$$

and of channel N is

$$N(z) = (1 + 0.5z)(1 + 0.5z^{-1})^{-1} \qquad (3.154)$$

Each channel can be made physically realizable by introducing a delay of one sampling interval.

What are the first six components of the sampled impulse-response of each channel and where in the z plane are the zeros and poles of its z-transform?

For each channel describe the effect on the z-transform and DFT of the sampled impulse-response of reversing the order of the components of the sampled impulse-response, the reversal being pivoted about the component at time $t=0$. What conclusions can be drawn in each case about the DFT of the sampled impulse-response of the channel and the signal distortion introduced?

Evaluate, for each channel,

(a) The aperiodic autocorrelation function of the sampled impulse-response of the channel, and

(b) the quantity d, which is a measure of the irreducible degradation in tolerance to additive white Gaussian noise, likely to be experienced in a data-transmission system on account of the distortion introduced by the channel.

Over which channel would you expect to achieve a better tolerance to additive white Gaussian noise, with a good detector?

Solution
The z-transform of channel M is

$$\begin{aligned} M(z) &= 0.6963(1 + 0.5z)(1 + 0.5z^{-1}) \\ &= 0.3482z + 0.8704 + 0.3482z^{-1} \end{aligned} \qquad (3.155)$$

The z-transform of channel N is

$$\begin{aligned} N(z) &= (1 + 0.5z)(1 + 0.5z^{-1})^{-1} \\ &= (1 + 0.5z)(1 - 0.5z^{-1} + 0.25z^{-2} - 0.125z^{-3} + \cdots) \\ &= 0.5z + 0.75 - 0.375z^{-1} + 0.1875z^{-2} \\ &\quad - 0.0938z^{-3} + 0.0469z^{-4} - \cdots \end{aligned} \qquad (3.156)$$

THEORY OF SIGNAL DISTORTION   149

Thus the first six components of the sampled impulse-response of channel M are

$$0.3482, 0.8704, 0.3482, 0.0, 0.0, 0.0$$

and the first six components of the sampled impulse-response of channel N are

$$0.5, 0.75, -0.375, 0.1875, -0.0938, 0.0469$$

Channel M has a zero at $z = -2.0$ and another zero at $z = -0.5$, and it has no poles.

Channel N has a zero at $z = -2.0$ (as for channel M) and it has a pole at $z = -0.5$.

If the order of the components of the sampled impulse-response of channel M is reversed, the reversal being pivoted about the component at time $t = 0$, then so also are the two constituent sequences with z-transforms $1 + 0.5z$ and $1 + 0.5z^{-1}$. This is because the reversal in time of the convolution of two sequences is the same as the convolution of the two individual sequences reversed in time. The reversal of the first of the two sequences causes its z-transform to be changed from $1 + 0.5z$ to $1 + 0.5z^{-1}$, and the reversal of the second sequence causes its z-transform to be changed from $1 + 0.5z^{-1}$ to $1 + 0.5z$. Thus the resultant z-transform is still $(1 + 0.5z)(1 + 0.5z^{-1})$ which means that the resultant sequence or sampled impulse-response remains *unchanged*. The sequence is clearly symmetrical about the component at time $t = 0$, as is evident from a direct inspection of $M(z)$. But the DFT of any such real-valued sequence or sampled impulse response has only real-valued components and therefore represents the absence of any phase distortion or delay at the discrete frequencies of the DFT. Any signal distortion satisfying this condition is taken to be pure amplitude distortion.

If the order of the components of the sampled impulse-response of channel N is reversed, the reversal being pivoted about the component at time $t = 0$, then so also are the two constituent sequences with z-transforms $1 + 0.5z$ and $(1 + 0.5z^{-1})^{-1}$. The reversal of the first of the two sequences causes its z-transform to be changed from $1 + 0.5z$ to $1 + 0.5z^{-1}$, and the reversal of the second sequence causes its z-transform to be changed from $(1 + 0.5z^{-1})^{-1}$ to $(1 + 0.5z)^{-1}$. Thus the z-transform of the reversed sequence is $(1 + 0.5z)^{-1} \times (1 + 0.5z^{-1})$ and this is the *inverse* of the z-transform of the original sequence. It follows that the reversal of the order of the components of the sampled impulse-response of channel N gives the sampled impulse-response of a channel that is the inverse of channel N. But, if a sequence of real values is reversed about its component at time $t = 0$, the components of its DFT are replaced by their complex conjugates. If the reversed sequence is the inverse of the original sequence, each component of its DFT must also be the reciprocal (inverse) of the corresponding component of the DFT of the original sequence. For the two conditions to be satisfied simultaneously, all components of each DFT must have an absolute value (modulus) of unity, the reciprocal of a complex number with an absolute value of unity being the complex-conjugate of that number. But any sequence or sampled impulse-

150 THEORY OF SIGNAL DISTORTION

response with a DFT having only components of unit absolute value represents the absence of any amplitude distortion or attenuation at the discrete frequencies of the DFT. Any signal distortion satisfying this condition is taken to be pure phase distortion.

It is clear than channel M introduces pure amplitude distortion and channel N introduces pure phase distortion.

If the sampled impulse-response of a channel is given by the n-component row vector

$$Y = [\, y_0 \quad y_1 \quad \ldots \quad y_g \quad 0 \quad 0 \quad \ldots \quad 0\,] \tag{3.157}$$

where

$$n \geqslant 2g + 1 \tag{3.158}$$

and the $\{y_i\}$ are all real valued, then the aperiodic autocorrelation function of the sampled impulse-response of the channel is given by the n-component row vector

$$B = [\, b_0 \quad b_1 \quad \ldots \quad b_g \quad 0 \quad \ldots \quad 0 \quad b_{-g} \quad \ldots \quad b_{-1}\,] \tag{3.159}$$

where

$$b_i = \sum_{h=-\infty}^{\infty} y_h y_{h+i} \tag{3.160}$$

and $y_h = 0$ for $h < 0$ and $h > g$. The quantity d is now given by

$$d = \frac{1}{b_0^2} \sum_{i=1}^{g} b_i^2 \tag{3.161}$$

For channel M,

$$B = [1.0000 \quad 0.6061 \quad 0.1212 \quad 0.0 \quad \ldots \quad 0.0 \quad 0.1212 \quad 0.6061] \tag{3.162}$$

and

$$d = 0.382 \tag{3.163}$$

For channel N,

$$B = [1.0 \quad 0.0 \quad 0.0 \quad \ldots \quad 0.0] \tag{3.164}$$

and

$$d = 0.0 \tag{3.165}$$

Since b_0 is a measure of the attenuation introduced by a channel, and since $b_0 = 1.0$ for each channel, neither of these gains any advantage over the other on account of the signal attenuation introduced. But, channel M, with $d = 0.382$, introduces significant amplitude distortion, whereas channel N, with $d = 0.0$, introduces no amplitude distortion. Thus a better tolerance to additive white Gaussian noise would be achieved over channel N than over channel M, when a good detector was used.

Problem 2

The sampled impulse-response of a channel is given by the vector

$$Y = [0.10 \quad 0.99 \quad -0.10 \quad 0.01] \tag{3.166}$$

Evaluate

(a) the sampled impulse-response of the digital filter matched to Y,

(b) the aperiodic autocorrelation function of Y (expressing each com-

THEORY OF SIGNAL DISTORTION   151

ponent of the autocorrelation function to just two decimal places, that is, to an accuracy of ± 0.005, and using these results for all subsequent calculations),
(c) the 7-component discrete energy-density spectrum of Y, and
(d) the sampled impulse-response of the linear feedforward-transversal-filter equalizer for the given channel.

Solution
(a) The sampled impulse-response of the digital filter matched to Y is

$$Z = [0.01 \quad -0.10 \quad 0.99 \quad 0.10] \tag{3.167}$$

this being just the time reverse of Y.

(b) The aperiodic autocorrelation function of Y is the convolution of the vectors Y and Z appropriately shifted in time. The convolution of Y and Z is given by the 7-component vector

$$[0.10 \times 0.01 \quad (-0.10 \times 0.10 + 0.99 \times 0.01)$$
$$(0.10 \times 0.99 - 0.99 \times 0.10 - 0.10 \times 0.01)$$
$$(0.10^2 + 0.99^2 + 0.10^2 + 0.01^2)$$
$$(0.99 \times 0.10 - 0.10 \times 0.99 - 0.01 \times 0.10)$$
$$(-0.10 \times 0.10 + 0.01 \times 0.99) \quad 0.01 \times 0.10]$$
$$= [0.001 \quad (-0.01 + 0.0099)$$
$$(0.099 - 0.099 - 0.001)$$
$$(0.01 + 0.9801 + 0.01 + 0.0001)$$
$$(0.099 - 0.099 - 0.001)$$
$$(-0.01 + 0.0099) \quad 0.001]$$
$$= [0.001 \quad -0.0001 \quad -0.001 \quad 1.0002$$
$$-0.001 \quad -0.0001 \quad 0.001]$$
$$\simeq [0.00 \quad 0.00 \quad 0.00 \quad 1.00 \quad 0.00 \quad 0.00 \quad 0.00] \tag{3.168}$$

Thus the aperiodic autocorrelation function of Y is approximately

$$[1.00 \quad 0.00 \quad 0.00 \quad 0.00 \quad 0.00 \quad 0.00 \quad 0.00] \tag{3.169}$$

(c) Let

$$B = [1 \quad 0 \quad 0 \quad 0 \quad 0 \quad 0 \quad 0] \tag{3.170}$$

Then the 7-component discrete energy density spectrum of Y is

$$BF = [1 \quad 1 \quad 1 \quad 1 \quad 1 \quad 1 \quad 1] \tag{3.171}$$

where F is the 7×7 DFT matrix.

(d) It is clear from (b) and (c) that the channel introduces pure phase distortion (at least for practical purposes) and that the digital filter matched to Y is also the inverse of Y and so equalizes the channel.

Thus, the sampled impulse-response of the linear feedforward transversal filter equalizer for the channel is given approximately by Z in Equation 3.167.

Problem 3
The four components of the sequence

$$X = [0.5 \quad 0.5 \quad 0.5 \quad -0.5] \tag{3.172}$$

occur at the time instants 0, T, $2T$ and $3T$, respectively.
Determine the 4-component DFT of the sequence,

(a) as given,
(b) when advanced by T,
(c) when advanced by $2T$, and
(d) when delayed by T.

Comment on the various properties of the four sequences and their DFT's.

Solution
From Equation 2.42, the 4×4 DFT matrix is

$$F = \begin{bmatrix} 1 & 1 & 1 & 1 \\ 1 & -j & -1 & j \\ 1 & -1 & 1 & -1 \\ 1 & j & -1 & -j \end{bmatrix} \tag{3.173}$$

(a) Thus the DFT of X is

$$XF = [1 \quad -j \quad 1 \quad j] \tag{3.174}$$

(b) The DFT of the sequence X, advanced by one sampling interval of T, is given by the DFT of the sequence

$$X_{-1} = [0.5 \quad 0.5 \quad -0.5 \quad 0.5] \tag{3.175}$$

which is obtained from X by shifting its components *cyclically* by *one* place to the left. Thus the DFT of the sequence X advanced by T is

$$X_{-1}F = [1 \quad 1 \quad -1 \quad 1] \tag{3.176}$$

Alternatively, the required DFT may be obtained by postmultiplying the DFT of X (Equation 3.174) by the diagonal matrix D_{-1}, whose $(i+1)$th component along the main diagonal is ω^i, where

$$\omega^i = \exp(j2\pi i/4) \tag{3.177}$$

Clearly,

$$D_{-1} = \begin{bmatrix} 1 & 0 & 0 & 0 \\ 0 & j & 0 & 0 \\ 0 & 0 & -1 & 0 \\ 0 & 0 & 0 & -j \end{bmatrix} \tag{3.178}$$

so that the DFT of the sequence X advanced by T is

$$XFD_{-1} = [1 \quad 1 \quad -1 \quad 1] = X_{-1}F \tag{3.179}$$

(c) The DFT of the sequence X, advanced by two sampling intervals of T, is given by the DFT of the sequence

$$X_{-2} = [0.5 \quad -0.5 \quad 0.5 \quad 0.5] \tag{3.180}$$

which is obtained from X by shifting its components cyclically by *two* places to the left. Thus the DFT of the sequence X advanced by $2T$ is

$$X_{-2}F = [1 \quad j \quad 1 \quad -j] \tag{3.181}$$

Alternatively, the required DFT may be obtained by postmultiplying the DFT of X by the diagonal matrix D_{-2}, whose $(i+1)$th component along the main diagonal is ω^{2i}, to give

$$D_{-2} = \begin{bmatrix} 1 & 0 & 0 & 0 \\ 0 & -1 & 0 & 0 \\ 0 & 0 & 1 & 0 \\ 0 & 0 & 0 & -1 \end{bmatrix} \tag{3.182}$$

and

$$XFD_{-2} = [1 \quad j \quad 1 \quad -j] = X_{-2}F \tag{3.183}$$

(d) The DFT of the sequence X, delayed by one sampling interval of T, is given by the DFT of

$$X_1 = [-0.5 \quad 0.5 \quad 0.5 \quad 0.5] \tag{3.184}$$

to give

$$X_1F = [1 \quad -1 \quad -1 \quad -1] \tag{3.185}$$

Alternatively, the required DFT is XFD_1, where D_1 is the diagonal matrix whose $(i+1)$th component along the main diagonal is ω^{-i}, so that

$$D_1 = \begin{bmatrix} 1 & 0 & 0 & 0 \\ 0 & -j & 0 & 0 \\ 0 & 0 & -1 & 0 \\ 0 & 0 & 0 & j \end{bmatrix} \tag{3.186}$$

to give

$$XFD_1 = [1 \quad -1 \quad -1 \quad -1] = X_1F \tag{3.187}$$

(1) In each case the absolute value of every component of the DFT of the sequence is equal to unity.
(2) In cases (b) and (d) every component of the DFT of the sequence is real valued.
(3) The sequence X or any shifted version of this sequence does *not* represent either pure phase distortion or pure amplitude distortion.

The reason for these apparent anomalies is that, before taking the DFT of any $(g+1)$-component sequence in order to determine whether it represents pure phase distortion or pure amplitude distortion, it is necessary first to extend the sequence by adding at least g zeros and then to determine the corresponding DFT, having the appropriately increased number of components. If this is done, the above anomalies disappear. It is interesting to observe that the more zeros that are added to the sequence, to give more components in the DFT, the smaller is the frequency difference between adjacent discrete frequencies of the DFT and so the finer is the sampling (in frequency) of the Fourier transform of the corresponding sequence of impulses. Thus the more components there are in the DFT, the less likely it is that any significant features of the Fourier transform will be missed and so give a misleading result.

From Equations 3.175 and 3.176,

$$X_{-1}F = 2X_{-1} \qquad (3.188)$$

so that X_{-1} is an eigenvector of F (Equation 3.173) associated with the eigenvalue 2.

Again, from Equations 3.184 and 3.185,

$$X_1 F = -2X_1 \qquad (3.189)$$

so that X_1 is an eigenvector of F associated with the eigenvalue -2.

From Equations 3.174 and 3.181,

$$X_{-2}F = \overline{(XF)} \qquad (3.190)$$

But, when the sequence X is reversed in time, the reversal being pivoted about its component at time $t=0$, the sequence

$$[-0.5 \quad 0.5 \quad 0.5 \quad 0.5] \qquad (3.191)$$

is obtained, whose components occur at the time instants $-3T, -2T, -T$ and 0, respectively. To obtain the DFT of this sequence, its components must be shifted *cyclically* to make the component at time $t=0$ the *first* component of the sequence, *as written*. The resulting sequence is

$$X_r = [0.5 \quad -0.5 \quad 0.5 \quad 0.5] = X_{-2} \qquad (3.192)$$

By the Reversal theorem, the DFT of this sequence must be the *complex conjugate* of that of X, as is, in fact, shown by Equation 3.190.

By an equivalent line of reasoning to that just used, it can be seen, that, if the sequence X_{-1} is reversed in time, the reversal being pivoted about the component at time $t=0$, the DFT of the resulting sequence is the *same* as that of X_{-1} itself. Similarly, if X_1 is reversed in time, the reversal being pivoted about the component at time $t=0$, the DFT of the resulting sequence is the *same* as that of X_1 itself. This is because, in each case, when the reversed sequence is cyclically shifted to make the component at time $t=0$ the first component, as written, the resulting sequence is the *same* as the original sequence. Now, since the DFT of the reversed sequence is the *complex conjugate* of that of the original sequence, it is clear that, in each case, all components of the DFT must also be *real valued*, as is, in fact, the case.

A sequence which is *orthogonal* to every cyclic shift of itself is said to be a *self-orthogonal* sequence[11]. It can be seen that X, together with every cyclic shift of X and the DFT of each resulting sequence, are *all* self-orthogonal sequences. In any one of these sequences, all components have the *same* absolute value, but this condition need not by any means be satisfied by a self-orthogonal sequence. As is illustrated by the particular example considered here, the DFT's of cyclic shifts of a self-orthogonal sequence are not necessarily or usually cyclic shifts of each other. It is evident that any scalar multiple of a self-orthogonal sequence, together with its complex conjugate, reverse and complex-conjugate reverse, are all self-orthogonal sequences. Self-orthogonal sequences, all of whose components have the same absolute value, have a number of important practical applications[11].

REFERENCES

1. Dwight, H.B. *Tables of Integrals and other Mathematical Data*, Fourth Edition, Macmillan, New York (1961)
2. Huffman, D.A. 'The generation of impulse-equivalent pulse trains', *IRE Trans. Inform. Theory*, **IT-8**, S10–S16 (1962)
3. Ackroyd, M.H. 'Huffman sequences with approximately uniform envelopes or cross-correlation functions', *IEEE Trans. Inform. Theory*, **IT-23**, 620–623 (1977)
4. Hunt, J.N. and Ackroyd, M.H. 'Some integer Huffman sequences', *IEEE Trans. Inform. Theory*, **IT-26**, 105–107 (1980)
5. Rabiner, L.R. and Gold, B. *'Theory and Application of Digital Signal Processing'*, Prentice-Hall, Englewood Cliffs, New Jersey (1975)
6. Clark, A.P. *Advanced Data-Transmission Systems*, Pentech Press, London (1977)
7. Clark, A.P. 'Signal distortion in a sampled digital signal', *IERE Conference Proceedings*, No. 37, 115–124 (1977)
8. Irwin, G.W. 'Predictors of maximum-likelihood performance of an unknown channel', *Proc. IEE*, Pt. F, **128**, 23–27 (1981)
9. Wozencraft, J.M. and Jacobs, I.M. *Principles of Communication Engineering*, Wiley, New York (1965)
10. Franks, L.E. *Signal Theory*, Prentice Hall, Englewood Cliffs, New Jersey (1969)
11. Clark, A.P., Zhu, Z.C. and Joshi, J.K. 'Fast start-up channel estimation', *Proc. IEE*, Pt. F, **131**, 375–382 (1984)

4

Linear equalizers

4.1 INTRODUCTION

The most widely used transmission systems for *digital data* and those to be assumed throughout this discussion are synchronous *serial* systems[1]. In transmitting digital data at relatively high rates over a bandlimited channel, the received signal is distorted, leading to *intersymbol interference* in the sampled baseband signal from which the transmitted data symbols are to be detected. Over many practical channels the most important signal distortion is *linear*, being caused partly by the bandlimiting of the signal in transmission and partly also by multipath propagation effects and other non-ideal characteristics of the transmission path. If not suitably corrected for, intersymbol interference can seriously degrade the performance of the detector[2-26], so that care must always be taken to reduce as far as possible the intersymbol interference at the output of the receiver filter and at the same time to maximize the signal/noise ratio here[27-48]. Chapter 3 describes and analyses the two different types of *linear* distortion that may be introduced into a baseband signal by the transmission path, and Chapters 4–7 are concerned with various techniques for detecting digital signals that have been subjected to this distortion.

Detection processes for distorted digital signals may be classified into two separate groups. In the first of these, the received sampled digital signal is fed through an *equalizer* that corrects the distortion introduced by the channel and restores the received signal into a copy of the transmitted signal, neglecting for the moment the effects of noise. The resultant received signal is then detected in the conventional manner, as normally applied to a serial digital signal in the absence of intersymbol interference. In other words, the equalizer acts as the *inverse* of the channel, so that the channel and equalizer together introduce no signal distortion, and each data symbol is detected as it arrives, independently of the others, by comparing the corresponding sample value with the appropriate threshold level (or levels)[49-154]. Equalizers may be *linear* or *nonlinear*. In the second group of detection processes, the decision process itself is modified to take account of the signal distortion that has been introduced by the channel, and no attempt need, in fact, be made here to reduce the signal distortion prior to the actual decision process. Although no equalizer is now required, the decision process may be considerably more complex than that when an equalizer is used[155].

In this chapter we are concerned with *linear equalizers* for a linear baseband channel[49-154]. We shall study various design techniques for such equalizers and compare their relative performances. It is assumed throughout that the receiver has *exact* prior knowledge of the sampled impulse-response of the channel, and that the channel is either time invariant or else varies only very slowly with time. This means that, for practical purposes, the sampled impulse-response of the channel is *constant* and *known* over the period involved in the equalization of any one received signal-element, so that, with the appropriate techniques, the channel can be equalized accurately. Conventional methods for holding an equalizer correctly adjusted for a channel, are described in Section 6.1.

The emphasis throughout this and the following chapters is on the transmission of *digital data*, where a *low error rate* of around 10^{-5} or 10^{-6} is desired in the detected data-symbols, implying a *high signal/noise ratio* in the received signal. Thus, except where specifically stated to the contrary, this condition is assumed. Nevertheless, the equalizers studied here are equally applicable to the transmission of digitally-coded speech or video signals, where much higher error rates can be tolerated. Under the latter conditions, however, the approximations that are derived on the assumption of a low error rate are no longer always valid.

4.2 BASIC ASSUMPTIONS

Consider the synchronous serial binary data-transmission system shown in Fig. 4.1. The signal at the input to the transmitter filter is a sequence of regularly spaced impulses $\{s_i \delta(t - iT)\}$, the $(i+1)$th of which occurs at time $t = iT$ seconds and has the value (area)

$$s_i = \pm k \qquad (4.1)$$

where k is a positive constant. It is assumed that $s_i = 0$ for $i < 0$. For $i \geqslant 0$, each impulse $s_i \delta(t - iT)$ is a binary polar signal-element, and a typical sequence of such elements is shown in Fig. 4.2. In practice, of course, a rectangular or rounded waveform would be used, with the appropriate change in the transmitter filter. The $\{s_i\}$ are statistically independent and equally likely to have either binary value. Where this

158 LINEAR EQUALIZERS

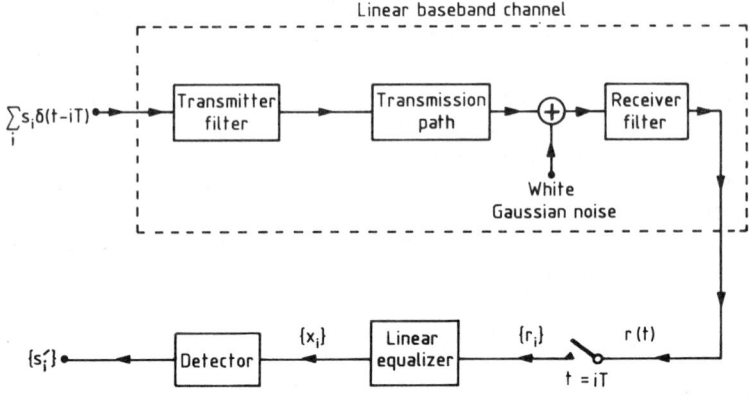

Fig. 4.1 Data-transmission system

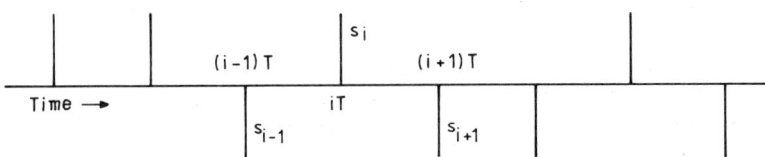

Fig. 4.2 A typical sequence of signal elements at the input to the baseband channel

condition is not satisfied by the $\{s_i\}$, it can normally be achieved, for practical purposes, by *scrambling* the transmitted sequence of data symbols, and then appropriately *descrambling* the corresponding detected data-symbols at the receiver, to obtain a copy of the original transmitted sequence[156-159]. Multilevel data-symbols could be used in place of binary symbols without affecting any of the important results in this analysis.

The transmission path in Fig. 4.1 is a linear baseband channel and its structure and properties are considered in some detail in Section 1.2. In particular, the transmission path could be a bandpass channel, such as a voiceband telephone circuit, together with a linear modulator at the transmitter and a linear demodulator at the receiver. The transmitter filter, transmission path and receiver filter in Fig. 4.1 together form a *linear baseband channel* whose impulse response is $y(t)$ and has, for practical purposes, a finite duration. It is assumed that $y(t)$ is real valued and that $y(t-iT)$ is exactly equal to a time-shifted version of $y(t-jT)$, at least where $|i-j|$ does not exceed the number of received signal elements involved in the detection of any one data symbol. The symbol j here is, of course, a positive integer.

White Gaussian noise with zero mean and a two-sided power

spectral density of $\frac{1}{2}N_0$ is added to the data signal at the output of the transmission path, giving the Gaussian waveform $w(t)$ at the output of the receiver filter, that is, at the output of the baseband channel in Fig. 4.1. Although many practical channels do not introduce significant levels of Gaussian noise, the relative tolerances of different data-transmission systems to additive white Gaussian noise is a good measure of their relative overall tolerances to practical types of additive noise (see Section 1.1)[1].

The signal waveform at the output of the baseband channel in Fig. 4.1 is

$$r(t) = \sum_i s_i y(t - iT) + w(t) \qquad (4.2)$$

where $\sum_i s_i y(t - iT)$ is the received data signal and $w(t)$ is the band-limited Gaussian noise waveform, originating from the white Gaussian noise at the input to the receiver filter. The received waveform $r(t)$ is sampled once per signal element at the time instants $\{iT\}$, for $i = 0, 1, 2, \ldots$. Various techniques are available for holding the sampling instants correctly synchronized to the received signal[1,160-162], but are not considered further here. The received signal at time $t = iT$, at the output of the sampler, is

$$r_i = \sum_{j=0}^{g} s_{i-j} y_j + w_i \qquad (4.3)$$

where $r_i = r(iT)$, $y_j = y(jT)$ and $w_i = w(iT)$. All signals here are *real valued*.

When the transmission path in Fig. 4.1 introduces no distortion, attenuation or delay, the impulse response $y(t)$ of the linear baseband channel becomes that of the transmitter and receiver filters in cascade and is such that

$$y_0 = 1 \qquad (4.4)$$

and
$$y_j = 0 \qquad (4.5)$$

for all nonzero integers $\{j\}$. The transmitter and receiver filters here are not physically realizable, but can be made so by introducing a sufficient delay into their impulse response. They are considered in more detail in Section 1.2. Under the conditions just described,

$$r_i = s_i + w_i \qquad (4.6)$$

The baseband channel in Fig. 4.1 now introduces no intersymbol interference or attenuation into the received sampled signal, and the best detection process is to detect each s_i from the corresponding r_i. When $r_i > 0$, s_i is detected as k, and when $r_i < 0$, s_i is detected as $-k$.

Notice also that there is no delay in transmission, the data-symbol s_i, carried by the signal-element $s_i \delta(t - iT)$ at the input to the baseband channel in Fig. 4.1, being detected at the receiver from the sample r_i at the *same* time instant $t = iT$.

Consider next the general case where the transmission path introduces signal distortion. If now a unit impulse at time $t = 0$ is fed to the baseband channel in the absence of any other signals, the resultant waveform at the input to the sampler in Fig. 4.1 is $y(t)$, which has for practical purposes a finite time duration that is less than $(g + 1)T$ seconds but may or may not be greater than gT seconds, where g is an appropriate positive integer. The corresponding samples at the sampler output form the *sampled impulse-response* of the baseband channel and are given by the $(g + 1)$-component row vector

$$Y = [\, y_0 \quad y_1 \quad \ldots \quad y_g \,] \tag{4.7}$$

where $y_i = y(iT)$. The delay in transmission, other than that involved in the time dispersion of the transmitted signal, is neglected here, so that $y_0 \neq 0$ and $y_i = 0$ for $i < 0$ and $i > g$. Clearly, the baseband channel now disperses (or 'spreads out') each signal-element in time, the *shape* of each individual received signal-element being identical to that of any other element with the same value of s_i.

If the $(i + 1)$th signal-element $s_i y(t - iT)$ is received alone and in the absence of noise, the corresponding samples $\{r_j\}$, starting with r_i at time $t = iT$, are

$$s_i y_0, s_i y_1, \ldots, s_i y_g, 0, 0, \ldots \tag{4.8}$$

all preceding $\{r_j\}$ being zero. It is assumed that the receiver has prior knowledge of both k and Y, so that it knows the two *possible* sets of sample values corresponding to any individual received signal-element, although, *before* the detection of a particular received signal-element, it has no knowledge of *which* of its two possible sets of sample values are in fact received. Since there is no variation in the impulse-response of the channel during the processing of any individual received signal-element, the sampled impulse-response Y is also time invariant.

With the reception of an uninterrupted sequence of signal-elements in the presence of noise, which will now be assumed, the sample of the received signal at the output of the baseband channel, at time $t = iT$, is, of course, given by r_i in Equation 4.3.

As is shown in Section 1.2, the transmitter filter in Fig. 4.1 is such that the average transmitted energy per signal element is equal to the mean-square value of s_i, which is

$$\mathscr{E}[s_i^2] = k^2 \tag{4.9}$$

from Equation 1.22, and, from Equation 1.24, the receiver filter is such that the Gaussian noise components $\{w_i\}$ are statistically independent Gaussian random variables with zero mean and variance

$$\sigma^2 = \tfrac{1}{2}N_0 \qquad (4.10)$$

where $\tfrac{1}{2}N_0$ is the two-sided power spectral density of the white Gaussian noise at the input to the receiver filter. The noise components are here designated $\{w_i\}$ to emphasize the fact that they are statistically independent (or 'white'). Each transmitted binary signal-element carries one bit of information, so that the *signal/noise ratio*, measured as the average transmitted energy per *bit* at the *input* to the transmission path, divided by the *two-sided* noise power spectral density at the *output* of the transmission path, is k^2/σ^2.

Suppose next that it is required to detect s_i from r_{i+h}, where h is a nonnegative integer in the range 0 to g. Now, from Equation 4.3,

$$r_{i+h} = s_i y_h + \sum_{\substack{j=0 \\ j \neq h}}^{g} s_{i+h-j} y_j + w_{i+h} \qquad (4.11)$$

where $s_i y_h$ is the *wanted signal*,

$$\sum_{\substack{j=0 \\ j \neq h}}^{g} s_{i+h-j} y_j$$

is the *intersymbol interference*, and w_{i+h} is the *additive noise*. Clearly, it may not be possible to detect s_i correctly from r_{i+h}, even in the absence of noise, unless the maximum value of the intersymbol-interference component has a magnitude less than $|ky_h|$. Under these conditions, the intersymbol interference on its own cannot prevent the correct detection of s_i from r_{i+h}, but it may seriously reduce the tolerance of the system to the additive noise. The evaluation of the precise effect of intersymbol interference on the tolerance of the system to noise is in general quite difficult[2-26].

The equalizer in Fig. 4.1 operates on the received samples $\{r_i\}$ in such a way that the output signal from the equalizer, at time $t = (i+h)T$, is

$$x_{i+h} \simeq s_i + u_{i+h} \qquad (4.12)$$

where h is an appropriate nonnegative integer and u_{i+h} is the Gaussian noise component at the equalizer output, at time $t = (i+h)T$. Clearly, all intersymbol interference has been effectively eliminated in x_{i+h}.

When exact equality holds in Equation 4.12, the optimum detection process for s_i from the equalized signal x_{i+h} is to detect s_i as its *possible* value *closest* to x_{i+h}. Thus, in the detector, s_i is detected by comparing x_{i+h} with a decision threshold of zero. If $x_{i+h} > 0$, s_i is detected as k, and if $x_{i+h} < 0$, s_i is detected as $-k$. The detected value of s_i is designated s'_i.

A most powerful tool for the analysis of an equalizer that operates on a sampled signal is the z-transform. This is described in Section 2.2. The particular importance of the z-transform of a sampled signal is firstly that it is very simply related to the sample values of the signal, and secondly that the resultant z-transform of two sampled networks in cascade is the *product* of their respective z-transforms.

As is shown in Section 2.2, the *z-transform* of the sampled impulse-response Y of the baseband channel is

$$Y(z) = y_0 + y_1 z^{-1} + y_2 z^{-2} + \cdots + y_g z^{-g} \qquad (4.13)$$

where z^{-i} represents the time instant $t = iT$. The sampled impulse-response Y may alternatively be represented as the corresponding sequence of *impulses*

$$y_0 \delta(t) + y_1 \delta(t - T) + \cdots + y_g \delta(t - gT) \qquad (4.14)$$

whose Fourier transform gives $Y(z)$. Each impulse here identifies precisely the time of occurrence of the corresponding sample, the sample *value* y_i, at time $t = iT$, being equal to the *area* of the impulse $y_i \delta(t - iT)$.

The z-transform of the $(i+1)$th transmitted signal-element, at the input to the baseband channel in Fig. 4.1, is $s_i z^{-i}$, and the z-transform of the $(i+1)$th received signal-element at the output of the sampler is

$$s_i z^{-i} Y(z) = s_i y_0 z^{-i} + s_i y_1 z^{-i-1} + \cdots + s_i y_g z^{-i-g} \qquad (4.15)$$

$Y(z)$ represents the *distortion* introduced by the channel. For no distortion, attenuation or delay,

$$Y(z) = 1 \qquad (4.16)$$

For no distortion or attenuation, and a delay of h sampling intervals,

$$Y(z) = z^{-h} \qquad (4.17)$$

For no distortion, but some attenuation (or gain) and delay,

$$Y(z) = y_h z^{-h} \qquad (4.18)$$

where y_h may take on any positive or negative real value. Normally, however, the delay in transmission is neglected, so that $y_0 \neq 0$.

The function of the equalizer is to convert $Y(z)$, the z-transform of the baseband channel and sampler, to z^{-h}, which represents a simple delay of hT seconds with no signal distortion or attenuation, where h is an appropriate nonnegative integer. z^{-h} is the z-transform of the baseband channel, sampler and equalizer. Thus, if the z-transform of the equalizer is $C(z)$, the z-transform of the $(i+1)$th received signal-element at the output of the equalizer is

$$s_i z^{-i} Y(z) C(z) \simeq s_i z^{-i-h} \tag{4.19}$$

Equation 4.12 is now satisfied, so that s_i is detected from the signal (sample) x_{i+h} at the output of the equalizer, at time $t=(i+h)T$. Clearly,

$$C(z) \simeq z^{-h} Y^{-1}(z) \tag{4.20}$$

since, of course,

$$Y(z) Y^{-1}(z) = 1 \tag{4.21}$$

so that the equalizer is equivalent to the *inverse* of the channel together with a *delay* of hT seconds.

Except where otherwise stated, a high signal/noise ratio is assumed. Under these conditions, the best tolerance to additive Gaussian noise is achieved through the effective elimination of *all* intersymbol interference, which now tends to predominate over the noise in the received signal[59,155]. Again, except where specifically stated to the contrary, the equalizer is taken to equalize the channel sufficiently accurately to reduce the residual intersymbol interference to a negligible level, and the channel is taken to be such that it can be equalized accurately by a linear filter. The conditions for which this is possible are that $Y(z)$ has no zeros (roots) on the unit circle in the z plane, as is shown in Section 3.6.

4.3 FEEDFORWARD TRANSVERSAL FILTER

The most widely studied linear equalizer is the feedforward transversal filter[49-154]. It has been shown that the optimum linear receiver, using only linear processing of the signal prior to detection, includes such a transversal filter[27-48]. Assume now that the equalizer in Fig. 4.1 is the linear feedforward transversal filter with $m+1$ taps, shown in Fig. 4.3. The signals shown here are, of course, all *real valued*.

The equalizer operates on sample values (or numbers) which may be in analogue or digital (binary coded) form, usually the latter. The binary coded numbers have typically from 8 to 32 binary digits. The signals (samples) shown in Fig. 4.3 are those present at the time instant $t=iT$. Each square marked T is here a storage element that holds the corresponding signal (sample value) r_j, in binary coded form. The storage elements are triggered at the time instants $\{iT\}$, for

164 LINEAR EQUALIZERS

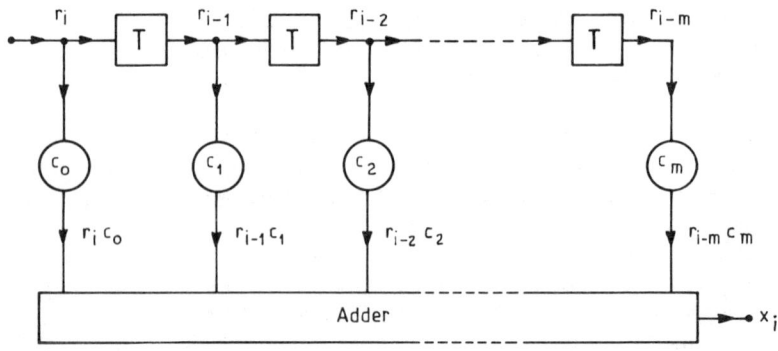

Fig. 4.3 An $(m+1)$-tap linear feedforward transversal equalizer for the baseband channel

all positive integers $\{i\}$, and each time the transversal filter is triggered, the stored signals $\{r_j\}$ are shifted one place to the right. Thus the extreme left-hand signal at the input to the equalizer is that received at time $t=iT$, the next signal is that received at time $t=(i-1)T$, and so on, each storage element introducing a delay of T seconds. A circle marked c_j is a multiplier that multiplies its input signal by c_j. The output signals from the multipliers, $\{r_{i-j}c_j\}$, are added together to give the output signal

$$x_i = \sum_{j=0}^{m} r_{i-j} c_j \qquad (4.22)$$

at time $t=iT$.

A practical linear feedforward transversal equalizer has between 8 and 64 taps, and is often implemented in a rather different form from that suggested by Fig. 4.3. For instance, only a single multiplier may be used, the multiplier being shared between the different taps.

The transversal filter in Fig. 4.3 has a sampled impulse-response given by the $(m+1)$-component row vector

$$C = [c_0 \quad c_1 \quad \ldots \quad c_m] \qquad (4.23)$$

and the z-transform of this sampled impulse-response is

$$C(z) = c_0 + c_1 z^{-1} + \cdots + c_m z^{-m} \qquad (4.24)$$

Notice that the sampled impulse-response of the filter is given by the sequence of its *tap gains* $\{c_j\}$, which also form the coefficients in the corresponding z-transform.

For the accurate equalization of the channel, the z-transform of the channel and equalizer is

$$Y(z)C(z) \simeq z^{-h} \tag{4.25}$$

where h is a nonnegative integer in the range 0 to $m+g$.

An alternative technique to z-transforms for analysing the operation of an equalizer is that of *convolution matrices*, as described in Section 2.3. The sampled impulse-response of two sampled networks in cascade is the *convolution* of their individual sampled impulse-responses, and is given by the product of one of the two sampled impulse-responses and the convolution matrix corresponding to the other.

Let Y_c be the $(m+1) \times (m+g+1)$ convolution matrix whose $(i+1)$th row is given by the $(m+g+1)$-component row vector

$$Y_i = [\overbrace{0 \ \ldots \ 0}^{i} \ y_0 \ y_1 \ \ldots \ y_g \ 0 \ \ldots \ 0] \tag{4.26}$$

so that the *first* component of Y_0 (the first row of Y_c) is y_0, and the last component of Y_m (the last row of Y_c) is y_g. Y_c is in fact formed by the first $m+1$ rows of the corresponding $(m+g+1) \times (m+g+1)$ *circulant* matrix (see Equation 2.25). From Equation 4.25, the sampled impulse-response of the channel and equalizer is

$$\sum_{i=0}^{m} c_i Y_i = C Y_c \simeq E_h \tag{4.27}$$

where E_h is the $(m+g+1)$-component row vector given by

$$E_h = [\overbrace{0 \ \ldots \ 0}^{h} \ 1 \ 0 \ \ldots \ 0] \tag{4.28}$$

E_h is the *ideal* or desired sampled impulse-response of the equalized channel, whereas CY_c is the *actual* sampled impulse-response obtained. Y_c is the convolution matrix corresponding to the sampled impulse-response of the channel, and CY_c is the $(m+g+1)$-component row vector obtained by the convolution of the vectors (sequences) Y and C (Equations 4.7 and 4.23). Y_c is not a circulant.

Various design techniques are available for evaluating the tap gains of the linear equalizer to satisfy Equation 4.27. With a finite feedforward transversal filter, Equation 4.27 will not in general be satisfied exactly, but instead

$$CY_c = E = [e_0 \ e_1 \ \ldots \ e_{m+g}] \tag{4.29}$$

where $e_h \simeq 1$ and $e_i \simeq 0$, for $i = 0, 1, \ldots, m+g$ and $i \neq h$.

166 LINEAR EQUALIZERS

The *peak distortion* in the equalized signal is defined to be[59,65]

$$D_p = \frac{1}{|e_h|} \sum_{\substack{i=0 \\ i \neq h}}^{m+g} |e_i| \qquad (4.30)$$

and the *mean-square distortion* is[59,65]

$$D_m = \frac{1}{e_h^2} \sum_{\substack{i=0 \\ i \neq h}}^{m+g} e_i^2 \qquad (4.31)$$

The definitions of D_p and D_m apply only to the *binary* polar data-symbols $\{s_i\}$ assumed here.

It can be seen from Equations 4.12 and 4.29 that, in the *absence of noise*,

$$x_{i+h} \simeq s_i \qquad (4.32)$$

where

$$x_{i+h} = \sum_{j=0}^{m+g} s_{i+h-j} e_j \qquad (4.33)$$

Thus the *mean-square error* in the equalized signal, due to *intersymbol interference*, is the expected value $\mathscr{E}[\cdot]$ of the error squared, and is

$$\mathscr{E}[(x_{i+h} - s_i)^2] = \mathscr{E}\left[\left(\sum_{j=0}^{m+g} s_{i+h-j} e_j - s_i\right)^2\right]$$

$$= \mathscr{E}\left[\left(\sum_{j=0}^{m+g} s_{i+h-j} e_j\right)^2 - 2s_i \left(\sum_{j=0}^{m+g} s_{i+h-j} e_j\right) + s_i^2\right]$$

$$= \mathscr{E}\left[\sum_{j=0}^{m+g} s_{i+h-j}^2 e_j^2 - 2s_i^2 e_h + s_i^2\right]$$

$$= \sum_{j=0}^{m+g} \mathscr{E}[s_{i+h-j}^2 e_j^2] - 2\mathscr{E}[s_i^2 e_h] + \mathscr{E}[s_i^2]$$

$$= k^2 \sum_{j=0}^{m+g} e_j^2 - 2k^2 e_h + k^2 = k^2 \sum_{\substack{j=0 \\ j \neq h}}^{m+g} e_j^2 + k^2 (e_h - 1)^2$$

$$= k^2 (E - E_h)(E - E_h)^T = k^2 |E - E_h|^2 \qquad (4.34)$$

The above derivation uses the fact that the mean-square value of s_i is

$$\mathscr{E}[s_i^2] = k^2 \qquad (4.35)$$

Also, since the $\{s_i\}$ are statistically independent with zero mean, they are *statistically orthogonal*, such that

$$\mathscr{E}[s_i s_j] = 0 \qquad (4.36)$$

for $i \neq j$[163-165]. In Equation 4.34, $(E - E_h)^T$ and $|E - E_h|$ are the transpose and Euclidean length, respectively, of the $(m+g+1)$-component row vector $E-E_h$. In every case, the ideal (or desired) value of the sampled impulse-response of the equalized channel is taken to be E_h, so that $e_h \simeq 1$ in Equation 4.29. Again, only the effects of *intersymbol interference* are considered here, the effects of the noise being for the moment neglected. For a given equalizer it may be required to minimize any one of the three quantities defined by the Equations 4.30, 4.31 and 4.34.

4.4 ZERO FORCING

In the early work on linear equalizers these were generally designed to minimize the peak distortion[49-61]. This is achieved, under certain conditions, by a technique known as *zero forcing*[52,53,59].

Assume that the component of Y (Equation 4.7) of greatest magnitude is y_j and that the number of taps of the linear equalizer in Fig. 4.3 is

$$m+1 = 2l+1 \tag{4.37}$$

where l is a positive integer. For reasons which will become clear later, the integer h in Equation 4.28 is now chosen to have the value

$$h = j+l \tag{4.38}$$

In the technique of zero forcing, the tap gains of the linear equalizer are chosen to set e_h to 1 and $2l$ of the remaining $\{e_i\}$ in Equation 4.29 to zero. It has been shown[52,59] that, when the peak distortion of the *received* signal, given by

$$\frac{1}{|y_j|} \sum_{\substack{i=0 \\ i \neq j}}^{g} |y_i|$$

is less than unity, the peak distortion in the equalized signal, for the given value of h, is *minimized* when $e_i = 0$ for $h-l \leqslant i < h$ and $h < i \leqslant h+l$, with $e_h = 1$. From Equations 4.29, 4.37 and 4.38 we now have

$$CY = [e_0 \ \ldots \ e_{j-1} \ \overbrace{0 \ \ldots \ 0}^{l} \ 1 \ \overbrace{0 \ \ldots \ 0}^{l} \ e_{j+m+1} \ \ldots \ e_{m+g}] \tag{4.39}$$

Let Z be the $(m+1) \times (m+1)$ matrix formed from Y_c by removing its first j and last $g-j$ columns. The first and last rows of Z are now given by the $(m+1)$-component vectors

$$Z_0 = [y_j \ y_{j+1} \ \ldots \ y_g \ 0 \ 0 \ \ldots \ 0] \tag{4.40}$$

and
$$Z_m = [0 \ 0 \ \ldots \ 0 \ y_0 \ y_1 \ \ldots \ y_j] \quad (4.41)$$
respectively, and each component along the main diagonal of Z has the value y_j. From Equation 4.39,

$$CZ = [\overbrace{0 \ \ldots \ 0}^{l} \ 1 \ \overbrace{0 \ \ldots \ 0}^{l}] \quad (4.42)$$

When the peak distortion in the received signal is less than unity, the matrix Z is diagonally dominant and must therefore be nonsingular. Under these conditions,

$$C = [\overbrace{0 \ \ldots \ 0}^{l} \ 1 \ \overbrace{0 \ \ldots \ 0}^{l}] Z^{-1} \quad (4.43)$$

so that C is given by the central row of Z^{-1} (m is even). Clearly, the vector C, that satisfies Equation 4.42 and so minimizes the peak distortion in the equalized signal, is now *uniquely* determined[52].

For the exact equalization of the channel, $e_i = 0$ in Equation 4.39 for all values of i other than h. However, this requires the $m+1$ tap gains to satisfy $m+g+1$ linear simultaneous equations, which is not normally possible.

Not only does the technique of zero forcing minimize the peak distortion in the equalized signal, when the peak distortion of the received signal is less than unity, but it also leads to some simple iterative processes for holding the equalizer correctly adjusted for a slowly time-varying channel[49-61]. The weakness of the technique is that it cannot be used in its simple form for equalizing severe time-varying signal distortion.

4.5 SEPARATE EQUALIZATION OF THE TWO FACTORS OF THE CHANNEL RESPONSE

Suppose that the z-transform of the baseband channel in Fig. 4.1 is

$$Y(z) = 1 + yz^{-1} \quad (4.44)$$

which means, of course, that the sampled impulse-response of the channel is given by the two-component vector $[1 \ y]$.

The linear equalizer should now ideally have the z-transform

$$Y^{-1}(z) = (1 + yz^{-1})^{-1} \quad (4.45)$$

where,
$$Y(z)Y^{-1}(z) = 1 \quad (4.46)$$

by definition.

LINEAR EQUALIZERS 169

Consider first the case where $|y|<1$. Now $|yz^{-1}|<1$, since

$$z^{-1}=\exp(-j2\pi fT) \qquad (4.47)$$

where $j=\sqrt{-1}$, bearing in mind that the z-transform of a sequence of samples is here taken to be the *Fourier transform* of the corresponding sequence of *impulses* (Section 2.2). Under these conditions, the binomial expansion can be applied to $(1+yz^{-1})^{-1}$ to give

$$(1+yz^{-1})^{-1}=1-yz^{-1}+y^2z^{-2}-y^3z^{-3}+\cdots \qquad (4.48)$$

which is a convergent series[166] (see also Section 3.3).

The value of $(1+yz^{-1})^{-1}$ can alternatively be obtained through the division of 1 by $(1+yz^{-1})$, using a normal process of long division, as follows[166-169]:

$$
\begin{array}{r}
1-yz^{-1}+y^2z^{-2}-y^3z^{-3}+\cdots \\
1+yz^{-1}\overline{\smash{\big)}\,1} \\
\underline{1+yz^{-1}} \\
-yz^{-1} \\
\underline{-yz^{-1}-y^2z^{-2}} \\
y^2z^{-2} \\
\underline{y^2z^{-2}+y^3z^{-3}} \\
-y^3z^{-3} \\
\cdots
\end{array}
$$

The linear transversal equalizer for a channel with z-transform $1+yz^{-1}$ has the given z-transform $(1+yz^{-1})^{-1}$ and therefore has ideally an infinite number of taps. It is shown in Fig. 4.4.

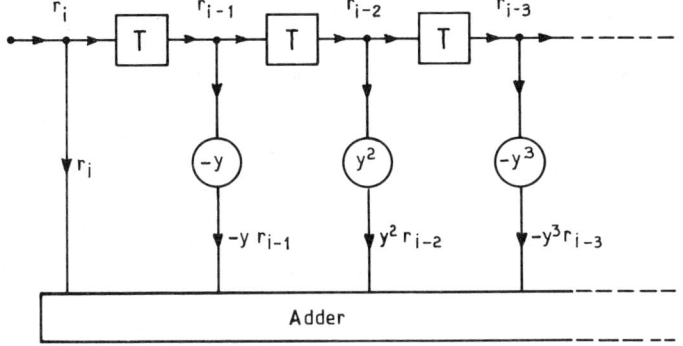

Fig. 4.4 Linear equalizer for a channel with z-transform $1+yz^{-1}$, where $|y|<1$

170 LINEAR EQUALIZERS

If $y \ll 1$, a good approximation to the ideal response for the equalized channel can be obtained with only the first few stages (storage elements and multipliers) of the filter in Fig. 4.4, since the gain introduced by the ith multiplier is $(-y)^i$ which decreases rapidly towards zero as i increases. Thus a practical equalizer for the given channel has a finite number of $m+1$ taps, as in Fig. 4.3, and only approximately equalizes the channel. Clearly, as $|y|$ increases towards unity, a longer transversal equalizer is needed, until, when $|y|=1$, equalization is no longer possible with a finite transversal filter. The significance of this is considered in more detail towards the end of the section.

The z-transform of an $(m+1)$-tap linear transversal filter, that approximately equalizes the channel with z-transform $1 + yz^{-1}$, where $|y| < 1$, is obtained by taking the first $m+1$ terms of $(1+yz^{-1})^{-1}$ and is

$$C(z) = 1 - yz^{-1} + y^2 z^{-2} - \cdots + (-y)^m z^{-m} \qquad (4.49)$$

Clearly, for a sufficiently large value of m, $(1 + yz^{-1})C(z) \simeq 1$ and $C(z) \simeq (1+yz^{-1})^{-1}$.

Suppose now that $y = \tfrac{1}{2}$ and the transversal equalizer has six taps. The z-transform of the baseband channel is

$$Y(z) = 1 + \tfrac{1}{2} z^{-1} \qquad (4.50)$$

and the z-transform of the equalizer is

$$C(z) = 1 - \tfrac{1}{2} z^{-1} + \tfrac{1}{4} z^{-2} - \tfrac{1}{8} z^{-3} + \tfrac{1}{16} z^{-4} - \tfrac{1}{32} z^{-5} \qquad (4.51)$$

the transversal equalizer being as shown in Fig. 4.5.

The z-transform of the channel and linear equalizer is now

$$Y(z)C(z) = (1 + \tfrac{1}{2} z^{-1})(1 - \tfrac{1}{2} z^{-1} + \tfrac{1}{4} z^{-2} - \tfrac{1}{8} z^{-3} + \tfrac{1}{16} z^{-4} - \tfrac{1}{32} z^{-5})$$

$$= 1 - \tfrac{1}{64} z^{-6} \qquad (4.52)$$

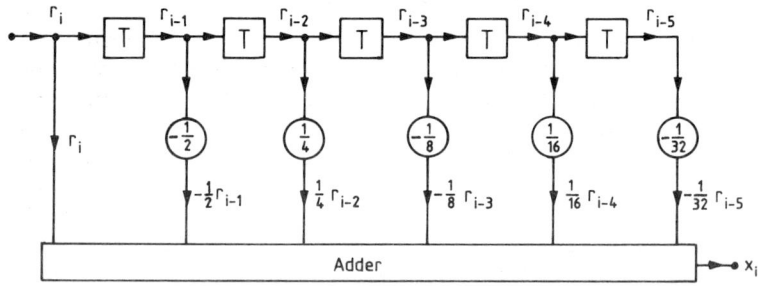

Fig. 4.5 6-tap linear equalizer for a channel with z-transform $1 + \tfrac{1}{2} z^{-1}$

LINEAR EQUALIZERS

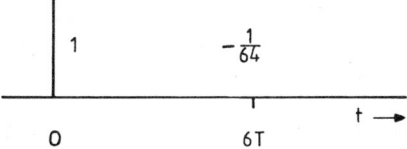

Fig. 4.6   Sampled impulse-response of a channel, with z-transform $1+\frac{1}{2}z^{-1}$, in cascade with the corresponding 6-tap transversal equalizer

so that

$$Y(z)C(z) \simeq 1 \tag{4.53}$$

and

$$C(z) \simeq Y^{-1}(z) \tag{4.54}$$

If the residual intersymbol interference at the output of the equalizer is not neglected, the sampled impulse response of the channel and linear equalizer, represented in terms of the corresponding impulses, is

$$\delta(t) - \tfrac{1}{64}\delta(t-6T) \tag{4.55}$$

as shown in Fig. 4.6. Clearly, this is an arrangement of zero forcing (Section 4.4).

When the $(i+1)$th individual signal-element is received alone and in the absence of noise, the signal at the output of the equalizer is

$$s_i\delta(t-iT) - \tfrac{1}{64}s_i\delta(t-6T-iT) \tag{4.56}$$

if represented in terms of impulses. Thus, when a continuous stream of data elements is received in the presence of Gaussian noise, the signal (sample) at the output of the equalizer, at time $t=iT$, is

$$x_i = -\tfrac{1}{64}s_{i-6} + s_i + u_i \tag{4.57}$$

where u_i is a Gaussian random variable with zero mean and variance dependent upon the transversal equalizer. Since $|-\tfrac{1}{64}s_{i-6}| \ll |s_i|$,

$$x_i \simeq s_i + u_i \tag{4.58}$$

and s_i is detected from x_i, at time $t=iT$. When $x_i>0$, s_i is detected as k, and when $x_i<0$, s_i is detected as $-k$. The equalizer introduces no delay into the equalized signal, and is a minimum delay (minimum-phase) network.

Suppose now that the z-transform of the channel is given by

$$Y(z) = 1 + yz^{-1} \tag{4.59}$$

where $|y|>1$. Clearly, $|yz^{-1}|>1$ since $|z^{-1}|=1$. Thus $(1+yz^{-1})^{-1}$ does not here satisfy Equation 4.48, since the binomial expansion cannot now be applied to $(1+yz^{-1})^{-1}$. This can be seen from the fact

172 LINEAR EQUALIZERS

that the expansion of $(1+yz^{-1})^{-1}$ given by Equation 4.48 has become a *divergent* series. The *same* divergent series is obtained if 1 is divided by $1+yz^{-1}$, using a normal process of long division.

It is interesting to observe that the zero (root) of $1+yz^{-1}$ is at $z=-y$, so that when $|y|<1$ the zero lies *inside* the unit circle in the z plane, and when $|y|>1$ the zero lies outside the unit circle.

To evaluate the $(m+1)$-tap linear traversal equalizer for the channel whose z-transform is given by Equation 4.59, consider the channel with z-transform

$$M(z) = y + z^{-1} \tag{4.60}$$

which is obtained from $Y(z)$ by *reversing the order* of its coefficients. The zero of $M(z)$ is $-1/y$ which is the *reciprocal* of the zero of $Y(z)$ and lies inside the unit circle in the z plane. Thus

$$\begin{aligned} M^{-1}(z) &= (y+z^{-1})^{-1} = y^{-1}(1+y^{-1}z^{-1})^{-1} \\ &= y^{-1}(1 - y^{-1}z^{-1} + y^{-2}z^{-2} - y^{-3}z^{-3} + \cdots) \\ &= y^{-1} - y^{-2}z^{-1} + y^{-3}z^{-2} - y^{-4}z^{-3} + \cdots \end{aligned} \tag{4.61}$$

since $|y^{-1}|<1$, which is the condition that must be satisfied[166] for the binomial expansion of $(1+y^{-1}z^{-1})^{-1}$. Equation 4.61 represents $M^{-1}(z)$ as a convergent series.

The value of $M^{-1}(z)$ can alternatively be obtained through the division of 1 by $M(z)$, using a normal process of long division, as follows[166-169]:

$$\begin{array}{r} y^{-1} - y^{-2}z^{-1} + y^{-3}z^{-2} - y^{-4}z^{-3} + \cdots \\ y+z^{-1} \overline{\smash{\big)}\ 1 } \\ \underline{1 + y^{-1}z^{-1}} \\ -y^{-1}z^{-1} \\ \underline{-y^{-1}z^{-1} - y^{-2}z^{-2}} \\ y^{-2}z^{-2} \\ \underline{y^{-2}z^{-2} + y^{-3}z^{-3}} \\ -y^{-3}z^{-3} \end{array}$$

$$\cdots$$

The linear transversal equalizer for a channel with z-transform $M(z)$ has the z-transform $M^{-1}(z)$ and therefore has ideally an infinite number of taps. The z-transform of an $(m+1)$-tap linear transversal filter, that approximately equalizes the channel with z-transform $M(z)$, is obtained by taking the first $m+1$ terms of $M^{-1}(z)$ and is

$$N(z) = y^{-1} - y^{-2}z^{-1} + y^{-3}z^{-2} - \cdots - (-y)^{-m-1}z^{-m} \tag{4.62}$$

Thus, for a sufficiently large value of m, the z-transform of the channel and linear equalizer is

$$M(z)N(z) \simeq 1 \qquad (4.63)$$

Consider now the z-transform $C(z)$ which is obtained from $N(z)$ by *reversing the order* of its coefficients, so that

$$C(z) = -(-y)^{-m-1} - (-y)^{-m}z^{-1} - (-y)^{-m+1}z^{-2} - \cdots + y^{-1}z^{-m} \qquad (4.64)$$

Bearing in mind that $Y(z)$ is obtained from $M(z)$ by reversing the order of its coefficients, it can be seen from Equation 4.63 that

$$Y(z)C(z) \simeq z^{-m-1} \qquad (4.65)$$

This follows from the fact that, if the orders of the coefficients in two polynomials in z are reversed, then so also is the order of the coefficients in their *product*, there being no other change in this product. Similarly, if two sequences are reversed in time, then so also is their *convolution*. Thus, if the sampled impulse-responses of the channel and the corresponding linear equalizer are reversed in time, then so also is the sampled impulse-response of the two in cascade, which implies that correct equalization of the channel is maintained, although there may be a change in the delay introduced by the equalizer. Clearly, when $|y| > 1$, $C(z)$ in Equation 4.64 is the z-transform of the required linear equalizer for the channel with z-transform $Y(z)$ in Equation 4.59.

The greater is the magnitude of y, the more rapidly do the magnitudes of the tap gains increase with their positions along the equalizer (measured from the input end), so that the smaller is the filter needed for a given accuracy of equalization.

Consider now the case where the z-transform of the channel is

$$Y(z) = 1 + 2z^{-1} \qquad (4.66)$$

and the transversal equalizer has six taps. Let

$$M(z) = 2 + z^{-1} \qquad (4.67)$$

The long division of 1 by $M(z)$ is now carried out, as follows:

$$
\begin{array}{r}
\frac{1}{2} - \frac{1}{4}z^{-1} + \frac{1}{8}z^{-2} - \frac{1}{16}z^{-3} + \cdots \\
2 + z^{-1} \overline{) 1 } \\
\underline{1 + \frac{1}{2}z^{-1}} \\
-\frac{1}{2}z^{-1} \\
\underline{-\frac{1}{2}z^{-1} - \frac{1}{4}z^{-2}} \\
\frac{1}{4}z^{-2} \\
\underline{\frac{1}{4}z^{-2} + \frac{1}{8}z^{-3}} \\
-\frac{1}{8}z^{-3}
\end{array}
$$

174 LINEAR EQUALIZERS

so that
$$M^{-1}(z) = \tfrac{1}{2} - \tfrac{1}{4}z^{-1} + \tfrac{1}{8}z^{-2} - \tfrac{1}{16}z^{-3} + \cdots \quad (4.68)$$

The first six terms of $M^{-1}(z)$ are given by
$$N(z) = \tfrac{1}{2} - \tfrac{1}{4}z^{-1} + \tfrac{1}{8}z^{-2} - \tfrac{1}{16}z^{-3} + \tfrac{1}{32}z^{-4} - \tfrac{1}{64}z^{-5} \quad (4.69)$$

and the z-transform $C(z)$ of the required linear equalizer for $Y(z)$ is obtained from $N(z)$ by reversing the order of its coefficients, to give
$$C(z) = -\tfrac{1}{64} + \tfrac{1}{32}z^{-1} - \tfrac{1}{16}z^{-2} + \tfrac{1}{8}z^{-3} - \tfrac{1}{4}z^{-4} + \tfrac{1}{2}z^{-5} \quad (4.70)$$

Thus the z-transform of the sampled impulse-response of the baseband channel and linear equalizer is
$$Y(z)C(z) = (1 + 2z^{-1})(-\tfrac{1}{64} + \tfrac{1}{32}z^{-1} - \cdots - \tfrac{1}{4}z^{-4} + \tfrac{1}{2}z^{-5})$$
$$= -\tfrac{1}{64} + z^{-6} \quad (4.71)$$

so that
$$Y(z)C(z) \simeq z^{-6} \quad (4.72)$$

Since, by definition, $Y(z)Y^{-1}(z) = 1$, it can be seen from Equation 4.72 that
$$C(z) \simeq z^{-6} Y^{-1}(z) \quad (4.73)$$

which means that the transversal filter equalizes the channel and in addition introduces a delay of $6T$ seconds (six sampling intervals).

If the residual intersymbol interference at the output of the equalizer is not neglected, the sampled impulse-response of the channel and linear equalizer, represented in terms of the corresponding impulses, is
$$-\tfrac{1}{64}\delta(t) + \delta(t - 6T) \quad (4.74)$$

as shown in Fig. 4.7. Clearly, this is an arrangement of zero forcing.

When the $(i+1)$th individual signal-element is received alone and in the absence of noise, the signal at the output of the equalizer is
$$-\tfrac{1}{64}s_i\delta(t - iT) + s_i\delta(t - 6T - iT) \quad (4.75)$$

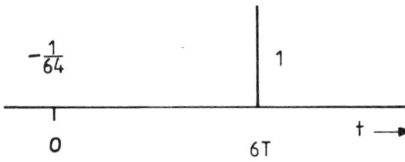

Fig. 4.7 Sampled impulse-response of a channel, with z-transform $1 + 2z^{-1}$, in cascade with the corresponding 6-tap transversal equalizer

if represented in terms of impulses. Thus, when a continuous stream of data elements is received in the presence of Gaussian noise, the signal at the output of the equalizer, at time $t = (i+6)T$, is

$$x_{i+6} = s_i - \tfrac{1}{64} s_{i+6} + u_{i+6} \tag{4.76}$$

where u_{i+6} is a Gaussian random variable with zero mean and variance dependent upon the transversal equalizer. Since $|-\tfrac{1}{64} s_{i+6}| \ll |s_i|$,

$$x_{i+6} \simeq s_i + u_{i+6} \tag{4.77}$$

and s_i is detected from the sign of x_{i+6}, at time $t = (i+6)T$. The equalizer introduces the maximum possible delay into the equalized signal, for the given number of taps, and so is a maximum-delay (maximum phase) network.

It is interesting now to consider the relationship between the ideal and practical equalizers for a channel whose z-transform is $Y(z) = 1 + yz^{-1}$, where $|y| > 1$. The ideal equalizer has the z-transform

$$\begin{aligned} Y^{-1}(z) &= (1 + yz^{-1})^{-1} = y^{-1} z (1 + y^{-1} z)^{-1} \\ &= y^{-1} z (1 - y^{-1} z + y^{-2} z^2 - y^{-3} z^3 + \cdots) \\ &= y^{-1} z - y^{-2} z^2 + y^{-3} z^3 - \cdots \end{aligned} \tag{4.78}$$

which is a convergent series. The z-transform given by Equation 4.78 is unfortunately not physically realizable, since z^i represents the time instant $-iT$ seconds or an *advance* in time of iT seconds. However, for a sufficiently large integer m,

$$Y^{-1}(z) \simeq y^{-1} z - y^{-2} z^2 + y^{-3} z^3 - \cdots - (-y)^{-m-1} z^{m+1}$$

and

$$\begin{aligned} z^{-m-1} Y^{-1}(z) &\simeq y^{-1} z^{-m} - y^{-2} z^{-m+1} \\ &\quad + y^{-3} z^{-m+2} - \cdots - (-y)^{-m-1} \\ &= -(-y)^{-m-1} - (-y)^{-m} z^{-1} \\ &\quad -(-y)^{-m+1} z^{-2} - \cdots + y^{-1} z^{-m} \end{aligned} \tag{4.79}$$

which is physically realizable and is in fact the z-transform $C(z)$ of the $(m+1)$-tap transversal equalizer given by Equation 4.64. This equalizer approximately equalizes the channel and at the same time introduces a delay of $(m+1)T$ seconds into the equalized signal (Equation 4.65). The equalizer for the given channel (Equation 4.59) could, if required, be determined by the method just described, but no advantage would in fact be gained by this. The ideal equalizer with z-transform $Y^{-1}(z)$ of course introduces *no* delay into the equalized

176 LINEAR EQUALIZERS

signal. The practical filter *must* introduce a delay in order that it may be physically realizable.

Consider next the case where the z-transform of the channel is

$$Y(z) = (1 + \tfrac{1}{2}z^{-1})(1 + 2z^{-1}) = 1 + 2\tfrac{1}{2}z^{-1} + z^{-2} \quad (4.80)$$

where $2\tfrac{1}{2}$ is, for convenience, written in place of 2.5. The z-transforms of the 6-tap equalizers for $1 + \tfrac{1}{2}z^{-1}$ and $1 + 2z^{-1}$ are given by Equations 4.51 and 4.70, respectively, and the z-transforms of the corresponding equalized channels are $1 - \tfrac{1}{64}z^{-6}$ and $-\tfrac{1}{64} + z^{-6}$, respectively, as can be seen from Equations 4.52 and 4.71. Thus the z-transform of the channel, given by Equation 4.80, in cascade with the two 6-tap equalizers is

$$(1 - \tfrac{1}{64}z^{-6})(-\tfrac{1}{64} + z^{-6}) \simeq z^{-6} \quad (4.81)$$

The equalizer is as shown in Fig. 4.8. Each of the two transversal filters here equalizes or removes the corresponding *factor* of $Y(z)$. The equalizer in Fig. 4.8 has the z-transform

$$(1 - \tfrac{1}{2}z^{-1} + \tfrac{1}{4}z^{-2} - \tfrac{1}{8}z^{-3} + \tfrac{1}{16}z^{-4} - \tfrac{1}{32}z^{-5}) \times$$
$$(-\tfrac{1}{64} + \tfrac{1}{32}z^{-1} - \tfrac{1}{16}z^{-2} + \tfrac{1}{8}z^{-3} - \tfrac{1}{4}z^{-4} + \tfrac{1}{2}z^{-5})$$
$$= -\tfrac{1}{64}(1 - 2\tfrac{1}{2}z^{-1} + 5\tfrac{1}{4}z^{-2} - 10\tfrac{5}{8}z^{-3} + 21\tfrac{5}{16}z^{-4}$$
$$- 42\tfrac{21}{32}z^{-5} + 21\tfrac{5}{16}z^{-6} - 10\tfrac{5}{8}z^{-7} + 5\tfrac{1}{4}z^{-8} - 2\tfrac{1}{2}z^{-9} + z^{-10}) \quad (4.82)$$

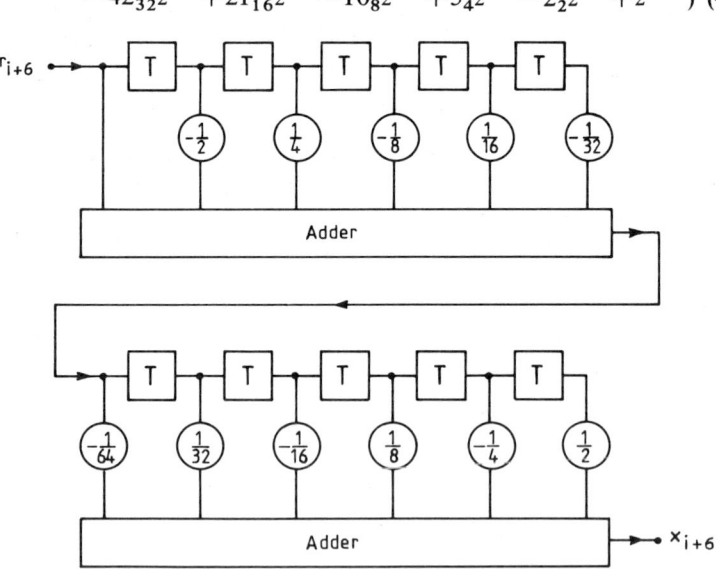

Fig. 4.8 Separate equalization of the two factors $1 + \tfrac{1}{2}z^{-1}$ and $1 + 2z^{-1}$ of the z-transform $1 + 2\tfrac{1}{2}z^{-1} + z^{-2}$

LINEAR EQUALIZERS

so that the equalizer is equivalent to a single transversal filter with 11 taps whose gains are given by the respective coefficients of the $\{z^{-i}\}$ on the right-hand side of Equation 4.82. Improper fractions are used for the coefficients here so that the *exact* values of these can be given. The single transversal filter may clearly be used in place of the two filters in cascade, with some reduction in the equipment complexity but without affecting the performance of the system.

The z-transform of the channel and linear equalizer is now

$$(1+2\tfrac{1}{2}z^{-1}+z^{-2})(-\tfrac{1}{64})(1-2\tfrac{1}{2}z^{-1}+5\tfrac{1}{4}z^{-2}-10\tfrac{5}{8}z^{-3}$$
$$+21\tfrac{5}{16}z^{-4}-42\tfrac{21}{32}z^{-5}+21\tfrac{5}{16}z^{-6}-10\tfrac{5}{8}z^{-7}$$
$$+5\tfrac{1}{4}z^{-8}-2\tfrac{1}{2}z^{-9}+z^{-10})$$
$$=-\tfrac{1}{64}(1-64\tfrac{1}{64}z^{-6}+z^{12})$$
$$\simeq -\tfrac{1}{64}+z^{-6}-\tfrac{1}{64}z^{-12}\simeq z^{-6} \qquad (4.83)$$

If the residual intersymbol interference at the output of the equalizer is not neglected, the sampled impulse-response of the channel and linear equalizer, represented in terms of the corresponding impulses, is approximately

$$-\tfrac{1}{64}\delta(t)+\delta(t-6T)-\tfrac{1}{64}\delta(t-12T) \qquad (4.84)$$

as shown in Fig. 4.9. As before, this is an arrangement of zero forcing.

When the $(i+1)$th individual signal-element is received alone and in the absence of noise, the signal at the output of the equalizer is

$$-\tfrac{1}{64}s_i\delta(t-iT)+s_i\delta(t-6T-iT)-\tfrac{1}{64}s_i\delta(t-12T-iT) \qquad (4.85)$$

if represented in terms of impulses. Thus, when a continuous stream of data elements is received in the presence of Gaussian noise, the signal at the output of the equalizer, at time $t=(i+6)T$, is

$$x_{i+6}\simeq -\tfrac{1}{64}s_{i-6}+s_i-\tfrac{1}{64}s_{i+6}+u_{i+6} \qquad (4.86)$$

so that

$$x_{i+6}\simeq s_i+u_{i+6} \qquad (4.87)$$

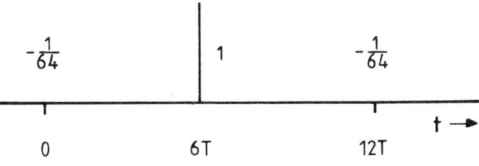

Fig. 4.9 Sampled impulse-response of a channel, with z-transform $1+2\tfrac{1}{2}z^{-1}+z^{-2}$, in cascade with the corresponding 11-tap transversal equalizer

and s_i is detected from the sign of x_{i+6}, at time $t=(i+6)T$. Clearly the residual intersymbol interference in x_{i+6} now involves the *two* data-symbols s_{i-6} and s_{i+6}, but is still negligible.

The z-transform of the equalizer for the given channel approximates to

$$z^{-h}(1+2\tfrac{1}{2}z^{-1}+z^{-2})^{-1} = z^{-h}(1+\tfrac{1}{2}z^{-1})^{-1}(1+2z^{-1})^{-1} \quad (4.88)$$

where h is an appropriate nonnegative integer. It might be thought that this z-transform could be evaluated directly through the division of 1 by $1+2\tfrac{1}{2}z^{-1}+z^{-2}$, using a normal process of long division. However this does not in fact give the required z-transform. The reason for the failure of the method is that the long division of 1 by $1+2\tfrac{1}{2}z^{-1}+z^{-2}$ gives a polynomial which is equal to the *product* of the polynomial obtained through the long division of 1 by $1+\tfrac{1}{2}z^{-1}$ and the polynomial obtained through the long division of 1 by $1+2z^{-1}$. The former polynomial approximates to $(1+\tfrac{1}{2}z^{-1})^{-1}$, but the latter polynomial forms a divergent series which does not therefore approximate to $z^{-h}(1+2z^{-1})^{-1}$, for *any* nonnegative integer h. Thus the long division of 1 by $1+2\tfrac{1}{2}z^{-1}+z^{-2}$ does not approximate to $z^{-h}(1+2\tfrac{1}{2}z^{-1}+z^{-2})^{-1}$. Furthermore, the reversal of the order of the coefficients in $Y(z)$ (Equation 4.80) leaves $Y(z)$ *unchanged*. For this reason, the equalizer for the given channel must be determined by the method previously described. It is important to observe that $1+2\tfrac{1}{2}z^{-1}+z^{-2}$ has one zero *inside* the unit circle in the z plane and one zero *outside* the unit circle.

The principles and techniques that have just been considered may now be generalized quite simply so that they may be applied to any given channel.

Firstly, when y is a *complex* number such that $|y|<1$, $(1+yz^{-1})^{-1}$ is obtained through the division of 1 by $1+yz^{-1}$, using a normal process of long division, exactly as previously described for the case where y is a *real* number. Similarly, of course, when $|y|>1$, $(1+yz^{-1})^{-1}$ is obtained through the division of 1 by $y+z^{-1}$, followed by the reversal of the order of the coefficients over the given number of terms generated. Secondly, if $Y(z) = Y_1(z)Y_2(z) \ldots Y_g(z)$ where each $Y_i(z)$ is a linear (elementary) factor of $Y(z)$, the polynomial obtained through the long division of 1 by $Y(z)$ is the *product* of the g polynomials obtained from the separate long division of 1 by the g factors $\{Y_i(z)\}$. Since $Y^{-1}(z) = Y_1^{-1}(z)Y_2^{-1}(z) \ldots Y_g^{-1}(z)$, it follows immediately that when *all* the zeros of $Y(z)$ lie *inside* the unit circle in the z plane, the long division of 1 by $Y(z)$ gives $Y^{-1}(z)$. Suppose now that one or more zeros of $Y(z)$ lie *outside* the unit circle, and let $Y_i(z)$ be one of the corresponding linear factors of $Y(z)$. The long division of 1 by $Y_i(z)$ gives a polynomial that forms a divergent series and is not therefore equal to $Y_i^{-1}(z)$. Since, as we have seen, the polynomial obtained from the long division of 1 by $Y(z)$ is the product of the polynomials obtained from the separate long division of 1 by the individual factors of $Y(z)$, it is clear that the long division of 1 by $Y(z)$ does *not* now now give $Y^{-1}(z)$. Using these facts, the general design technique for a linear equalizer has been developed, as follows.

Suppose first that the z-transform of the baseband channel in Fig. 4.1 is given by

$$Y(z) = y_0 + y_1 z^{-1} + y_2 z^{-2} + \cdots + y_g z^{-g} \qquad (4.89)$$

where *all* the zeros of $Y(z)$ lie *inside* the unit circle in the z plane.

To obtain the z-transform $C(z)$ of the $(m+1)$-tap linear equalizer for this channel, use a normal process of long division to divide 1 by $Y(z)$, as described for $Y(z) = 1 + yz^{-1}$ where $|y| < 1$, and taken the first $m+1$ terms of the resultant polynomial to give $C(z)$. Now

$$Y(z)C(z) \simeq 1 \qquad (4.90)$$

The equalizer is a minimum-delay (minimum-phase) network and applies zero forcing to the sampled impulse-response of the equalized channel. The main tap of the equalizer, that passes the wanted signal component through to the output, is here the first (or extreme left-hand) tap of the equalizer.

Suppose next that *all* the zeros of $Y(z)$ lie *outside* the unit circle in the z plane. Reverse the order of the coefficients in $Y(z)$ to give

$$M(z) = y_g + y_{g-1} z^{-1} + y_{g-2} z^{-2} + \cdots + y_0 z^{-g} \qquad (4.91)$$

The zeros of $M(z)$ are the reciprocals of the zeros of $Y(z)$ and therefore lie *inside* the unit circle in the z plane. Use a normal process of long division to divide 1 by $M(z)$, and take the first $m+1$ terms of the resultant polynomial to give the z-transform $N(z)$ of the equalizer for $M(z)$, such that

$$M(z)N(z) \simeq 1 \qquad (4.92)$$

Reverse the order of the coefficients in $N(z)$, to give $C(z)$, which is now the z-transform of the $(m+1)$-tap linear equalizer for $Y(z)$, such that

$$Y(z)C(z) \simeq z^{-g-m} \qquad (4.93)$$

The equalizer is a maximum-delay (maximum-phase) network and applies zero forcing to the sampled impulse-response of the equalized channel. The main tap of the equalizer, that passes the wanted signal component through to the output, is here the last (or extreme right-hand) tap of the equalizer.

Suppose finally that

$$Y(z) = Y_1(z) Y_2(z) \qquad (4.94)$$

180 LINEAR EQUALIZERS

where $Y_1(z)$ and $Y_2(z)$ may each have several zeros, with all the zeros of $Y_1(z)$ *inside* the unit circle in the z plane and all the zeros of $Y_2(z)$ *outside* the unit circle. Let

$$Y_1(z) = 1 + p_1 z^{-1} + p_2 z^{-2} + \cdots + p_{g-f} z^{-g+f} \quad (4.95)$$

and $\quad Y_2(z) = q_0 + q_1 z^{-1} + q_2 z^{-2} + \cdots + q_f z^{-f} \quad (4.96)$

where f is an integer in the range 0 to g. All coefficients in $Y_1(z)$ and $Y_2(z)$ are *real valued*.

The z-transform $C_1(z)$ of the $(m_1 + 1)$-tap linear equalizer for $Y_1(z)$ is determined as follows. Use a normal process of long division to divide 1 by $Y_1(z)$, and take the first $m_1 + 1$ terms of the resultant polynomial to give $C_1(z)$. Thus

$$Y_1(z) C_1(z) \simeq 1 \quad (4.97)$$

The z-transform $C_2(z)$ of the $(m_2 + 1)$-tap linear equalizer for $Y_2(z)$ is determined as follows. Let $M(z)$ be the polynomial determined from $Y_2(z)$ by reversing the order of its coefficients. Use a normal process of long division to divide 1 by $M(z)$. Take the first $m_2 + 1$ terms of the resultant polynomial and reverse the order of the coefficients, to give $C_2(z)$. Then

$$Y_2(z) C_2(z) \simeq z^{-f-m_2} \quad (4.98)$$

The z-transform of the linear equalizer for the baseband channel is now

$$C(z) = C_1(z) C_2(z) \quad (4.99)$$

so that the z-transform of the channel and equalizer is

$$Y(z) C(z) \simeq z^{-f-m_2} \quad (4.100)$$

as can be seen from Equations 4.97 and 4.98. The equalizer here is neither a minimum nor a maximum delay network.

The equalization of either $Y_1(z)$ or $Y_2(z)$, in the manner just described, leads to a sampled impulse-response with a consecutive set of zero-valued components between the wanted and unwanted components, so that $C(z)$ applies zero forcing to each of the two factors of $Y(z)$, and for practical purposes (but not exactly) $C(z)$ applies zero forcing also to $Y(z)$ itself. It follows that zero forcing can effectively be applied to *any* linear baseband channel that can be equalized linearly. However, with a time-varying channel there is not necessarily a simple algorithm for holding the tap gains correctly adjusted for the channel[53,59].

The general arrangement just described achieves the separate linear equalization of the two factors $Y_1(z)$ and $Y_2(z)$ of the channel z-

transform. When the peak distortion in the received signal is less than unity, the arrangement tends to minimize the peak distortion in the equalized signal. It is in general the most suitable design technique for the linear equalization of a known time-variant baseband channel, where a low peak distortion is required in the equalized signal.

The technique that has just been described may be further generalized as follows. Let

$$Y(z) = Y_1(z) Y_2(z) \ldots Y_n(z) \tag{4.101}$$

where $0 < n \leqslant g$, and $Y_i(z)$, for each i, is a polynomial with real- or complex-valued coefficients and *all* of whose zeros are either inside or outside the unit circle in the z plane. Determine, for each i, the z-transform $C_i(z)$ of the linear equalizer for $Y_i(z)$, as previously described. The z-transform of the linear equalizer for $Y(z)$ is now

$$C(z) = C_1(z) C_2(z) \ldots C_n(z) \tag{4.102}$$

and the z-transform of the channel and linear equalizer is

$$Y(z) C(z) \simeq z^{-h} \tag{4.103}$$

where h is an appropriate integer.

The equalizer here does *not* apply zero forcing to the sampled impulse-response of the equalized channel, except when it becomes identical to the arrangement previously described, nor (with this exception) does it normally achieve as low a peak distortion in the equalized signal, for a given number of taps in the transversal filter, as does the other arrangement. It is also more difficult to evaluate and there is no simple algorithm for holding the tap gains correctly adjusted for a time-varying channel, that operates according to this technique. The arrangement does not therefore appear to be of much practical value. It does however help to clarify an important property of linear equalizers.

It follows from Equation 4.101 that

$$Y^{-1}(z) = Y_1^{-1}(z) Y_2^{-1}(z) \ldots Y_n^{-1}(z) \tag{4.104}$$

for *any* selection of the n factors of $Y(z)$. Furthermore, if a practical equalizer is designed to satisfy Equation 4.103 *fairly accurately* for *given* and appropriate values of h and m, then it does not matter which of the possible combinations of the factors of $Y(z)$ are used in the design of this equalizer; the resultant equalizer will, for practical purposes, be the *same*. In other words, the equalizer is effectively determined by $Y(z)$, h and m, and not by the particular design technique.

182 LINEAR EQUALIZERS

Suppose next that, in Equation 4.101,

$$Y_i(z) = 1 + yz^{-1} \qquad (4.105)$$

where $|y| = 1$. It has been shown (near the beginning of this section) that, when $y = 1$, $Y_i(z)$ cannot be equalized by a *finite* linear feedforward transversal filter, in the sense that the *peak distortion* in the equalized signal cannot be reduced to an acceptably low value. This result may readily be extended to the more general case considered here, where y may be any complex number such that $|y| = 1$, which means, of course, that $Y_i(z)$ has a zero (root) at any required point *on* the unit circle in the z plane. Under these conditions, $Y_i(z)$ again cannot be equalized by a linear feedforward transversal filter to reduce the *peak distortion* in the equalized signal to an acceptably low value. Clearly, if $Y_i(z)$ in Equation 4.101 cannot be equalized correctly, then neither can $Y(z)$. Furthermore, $Y(z)$ cannot now be equalized correctly through the choice of any *other* set of factors in Equation 4.101.

It is shown in Section 3.6 that, if $Y(z)$ contains one or more zeros on the unit circle, the channel introduces infinite attenuation at certain discrete frequencies, which means that the corresponding frequency components of the transmitted signal are lost completely. When this happens, no amount of *linear* signal processing at the receiver can restore the lost frequencies. The linear equalization of the channel in fact requires a linear filter that introduces *infinite gain* at these frequencies, which is not physically realizable. In the case of a linear feedforward transversal filter, the infinite gain needed must be achieved through an infinite number of taps. The channel cannot therefore be equalized linearly, no matter what type of linear filter is used, if *exact* equalization is required. It is not just linear feedforward transversal equalizers that cannot be used. However, if a linear feedforward transversal equalizer is designed to minimize the *mean-square error* (Equation 4.34) or *mean-square distortion* (Equation 4.31) in the equalized signal and if the corresponding one of these two quantities is then taken as a measure of the *accuracy* of equalization (see Section 4.11), it is possible, with a sufficient number of taps, to approach quite closely to the reasonably accurate equalization of such a channel. Unfortunately, the very high gain introduced in the neighbourhood of the severely attenuated frequencies correspondingly *enhances* any *noise* present at these frequencies, and it is this effect that, in practice, prevents the use of a linear equalizer here. The effects of the noise on the equalized signal are considered further in Section 4.9.

4.6 EXAMPLES

Problem 1

The sampled impulse-response of channel A is

$$Y_A = [1 \quad -0.2] \quad (4.106)$$

and the sampled impulse response of channel B is

$$Y_B = [0.2 \quad 0.96 \quad -0.2] \quad (4.107)$$

Evaluate the tap gains of a 5-tap linear feedforward transversal equalizer for each channel.

Solution

The z-transform of the sampled impulse-response of channel A is

$$Y_A(z) = 1 - 0.2z^{-1} \quad (4.108)$$

whose one zero is at $z = 0.2$ and so lies inside the unit circle in the z plane. Thus $Y_A^{-1}(z)$ can be determined through the long division of 1 by $Y_A(z)$, in the manner previously described. This leads to the result that the z-transform of the required 5-tap linear feedforward transversal equalizer for the channel is

$$C(z) = 1 + 0.2z^{-1} + 0.04z^{-2} + 0.008z^{-3} + 0.0016z^{-4} \quad (4.109)$$

which means that the tap gains of the equalizer are 1, 0.2, 0.04, 0.008 and 0.0016.

The z-transform of the sampled impulse-response of channel B is

$$Y_B(z) = 0.2 + 0.96z^{-1} - 0.2z^{-2} = (0.2 + z^{-1})(1 - 0.2z^{-1}) \quad (4.110)$$

whose zeros are at $z = -5$ and $z = 0.2$. Clearly, $Y_B(z)$ has one zero inside the unit circle and one zero outside. Thus $Y_B^{-1}(z)$ cannot be determined through the long division of 1 by $Y_B(z)$ and the equalizers for the two factors of $Y_B(z)$ must be determined separately. The equalizer for $1 - 0.2z^{-1}$ has already been determined and the equalizer for $0.2 + z^{-1}$ is obtained by considering first the equalizer for

$$M(z) = 1 + 0.2z^{-1} \quad (4.111)$$

Since the zero of $M(z)$ is at $z = -0.2$ and so lies inside the unit circle, $M^{-1}(z)$ can be determined through the long division of 1 by $M(z)$. The process of long division is carried out as previously described and shows that the z-transform of a 5-tap linear feedforward transversal equalizer for a channel with z-transform $M(z)$ is

$$N(z) = 1 - 0.2z^{-1} + 0.04z^{-2} - 0.008z^{-3} + 0.0016z^{-4} \quad (4.112)$$

Since $0.2 + z^{-1}$ is obtained from $M(z)$ by reversing the order of its coefficients, the z-transform of the equalizer for $0.2 + z^{-1}$ is obtained from $N(z)$ by reversing the order of its coefficients, to give

$$D(z) = 0.0016 - 0.008z^{-1} + 0.04z^{-2} - 0.2z^{-3} + z^{-4} \quad (4.113)$$

this equalizer being a 5-tap linear feedforward transversal filter, with tap gains 0.0016, -0.008, 0.04, -0.2 and 1.

184 LINEAR EQUALIZERS

The z-transform of the linear equalizer for channel B is therefore

$$C(z)D(z) = 0.0016 - 0.0077z^{-1} + 0.0385z^{-2} - 0.1923z^{-3} + 0.9615z^{-4}$$
$$+ 0.1923z^{-5} + 0.0385z^{-6} + 0.0077z^{-7} + 0.0016z^{-8} \qquad (4.114)$$

To obtain the required 5-tap linear feedforward transversal equalizer for channel B, the four smallest terms in $C(z)D(z)$ are omitted to give the z-transform

$$E(z) = 0.0385 - 0.1923z^{-1} + 0.9615z^{-2} + 0.1923z^{-3} + 0.0385z^{-4} \qquad (4.115)$$

which means that the tap gains of the qualizer are 0.0385, -0.1923, 0.9615, 0.1923 and 0.0385.

Alternatively, and much more simply, express $C(z)$ and $D(z)$ to only three terms each, to give

$$C(z) = 1 + 0.2z^{-1} + 0.04z^{-2} \qquad (4.116)$$

and
$$D(z) = 0.04 - 0.2z^{-1} + z^{-2} \qquad (4.117)$$

The z-transform of the 5-tap linear feedforward transversal equalizer for channel B now becomes

$$E(z) = C(z)D(z) = 0.04 - 0.192z^{-1} + 0.9616z^{-2} + 0.192z^{-3} + 0.04z^{-4}$$
$$(4.118)$$

which means that the tap gains of the equalizer are 0.04, -0.192, 0.9616, 0.192 and 0.04. This agrees well with the previous result.

Problem 2
A binary polar data signal is transmitted over a linear baseband channel whose sampled impulse-response has the z-transform

$$Y(z) = 1 + 0.2z^{-1} + 0.1z^{-2} \qquad (4.119)$$

Use the method of long division to determine the tap gains of a 4-tap linear feedforward transversal equalizer for the channel.

Determine the sampled impulse-response of the channel and equalizer, and hence evaluate the peak distortion in the equalized signal.

What are the tap gains of a 4-tap linear feedforward transversal equalizer for the channel with z-transform

$$V(z) = 0.1 + 0.2z^{-1} + z^{-2} \qquad (4.120)$$

and what is the peak distortion of the equalized signal here?

Solution
Since each of the coefficients of z^{-1} and z^{-2} has a magnitude (absolute value) that is much less than unity, it is clear that the two zeros of $Y(z)$ must lie *inside* the unit circle in the z plane, so that $Y^{-1}(z)$ may be determined through the long division of 1 by $Y(z)$. This is achieved as follows.

LINEAR EQUALIZERS

$$1+0.2z^{-1}+0.1z^{-2} \overline{\smash{\big)}\begin{array}{l} 1-0.2z^{-1}-0.06z^{-2}+0.032z^{-3}-\cdots \\ 1 \\ \underline{1+0.2z^{-1}+0.1z^{-2}} \\ -0.2z^{-1}-0.1z^{-2} \\ \underline{-0.2z^{-1}-0.04z^{-2}-0.02z^{-3}} \\ -0.06z^{-2}+0.02z^{-3} \\ \underline{-0.06z^{-2}-0.012z^{-3}-0.006z^{-4}} \\ 0.032z^{-3}+0.006z^{-4} \end{array}}$$

Thus the z-transform of the required 4-tap linear feedforward transversal equalizer for the channel is

$$C(z) = 1 - 0.2z^{-1} - 0.06z^{-2} + 0.032z^{-3} \tag{4.121}$$

which means that the tap gains of the equalizer are 1, -0.2, -0.06 and 0.032. The z-transform of the channel and equalizer in cascade is

$$Y(z)C(z) = (1 + 0.2z^{-1} + 0.1z^{-2})(1 - 0.2z^{-1} - 0.06z^{-2} + 0.032z^{-3})$$
$$= 1 + 0.0004z^{-4} + 0.0032z^{-5} \tag{4.122}$$

Thus the sampled impulse-response of the channel and equalizer is given by the vector

$$E = [e_0 \quad e_1 \quad \ldots \quad e_5] \tag{4.123}$$

where
$$e_0 = 1$$
$$e_1 = e_2 = e_3 = 0$$
$$e_4 = 0.0004$$
$$e_5 = 0.0032 \tag{4.124}$$

and the peak distortion in the equalized signal is

$$\frac{1}{|e_0|} \sum_{i=1}^{5} |e_i| = 0.0004 + 0.0032 = 0.0036 \tag{4.125}$$

Since
$$V(z) = 0.1 + 0.2z^{-1} + z^{-2} \tag{4.126}$$

is obtained from $Y(z)$ by reversing the order of the coefficients, the z-transform $D(z)$, of the 4-tap linear feedforward transversal equalizer for the channel with z-transform $V(z)$, is obtained from $C(z)$ by reversing the order of its coefficients. Thus

$$D(z) = 0.032 - 0.06z^{-1} - 0.2z^{-2} + z^{-3} \tag{4.127}$$

which means that the tap gains of the equalizer are 0.032, -0.06, -0.2 and 1. The z-transform of the channel and equalizer in cascade is now

$$V(z)D(z) = (0.1 + 0.2z^{-1} + z^{-2})(0.032 - 0.06z^{-1} - 0.2z^{-2} + z^{-3})$$
$$= 0.0032 + 0.0004z^{-1} + z^{-5} \tag{4.128}$$

186 LINEAR EQUALIZERS

where again $V(z)D(z)$ is derived from $Y(z)C(z)$ simply by reversing the order of the coefficients.

The sampled impulse-response of the channel and equalizer is here given by the vector

$$E = [e_0 \; e_1 \; \ldots \; e_5] \qquad (4.129)$$

where
$$e_5 = 1$$
$$e_4 = e_3 = e_2 = 0$$
$$e_1 = 0.0004$$
$$e_0 = 0.0032 \qquad (4.130)$$

The component values of E in Equation 4.129 are obtained from those of E in Equation 4.123 simply be reversing their order, and the peak distortion in the equalized signal is

$$\frac{1}{|e_5|} \sum_{i=0}^{4} |e_i| = 0.0032 + 0.0004 = 0.0036 \qquad (4.131)$$

as in Equation 4.125.

Problem 3

A baseband channel has a sampled impulse-response whose z-transform is

$$Y(z) = 1 - \tfrac{1}{2}z^{-1} + 4z^{-2} - 2z^{-3} \qquad (4.132)$$

Design a 9-tap linear feedforward transversal equalizer for this channel.

Solution

$$Y(z) = (1 - \tfrac{1}{2}z^{-1})(1 + 4z^{-2})$$
$$= (1 - \tfrac{1}{2}z^{-1})(1 + 2jz^{-1})(1 - 2jz^{-1}) \qquad (4.133)$$

where $j = \sqrt{-1}$. Let

$$Y_1(z) = 1 - \tfrac{1}{2}z^{-1} \qquad (4.134)$$

and
$$Y_2(z) = 1 + 4z^{-2} \qquad (4.135)$$

Clearly, the zero of $Y_1(z)$ lies inside the unit circle in the z plane, whereas the two zeros of $Y_2(z)$ both lie outside the unit circle. The equalizers for these two factors must therefore be determined separately.

The z-transform $C_1(z)$ of the 5-tap linear equalizer for $Y_1(z)$ is determined through the long division of 1 by $Y_1(z)$, to give

$$C_1(z) = 1 + \tfrac{1}{2}z^{-1} + \tfrac{1}{4}z^{-2} + \tfrac{1}{8}z^{-3} + \tfrac{1}{16}z^{-4} \qquad (4.136)$$

which is such that

$$Y_1(z)C_1(z) \simeq 1 \qquad (4.137)$$

The z-transform $C_2(z)$ of the linear equalizer for $Y_2(z)$ is determined as follows. The z-transform obtained from $Y_2(z)$ by reversing the order of its coefficients is

$$M(z) = 4 + z^{-2} \qquad (4.138)$$

LINEAR EQUALIZERS

and the two zeros of $M(z)$ lie inside the unit circle in the z plane. Thus the z-transform of the linear equalizer for $M(z)$ is determined through the long division of 1 by $M(z)$, to give

$$M^{-1}(z) = \tfrac{1}{4} - \tfrac{1}{16}z^{-2} + \tfrac{1}{64}z^{-4} - \cdots \tag{4.139}$$

The z-transform obtained from $M^{-1}(z)$ by taking its first three terms and reversing the order of the coefficients is

$$C_2(z) = \tfrac{1}{64} - \tfrac{1}{16}z^{-2} + \tfrac{1}{4}z^{-4} \tag{4.140}$$

and this is the z-transform of the 3-tap linear equalizer for $Y_2(z)$, such that

$$Y_2(z)C_2(z) \simeq z^{-6} \tag{4.141}$$

Notice that both $C_1(z)$ and $C_2(z)$ have been selected to have z^{-4} as the highest power of z^{-1}, which means that each of the corresponding transversal filters has effectively 5 taps, although, in the case of the filter with z-transform $C_2(z)$, two of these taps have zero gain. The z-transform $C(z)$ of the 9-tap transversal equalizer for the given baseband channel is now

$$C(z) = C_1(z)C_2(z) = \tfrac{1}{64}(1 + \tfrac{1}{2}z^{-1} - 3\tfrac{3}{4}z^{-2} - 1\tfrac{7}{8}z^{-3} + 15\tfrac{1}{16}z^{-4} + 7\tfrac{1}{2}z^{-5}$$
$$+ 3\tfrac{3}{4}z^{-6} + 2z^{-7} + z^{-8}) \tag{4.142}$$

The z-transform of the channel and linear equalizer is

$$Y(z)C(z) = (1 - \tfrac{1}{2}z^{-1} + 4z^{-2} - 2z^{-3})\tfrac{1}{64}(1 + \tfrac{1}{2}z^{-1} - 3\tfrac{3}{4}z^{-2}$$
$$- 1\tfrac{7}{8}z^{-3} + 15\tfrac{1}{16}z^{-4} + 7\tfrac{1}{2}z^{-5} + 3\tfrac{3}{4}z^{-6} + 2z^{-7} + z^{-8})$$
$$= \tfrac{1}{64}(1 - \tfrac{1}{32}z^{-5} + 64z^{-6} - 2z^{-11})$$
$$\simeq \tfrac{1}{64} + z^{-6} - \tfrac{1}{32}z^{-11} \tag{4.143}$$

Thus the $(i+1)$th received signal-element has a z-transform that approximates to

$$s_i z^{-i}(\tfrac{1}{64} + z^{-6} - \tfrac{1}{32}z^{-11}) = s_i(\tfrac{1}{64}z^{-i} + z^{-i-6} - \tfrac{1}{32}z^{-i-11}) \tag{4.144}$$

and is detected from the sign of the signal

$$x_{i+6} \simeq -\tfrac{1}{32}s_{i-5} + s_i + \tfrac{1}{64}s_{i+6} + u_{i+6} \simeq s_i + u_{i+6} \tag{4.145}$$

at the output of the equalizer, at time $t = (i+6)T$. u_{i+6} is a Gaussian noise component.

It would, of course, have been possible to take

$$C_1(z) = 1 + \tfrac{1}{2}z^{-1} + \tfrac{1}{4}z^{-2} + \tfrac{1}{8}z^{-3} + \tfrac{1}{16}z^{-4} + \tfrac{1}{32}z^{-5} + \tfrac{1}{64}z^{-6} \tag{4.146}$$
$$C_2(z) = -\tfrac{1}{16} + \tfrac{1}{4}z^{-2} \tag{4.147}$$

or else

$$C_1(z) = 1 + \tfrac{1}{2}z^{-1} + \tfrac{1}{4}z^{-2} \tag{4.148}$$
$$C_2(z) = -\tfrac{1}{256} + \tfrac{1}{64}z^{-2} - \tfrac{1}{16}z^{-4} + \tfrac{1}{4}z^{-6} \tag{4.149}$$

since each of these combinations of $C_1(z)$ and $C_2(z)$ gives 9 taps for the resultant equalizer with z-transform $C(z)$. However, the evaluation of the z-transform of the equalized channel for each of these combinations shows a

188 LINEAR EQUALIZERS

greater value of the peak distortion in the equalized signal than for the chosen combination. Whenever the tap gains of an equalizer have been evaluated, it is important to check the sampled impulse-response of the equalized channel in order to ensure that the required accuracy of equalization has been achieved.

4.7 FEEDBACK TRANSVERSAL FILTER

The transversal equalizers considered so far are linear feedforward transversal filters of the type normally used in practice. Equalization of a channel may alternatively be achieved by means of a *feedback transversal filter*, provided that certain conditions are satisfied by the channel response. Let the z-transform of the sampled impulse-response of the channel be

$$Y(z) = y_0 + y_1 z^{-1} + y_2 z^{-2} + \cdots + y_g z^{-g} \quad (4.150)$$

where $y_0 \neq 0$, and let $V(z)$ be the z-transform obtained by scaling $Y(z)$ to make the first term equal to unity. Thus

$$V(z) = (1/y_0) Y(z) = 1 + v_1 z^{-1} + v_2 z^{-2} + \cdots + v_g z^{-g}$$
$$= (1 + \beta_1 z^{-1})(1 + \beta_2 z^{-1}) \ldots (1 + \beta_g z^{-1}) \quad (4.151)$$

where
$$v_i = y_i / y_0 \quad (4.152)$$

for $i = 1, 2, \ldots, g$. The $\{\beta_i\}$ may be real- or complex-valued and are the *negatives* of the g zeros of $V(z)$, and are therefore also the negatives of the g zeros of $Y(z)$.

The linear feedback transversal equalizer for the channel is as shown in Fig. 4.10. The signals shown here are those present at the

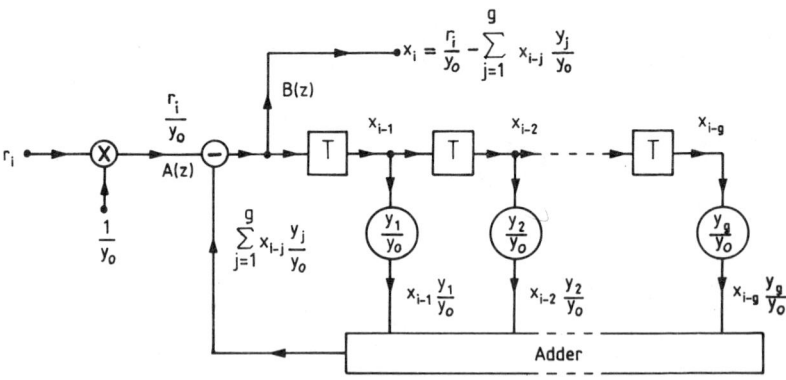

Fig. 4.10   Linear feedback transversal equalizer for a channel with z-transform Y(z)

time instant $t=iT$. It can be seen that if an individual signal-element $s_i\delta(t-iT)$ is fed to the input of the baseband channel in Fig. 4.1, the resultant sequence of signals (samples) at the output of the channel, in the absence of noise, are

$$s_i y_0, s_i y_1, \ldots, s_i y_g \qquad (4.153)$$

at the time instants $iT, (i+1)T, \ldots, (i+g)T$, respectively, the preceding and following signals being all zero. The operation of the feedback transversal filter on the received sequence of signals is to give at its output a single nonzero signal $x_i = s_i$, at the time instant $t=iT$, this being preceded and followed by zero signals (samples), assuming that all signals in the equalizer were initially set to zero. The mechanism whereby the above result is obtained can be explained as follows. At time $t=iT$, the input signal to the equalizer is $s_i y_0$, which, at the output of the multiplier ($\times 1/y_0$), becomes s_i. The signal subtracted from s_i in the subtractor is zero, to give the signal s_i at the output of the equalizer and at the input to the first store (the left-hand square marked T). At time $t=(i+1)T$, the signal s_i that was previously at the input of the first store of the equalizer is moved to the output of this store, where it is multiplied by y_1/y_0 to give the signal $s_i y_1/y_0$, which is subtracted from the signal at the input to the subtractor, that is now $s_i y_1/y_0$. This gives a zero output signal and a zero signal at the input to the first store, the remaining stored signals (excluding s_i) all being zero. At time $t=(i+2)T$, the signal s_i is moved to the output of the second store and the input zero signal is moved to the output of the first store. The signal s_i is now multiplied by y_2/y_0 to give the signal $s_i y_2/y_0$, which is subtracted from the signal $s_i y_2/y_0$ at the input to the subtractor, to give again a zero output signal. The process continues in this way until time $t=(i+g+1)T$ and subsequently, when *all* signals associated with the equalizer (input, output and stored) are zero. Thus the filter clearly equalizes the channel. Furthermore, the equalization is exact and requires only g taps, which is a far smaller number than that often required by the corresponding feedforward transversal equalizer.

Consider now the feedback transversal equalizer, shown in Fig. 4.11, for a channel with z-transform

$$Y(z) = 1 + \beta_i z^{-1} \qquad (4.154)$$

Let the z-transforms of the sequences at the input and output of the equalizer in Fig. 4.11 be $A(z)$ and $B(z)$, respectively. Then

$$B(z) = A(z) - \beta_i z^{-1} B(z) \qquad (4.155)$$

so that

$$B(z)(1 + \beta_i z^{-1}) = A(z) \qquad (4.156)$$

190 LINEAR EQUALIZERS

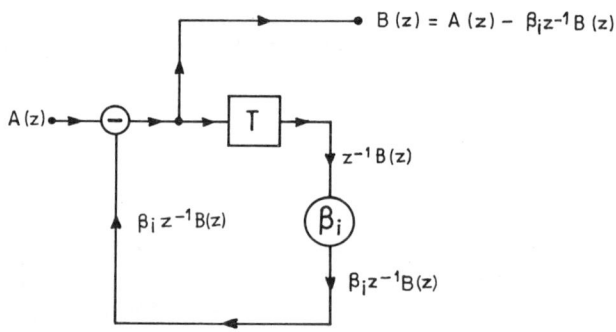

Fig. 4.11 Linear feedback transversal equalizer for a channel with z-transform $1 + \beta_i z^{-1}$

Now, when $|\beta_i| < 1$,

$$B(z) = A(z)(1 + \beta_i z^{-1})^{-1}$$
$$= A(z)(1 - \beta_i z^{-1} + \beta_i^2 z^{-2} - \beta_i^3 z^{-3} + \cdots) \quad (4.157)$$

the result being obtained either by applying the binomial expansion to $(1 + \beta_i z^{-1})^{-1}$ (which is valid since $|\beta_i z^{-1}| < 1$) or else through the division of 1 by $1 + \beta_i z^{-1}$ using a normal process of long division. However, when $|\beta_i| > 1$,

$$(1 + \beta_i z^{-1})^{-1} = \beta_i^{-1} z (1 + \beta_i^{-1} z)^{-1}$$
$$= \beta_i^{-1} z (1 - \beta_i^{-1} z + \beta_i^{-2} z^2 - \beta_i^{-3} z^3 + \cdots)$$
$$= \beta_i^{-1} z - \beta_i^{-2} z^2 + \beta_i^{-3} z^3 - \beta_i^{-4} z^4 + \cdots \quad (4.158)$$

But, clearly, now

$$B(z) \neq A(z)(1 + \beta_i z^{-1})^{-1} \quad (4.159)$$

since, from Equation 4.158, the output signal *precedes* the input signal, which is not physically realizable. In any case it does *not* describe the actual operation of the equalizer. It is evident from Fig. 4.11 that the sampled impulse-response of the equalizer is

$$[1 \quad -\beta_i \quad \beta_i^2 \quad -\beta_i^3 \quad \ldots] \quad (4.160)$$

with z-transform

$$1 - \beta_i z^{-1} + \beta_i^2 z^{-2} - \beta_i^3 z^{-3} + \cdots \quad (4.161)$$

so that, in fact,

$$B(z) = A(z)(1 - \beta_i z^{-1} + \beta_i^2 z^{-2} - \beta_i^3 z^{-3} + \cdots) \quad (4.162)$$

Thus the z-transform of the equalizer is obtained from the long division of 1 by $1+\beta_i z^{-1}$, *regardless* of the value of β_i, and it is written as $1/(1+\beta_i z^{-1})$. This is here taken to imply the process of *long division*. When $|\beta_i|<1$,

$$\frac{1}{1+\beta_i z^{-1}} = (1+\beta_i z^{-1})^{-1} \qquad (4.163)$$

and, when $|\beta_i|>1$,

$$\frac{1}{1+\beta_i z^{-1}} \neq (1+\beta_i z^{-1})^{-1} \qquad (4.164)$$

bearing in mind that, by definition,

$$(1+\beta_i z^{-1})(1+\beta_i z^{-1})^{-1} = 1 \qquad (4.165)$$

The above results apply for both real and complex values of β_i.

The fact that, when $|\beta_i|>1$, the operation of the feedback equalizer is given by Equation 4.162 and not by Equation 4.158, shows that the feedback equalizer is *not* now equivalent to the ideal linear equalizer for $1+\beta_i z^{-1}$, and in fact it performs a quite different operation on its input signal.

In the absence of noise and with the equalizer tap gain adjusted to its *exact* value, the equalizer achieves exact equalization regardless of whether $|\beta_i|<1$ or $|\beta_i|>1$, but just so long as the signal stored in the equalizer at the start of the equalization process has been set to zero. However, when $|\beta_i|>1$, any error in the equalizer tap gain (that is, any departure from β_i) or any noise added to the data signal at the input to the equalizer, however small its magnitude, results in a sequence of error or noise components of steadily increasing magnitudes, in the output signal from the equalizer, so that the system is *unstable*.

Consider now the feedback transversal equalizer in Fig. 4.10 for a channel with z-transform $Y(z)$ in Equation 4.150, where, as before,

$$Y(z) = y_0 V(z) \qquad (4.166)$$

and $V(z)$ is given by Equations 4.151 and 4.152. Let the z-transforms of the sequences of signals (samples) at the input and output of the linear feedback transversal filter in Fig. 4.10 be $A(z)$ and $B(z)$, respectively, as shown. Thus

$$B(z) = A(z) - \sum_{i=1}^{g} v_i z^{-i} B(z) \qquad (4.167)$$

or

$$A(z) = B(z)\left(1 + \sum_{i=1}^{g} v_i z^{-i}\right) = B(z)V(z) \qquad (4.168)$$

so that

$$B(z) = A(z)/V(z) \qquad (4.169)$$

and the z-transform of the linear feedback transversal equalizer in Fig. 4.10 (excluding the multiplier) is

$$\frac{1}{V(z)} = \frac{1}{1 + v_1 z^{-1} + v_2 z^{-2} + \cdots + v_g z^{-g}} \quad (4.170)$$

Now, from Equation 4.151[167-169],

$$\frac{1}{V(z)} = \frac{1}{(1 + \beta_1 z^{-1})(1 + \beta_2 z^{-1}) \ldots (1 + \beta_g z^{-1})}$$

$$= \frac{1}{1 + \beta_1 z^{-1}} \cdot \frac{1}{1 + \beta_2 z^{-1}} \cdots \frac{1}{1 + \beta_g z^{-1}} \quad (4.171)$$

This means that the equalizer in Fig. 4.10 may alternatively be implemented as a sequence of g equalizers, with z-transforms $\{1/(1 + \beta_i z^{-1})\}$, connected in cascade, the resultant equalizer having exactly the same performance as that in Fig. 4.10. The $\{\beta_i\}$ may, of course, be real- or complex-valued. It follows that the equalizer in Fig. 4.10 is stable if and only if the equalizer in Fig. 4.11 is stable for every value of i in the range 1 to g, that is, if and only if *all* the zeros of $Y(z)$ lie *inside* the unit circle in the z plane. When this condition is satisfied, the z-transform

$$1 \bigg/ \left(1 + \frac{y_1}{y_0} z^{-1} + \frac{y_2}{y_0} z^{-2} + \cdots + \frac{y_g}{y_0} z^{-g}\right) \quad (4.172)$$

of the equalizer in Fig. 4.10 becomes equal to

$$\left(1 + \frac{y_1}{y_0} z^{-1} + \frac{y_2}{y_0} z^{-2} + \cdots + \frac{y_g}{y_0} z^{-g}\right)^{-1} \quad (4.173)$$

and the equalizer is both stable and achieves the accurate equalization of the channel.

To summarize, if one or more zeros of the z-transform $Y(z)$ of the baseband channel in Fig. 4.1 lie *outside* or *on* the unit circle in the z plane, a linear feedback transversal equalizer is unstable and so cannot be used, whether it is implemented as in Fig. 4.10 or as a sequence of separate filters, the ith of which is shown in Fig. 4.11. When all the zeros of $Y(z)$ lie inside the unit circle, the linear feedback transversal equalizer is always stable. It now gives *exact* equalization and is often much smaller than the equivalent feedforward filter.

When the baseband channel introduces appreciable amplitude distortion and is time invariant and known, the linear equalizer giving the most accurate equalization for a given number of taps is usually that where the factor $Y_1(z)$ of the z-transform of the channel (Equation 4.94) whose zeros lie inside the unit circle in the z plane, is equalized by means of a linear feedback transversal filter, while the factor $Y_2(z)$, whose zeros lie outside the unit circle, is equalized by a linear feedforward transversal filter.[60] Linear feedback transversal filters are not, however, generally suitable for use as adaptive equalizers for time-varying channels, because of the risk of instability.

4.8 EXAMPLES

Problem 1
Describe a modification to the equalizer in Fig. 4.8 for the channel with z-transform

$$Y(z) = (1 + \tfrac{1}{2}z^{-1})(1 + 2z^{-1}) = 1 + 2\tfrac{1}{2}z^{-1} + z^{-2} \qquad (4.174)$$

that leads to a simpler equalizer but without degrading its performance.

Solution
Figure 4.8 shows a linear equalizer for this channel, comprising two linear feedforward transversal filters each of which equalizes the corresponding factor of $Y(z)$. The first of the two transversal filters equalizes the factor $1 + \tfrac{1}{2}z^{-1}$ whose zero is at $z = -\tfrac{1}{2}$ and so lies inside the unit circle. This factor of $Y(z)$ may be equalized instead by the linear feedback transversal filter shown in Fig. 4.12. The output of the feedback filter is fed to the input of the second of the two feedforward transversal filters in Fig. 4.8, to give the complete equalizer for the channel. The resulting linear equalizer has just 7 taps instead of the 11 taps required by a *single* linear feedforward transversal equalizer for the channel (Equation 4.82). Furthermore, since the feedback filter equalizes the factor $1 + \tfrac{1}{2}z^{-1}$ *exactly*, the sampled impulse-response of the channel with z-transform $1 + 2\tfrac{1}{2}z^{-1} + z^{-2}$ and the 7-tap linear equalizer just described, is as shown in Fig. 4.7, instead of that shown in Fig. 4.9, which holds for the given channel and the 11-tap linear feedforward transversal equalizer. Thus, not only have the number of taps of the equalizer been reduced from 11 to 7 but the peak distortion in the equalized signal has been halved.

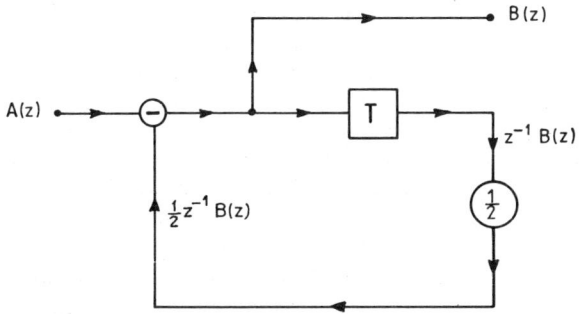

Fig. 4.12 Linear feedback transversal equalizer for the factor $1 + \tfrac{1}{2}z^{-1}$ of $Y(z)$

Problem 2

The z-transforms of two different channels are

$$Y_A(z) = 1 - 0.7z^{-1} + 0.1z^{-2} \qquad (4.175)$$

and
$$Y_B(z) = 0.1 - 0.7z^{-1} + z^{-2} \qquad (4.176)$$

Can either of the two channels be equalized by a linear feedback transversal equalizer, and, if so, what are the tap gains of the equalizer and what is the peak distortion in the equalized signal?

Solution

Now
$$Y_A(z) = (1 - 0.5z^{-1})(1 - 0.2z^{-1}) \qquad (4.177)$$

and
$$Y_B(z) = (0.5 - z^{-1})(0.2 - z^{-1}) \qquad (4.178)$$

so that the zeros of $Y_A(z)$ are at $z = 0.5$ and $z = 0.2$, and the zeros of $Y_B(z)$ are at $z = 2$ and $z = 5$.

The zeros of $Y_B(z)$ may alternatively be derived as follows. Since the coefficients in $Y_B(z)$ are obtained from those in $Y_A(z)$ by *reversing their order*, the zeros of $Y_B(z)$ must be the *reciprocals* of those of $Y_A(z)$ and so must be as given.

Since both zeros of $Y_A(z)$ lie *inside* the unit circle in the z plane, the channel with z-transform $Y_A(z)$ can be equalized by a linear feedback transversal equalizer. On the other hand, since both zeros of $Y_B(z)$ lie *outside* the unit circle, the channel with z-transform $Y_B(z)$ cannot be equalized by a linear feedback transversal equalizer.

The linear feedback transversal equalizer for the channel with z-transform $Y_A(z)$ is as shown in Fig. 4.13, and its tap gains are -0.7 and 0.1. The equalizer achieves the exact equalization of the channel, so that the peak distortion here is zero. This can be seen by considering the z-transforms $A(z)$ and $B(z)$ of the input and output signals, as shown in the diagram. Thus

$$B(z) = A(z) + 0.7z^{-1}B(z) - 0.1z^{-2}B(z) \qquad (4.179)$$

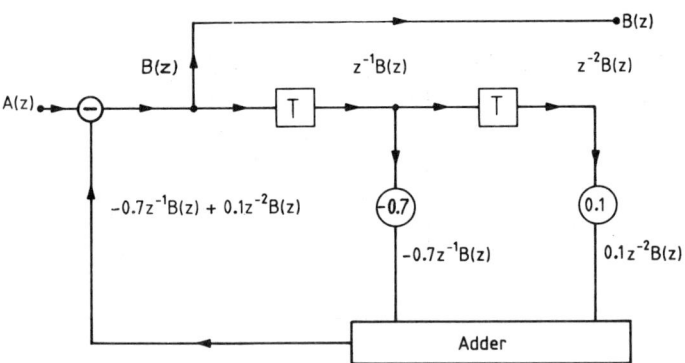

Fig. 4.13 Linear feedback transversal equalizer for the channel with z-transform $Y_A(z) = 1 - 0.7z^{-1} + 0.1z^{-2}$

or
$$B(z)(1-0.7z^{-1}+0.1z^{-2})=A(z) \quad (4.180)$$
or
$$\frac{B(z)}{A(z)}=\frac{1}{1-0.7z^{-1}+0.1z^{-2}}=(1-0.7z^{-1}+0.1z^{-2})^{-1} \quad (4.181)$$

Since $B(z)/A(z)$ is the z-transform of the equalizer, the equalizer is the exact inverse of the channel and so equalizes the channel exactly.

4.9 TOLERANCE TO ADDITIVE GAUSSIAN NOISE

Consider the linear feedforward transversal filter with $m+1$ taps, as shown in Fig. 4.14. Assume that the filter equalizes the baseband channel in Fig. 4.1 and that the signal at the output of the equalizer, at time $t=(i+h)T$, is

$$x_{i+h} \simeq s_i + u_{i+h} \quad (4.182)$$

where, as before,

$$s_i = \pm k \quad (4.183)$$

and u_{i+h} is a noise component. s_i is detected from the sign of x_{i+h}, so that when $x_{i+h}>0$, s_i is detected as k, and when $x_{i+h}<0$, s_i is detected as $-k$. The detected value of s_i is designated s'_i.

The noise components at different points in the transversal equalizer, at the time instant $t=(i+h)T$, are shown in Fig. 4.14, from which it can be seen that the output noise signal from the equalizer, at time $t=(i+h)T$, is

$$u_{i+h} = \sum_{j=0}^{m} w_{i+h-j} c_j \quad (4.184)$$

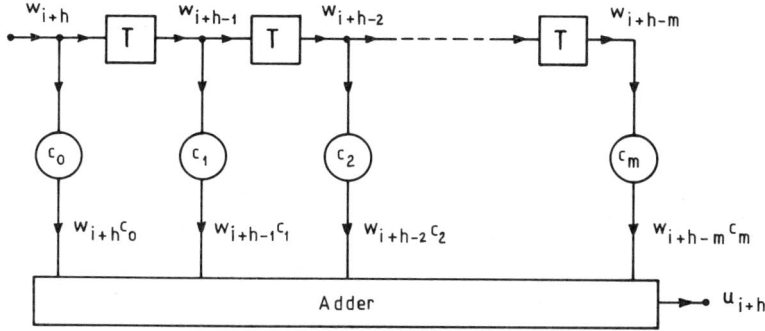

Fig. 4.14 Noise signals in an $(m+1)$-tap linear feedforward transversal equalizer at the time instant $t=(i+h)T$

196 LINEAR EQUALIZERS

The noise components $\{w_i\}$ are statistically independent Gaussian random variables with zero mean and variance σ^2, so that the $\{w_{i+h-j}c_j\}$ are uncorrelated Gaussian random variables with variances $\{\sigma^2 c_j^2\}$, and the variance of their sum is equal to the sum of their variances[163-165]. Thus u_{i+h} is a Gaussian random variable with zero mean and variance

$$\eta^2 = \sum_{j=0}^{m} \sigma^2 c_j^2 = \sigma^2 C C^T = \sigma^2 |C|^2 \qquad (4.185)$$

where $|C|$ is the *Euclidean length* of the $(m+1)$-component row vector

$$C = [c_0 \quad c_1 \quad \ldots \quad c_m] \qquad (4.186)$$

η^2 is, of course, also the *mean-square value* of u_{i+h}, and so is the *mean-square error* due to *noise* in x_{i+h}.

An error occurs in the detection of s_i from x_{i+h}, whenever x_{i+h} has the opposite sign to s_i, and therefore whenever u_{i+h} has a magnitude greater than k and the opposite sign to s_i.

The probability density function of the noise component u_{i+h}, at the sample-value u, is[163-165]

$$p(u) = \frac{1}{\sqrt{(2\pi\eta^2)}} \exp\left(-\frac{u^2}{2\eta^2}\right) \qquad (4.187)$$

and this is clearly an *even* function of u. It is now, for convenience, assumed that the channel is equalized *exactly*, such that exact equality holds in Equation 4.182. It is also assumed here and throughout this book that the data-symbol s_i and the noise component u_{i+h} are *statistically independent* for every i and h, as is normally the case in practice.

If $s_i = -k$, an error occurs in s_i' when $u_{i+h} > k$, and the probability of this occurring is

$$\int_k^\infty p(u) \, du = \int_k^\infty \frac{1}{\sqrt{2\pi\eta^2}} \exp\left(-\frac{u^2}{2\eta^2}\right) du$$

$$= \int_{k/\eta}^\infty \frac{1}{\sqrt{2\pi}} \exp\left(-\tfrac{1}{2}u^2\right) du = Q\left(\frac{k}{\eta}\right) \qquad (4.188)$$

where

$$Q(v) = \int_v^\infty \frac{1}{\sqrt{2\pi}} \exp\left(-\tfrac{1}{2}u^2\right) du \qquad (4.189)$$

If $s_i = k$, an error occurs in s_i' when $u_{i+h} < -k$, and the probability of this occurring is

$$\int_{-\infty}^{-k} p(u)\, du = \int_{-\infty}^{-k} \frac{1}{\sqrt{2\pi\eta^2}} \exp\left(-\frac{u^2}{2\eta^2}\right) du$$

$$= \int_{k}^{\infty} \frac{1}{\sqrt{2\pi\eta^2}} \exp\left(-\frac{u^2}{2\eta^2}\right) du = Q\left(\frac{k}{\eta}\right) \quad (4.190)$$

The above result follows from the fact that $p(u)$ is an *even* function of u.

It is evident from Equations 4.188 and 4.190 that the error probability is *independent* of the binary value of s_i. Thus, regardless of the binary value of s_i and therefore on average, even when s_i is not equally likely to have either value $\pm k$, the probability of error in s'_i is

$$P_e = Q\left(\frac{k}{\eta}\right) = Q\left(\frac{k}{\sigma|C|}\right) \quad (4.191)$$

using the result of Equation 4.185.

The tolerance of any system to additive Gaussian noise is taken to be the signal/noise ratio for some given error rate in the $\{s'_i\}$. The signal/noise ratio is here

$$20 \log_{10}(k/\sigma) \quad (4.192)$$

dB, as can be seen from Equations 4.9 and 4.10, and the error rate is P_e, which is normally taken to be quite low, say 10^{-5} or 10^{-6}. The *smaller* the signal/noise ratio for the particular error rate, the *greater* the tolerance of the system to noise. Thus the *advantage* in tolerance to noise of a given system over another is the *reduction* in signal/noise ratio for a particular error rate, when the given system replaces the other.

Consider now the case where the z-transform of the baseband channel in Fig. 4.1 is

$$Y(z) = 1 + 2\tfrac{1}{2}z^{-1} + z^{-2} \quad (4.193)$$

and the linear transversal equalizer for the channel has the z-transform

$$C(z) = -\tfrac{1}{64}(1 - 2\tfrac{1}{2}z^{-1} + 5\tfrac{1}{4}z^{-2} - 10\tfrac{5}{8}z^{-3} + 21\tfrac{5}{16}z^{-4} - 42\tfrac{21}{32}z^{-5}$$
$$+ 21\tfrac{5}{16}z^{-6} - 10\tfrac{5}{8}z^{-7} + 5\tfrac{1}{4}z^{-8} - 2\tfrac{1}{2}z^{-9} + z^{-10}) \quad (4.194)$$

so that the z-transform of the equalized channel is

$$Y(z)C(z) \simeq z^{-6} \quad (4.195)$$

as in Equation 4.83.

The data-symbol s_i is detected from the sign of the signal

$$x_{i+6} \simeq s_i + u_{i+6} \quad (4.196)$$

198 LINEAR EQUALIZERS

at the output of the equalizer, at time $t=(i+6)T$. The noise component u_{i+6} is a Gaussian random variable with zero mean and variance

$$\eta^2 = \frac{\sigma^2}{64^2}[1^2+(2\tfrac{1}{2})^2+(5\tfrac{1}{4})^2+(10\tfrac{5}{8})^2+(21\tfrac{5}{16})^2$$
$$+(42\tfrac{21}{32})^2+(21\tfrac{5}{16})^2+(10\tfrac{5}{8})^2+(5\tfrac{1}{4})^2+(2\tfrac{1}{2})^2+1^2]$$
$$=0.738\sigma^2 \qquad (4.197)$$

so that

$$\eta = 0.859\sigma \qquad (4.198)$$

From Equation 4.191, the probability of error in the detection of s_i from x_{i+6} is approximately

$$P_e = Q\left(\frac{k}{\eta}\right) = Q\left(\frac{k}{0.859\sigma}\right) = Q\left(\frac{1.164k}{\sigma}\right) \qquad (4.199)$$

If now *no* equalization is used, the z-transform of the $(i+1)$th received signal-element is

$$s_i z^{-i} Y(z) = s_i(z^{-i}+2\tfrac{1}{2}z^{-i-1}+z^{-i-2}) \qquad (4.200)$$

from Equation 4.193, so that the components (samples) of the $(i+1)$th signal-element are s_i, $2\tfrac{1}{2}s_i$ and s_i, at times iT, $(i+1)T$ and $(i+2)T$, respectively.

Clearly, s_i is best detected from the received sample containing the component $2\tfrac{1}{2}s_i$, that is, from the received sample

$$r_{i+1} = s_{i-1}+2\tfrac{1}{2}s_i+s_{i+1}+w_{i+1} \qquad (4.201)$$

at time $t=(i+1)T$. When $r_{i+1}>0$, s_i is detected as k, and when $r_{i+1}<0$, s_i is detected as $-k$.

At high signal/noise ratios with additive Gaussian noise, as assumed here, even a small increase in the distance to the decision threshold, in the detection of a signal, produces a very large reduction in the corresponding error probability. Thus practically all the errors occur when

$$s_{i-1} = s_{i+1} = -s_i \qquad (4.202)$$

since now the signal component in r_{i+1} has its *smallest* magnitude and is such that

$$r_{i+1} = \tfrac{1}{2}s_i + w_{i+1} \qquad (4.203)$$

An error occurs here in the detection of s_i when w_{i+1} has a magnitude greater than $\tfrac{1}{2}k$ and the opposite sign to s_i. Of course, w_{i+1} is a

Gaussian random variable with zero mean and variance σ^2, and w_{i+1} is statistically independent of s_i.

Since there is a probability of $\frac{1}{4}$ that Equation 4.202 is satisfied, the probability of error in the detection of s_i from r_{i+1} is approximately equal to a quarter of the conditional probability of error, given that Equation 4.202 is satisfied. Thus the probability of error in the detection of s_i from r_{i+1} is approximately

$$\frac{1}{4}\int_{k/2}^{\infty} \frac{1}{\sqrt{2\pi\sigma^2}} \exp\left(-\frac{w^2}{2\sigma^2}\right) dw = \frac{1}{4}\int_{k/2\sigma}^{\infty} \frac{1}{\sqrt{2\pi}} \exp(-\tfrac{1}{2}w^2) \, dw$$

$$= \tfrac{1}{4} Q\left(\frac{0.5k}{\sigma}\right) \qquad (4.204)$$

At error probabilities of around 1 in 10^5 or 1 in 10^6, a reduction in the error probability by a factor of four times is equivalent to an increase in tolerance to additive Gaussian noise of only a little over $\tfrac{1}{2}$ dB. Although this must obviously be taken into account in an accurate comparison of error probabilities, it may be neglected for our purposes here. Thus the probability of error in the detection of s_i from r_{i+1} can be taken to be

$$P_e = Q\left(\frac{0.5k}{\sigma}\right) \qquad (4.205)$$

Comparing Equations 4.199 and 4.205, it can be seen that for *given* values of P_e and k, the value of σ in the case where the linear equalizer is used is 2.33 times that where no equalizer is used. This means that the linear equalizer gains an advantage in tolerance to additive white Gaussian noise of 7.3 dB over the arrangement with no equalizer. Making an allowance for the omission of the factor 1/4 in the error probability for no equalizer, it can be seen that the equalizer increases the tolerance to additive white Gaussian noise by about 7 dB. This is a considerable improvement and is quite typical of the advantage that may be gained with a linear equalizer.

4.10 EXAMPLES

Problem 1

The $(i+1)$th sample of a binary baseband signal received over channel A is

$$q_i = s_i - 0.2s_{i-1} + w_i \qquad (4.206)$$

and the corresponding sample received over channel B is

$$r_i = 0.2s_i + 0.96s_{i-1} - 0.2s_{i-2} + w_i \qquad (4.207)$$

200 LINEAR EQUALIZERS

where the data-symbols $\{s_i\}$ are statistically independent and equally likely to have either value $\pm k$, and the $\{w_i\}$ are statistically independent Gaussian random variables with zero mean and variance σ^2.

Which channel gives the more reliable transmission of data when a 5-tap linear equalizer is used at the receiver?

Solution

The sampled impulse-response of channel A is

$$Y_A = [1 \quad -0.2] \tag{4.208}$$

with z-transform

$$Y_A(z) = 1 - 0.2z^{-1} \tag{4.209}$$

and the sampled impulse-response of channel B is

$$Y_B = [0.2 \quad 0.96 \quad -0.2] \tag{4.210}$$

with z-transform

$$Y_B(z) = 0.2 + 0.96z^{-1} - 0.2z^{-2} \tag{4.211}$$

It has been shown in Section 4.6 that the tap gains of the 5-tap linear feedforward transversal equalizer for channel A are

$$1, 0.2, 0.04, 0.008, 0.0016$$

and for channel B are

$$0.0385, -0.1923, 0.9615, 0.1923, 0.0385$$

In each case the noise variance at the output of the equalizer is equal to the variance of the input noise samples multiplied by the sum of the squares of the tap gains of the equalizer. Thus the noise variance at the output of the equalizer for channel A is

$$\eta_A^2 = \sigma^2(1^2 + 0.2^2 + 0.04^2 + 0.008^2 + 0.0016^2) = 1.0417\sigma^2 \tag{4.212}$$

and the noise variance at the output of the equalizer for channel B is

$$\eta_B^2 = \sigma^2(0.0385^2 + 0.1923^2 + 0.9615^2 + 0.1923^2 + 0.0385^2) = 1.0014\sigma^2 \tag{4.213}$$

For channel A, the equalized signal containg s_i is

$$x_i \simeq s_i + u_i \tag{4.214}$$

where u_i is a Gaussian random variable with zero mean and variance η_A^2. Thus the probability density function of u_i, at the sample value u, is the even function

$$p(u) = \frac{1}{\sqrt{2\pi\eta_A^2}} \exp\left(-\frac{u^2}{2\eta_A^2}\right) \tag{4.215}$$

.Now s_i', the detected value of s_i, is determined from the sign of x_i, such that, when $x_i > 0$, $s_i' = k$, and when $x_i < 0$, $s_i' = -k$. Thus an error occurs in s_i' when u_i has a magnitude greater than k and the opposite sign to s_i. Hence, regardless of the sign of s_i, the probability of error in s_i' is approximately

$$P_A = \int_k^\infty p(u)\,du = \int_k^\infty \frac{1}{\sqrt{2\pi\eta_A^2}} \exp\left(-\frac{u^2}{2\eta_A^2}\right) du$$

$$= \int_{k/\eta_A}^\infty \frac{1}{\sqrt{2\pi}} \exp\left(-\tfrac{1}{2}u^2\right) du$$

$$= Q\left(\frac{k}{\eta_A}\right) = Q\left(\frac{k}{1.0206\sigma}\right) = Q\left(\frac{0.9798k}{\sigma}\right) \qquad (4.216)$$

There is a small (but for our purposes negligible) error in this result caused by the fact that the channel is not equalized *exactly* by the linear equalizer.

For channel B, the equalized signal containing s_i is (from Section 4.6)

$$x_{i+3} \simeq s_i + u_{i+3} \qquad (4.217)$$

and, by a derivation similar to that above, it may be shown that the probability of error in s_i' is now approximately

$$P_B = Q\left(\frac{k}{\eta_B}\right) = Q\left(\frac{k}{1.0007\sigma}\right) = Q\left(\frac{0.9993k}{\sigma}\right) \qquad (4.218)$$

Comparing Equations 4.216 and 4.218, it can be seen that at a *given* low error probability and a *given* value of k, the value of σ for channel B is 0.9993/0.9798 times that for channel A. Thus, neglecting the small inaccuracies in the equalized signals, channel B gains an advantage of

$$20\log_{10}\left(\frac{0.9993}{0.9798}\right) = 0.17 \qquad (4.219)$$

dB in tolerance to noise over channel A and so gives the more reliable transmission of data. For practical purposes, however, the advantage gained is not very significant. It is however interesting to observe than channel B introduces *more* intersymbol interference into the received (unequalized) signal than does channel A. Furthermore, channel B introduces nearly pure phase distortion into the data signal, whereas channel A introduces similar levels of amplitude and phase distortions (see Chapter 3).

Problem 2

A binary polar data signal is transmitted over a linear baseband channel whose sampled impulse-response has the z-transform

$$Y(z) = 0.4 + z^{-1} \qquad (4.220)$$

Use the method of long division to determine the tap gains of a 6-tap linear feedforward transversal equalizer for the channel.

Determine the sampled impulse-response of the channel and equalizer and hence evaluate the peak distortion in the equalized signal.

What is the delay in detection of a received data symbol?

If the noise components $\{w_i\}$ in the received samples $\{r_i\}$ at the equalizer input are statistically independent Gaussian random variables with zero mean and variance σ^2, what is the noise variance in the equalized signal?

202 LINEAR EQUALIZERS

Suppose now that the given channel is replaced by one with the z-transform

$$X(z) = (1 + 0.4z^{-1})^{-1} \qquad (4.221)$$

without changing the noise statistics at the receiver input.

What is the z-transform, and hence what are the tap gains, of the linear feedforward transversal equalizer for this channel?

What is (a) the sampled impulse-response of the channel and equalizer, (b) the peak distortion in the equalized signal, (c) the delay in detection, and (d) the noise variance in the equalized signal?

Which cannel gives the more reliable transmission of data?

Solution

The zero of $Y(z)$ is at $z = -2.5$ and so lies *outside* the unit circle in the z plane. This means that $Y^{-1}(z)$ cannot be obtained through the long division of 1 by $Y(z)$. Let

$$M(z) = 1 + 0.4z^{-1} \qquad (4.222)$$

where $M(z)$ is obtained from $Y(z)$ by reversing the order of the coefficients. The zero of $M(z)$ is at $z = -0.4$, which is the reciprocal of the zero of $Y(z)$ and lies *inside* the unit circle. Thus $M^{-1}(z)$ can be obtained through the long division of 1 by $M(z)$, to give

$$M^{-1}(z) = 1 - 0.4z^{-1} + 0.16z^{-2} - 0.064z^{-3} + 0.0256z^{-4} - 0.01024z^{-5}$$
$$(4.223)$$

The z-transform of the 6-tap linear feedforward transversal equalizer for $Y(z)$ is now obtained from $M^{-1}(z)$ by taking its first six terms and reversing the order of the coefficients, to give

$$C(z) = -0.01024 + 0.0256z^{-1} - 0.064z^{-2} + 0.16z^{-3} - 0.4z^{-4} + z^{-5}$$
$$(4.224)$$

Thus the tap gains of the required equalizer are

$$-0.01024, \ 0.0256, \ -0.064, \ 0.16, \ -0.4 \quad \text{and} \quad 1$$

The z-transform of the channel and equalizer is

$$Y(z)C(z) = -0.004096 + z^{-6} \qquad (4.225)$$

which means that the sampled impulse-response of the channel and equalizer is

$$E = [-0.004096 \ 0 \ 0 \ 0 \ 0 \ 1] \qquad (4.226)$$

and the peak distortion in the equalized signal is

$$\frac{1}{|e_6|} \sum_{i=0}^{5} |e_i| = 0.004096 \qquad (4.227)$$

where e_i is the $(i+1)$th component of E.

Since $Y(z)C(z) \simeq z^{-6}$, the delay in detection of a received data-symbol is 6 sampling intervals.

LINEAR EQUALIZERS

The received sample at time $t = iT$ is

$$r_i = 0.4s_i + s_{i-1} + w_i \tag{4.228}$$

where s_i is the data symbol transmitted at time $t = iT$, and $s_i = \pm k$, k being a positive constant.

If the $\{w_i\}$ are statistically independent Gaussian random variables with zero mean and variance σ^2, the noise components $\{u_i\}$ at the output of the equalizer are Gaussian random variables with zero mean and variance

$$\sigma^2(0.01024^2 + 0.0256^2 + 0.064^2 + 0.16^2 + 0.4^2 + 1^2) = 1.1905\sigma^2 \tag{4.229}$$

the output noise variance being the input noise variance, σ^2, times the sum of the squares of the tap gains of the linear feedforward transversal equalizer.

If the given channel is replaced by one with the z-transform

$$X(z) = (1 + 0.4z^{-1})^{-1} \tag{4.230}$$

the z-transform of the linear feedforward transversal equalizer for the channel becomes

$$X^{-1}(z) = 1 + 0.4z^{-1} \tag{4.231}$$

so that the tap gains of the equalizer are 1 and 0.4, the equalizer having just two taps.

Since $X(z)X^{-1}(z) = 1$,

(a) the sampled impulse-response of the channel and equalizer is

$$E = [1 \ 0 \ 0 \ \ldots \ 0] \tag{4.232}$$

(b) the peak distortion in the equalized signal is zero,
(c) the delay in detection is zero, and
(d) the noise variance in the equalized signal is

$$\sigma^2(1^2 + 0.4^2) = 1.16\sigma^2 \tag{4.233}$$

The noise variance in the equalized signal here is reduced from $1.19\sigma^2$ to $1.16\sigma^2$, relative to the channel with z-transform $Y(z)$ (Equation 4.220), and, at the same time, the peak distortion is reduced from 0.004 to zero. The channel with z-transform $X(z)$ therefore gives the more reliable transmission of data.

Problem 3

The $(i+1)$th received sample of a noisy digital data signal is

$$r_i = s_i + 0.2s_{i-1} + 0.1s_{i-2} + w_i \tag{4.234}$$

where the data-symbols $\{s_i\}$ are statistically independent and equally likely to have either value $\pm k$, and the $\{w_i\}$ are statistically independent Gaussian random variables with zero mean and variance σ^2.

What is the signal/noise ratio at the output of a linear *feedback* transversal equalizer for the given channel, and how does this signal/noise ratio compare with that at the input to the equalizer?

Solution

The sampled impulse-response of the channel is

$$Y = [1 \ 0.2 \ 0.1] \tag{4.235}$$

204 LINEAR EQUALIZERS

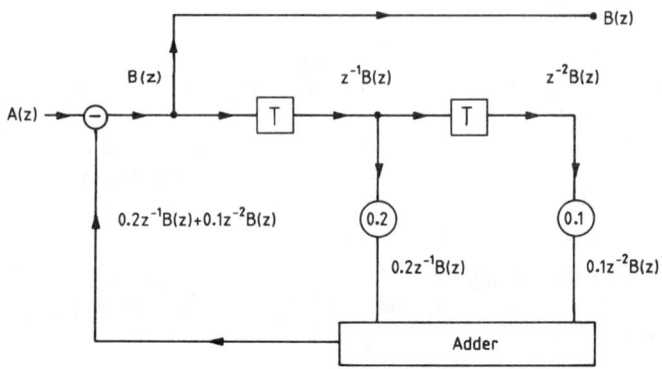

Fig. 4.15 Linear feedback transversal equalizer for the channel with z-transform $1 + 0.2z^{-1} + 0.1z^{-2}$

with z-transform

$$Y(z) = 1 + 0.2z^{-1} + 0.1z^{-2} \tag{4.236}$$

both of whose zeros are *inside* the unit circle.

The linear feedback transversal equalizer for this channel is as shown in Fig. 4.15, and the equalized signal at the output of the equalizer is

$$x_i = s_i + u_i \tag{4.237}$$

where u_i is a Gaussian random variable with zero mean.

The variance of u_i can be determined by considering the linear feedforward transversal equalizer (having ideally an infinite number of taps) that is equivalent to the given feedback equalizer. In Section 4.6 the 4-tap linear feedforward transversal equalizer for the given channel is shown to have the tap gains

$$1, -0.2, -0.06 \quad \text{and} \quad 0.032$$

Thus an approximate value of the variance of u_i is

$$\eta^2 = \sigma^2(1^2 + 0.2^2 + 0.06^2 + 0.032^2) = 1.0446\sigma^2 \tag{4.238}$$

Since u_i has zero mean, the mean-square value of u_i is also $1.0446\sigma^2$.

The mean-square value of the data signal in the equalized signal x_i is the mean-square value of s_i, which is clearly k^2.

Thus the signal/noise ratio at the output of the linear feedback transversal equalizer is approximately

$$\frac{k^2}{\eta^2} = \frac{k^2}{1.0446\sigma^2} = \frac{0.9573k^2}{\sigma^2} \tag{4.239}$$

When expressed in dB this becomes

$$10 \log_{10}\left(\frac{k^2}{\eta^2}\right) = 10 \log_{10}\left(\frac{0.9573k^2}{\sigma^2}\right) \tag{4.240}$$

From Equation 4.234, the mean-square value of the data signal at the input to the equalizer is

$$\mathscr{E}[(s_i+0.2s_{i-1}+0.1s_{i-2})^2] = \mathscr{E}[s_i^2 + 0.04s_{i-1}^2 + 0.01s_{i-2}^2 + 0.4s_i s_{i-1}$$
$$+ 0.2s_i s_{i-2} + 0.04s_{i-1}s_{i-2}]$$
$$= \mathscr{E}[s_i^2] + 0.04\mathscr{E}[s_{i-1}^2] + 0.01\mathscr{E}[s_{i-2}^2]$$
$$= k^2(1+0.04+0.01) = 1.05k^2 \qquad (4.241)$$

This follows because the $\{s_i\}$ are statistically independent and have zero mean, so that they are statistically *orthogonal* and

$$\mathscr{E}[s_i s_j] = 0 \qquad (4.242)$$

for any $i \neq j$.

Since the noise at the input to the equalizer has zero mean, its mean-square value is equal to its variance, which is σ^2.

Thus the signal/noise ratio at the input to the equalizer, expressed in dB, is

$$10 \log_{10}\left(\frac{1.05k^2}{\sigma^2}\right) \qquad (4.243)$$

from which it follows that the signal/noise ratio at the output of the equalizer is

$$10 \log_{10}\left(\frac{1.05}{0.9573}\right) = 10 \log_{10}(1.0968) = 0.40 \qquad (4.244)$$

dB *below* that at the input to the equalizer. The signal/noise ratio is therefore *reduced* slightly by the equalizer. The reduction in signal/noise ratio, in fact, tends to increase quite rapidly with the amplitude distortion in the received signal.

4.11 EQUALIZER THAT MINIMIZES THE MEAN-SQUARE ERROR IN THE EQUALIZED SIGNAL

It is now recognized that a linear transversal equalizer that minimizes the mean-square error in its output signal[59-154] generally gives a more useful or effective degree of equalization, for a given number of taps, than an equalizer that minimizes the peak distortion[49-61]. It is also more easily held in the correct adjustment for a slowly time-varying channel, in an adaptive system, particularly when the channel introduces severe signal distortion[59-154]. Such an equalizer *minimizes the mean-square difference* between the *actual* and *ideal* sample values (signals) at its output, for the given number of taps, when a continuous stream of data elements is being received in the presence of noise.

Suppose that the sampled impulse-response of the baseband

channel in Fig. 4.1 is given by the $(g+1)$-component row vector

$$Y = [y_0 \quad y_1 \quad \ldots \quad y_g] \tag{4.245}$$

The sampled impulse-response of the $(m+1)$-tap linear feedforward transversal equalizer is given by the $(m+1)$-component row vector

$$C = [c_0 \quad c_1 \quad \ldots \quad c_m] \tag{4.246}$$

and the sampled impulse-response of the channel and equalizer is given by the $(m+g+1)$-component row vector

$$E = [e_0 \quad e_1 \quad \ldots \quad e_{m+g}] \tag{4.247}$$

The *ideal* value of the sampled impulse-response of the channel and equalizer, for a total transmission delay of hT seconds, is given by the $(m+g+1)$-component row vector

$$E_h = [\overbrace{0 \quad \ldots \quad 0}^{h} \quad 1 \quad 0 \quad \ldots \quad 0] \tag{4.248}$$

where h is an integer in the range 0 to $m+g$. With the accurate equalization of the channel, $E \simeq E_h$.

Under these conditions, the output signal from the linear equalizer, at time $t = (i+h)T$, is

$$x_{i+h} \simeq s_i + u_{i+h} \tag{4.249}$$

where u_{i+h} is the Gaussian noise component. Ideally, exact equality should hold in Equation 4.249, but in fact

$$x_{i+h} = \sum_{j=0}^{m+g} s_{i+h-j} e_j + u_{i+h} \tag{4.250}$$

as can be seen from Equation 4.247.

Now, in Equation 4.250, the $\{s_{i+h-j}\}$ for $j = 0, 1, \ldots, m+g$, are *statistically independent* and equally likely to have either value $\pm k$, so that the $\{s_{i+h-j}\}$ have *zero mean* and are *statistically orthogonal*, which means that

$$\mathscr{E}[s_{i+h-j} s_{i+h-l}] = 0 \tag{4.251}$$

for any $j \neq l$[163-165], where $\mathscr{E}[x]$ is the expected or mean value of x. Also,

$$\mathscr{E}[s_{i+h-j}^2] = k^2 \tag{4.252}$$

Again, the $\{s_{i+h-j}\}$ are statistically independent of u_{i+h} which has zero mean, so that s_{i+h-j} and u_{i+h} are *statistically orthogonal*, and

$$\mathscr{E}[s_{i+h-j} u_{i+h}] = 0 \tag{4.253}$$

for any j[163-165].

LINEAR EQUALIZERS 207

A linear equalizer that minimizes the mean-square error in its output signal, minimizes the mean-square value of $x_{i+h} - s_i$, which is

$$\mathcal{E}[(x_{i+h} - s_i)^2] = \mathcal{E}\left[\left(\sum_{j=0}^{m+g} s_{i+h-j} e_j + u_{i+h} - s_i\right)^2\right]$$

$$= \mathcal{E}\left[\left(\sum_{j=0}^{m+g} s_{i+h-j} e_j\right)^2 - 2(s_i - u_{i+h})\right.$$

$$\left. \times \sum_{j=0}^{m+g} s_{i+h-j} e_j + (s_i - u_{i+h})^2\right]$$

$$= \sum_{j=0}^{m+g} \mathcal{E}[s_{i+h-j}^2 e_j^2] - \mathcal{E}[2s_i^2 e_h]$$

$$+ \mathcal{E}[s_i^2 - 2s_i u_{i+h} + u_{i+h}^2]$$

$$= k^2 \sum_{j=0}^{m+g} e_j^2 - 2k^2 e_h + k^2 + 0 + \sigma^2 |C|^2$$

$$= k^2 \sum_{\substack{j=0 \\ j \neq h}}^{m+g} e_j^2 + k^2 (e_h - 1)^2 + \sigma^2 |C|^2$$

$$= k^2 |E - E_h|^2 + \sigma^2 |C|^2 \quad (4.254)$$

where $|E - E_h|$ and $|C|$ are the *Euclidean lengths* of the vectors $E - E_h$ and C, respectively. $k^2|E - E_h|^2$ is the mean-square error in x_{i+h} due to *intersymbol interference* (Equation 4.34), and $\sigma^2|C|^2$ is the mean square error due to the *Gaussian noise* (Equation 4.185).

The equalizer here minimizes the mean-square error due to *both* intersymbol interference and noise, and not just the mean-square error $k^2|E - E_h|^2$ due to intersymbol interference alone (Equation 4.34). The relationship between the two minimization criteria will be considered presently.

Let Y_c be the $(m+1) \times (m+g+1)$ *convolution matrix* whose $(i+1)$th row is given by the $(m+g+1)$-component row vector

$$Y_i = [\overbrace{0 \quad \ldots \quad 0}^{i} \quad y_0 \quad y_1 \quad \ldots \quad y_g \quad 0 \quad \ldots \quad 0] \quad (4.255)$$

The first row of Y_c is clearly Y_0, whose *first* component is y_0, and the last row of Y_c is Y_m, whose *last* component is y_g. Y_c is always taken to be the *convolution matrix* itself, whereas Y_0, Y_1, Y_i, Y_m, and so on are the corresponding *rows* of this matrix. Notice also that the matrix Y_c here is not *quite* the same as that in Equation 2.25, the latter being a *square* convolution matrix and therefore also a *circulant* matrix, but being otherwise equivalent in that it gives the *same* resultant convolution. The sampled impulse-response of the channel and equalizer is now given by the $(m+g+1)$-component vector

$$E = \sum_{i=0}^{m} c_i Y_i = C Y_c \tag{4.256}$$

From Equation 4.254, the linear equalizer is required to minimize
$k^2|CY_c - E_h|^2 + \sigma^2|C|^2$

$$= k^2(CY_c - E_h)(CY_c - E_h)^T + \sigma^2 CC^T$$
$$= k^2 CY_c Y_c^T C^T - 2k^2 E_h Y_c^T C^T + k^2 E_h E_h^T + \sigma^2 CC^T$$
$$= C(k^2 Y_c Y_c^T + \sigma^2 I)C^T - 2k^2 E_h Y_c^T C^T + k^2 \tag{4.257}$$

where I is an $(m+1) \times (m+1)$ identity matrix, and E_h is given by Equation 4.248.

The $(m+1) \times (m+1)$ matrix $k^2 Y_c Y_c^T + \sigma^2 I$ is real, symmetric and positive definite, since the matrices $Y_c Y_c^T$ and I are both real, symmetric and positive definite. It follows that there is an $(m+1) \times (m+1)$ real nonsingular matrix G such that[170]

$$GG^T = k^2 Y_c Y_c^T + \sigma^2 I \tag{4.258}$$

Thus, from Equation 4.257,

$k^2|CY_c - E_h|^2 + \sigma^2|C|^2$
$$= CGG^T C^T - 2k^2 E_h Y_c^T C^T + k^2$$
$$= [CG - k^2 E_h Y_c^T (G^T)^{-1}][CG - k^2 E_h Y_c^T (G^T)^{-1}]^T$$
$$\quad + k^2[1 - k^2 E_h Y_c^T (G^T)^{-1} G^{-1} Y_c E_h^T]$$
$$= |CG - k^2 E_h Y_c^T (G^T)^{-1}|^2 + k^2[1 - k^2 E_h Y_c^T (GG^T)^{-1} Y_c E_h^T]$$
$$\tag{4.259}$$

It is required to select C to minimize $k^2|CY_c - E_h|^2 + \sigma^2|C|^2$. In Equation 4.259

$$k^2[1 - k^2 E_h Y_c^T (GG^T)^{-1} Y_c E_h^T] \tag{4.260}$$

is independent of C and

$$|CG - k^2 E_h Y_c^T (G^T)^{-1}|^2 \tag{4.261}$$

is nonnegative, so that

$$k^2|CY_c - E_h|^2 + \sigma^2|C|^2 \tag{4.262}$$

is minimum when

$$|CG - k^2 E_h Y_c^T (G^T)^{-1}|^2 = 0 \tag{4.263}$$

that is, when
$$CG - k^2 E_h Y_c^T (G^T)^{-1} = 0 \qquad (4.264)$$
or
$$C = k^2 E_h Y_c^T (G^T)^{-1} G^{-1} = k^2 E_h Y_c^T (GG^T)^{-1}$$
$$= k^2 E_h Y_c^T (k^2 Y_c Y_c^T + \sigma^2 I)^{-1}$$
$$= E_h Y_c^T \left(Y_c Y_c^T + \frac{\sigma^2}{k^2} I \right)^{-1} \qquad (4.265)$$

Equation 4.265 gives the values of the tap gains of the $(m+1)$-tap linear feedforward transversal equalizer that minimizes the mean-square error due to both intersymbol interference and noise, in the output signal from the equalizer. From Equation 4.259, the minimum value of this mean-square error, and therefore the mean-square error obtained with the equalizer of Equation 4.265, is

$$k^2 [1 - k^2 E_h Y_c^T (GG^T)^{-1} Y_c E_h^T]$$
$$= k^2 \left[1 - E_h Y_c^T \left(Y_c Y_c^T + \frac{\sigma^2}{k^2} I \right)^{-1} Y_c E_h^T \right] \qquad (4.266)$$

The matrix $k^2 Y_c Y_c^T + \sigma^2 I$ (see Equation 4.265) can be considered as the $(m+1) \times (m+1)$ *correlation matrix* for the received samples $\{r_i\}$. The component in the ith row and jth column of this matrix (for any integer $n \geqslant g$) is $\mathscr{E}[r_n r_{n+i-j}]$ which, by means of a derivation similar to that in Equations 4.251–4.254, can be shown to be equal to

$$k^2 \sum_{l=-\infty}^{\infty} y_l y_{l+i-j} + \sigma^2 \delta_{ij}$$

where $\delta_{ij} = 1$ for $i = j$ and $\delta_{ij} = 0$ for $i \neq j$. Furthermore, the quantity

$$\sum_{l=-\infty}^{\infty} y_l y_{l+i-j}$$

here is the component b_{i-j} of the *aperiodic autocorrelation function* of the sequence Y or Y_0 (Equations 3.14 and 3.15). The matrix $Y_c Y_c^T + (\sigma^2/k^2) I$ can therefore be considered as a *normalized correlation matrix*.

At high signal/noise ratios, $k^2 \gg \sigma^2$ (assuming, for convenience, that $|Y| \simeq 1$ and hence $|Y_i| \simeq 1$). Thus, from Equation 4.265, the equalizer that minimizes the mean-square error is now given by

$$C \simeq E_h Y_c^T (Y_c Y_c^T)^{-1} \qquad (4.267)$$

This equalizer minimizes $k^2 |E - E_h|^2$ in Equation 4.254, as will be shown in more detail later.

At very-low signal/noise ratios, $k^2 \ll \sigma^2$, so that the equalizer is now

210 LINEAR EQUALIZERS

given by

$$C \simeq \frac{k^2}{\sigma^2} E_h Y_c^T = \frac{k^2}{\sigma^2} [y_h \quad y_{h-1} \quad \ldots \quad y_0 \quad 0 \quad \ldots \quad 0] \quad (4.268)$$

where $y_i = 0$ for $i > g$. The filter is here *matched* to the $(i+1)$th received signal-element, or at least to as much of it as has been received over the $m+1$ consecutive sampling instants ending at the time instant $t = (i+h)T$, when s_i is detected from the output signal x_{i+h}. The equalizer therefore maximizes the output signal/noise ratio and hence minimizes the effect of the noise on the signal. Consequently, it reduces $\sigma^2 |C|^2$ in Equation 4.254. It does not however necessarily reduce and may well increase the mean-square error $k^2 |E - E_h|^2$ due to intersymbol interference. This is, of course, to be expected since the noise now predominates over the intersymbol interference, so that it is more important to reduce the noise than the intersymbol interference.

It can be seen from Equation 4.265 that, even if there are a sufficient number of taps in the equalizer to ensure accurate equalization at *high* signal/noise ratios, when Equation 4.267 is satisfied, the following undesirable effect is experienced at *low* signal/noise ratios. In addition to the intersymbol interference now introduced into the equalized signal x_{i+h} (Equation 4.249), the wanted signal-component here, that is dependent on s_i, is *reduced in magnitude*, due to the fact that $\sigma^2/k^2 \gg 1$ in Equation 4.265. In other words, an appreciable *bias* is introduced into the wanted signal component.

Most adaptive equalizers are automatically adjusted to minimize the mean-square error in the equalized signal[59-154], which means that, when the data are received in the presence of additive white Gaussian noise, the equalizer is adjusted as just described. Unfortunately, over switched telephone circuits, the predominant type of noise is impulsive noise which occurs in short bursts, separated by relatively long intervals that are often comparatively noise-free. Under these conditions, the equalizer is normally adjusted so that, for practical purposes, it minimizes the mean-square error $k^2 |E - E_h|^2$ due to intersymbol interference alone. Again, when a fixed equalizer is designed for some given time-invariant channel that introduces additive white Gaussian noise, a prior knowledge of the signal/noise ratio k^2/σ^2 is required in order to minimize the mean-square error in the equalized signal, as can be seen from Equation 4.265. In practice this prior knowledge is often not available. The best that can be done under these conditions is again to minimize the mean-square error due to intersymbol interference alone.

Consider therefore the $(m+1)$-tap linear feedforward transversal

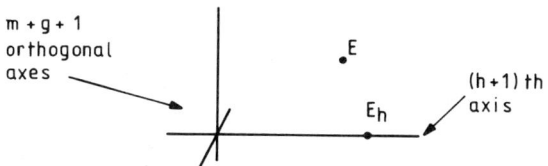

Fig. 4.16 Vectors E and E_h in the $(m+g+1)$-dimensional Euclidean vector space

equalizer that is designed to minimize $k^2|E-E_h|^2$, which is the mean-square error due to intersymbol interference in its output signal.

The vectors E and E_h may be represented as points in an $(m+g+1)$-dimensional Euclidean vector space, as illustrated in Fig. 4.16, and the equalizer must now be designed to minimize $|E-E_h|$, which is the *Euclidean distance* between the two vectors in the vector space. In the following analysis this distance will, for convenience, be written as $|E_h-E|$. Now, for given values of h and $m+g$, E_h is *fixed*, but, since $E=CY_c$ (from Equation 4.256), E can be moved in the vector space by changing C. Thus C must be chosen to minimize $|E_h-E|$. But, from Equation 4.256, E is a *linear combination* of the $m+1$ vectors $\{Y_i\}$ given by Equation 4.255. These $m+1$ vectors are *linearly independent*, since no one of them can be formed by a linear combination of any of the others, so that the $m+1$ vectors $\{Y_i\}$ span an $(m+1)$-dimensional subspace of the $(m+g+1)$-dimensional Euclidean vector space. This means, of course, that any vector in the given subspace must be a linear combination of the $m+1$ vectors $\{Y_i\}$, and, conversely, any linear combination of these vectors must lie in the subspace. Clearly, E must lie in the given $(m+1)$-dimensional subspace. Since *exact* equalization of the channel cannot be achieved by means of a linear feedforward transversal equalizer with a *finite* number of taps, and since exact equalization is, in fact, achieved only when $E=E_h$, it follows that, whenever the channel introduces *any* intersymbol interference in the sampled signal at the receiver input, E_h *cannot* lie in the $(m+1)$-dimensional subspace, at least not exactly.

Now $|E_h-E|$ is the distance between the fixed point E_h in Fig. 4.16 and the moveable point E which is confined to the given subspace. Since $E=CY_c$ and there are no constraints on C, the vector C may be selected such that CY_c is *any* required point in the $(m+1)$-dimensional subspace. It follows that $|E_h-E|$ is *minimum* when E is the point in the subspace at the minimum distance from E_h. By the Projection theorem, the required vector E is the *orthogonal projection* of E_h on to the $(m+1)$-dimensional subspace. This is illustrated in Fig. 4.17, and is verified as follows.

Let the $(m+g+1)$-component vector XY_c be the *orthogonal*

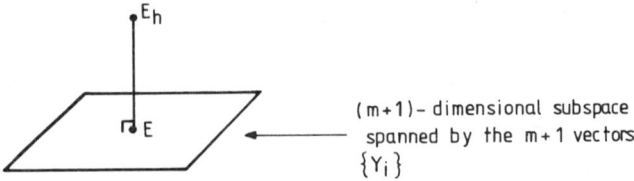

Fig. 4.17 *The vector E that minimizes the mean-square error due to intersymbol interference*

projection of E_h on to the $(m+1)$-dimensional subspace spanned by the $\{Y_i\}$, where X is the appropriate $(m+1)$-component row vector Since $E = CY_c$, the square of the distance between E and E_h is

$$\begin{aligned}
|E_h - E|^2 &= |E_h - CY_c|^2 \\
&= (E_h - CY_c)(E_h - CY_c)^T \\
&= (E_h - XY_c + XY_c - CY_c)(E_h - XY_c + XY_c - CY_c)^T \\
&= (E_h - XY_c)(E_h - XY_c)^T + (XY_c - CY_c)(XY_c - CY_c)^T \\
&\quad + (E_h - XY_c)(XY_c - CY_c)^T + (XY_c - CY_c)(E_h - XY_c)^T \\
&= (E_h - XY_c)(E_h - XY_c)^T + (XY_c - CY_c)(XY_c - CY_c)^T
\end{aligned}$$
(4.269)

Use is made here of the fact that $E_h - XY_c$ is orthogonal to $XY_c - CY_c$, because the vector $XY_c - CY_c$ lies in the $(m+1)$-dimensional subspace and $E_h - XY_c$ is orthogonal to this subspace. Now $(E_h - XY_c) \times (E_h - XY_c)^T$ is not dependent on C and $(XY_c - CY_c)(XY_c - CY_c)^T$ is nonnegative, so that $|E_h - E|^2$ is minimum when $CY_c = XY_c$ or $C = X$. The latter follows because the $\{Y_i\}$ are linearly independent, so that C is uniquely determined by CY_c.

The vector $E_h - E$ is now *orthogonal* to the $(m+1)$-dimensional subspace and therefore also to every vector that lies in the subspace. The subspace contains each Y_i, for $i = 0, 1, \ldots, m$, and, of course, it also contains the origin of the vector space, that is, the zero vector. Thus the vector $E_h - E$ is orthogonal to Y_i, for $i = 0, 1, \ldots, m$, and therefore to each *row* of the $(m+1) \times (m+g+1)$ matrix Y_c. Hence

$$(E_h - E)Y_i^T = 0 \qquad (4.270)$$

or

$$(E_h - E)Y_c^T = 0 \qquad (4.271)$$

where 0 is now the $(m+1)$-component zero row vector, so that

$$(E_h - CY_c)Y_c^T = 0 \qquad (4.272)$$

or

$$CY_c Y_c^T = E_h Y_c^T \qquad (4.273)$$

or
$$C = E_h Y_c^T (Y_c Y_c^T)^{-1} \quad (4.274)$$

since the $(m+1) \times (m+1)$ matrix $Y_c Y_c^T$ is nonsingular. Clearly, C is given by the $(h+1)$th *row* of the $(m+g+1) \times (m+1)$ matrix $Y_c^T (Y_c Y_c^T)^{-1}$. To obtain the optimum equalizer, C must be determined for each value of h in the range 0 to $m+g$, and the vector C for the required equalizer selected as that which gives the minimum value of $|E_h - E|$.

The *minimum mean-square error* in the equalized signal due to intersymbol interference alone is the minimum value of $k^2 |E_h - E|^2$, and is, of course, obtained with the equalizer satisfying Equation 4.274. Thus the mean-square error given by this equalizer is

$$k^2 |E_h - C Y_c|^2$$
$$= k^2 |E_h - E_h Y_c^T (Y_c Y_c^T)^{-1} Y_c|^2$$
$$= k^2 E_h [I - Y_c^T (Y_c Y_c^T)^{-1} Y_c][I - Y_c^T (Y_c Y_c^T)^{-1} Y_c]^T E_h^T$$
$$= k^2 E_h [I - Y_c^T (Y_c Y_c^T)^{-1} Y_c] E_h^T$$
$$= k^2 [1 - E_h Y_c^T (Y_c Y_c^T)^{-1} Y_c E_h^T] \quad (4.275)$$

where I is the $(m+g+1) \times (m+g+1)$ identity matrix and $Y_c^T (Y_c Y_c^T)^{-1} Y_c$ is an $(m+g+1) \times (m+g+1)$ real symmetric matrix. It can be seen from Equations 4.275 and 4.248 that the mean-square error due to intersymbol interference in the equalized signal, obtained with the required equalizer, is k^2 times the difference between unity and the $(h+1)$th component along the main diagonal of $Y_c^T (Y_c Y_c^T)^{-1} Y_c$.

Comparing Equations 4.267 and 4.274, it is clear that a linear equalizer, that minimizes the mean-square error due to intersymbol interference alone, also minimizes the mean-square error due to both intersymbol interference and noise, at high signal/noise ratios. This is, of course, to be expected since, at high signal/noise ratios, the intersymbol interference tends to predominate over the noise, so that it is now more important to reduce the intersymbol interference than the noise.

Consider the case where the z-transform of the channel is

$$Y(z) = 1 + 2.5z^{-1} + z^{-2} \quad (4.276)$$

and it is required to equalize the channel by means of an 11-tap linear feedforward transversal filter that minimizes the mean-square error due to intersymbol interference in its output signal.

Let Y_c be the 11×13 matrix whose $(i+1)$th row is

$$Y_i = [\overbrace{0 \; \ldots \; 0}^{i} \; 1 \; 2.5 \; 1 \; 0 \; \ldots \; 0] \quad (4.277)$$

so that the sampled impulse-response of the channel and equalizer is given by the 13-component row vector

$$E = CY_c \tag{4.278}$$

where C is the 11-component row vector whose component values are the tap gains of the required equalizer. From the symmetry of $Y(z)$ it seems reasonable to select the vector C so as to minimize the distance between E and E_6, where E_6 is the 13-component row vector

$$E_6 = [\overbrace{0 \;\ldots\; 0}^{6} \;\; 1 \;\; 0 \;\ldots\; 0] \tag{4.279}$$

that represents the *ideal* or desired sampled impulse-response. From Equation 4.274,

$$C = E_6 Y_c^T (Y_c Y_c^T)^{-1} \tag{4.280}$$

and the required vector C is now given by

$$C = [-0.012 \quad 0.035 \quad -0.079 \quad 0.164 \quad -0.331 \quad 0.665$$
$$-0.331 \quad 0.164 \quad -0.079 \quad 0.035 \quad -0.012] \tag{4.281}$$

From Equation 4.278, the sampled impulse-response of the channel and equalizer is

$$E = [-0.012 \quad 0.006 \quad -0.003 \quad 0.001 \quad -0.001 \quad 0.000 \quad 1.000$$
$$0.000 \quad -0.001 \quad 0.001 \quad -0.003 \quad 0.006 \quad -0.012] \tag{4.282}$$

The corresponding 11-tap equalizer, designed through the separate equalization of the two factors $1 + \tfrac{1}{2}z^{-1}$ and $1 + 2z^{-1}$, as described in Section 4.5, is given by

$$C = [-0.016 \quad 0.039 \quad -0.082 \quad 0.166 \quad -0.334 \quad 0.667$$
$$-0.334 \quad 0.166 \quad -0.082 \quad 0.039 \quad -0.016] \tag{4.283}$$

as can be seen from Equation 4.82, and now (from Expression 4.84)

$$E = [-0.016 \quad 0 \quad 0 \quad 0 \quad 0 \quad 1 \quad 0 \quad 0 \quad 0 \quad 0 \quad -0.016] \tag{4.284}$$

The latter equalizer achieves zero forcing in the sampled impulse-response of the equalized channel. Furthermore, it gives a lower peak distortion due to intersymbol interference in its output signal than does the other equalizer but a higher mean-square distortion and a higher mean-square error. There is not, however, much difference between the tap gains of the two equalizers.

For either equalizer, s_i is detected from the signal x_{i+6} at the output of the equalizer, at time $t = (i+6)T$, where

$$x_{i+6} \simeq s_i + u_{i+6} \tag{4.285}$$

and u_{i+6} is a Gaussian random variable with zero mean and variance $\sigma^2|C|^2$ (Equation 4.185).

In the case of the equalizer that minimizes the mean-square error, the probability of error in the detection of s_i is approximately

$$Q\left(\frac{k}{\sigma|C|}\right) = Q\left(\frac{k}{0.855\sigma}\right) = Q\left(\frac{1.170k}{\sigma}\right) \qquad (4.286)$$

whereas, for the equalizer that achieves zero forcing, it is approximately

$$Q\left(\frac{k}{\sigma|C|}\right) = Q\left(\frac{k}{0.859\sigma}\right) = Q\left(\frac{1.164k}{\sigma}\right) \qquad (4.287)$$

At very high signal/noise ratios, the slightly lower peak distortion obtained with the latter equalizer reduces the very small advantage gained by the former due to its lower noise variance, so that, for practical purposes, the two arrangements achieve the same tolerance to additive white Gaussian noise at high signal/noise ratios. Indeed, the *larger* the number of taps of an equalizer and with h appropriately related to the number of taps, the *more accurate* the equalization of the channel, and so, for the same delay of h sampling intervals introduced by each equalizer, the *more alike* the two equalizers become.

The important fact that has been illustrated by this example is that, with the accurate *linear* equalization of the channel, the tap gains of the transversal equalizer are effectively fixed by the *channel* and are not significantly affected by the particular design technique applied to the equalizer. This means that when a channel is equalized *linearly*, the tolerance to additive white Gaussian noise is essentially determined by the channel.

It can be shown[155] that a linear equalizer, that minimizes the *mean-square error* (due to intersymbol interference) in the equalized signal, also minimizes the *mean-square distortion* (Equation 4.31). The former equalizer is in fact a *special case* of the latter. Furthermore, for a given number of taps of the equalizer and with an optimum threshold-level detector operating on the equalized signal, at a given delay in detection, an equalizer that minimizes the mean-square distortion has tap gains whose values are in a constant ratio to the corresponding tap gains of the equalizer that minimizes the mean-square error, and it gives *exactly* the same probability of error in the detection of a data symbol as does the other equalizer.

When a QAM signal is transmitted over a bandpass channel, the signals in the data-transmission system of Fig. 4.1 become *complex valued* (see Section 3.7). Thus the received sample, at time $t = iT$, is

$$r_i = \sum_{j=0}^{g} s_{i-j} y_j + w_i \tag{4.288}$$

as in Equation 4.3, but r_i, s_{i-j}, y_j and w_i are now *complex valued*. Assume, for simplicity, a 4-level QAM signal, such that

$$s_i = s_{0,i} + j s_{1,i} \tag{4.289}$$

where
$$j = \sqrt{-1} \tag{4.290}$$

$$s_{0,i} = \pm k \tag{4.291}$$

and
$$s_{1,i} = \pm k \tag{4.292}$$

the $\{s_{0,i}\}$ and $\{s_{1,i}\}$ being statistically independent and equally likely to have either binary value. Thus the mean-square value of s_i is[163-165]

$$\mathscr{E}[|s_i|^2] = 2k^2 \tag{4.293}$$

Again, with a receiver filter corresponding to that for real-valued signals in Fig. 4.1, the real and imaginary parts of the noise components $\{w_i\}$ are statistically independent Gaussian random variables with zero mean and variance σ^2, so that the $\{w_i\}$ themselves are statistically independent complex-valued Gaussian random variables with zero mean and variance $2\sigma^2$. The signal/noise ratio is k^2/σ^2 as before (Section 4.2).

Under these conditions, the mean-square error in the equalized signal (Equation 4.254) becomes

$$\mathscr{E}[|x_{i+h} - s_i|^2] = 2k^2 |E - E_h|^2 + 2\sigma^2 |C|^2 \tag{4.294}$$

where
$$|E - E_h|^2 = (E - E_h)(E - E_h)^* \tag{4.295}$$

and is the square of the *unitary distance* between the vectors E and E_h in the $(m+g+1)$-dimensional *unitary* vector space containing these vectors. Also

$$|C|^2 = CC^* \tag{4.296}$$

and is the square of the *unitary length* of the $(m+1)$-component vector C. The vector V^* is, of course, the *conjugate transpose* of the vector V. The above relationships are considered in more detail in Section 1.3.

The linear feedforward transversal equalizer, that minimizes the mean-square error due to both intersymbol interference and noise in the equalized signal, may now be derived along exactly the same lines as before (Equations 4.255–4.265), but the $m+1$ tap gains of the equalizer are now given by the $(m+1)$-component complex row vector

$$C = E_h Y_c^* \left(Y_c Y_c^* + \frac{\sigma^2}{k^2} I \right)^{-1} \quad (4.297)$$

in place of C in Equation 4.265, and the mean-square error in the equalized signal obtained with this equalizer, is

$$2k^2 \left[1 - E_h Y_c^* \left(Y_c Y_c^* + \frac{\sigma^2}{k^2} I \right)^{-1} Y_c E_h^T \right] \quad (4.298)$$

in place of the value given by Equation 4.266. The matrices E_h, Y_c and I are as previously defined (Equations 4.248 and 4.255), and the matrix

$$Y_c Y_c^* + \frac{\sigma^2}{k^2} I \quad (4.299)$$

is a positive definite Hermitian matrix.

Similarly, the $m+1$ tap gains of the linear feedforward transversal equalizer, that minimizes the mean-square error due only to intersymbol interference in the equalized signal, are now given by the $m+1$ components of the *complex* row vector

$$C = E_h Y_c^* (Y_c Y_c^*)^{-1} \quad (4.300)$$

in place of C in Equation 4.274, and the mean-square error due to intersymbol interference in the equalized signal, obtained with this equalizer, is

$$2k^2 [1 - E_h Y_c^* (Y_c Y_c^*)^{-1} Y_c E_h^T] \quad (4.301)$$

in place of the value given by Equation 4.275.

4.12 LINEAR EQUALIZATION OF PHASE DISTORTION

Consider the data-transmission system in Fig. 4.1, where all signal are *real-valued*, and suppose that the baseband channel here introduces pure phase distortion (which may include some delay but no attenuation). Phase distortion is defined and described in some detail in Sections 3.2 and 3.3. Let the z-transform of the channel be

$$Y(z) = y_0 + y_1 z^{-1} + \cdots + y_g z^{-g} \quad (4.302)$$

where $y_i = 0$ for $i < 0$ and $i > g$, and suppose that the channel introduces no gain or attenuation, such that

$$y_0^2 + y_1^2 + \cdots + y_g^2 = 1 \quad (4.303)$$

218 LINEAR EQUALIZERS

It can be seen from Equations 3.20 and 3.51 that, when the channel introduces pure phase distortion, it not only satisfies Equation 4.303, but is also such that

$$\sum_{j=-\infty}^{\infty} y_j y_{j+h} \simeq 0 \qquad (4.304)$$

for any nonzero integer h. Equation 4.304 only holds exactly for *all* nonzero integer values of h, when $g \to \infty$, but the errors normally become negligible before g becomes unduly large.

Consider now two received signal-elements whose samples (components) are given by the n-component vectors

$$s_i Y_0 = [s_i y_0 \quad s_i y_1 \quad \ldots \quad s_i y_g \quad 0 \quad \ldots \quad 0] \qquad (4.305)$$

and

$$s_{i+h} Y_h = [0 \quad \ldots \quad 0 \quad s_{i+h} y_0 \quad s_{i+h} y_1 \quad \ldots \quad s_{i+h} y_g] \qquad (4.306)$$

where $n = g + h + 1$ and $h > 0$. All samples of the two signal elements preceding and following those given in Equations 4.305 and 4.306 are of value zero. The inner product of the two vectors is

$$s_i s_{i+h} Y_0 Y_h^T = s_i s_{i+h} \sum_{j=-\infty}^{\infty} y_j y_{j+h} \simeq 0 \qquad (4.307)$$

from Equation 4.304, so that the two received signal-elements are *orthogonal*. But, from Equation 4.303, the vectors Y_0 and Y_h have unit length, that is,

$$|Y_0| = |Y_h| = 1 \qquad (4.308)$$

and, since

$$s_i = \pm k \qquad (4.309)$$

it follows that

$$|s_i Y_0| = |s_{i+h} Y_h| = k \qquad (4.310)$$

Now, the n-component signal vectors at the *input* to the linear baseband channel in Fig. 4.1, corresponding to the received vectors $s_i Y_0$ and $s_{i+h} Y_h$, are

$$S_i = [s_i \quad 0 \quad 0 \quad \ldots \quad 0] \qquad (4.311)$$

and

$$S_{i+h} = [\overbrace{0 \quad 0 \quad \ldots \quad 0}^{h} \quad s_{i+h} \quad 0 \quad \ldots \quad 0] \qquad (4.312)$$

respectively. Clearly,

$$S_i S_{i+h}^T = S_{i+h} S_i^T = 0 \qquad (4.313)$$

for any nonzero integer h, and

$$|S_i| = |S_{i+h}| = k \tag{4.314}$$

so that the vectors S_i and S_{i+h} are orthogonal and of length k, just like the vectors $s_i Y_0$ and $s_{i+h} Y_h$. It is evident from the above analysis, together with that in Section 3.2, that pure phase distortion is an *orthogonal transformation* of the transmitted signal, such that the individual received signal-elements remain orthogonal to each other and unchanged in level (energy). Pure phase distortion can, in fact, be considered as a *rotation* of the orthogonal axes of the Euclidean vector space, in which the vectors corresponding to the individual signal-elements are given as *fixed points*. There are consequently no changes in the *relative positions* of these vectors in the vector space, with therefore no changes in any inner products, lengths, distances or angles associated with the vectors.

Suppose now that the linear feedforward transversal filter in the receiver of the data-transmission system of Fig. 4.1 is *matched* to the channel, whose z-transform is given by $Y(z)$ in Equation 4.302. Thus the z-transform of the linear filter is

$$C(z) = y_g + y_{g-1} z^{-1} + \cdots + y_0 z^{-g} \tag{4.315}$$

and the resultant z-transform of the channel and matched filter, connected in cascade, is

$$\begin{aligned} Y(z)C(z) &= (y_0 + y_1 z^{-1} + \cdots + y_g z^{-g}) \\ &\quad \times (y_g + y_{g-1} z^{-1} + \cdots + y_0 z^{-g}) \\ &= y_0 y_g + (y_0 y_{g-1} + y_1 y_g) z^{-1} + \cdots \\ &\quad + (y_0 y_1 + y_1 y_2 + \cdots + y_{g-1} y_g) z^{-g+1} \\ &\quad + (y_0^2 + y_1^2 + \cdots + y_g^2) z^{-g} \\ &\quad + (y_0 y_1 + y_1 y_2 + \cdots + y_{g-1} y_g) z^{-g-1} + \cdots \\ &\quad + (y_0 y_{g-1} + y_1 y_g) z^{-2g+1} + y_0 y_g z^{-2g} \\ &\simeq z^{-g} \end{aligned} \tag{4.316}$$

as can be seen from Equations 4.303 and 4.304. Thus the z-transform of the matched filter is

$$C(z) \simeq z^{-g} Y^{-1}(z) \tag{4.317}$$

which means that the matched filter is also a *linear equalizer* for the given channel, as in Fig. 4.1. It follows that the signal at the output of the equalizer, at time $t = (i+g)T$, is

$$x_{i+g} \simeq s_i + u_{i+g} \tag{4.318}$$

220 LINEAR EQUALIZERS

where u_{i+g} is a Gaussian random variable with zero mean and variance

$$\sigma^2|C|^2 = \sigma^2 \sum_{j=0}^{g} y_j^2 = \sigma^2 \qquad (4.319)$$

as can be seen from Equations 4.315 and 4.303. Clearly, the $\{u_i\}$ have the same variance as the $\{w_i\}$. The data-symbol s_i is detected from the sign of the equalized signal x_{i+g} and the probability of error in the detection of s_i is approximately

$$P_e = Q\left(\frac{k}{\sigma|C|}\right) = Q\left(\frac{k}{\sigma}\right) \qquad (4.320)$$

The error probability is here effectively the *same* as that when there is no distortion or attenuation of the received signal, in which case the received signal, at time $t = iT$, is

$$r_i = s_i + w_i \qquad (4.321)$$

neglecting the delay in transmission.

The linear filter is clearly matched both to the channel and to each received signal-element, which means that in the presence of additive white Gaussian noise, as assumed here, the output signal/noise ratio from the equalizer is maximized[171,172]. Since, furthermore, there is negligible intersymbol interference at the output of the equalizer, it follows from the analysis in Section 1.6 that the linear equalizer and detector achieve the best available tolerance to additive white Gaussian noise. There is no better detection process for the given received signal.

Consider now the two noise components u_i and u_{i+h}, at the time instants iT and $(i+h)T$, at the output of the linear equalizer with z-transform $C(z)$ in Equation 4.315. Clearly,

$$u_i = w_i y_g + w_{i-1} y_{g-1} + \cdots + w_{i-g} y_0 \qquad (4.322)$$

and $\quad u_{i+h} = w_{i+h} y_g + w_{i+h-1} y_{g-1} + \cdots + w_{i+h-g} y_0 \qquad (4.323)$

Since $\mathscr{E}[w_i] = 0$, where $\mathscr{E}[w_i]$ is the expected or mean value of w_i,

$$\mathscr{E}[w_i^2] = \sigma^2 \qquad (4.324)$$

where σ^2 is the variance of w_i, and

$$\mathscr{E}[u_i] = \mathscr{E}[u_{i+h}] = 0 \qquad (4.325)$$

Also, since the $\{w_i\}$ are statistically independent with zero mean, they are *statistically orthogonal*, such that[163-165]

$$\mathscr{E}[w_i w_{i+h}] = 0 \qquad (4.326)$$

for any nonzero integer h.

Thus, when $h \neq 0$,

$$\begin{aligned}
\mathscr{E}[u_i u_{i+h}] &= \mathscr{E}[(w_i y_g + w_{i-1} y_{g-1} + \cdots + w_{i-g} y_0) \\
&\quad \times (w_{i+h} y_g + w_{i+h-1} y_{g-1} + \cdots + w_{i+h-g} y_0)] \\
&= \mathscr{E}[w_i^2 y_g y_{g-h} + w_{i-1}^2 y_{g-1} y_{g-h-1} + \cdots + w_{i-g+h}^2 y_h y_0] \\
&= \sigma^2 y_g y_{g-h} + \sigma^2 y_{g-1} y_{g-h-1} + \cdots + \sigma^2 y_h y_0 \\
&= \sigma^2 \sum_{j=-\infty}^{\infty} y_j y_{j-h} \simeq 0
\end{aligned} \qquad (4.327)$$

so that the $\{u_i\}$ are *statistically orthogonal* and *uncorrelated*. It follows that the $\{u_i\}$ are *statistically independent* Gaussian random variables with zero mean and variance σ^2, just like the $\{w_i\}$[163-165].

The important results may now be summarized as follows. The signal distortion represented by $Y(z)$ is pure phase distortion, with no signal attenuation or gain, and the equalizer with z-transform $C(z)$ is both a pure phase equalizer and a matched filter, whose output noise components $\{u_i\}$ have exactly the same statistical properties as the input noise components $\{w_i\}$. The equalizer therefore introduces no change in level or other statistical properties of the input noise samples, nor does it change the level of the data signal. Thus the linear equalizer *removes* the phase distortion, without changing the signal/noise ratio or any noise statistics, which means that the presence of any phase distortion does not affect the tolerance of the system to additive white Gaussian noise.

Most practical channels introduce, in addition to phase distortion, some degree of *amplitude distortion*. Amplitude distortion is defined and described in some detail in Sections 3.4 and 3.5. In the presence of amplitude distortion, the linear equalizer for the channel is *not* matched to the channel or received signal, which necessarily implies that the signal/noise ratio at the output of the equalizer is *less* than that at the output of the corresponding matched filter (neglecting here the level of the intersymbol interference). This means that, with a linear equalizer followed by the appropriate threshold-level detector, the probability of error in the detection of a received signal-element is now *greater* than that where only a single element is received and is detected by a matched-filter detector. In other words, when a channel is equalized linearly at the receiver, *any* amplitude distortion introduced by the channel (with no signal attenuation or gain) always *reduces* the tolerance of the system to additive white Gaussian noise. Furthermore, in the presence of amplitude distortion, a linear equalizer no longer gives the best available tolerance to Gaussian noise, since it does *not* now correspond to a maximum-likelihood detector (Sections 1.4 and 1.5). The latter minimizes the probability of error in the detection of a received message and is here taken to be the *optimum* detector. Indeed, the tolerance to noise obtained with a linear equalizer may now fall far below that obtainable with the optimum detection process for the particular received signal, especially when $Y(z)$ has one or more zeros close to the unit circle in the z plane, so that a linear equalizer is no longer the most suitable arrangement. With the optimum detection process, the presence of amplitude distortion may or may not noticeably reduce the tolerance of the system to additive white Gaussian noise, depending upon the severity of the distortion[155].

REFERENCES

1. Clark, A.P. *Principles of Digital Data Transmission*, Second Edition, Pentech Press, London (1983)
2. Saltzberg, B.R. 'Error probabilities for a binary signal perturbed by intersymbol interference and Gaussian noise', *IEEE Trans. Commun. Syst.*, **CS-12**, 117–120 (1964)
3. Marinides, H.F. and Reijns, G.L. 'Influence of bandwidth restriction on the signal-to-noise performance of a PCM/NRZ signal', *IEEE Trans. Aerospace Electron. Syst.*, **AES-4**, 35–40 (1968)
4. Saltzberg, B.R. 'Intersymbol interference error bounds with application to ideal bandlimiting signalling', *IEEE Trans. Inform. Theory*, **IT-14**, 563–568 (1968)
5. Saltzberg, B.R. and Simon, M.K. 'Data transmission error probabilities in the presence of low-frequency removal and noise', *Bell Syst. Tech. J.*, **48**, 225–273 (1969)
6. Lugannani, R. 'Intersymbol interference and probability of error in digital systems', *IEEE Trans. Inform. Theory*, **IT-15**, 682–688 (1969)
7. Ho, E.Y. and Yeh, Y.S. 'A new approach for evaluating the error probability in the presence of intersymbol interference and additive Gaussian noise', *Bell Syst. Tech. J.*, **49**, 2249–2266 (1970)
8. Ho, E.Y. and Yeh, Y.S. 'Error probability of a multilevel digital system with intersymbol interference and Gaussian noise', *Bell Syst. Tech. J.*, **50**, 1017–1023 (1971)
9. Shimbo, O. and Celebiler, M.I. 'The probability of error due to intersymbol interference and Gaussian noise in digital communication systems', *IEEE Trans. Commun. Technol.*, COM-19, 113–119 (1971)
10. Hill, F.S. 'The computation of error probability for digital transmission', *Bell Syst. Tech. J.*, **50**, 2055–2077 (1971)
11. Yeh, Y.S. and Ho, E.Y. 'Improved intersymbol interference error bounds in digital systems', *Bell Syst. Tech. J.*, **50**, 2585–2598 (1971)
12. Prabhu, V.K. 'Some considerations of error bounds in digital systems', *Bell Syst. Tech. J.*, **50**, 3127–3151 (1971)
13. Forney, G.D. 'Lower bounds on error probability in the presence of large intersymbol interference', *IEEE Trans. Commun.*, **COM-20**, 76–77 (1972)
14. Falconer, D.D. and Gitlin, R.D. 'Bounds on error-pattern probabilities for digital communication systems', *IEEE Trans. Commun.*, **COM-20**, 132–139 (1972)
15. Glave, F.E. 'An upper bound to the probability of error due to intersymbol

interference for correlated digital signals', *IEEE Trans. Inform. Theory*, **IT-18**, 356–363 (1972)
16. Yao, K. 'On minimum average probability of error expression for a binary pulse communication system with intersymbol interference', *IEEE Trans. Inform. Theory*, **IT-18**, 528–531 (1972)
17. Benedetto, S., de Vincentis, G. and Luvison, A. 'Error probability in the presence of intersymbol interference and additive noise for multilevel digital signals, *IEEE Trans. Commun.*, **COM-21**, 181–190 (1973)
18. Hill, F.S. and Blanco, M.A. 'Random geometric series and intersymbol interference', *IEEE Trans. Inform. Theory*, **IT-19**, 326–335 (1973)
19. Matthews, J.W. 'Sharp error bounds for intersymbol interference', *IEEE Trans. Inform. Theory*, **IT-19**, 440–447 (1973)
20. McGee, W.F. 'A modified intersymbol interference error bound', *IEEE Trans. Commun.*, **COM-21**, 862–863 (1973)
21. Korn, I. 'Probability of error in binary communication systems with causal bandlimited filters—Part 1: Nonreturn-to-zero signal', *IEEE Trans. Commun.*, **COM-21**, 878–890 (1973)
22. Korn, I. 'Probability of error in binary communication systems with causal bandlimited filters—Part 2: Split-phase signal', *IEEE Trans. Commun.*, **COM-21**, 891–898 (1973)
23. Bozic, S.M. and Tahim, K.S. 'Dependence of intersymbol interference upon the signalling rate', *Int. J. Electron.*, **35**, 523–528 (1973)
24. Korn, I. 'Bounds to probability of error in binary communication systems with intersymbol interference and dependent or independent symbols', *IEEE Trans. Commun.*, **COM-22**, 251–254 (1974)
25. Vanelli, J.C. and Shehadeh, N.M. 'Computation of bit-error probability using the trapezoidal integration rule', *IEEE Trans. Commun.*, **COM-22**, 331–334 (1974)
26. Korn, I. 'Improvement to sharp error bounds for intersymbol interference', *Proc. IEE*, **122**, 265–267 (1975)
27. Aein, J.M. and Hancock, J.C. 'Reducing the effects of intersymbol interference with correlation receivers', *IEEE Trans. Inform. Theory*, **IT-9**, 167–175 (1963)
28. George, D.A. 'Matched filters for interfering symbols', *IEEE Trans. Inform. Theory*, **IT-11**, 153–154 (1965)
29. Tufts, D.W. 'Nyquist's problem—the joint optimization of transmitter and receiver in pulse amplitude modulation', *Proc. IEEE*, **53**, 248–259 (1965)
30. Smith, J.W. 'The joint optimization of transmitted signal and receiving filter for data transmission systems', *Bell Syst. Tech. J.*, **44**, 2363–2392 (1965)
31. Aaron, M.R. and Tufts, D.W. 'Intersymbol interference and error probability', *IEEE Trans. Inform. Theory*, **IT-12**, 26–34 (1966)
32. Deighton, P.D. 'Optimisation of realisable receiver in pulse amplitude modulation', *Electronics Letters*, **3**, 129–130 (1967)
33. Berger, T. and Tufts, D.W. 'Optimum pulse amplitude modulation, Part 1: Transmitter–receiver design and bounds from information theory', *IEEE Trans. Inform. Theory*, **IT-13**, 196–208 (1967)
34. Berger, T. and Tufts, D.W. 'Optimum pulse amplitude modulation, Part 2: Inclusion of timing jitter', *IEEE Trans. Inform. Theory*, **IT-13**, 209–216 (1967)
35. Deighton, P.D. 'Joint optimisation of realisable receiver and transmitter in datatransmission systems', *Electronics Letters*, **3**, 342–344 (1967)
36. Chang, R.W. and Freeny, S.L. 'Hybrid digital transmission systems—Part 1: Joint optimization of analogue and digital repeaters', *Bell Syst. Tech. J.*, **47**, 1663–1686 (1968)
37. Davisson, L.E. 'Steady state error in adaptive mean-square minimization', *IEEE Trans. Inform. Theory*, **IT-16**, 382–385 (1970)
38. Chang, R.W. 'Joint optimization of automatic equalization and carrier

acquisition for digital communication', *Bell Syst. Tech. J.*, **49**, 1069–1104 (1970)
39. Ericson, T. 'Structure of optimum receiving filters in data transmission systems', *IEEE Trans. Inform. Theory*, **IT-17**, 352–353 (1971)
40. Ho, E.Y. 'Optimum equalization and the effect of timing and carrier phase on synchronous data systems', *Bell Syst. Tech. J.*, **50**, 1671–1689 (1971)
41. Hansler, E. 'Some properties of transmission systems with minimum mean-square error', *IEEE Trans. Commun. Technol.*, **COM-19**, 576–579 (1971)
42. Ericson, T. and Johansson, U. 'Digital transmission over coaxial cables', *Ericsson Technics*, **27**, No. 4, 191–272 (1971)
43. Cho, Y.S. 'Optimal equalization of wideband coaxial cable channels using "bump" equalizers', *Bell Syst. Tech. J.*, **51**, 1327–1345 (1972)
44. Moore, J.B. and Hetrakul, P., 'Optimal demodulation of PAM signals', *IEEE Trans. Inform. Theory*, **IT-19**, 188–196 (1973)
45. Gardner, W.A. 'The structure of least mean square linear estimators for synchronous M-ary signals', *IEEE Trans. Inform. Theory*, **IT-19**, 240–243 (1973)
46. Ericson, T. 'Optimum PAM filters are always band limited', *IEEE Trans. Inform. Theory*, **IT-19**, 570–572 (1973)
47. Mark, J.W. 'Relationship between source coding, channel coding and equalization in data transmission', *Proc. IEEE*, **61**, 1657–1659 (1973)
48. Benedetto, S. and Biglieri, E. 'On linear receivers for digital transmission systems', *IEEE Trans. Commun.*, **COM-22**, 1205–1215 (1974)
49. Rappeport, M.A. 'Automatic equalization of data transmission facility distortion using transversal equalizers', *IEEE Trans. Commun. Technol.*, **COM-12**, 65–73 (1964)
50. Schreiner, K.E., Funk, H.L. and Hopner, E. 'Automatic distortion correction for efficient pulse transmission', *IBM J. Res. Develop.*, **9**, 20–30 (1965)
51. Becker, F.K., Holzman, L.N., Lucky, R.W. and Port, E. 'Automatic equalization for digital communication', *Proc. IEEE*, **53**, 96–97 (1965)
52. Lucky, R.W. 'Automatic equalization for digital communication', *Bell Syst. Tech. J.*, **44**, 547–588 (1965)
53. Lucky, R.W. 'Techniques for adaptive equalization of digital communication systems', *Bell Syst. Tech. J.*, **45**, 255-286 (1966)
54. Rudin, H.R. 'Automatic equalization using transversal filters', *IEEE Spectrum*, **4**, 53–59 (1967)
55. Lytle, D.W. 'Convergence criteria for transversal equalizers', *Bell Syst. Tech. J.*, **47**, 1775–1800 (1968)
56. Hirsh, D. and Wolf, W.J. 'A simple adaptive equalizer for efficient data transmission', *IEEE Trans. Commun. Technol.*, **COM-18**, 5–12 (1970)
57. Newhall, E.E., Qureshi, S.U.H. and Simone, C.F. 'A technique for finding approximate inverse systems and its application to equalization', *IEEE Trans. Commun. Technol.*, **COM-19**, 1116–1127 (1971)
58. Guida, A. 'Optimum tapped delay line for digital signals', *IEEE Trans. Commun.*, **COM-21**, 277–283 (1973)
59. Lucky, R.W., Salz, J. and Weldon, E.J. *Principles of Data Communication*, pp. 93–165, McGraw-Hill, New York (1968)
60. Salomonsson, G. 'An equalizer with feedback filter', *Ericsson Technics*, **28**, No. 2, 57–101 (1972)
61. Gitlin, R.D. and Mazo, J.E. 'Comparison of some cost functions for automatic equalization', *IEEE Trans. Commun.*, **COM-21**, 233–237 (1973)
62. Lucky, R.W. and Rudin, H.R. 'Generalized automatic equalization for communication channels', *Proc. IEEE*, **54**, 439–440 (1966)
63. Lucky, R.W. and Rudin, H.R. 'An automatic equalizer for general-purpose communication channels', *Bell Syst. Tech. J.*, **46**, 2179–2208 (1967)
64. Di Toro, M.J. 'Communication in time-frequency spread media using adaptive

equalization', *Proc. IEEE*, **56**, 1653–1679 (1968)
65. Gersho, A. 'Adaptive equalization of highly dispersive channels, for data transmission', *Bell Syst. Tech. J.*, **48**, 55–70 (1969)
66. Proakis, J.G. and Miller, J.H. 'An adaptive receiver for digital signalling through channels with intersymbol interference', *IEEE Trans. Inform. Theory*, **IT-15**, 484–497 (1969)
67. Rudin, H.R. 'A continuously adaptive equalizer for general purpose communication channels', *Bell Syst. Tech. J.*, **48**, 1865–1884 (1969)
68. Potter, J.B. 'The adaptive equalization of communication systems, particularly those using waveform fidelity criteria', *Proc. Inst. Radio Electron. Eng. Aust.*, **30**, 326–336 (1969)
69. Davisson, L.E. and Schwartz, S.C. 'Analysis of a decision-directed receiver with unknown priors', *IEEE Trans. Inform. Theory*, **IT-16**, 270–276 (1970)
70. Proakis, J.G. 'Adaptive digital filters for equalization of telephone channels', *IEEE, Trans. Audio Electroacoustics*, **AU-18**, 195–200 (1970)
71. Niessen, C.W. and Willim, D.K. 'Adaptive equalizer for pulse transmission', *IEEE Trans. Commun. Technol.*, **COM-18**, 377–395 (1970)
72. Lender, A. 'Decision-directed digital adaptive equalization technique for high-speed data transmission', *IEEE Trans. Commun. Technol.*, **COM-18**, 625–632 (1970)
73. De, S. and Davies, A.C. 'Convergence of adaptive equaliser for data transmission', *Electronics Letters*, **6**, 858–861 (1970)
74. Schonfield, T.J. and Schwartz, M. 'Rapidly converging first order training algorithm for an adaptive equalizer', *IEEE Trans. Inform. Theory*, **IT-17**, 431–439 (1971)
75. Chang, R.W. 'A new equalizer structure for fast start-up digital communication', *Bell Syst. Tech. J.*, **50**, 1969–2014 (1971)
76. Schonfield, T.J. and Schwartz, M. 'Rapidly converging second order tracking algorithm for adaptive equalization', *IEEE Trans. Inform. Theory*, **IT-17**, 572–579 (1971)
77. Lawrence, R.E. and Kaufman, H. 'The Kalman filter for the equalization of a digital communications channel', *IEEE Trans. Commun. Technol.*, **COM-19**, 1137–1141 (1971)
78. Mark, J.W. and Haykin, S.S. 'Adaptive equalization for digital communication', *Proc. IEE*, **118**, 1711–1720 (1971)
79. Brewster, R.L. 'A digital adaptive equaliser for high-speed voice channel modems', *IERE Conf. Proc.*, No. 23, 373–382 (1972)
80. Stuttard, E.B. 'Automatic adaptive equalisation in data modems', *IERE Conf. Proc.*, No. 23, 383–390 (1972)
81. Richman, S.H. and Schwartz, M. 'Dynamic programming training period for an MSE adaptive equalizer', *IEEE Trans. Commun.*, **COM-20**, 857–864 (1972)
82. Ungerboeck, G. 'Theory on the speed of convergence in adpative equalizers for digital communication', *IBM J. Res. Develop.*, **16**, 546–555 (1972)
83. Sha, R.T. and Tang, D.T. 'A new class of automatic equalizers', *IBM J. Res. Develop.*, **16**, 556–566 (1972)
84. Chang, R.W. and Ho, E.Y. 'On fast start-up data communication systems using pseudo-random training sequences', *Bell Syst. Tech. J.*, **51**, 2013–2027 (1972)
85. Walzman, T. and Schwartz, M. 'Automatic equalization using the discrete frequency domain', *IEEE Trans. Inform. Theory.*, **IT-19**, 59–68 (1973)
86. Potter, J.B. 'Application of time-series algebra to the adaptive equalization of band-limited waveform-transmission systems', *Proc. IEE*, **120**, 191–196 (1973)
87. Gitlin, R.D., Ho, E.Y. and Mazo, J.E. 'Passband equalization of differentially phase-modulated data signals', *Bell Syst. Tech. J.*, **52**, 219–238 (1973)
88. Karnaugh, M. 'Automatic equalizers having minimum adjustment time', *IBM J.*

Res. Develop., **17**, 176–179 (1973)
89. Mark, J.W. and Budihardjo, P.S. 'Joint optimization of receive filter and equalizer', *IEEE Trans. Commun.*, **COM-21**, 264–266 (1973)
90. Mark, J.W. 'A note on the modified Kalman filter for channel equalization', *Proc. IEEE*, **61**, 481–482 (1973)
91. Qureshi, S.U.H. 'Adjustment of the position of the reference tap of an adaptive equalizer', *IEEE Trans. Commun.*, **COM-21**, 1046–1052 (1973)
92. Walzman, T. and Schwartz, M. 'A projected gradient method for automatic equalization in the discrete frequency domain', *IEEE Trans. Commun.*, **COM-21**, 1442–1446 (1973)
93. Pulleyblank, R.W. 'A comparison of receivers designed on the basis of minimum mean-square error and probability of error for channels with intersymbol interference and noise', *IEEE Trans. Commun.*, **COM-21**, 1434–1438 (1973)
94. Taylor, M.G. 'A technique for using a time-multiplexed second order digital filter section for performing adaptive filtering', *IEEE Trans. Commun.*, **COM-22**, 326–330 (1974)
95. Godard, D. 'Channel equalization using a Kalman filter for fast data transmission', *IBM J. Res. Develop.*, **18**, 267–273 (1974)
96. Cho, Y.S. 'Mean square error equalization using manually adjusted equalizers', *Bell Syst. Tech. J.*, **53**, 847–865 (1974)
97. Lee, T.S. and Cunningham, D.R. 'Kalman filter equalizer for QPSK digital communication channel', *Conf. Rec. IEEE Int. Conf. Communications*, pp. 25A.1–25A.5 (1974)
98. Fleischer, P.E. 'Active adjustable loss and delay equalizers', *IEEE Trans. Commun.*, **COM-22**, 951–955 (1974)
99. Kleibanov, S.B., Privalskii, V.B. and Time, I.V. 'Kalman filter for equalization of digital communications channel', *Autom. Remote Control*, **35**, No. 7, Pt. 1, 1097–1102 (1974)
100. Eriksson, L.E. 'Transmitter and receiver filters for digital PAM using transversal filters with few taps', *IEEE Trans. Commun.*, **COM-22**, 1215–1225 (1974)
101. Allen, J.B. and Mazo, J.E. 'A decision-free equalization scheme for minimum phase channels', *IEEE Trans. Commun.*, **COM-22**, 1732–1733 (1974)
102. Kosovych, O.S. and Pickholtz, R.L. 'Automatic equalization using successive overrelaxation iterative technique', *IEEE Trans. Inform. Theory*, **IT-21**, 51–58 (1975)
103. Mueller, K.H. 'A new fast-converging mean square algorithm for adaptive equalizer with partial response signalling', *Bell Syst. Tech. J.*, **54**, 143–153 (1975)
104. Mueller, K.H. and Spaulding, D.A. 'Cyclic equalization—A new rapidly converging equalization technique for synchronous data communication', *Bell Syst. Tech. J.*, **54**, 369–406 (1975)
105. Luvison, A. and Pirani, G. 'A method to compute optimal gains in recursive linear filtering applications', *IEEE Trans. Commun.*, **COM-23**, 399–400 (1975)
106. Fitch, S.M. and Kurz, L. 'Recursive equalization in data transmission—A design procedure and performance evaluation', *IEEE Trans. Commun.*, **COM-23**, 546–550 (1975)
107. Butler, P. and Cantoni, A. 'Noniterative automatic equalization', *IEEE Trans. Commun.*, **COM-23**, 621–633 (1975)
108. Macleod, C.J., Ciapala, E. and Jelonek, Z.J. 'Quantisation in non-recursive equalisers for data transmission', *Proc. IEE*, **122**, 1105–1110 (1975)
109. Lee, T.S. and Cunningham, D.R. 'Kalman filter equalization for QPSK communications', *IEEE Trans. Commun.*, **COM-24**, 361–364 (1976)
110. Eriksson, L.E. and Van den Elzen, H.C. 'An equalizer structure with reduced sampling time reference sensitivity', *IEEE Trans. Commun.*, **COM-24**, 1337–1343 (1976)

111. Cantoni, A. and Butler, P. 'Linear minimum mean-square error estimators applied to channel equalization', *IEEE Trans. Commun.*, **COM-25**, 441–446 (1977)
112. Sakaki, H., Shintani, S. and Kuroda, H. 'A rapidly converging equalizer with variable tap weight adjusting coefficients', *Electron. Commun. Japan*, **60-B**, 77–84 (1977)
113. Gitlin, R.D. and Magee, F.R. 'Self-orthogonalizing adaptive equalization algorithms', *IEEE Trans. Commun.*, **COM-25**, 666–672 (1977)
114. Sakaki, H., Shintani, S. and Kuroda, H. 'Some consideration on automatic equalizer with coefficient matrices', *Electron. Commun. Japan*, **61-B**, 59–67 (1978)
115. Schmidt, W. 'An automatic adaptive equalizer for data transmission', *Proc. Int. Symp. Circuits and Systems*, New York, 436–440 (1978)
116. Cecchi, E., Martinelli, G., Orlandi, G. and Salerno, M. 'Possibility of automatically acquiring the optimal adjusting step in adaptive equalisers', *Proc. IEE*, **125**, 626–632 (1978)
117. Doan, H.B. and Cantoni, A. 'Fast adaptation of equaliser and desired response for digital data receivers', *Electron. Circuits and Systems*, **2**, 159–166 (1978)
118. Morgan, D.R. 'Adaptive multipath cancellation for digital data communications', *IEEE Trans. Commun.*, **COM-26**, 1380–1390 (1978)
119. Nicholson, G. and Norton, J.P. 'Bias and residual intersymbol interference of minimum-variance equalisers for digital communication', *Electronics Letters*, **14**, 678–679 (1978)
120. Falconer, D.D. and Ljung, L. 'Application of fast Kalman estimation to adaptive equalization', *IEEE Trans. Commun.*, **COM-26**, 1439–1446 (1978)
121. Glover, J.R. 'Higher order algorithms for adaptive filters', *IEEE Trans. Commun.*, **COM-27**, 216–221 (1979)
122. Nicholson, G. and Norton, J.P. 'Kalman filter equalization for a time-varying communication channel', *Australian Telecommun. Research*, **13**, 3–12 (1979)
123. Satorius, E.H. and Alexander, S.T. 'Channel equalization using adaptive lattice algorithms', *IEEE Trans. Commun.*, **COM-27**, 899–905 (1979)
124. Prasad, S. 'A novel theorem with potential application to automatic equalization', *IEEE Trans. Commun.*, **COM-27**, 1254–1257 (1979)
125. Jin, Y.I. 'New adaptive equalizer with infinite impulse response', *Electroncs Letters*, **15**, 557–559 (1979)
126. Luvison, A. and Pirani, G. 'Design and performance of an adaptive Kalman receiver for synchronous data transmission', *IEEE Trans. Aerospace and Electronic Systems*, **AES-15**, 635–648 (1979)
127. Weiss, A. and Mitra, D. 'Digital adaptive filters: Conditions for convergence, rates of convergence, effects of noise and errors arising from the implementation', *IEEE Trans. Inform. Theory*, **IT-25**, 637–652 (1979)
128. Speidel, J. 'A new automatic recursive equalizer with improved convergence properties', *Int. Zurich Seminar on Digital Transmission in Wireless Systems*, G7/1–6 (1980)
129. Millot, L.J. 'A general class of PAM equalizers'. *IEEE Trans. Commun.*, **COM-28**, 915–917 (1980)
130. Dalle Mese, E. and Corsini, G. 'Adaptive Kalman filter equaliser', *Electronics Letters*, **16**, 547–549 (1980)
131. Mazo, J.E. 'Analysis of decision-directed equalizer convergence', *Bell Syst. Tech. J.*, **59**, 1857–1876 (1980)
132. Marino-Acebal, J.B., Mayer-Pujadas, A., Masgrau, E. and Nadeu, C. 'Design of digital equalizers with minimax error', *Signal Processing 1: Theories and Applications*, M. Kunt and F. de Coulon (editors), North Holland, Amsterdam (1981)
133. Satorius, E.H. and Pack, J.D. 'Application of least-squares lattice algorithms to adaptive equalization', *IEEE Trans. Commun.*, **COM-29**, 136–142 (1981)

134. Westfall, F.A. 'Efficient digital signal processing realizations for PSK modem receivers', *IERE Conf. Proc.*, No. 49, 143–152 (1981)
135. Korobkov, D.L. 'Digital matrix correction in the frequency domain', *Telecommun. and Radio Eng.*, Pt. 2, USA, **36**, 60–63 (1981)
136. Hawksford, M.J. and Rezaee, N., 'Adaptive mean-square-error transversal equalizer', *Proc. IEE*, Pt. F, **128**, 296–304 (1981)
137. Kumar, R. and Moore, J.B. 'Adaptive equalization via fast quantized-state methods', *IEEE Trans. Commun.*, **COM-29**, 1492–1501 (1981)
138. Mueller, M.S. and Salz, J. 'A unified theory of data-aided equalization', *Bell Syst. Tech. J.*, **60**, 2023–2038 (1981)
139. Takatori, H., Suzuki, T., Tomooka, K. and Ogawa, M. 'A new equalizing scheme for digital subscriber loop', *Nat. Commun. Conf.*, New Orleans, USA, **3**, E1.5/1–6 (1981)
140. Mueller, M.S. 'On the rapid initial convergence of least-squares equalizer adjustment algorithms', *Bell Syst. Tech. J.*, **60**, 2345–2358 (1981)
141. Sari, H. 'Simplified algorithms for adaptive channel equalization', *Philips J. Res.*, **37**, 56–77 (1982)
142. Mueller, K.H. and Werner, J.J. 'A hardware efficient passband equalizer structure for data transmission', *IEEE Trans. Commun.*, **COM-30**, 538–540 (1982)
143. Mills, W.C. 'A microprocessor based 9600 bit/s modem', *Commun. and Broadcast.*, **7**, 35–40 (1982)
144. Treichler, J.R. and Larimore, M.G. 'Thinned impulse responses for adaptive FIR filters', *IEEE Int. Conf. on Acoustics, Speech and Signal Process.*, Paris, France, **2**, 631–634 (1982)
145. Sari, H. 'Performance evaluation of three adaptive equalization algorithms', *IEEE Int. Conf. on Acoustics, Speech and Signal Process.*, Paris, France, **3**, 1385–1389 (1982)
146. Eweda, E. and Macchi, O. 'Equalization of rapid selective fadings with unknown and time-varying forms', *IEEE Int. Conf. on Acoustics, Speech and Signal Process.*, Paris, France, **3**, 1390–1393 (1982)
147. Preis, D. and Bunks, C. 'Minimax equalizers for digital communication', *IEEE Int. Conf. on Acoustics, Speech and Signal Process.*, Paris, France, **3**, 1773–1776 (1982)
148. Ramesh, N.S. and Mitra, S.K. 'Block adaptive equalization', *Int. Symp. on Circuits and Systems*, Rome, Italy, **2**, 686–689 (1982)
149. Miller, C.K. 'Automatic adaptive equalization for medium speed modems', *Comput. Des. (USA)*, **21**, 233–238 (1982)
150. Sakaniwa, K. and Yokoyama, H. 'An automatic equalizer using block-iterative algorithm', *Trans. IECE Japan*, **J65-A**, 1011–1018 (1982)
151. Kuo, Y.L. and Aprille, T.J. 'A baseband adaptive equalizer for a 16-state QAM digital system over mastergroup band analogue networks', *IEEE Global Telecommun. Conf.*, Miami, FL, USA, **3**, 1246–1250 (1982)
152. Kabal, P. 'The stability of adaptive minimum mean square error equalizers using delayed adjustment', *IEEE Trans. Commun.*, **COM-31**, 430–432 (1983)
153. Yucel, M.D., Tepedelenlioglu, N. and Tanik, Y. 'A fast noniterative method for adaptive channel equalization', *IEEE Trans. Commun.*, **COM-31**, 922–927 (1983)
154. Picchi, G. and Prati, G. 'Self-orthogonalizing adaptive equalization in the discrete frequency domain', *IEEE Trans. Commun.*, **COM-32**, 371–379 (1984)
155. Clark, A.P. *Advanced Data-Transmission Systems*, Pentech Press, London (1977)
156. Savage, J.E. 'Some simple self-synchronizing digital data scramblers', *Bell Syst. Tech. J.*, **46**, 449–487 (1967)
157. Nakamura, K. and Iwadare, Y. 'Data scramblers for multilevel pulse sequences', *Electron. Commun. Japan*, **55-A**, No. 6, 8–16 (1972)

158. Leeper, D.G. 'A universal digital data scrambler', *Bell Syst. Tech. J.*, **52**, 1851–1865 (1973)
159. Kasai, H., Senmoto, S. and Matsushita, M. 'PCM jitter suppression by scrambling', *IEEE Trans. Commun.*, **COM-22**, 1114–1122 (1974)
160. Stiffler, J.J. *Theory of Synchronous Communications*, Prentice-Hall, Englewood Cliffs, New Jersey (1971)
161. Harvey, J.D. *Synchronisation of a Synchronous Modem*, SERC report GR/A/1200.7, SERC, Swindon (1980)
162. O'Reilly, J.J. 'Timing extraction for baseband digital transmission', Colloquium on Nonlinear Operations on Stochastic Processes, London, 42–53 (1981)
163. Papoulis, A. *Probability, Random Variables and Stochastic Processes*. McGraw-Hill, New York (1965)
164. Thomas, J.B. *An Introduction to Statistical Communication Theory*, Wiley, New York (1969)
165. Davenport, W.B. *Probability and Random Processes*, McGraw-Hill, New York (1970)
166. Dwight, H.B. *Tables of Integrals and other Mathematical Data*, fourth edition, Macmillan, New York (1961)
167. Hyslop, J.M. *Infinite series*, fifth edition, Oliver and Boyd, Edinburgh (1954)
168. Sokowski, E. *Fundamentals of College Algebra*, second edition, Prindle, Weber and Schmidt, Boston, Mass. (1971)
169. Paley, H. and Weichsel, P.M. *Elements of Abstract and Linear Algebra*, Holt, Rinehart and Winston, New York (1972)
170. Ayres, F. *Matrices*, McGraw-Hill, New York (1962)
171. Wozencraft, J.M. and Jacobs, I.M. *Principles of Communication Engineering*, Wiley, New York (1965)
172. Lathi, B.P. *An Introduction to Random Signals and Communication Theory*, Intertext Books, London (1968)

Chapter 5

Decision-feedback equalizers

5.1 DECISION DIRECTED CANCELLATION OF INTERSYMBOL INTERFERENCE

During the last few years a considerable amount of work has been carried out on decision-feedback (nonlinear) equalizers and it has been shown that these can sometimes achieve a much better tolerance to additive noise than linear equalizers[1-73]. To understand and hence to be able to analyse the method of operation of these equalizers, it is necessary first to study the basic principles of decision-directed cancellation of intersymbol interference.

Consider the data-transmission system shown in Fig. 5.1. The signal at the input to the transmitter filter is a sequence of regularly spaced impulses $\{s_i \delta(t - iT)\}$, the $(i+1)$th of which occurs at time $t = iT$ seconds and has the value (area)

$$s_i = \pm k \qquad (5.1)$$

where k is a positive constant. It is assumed that $s_i = 0$ for $i < 0$. For $i \geq 0$, each impulse $s_i \delta(t - iT)$ is a binary polar signal-element. An important assumption that is made throughout the following analysis is that the $\{s_i\}$ are *statistically independent* and *equally likely* to have either binary value. Where this condition is not satisfied by the $\{s_i\}$, it can normally be achieved by *scrambling* the transmitted sequence of data symbols and appropriately *descrambling* the corresponding detected data-symbols at the receiver, to obtain a copy of the original transmitted sequence[74-77]. Multilevel data-symbols could be used in place of binary symbols without unduly affecting any of the important results in this analysis.

The transmitter filter, transmission path and receiver filter in Fig. 5.1 together form a *linear baseband channel* whose impulse response is $y(t)$ and has, for practical purposes, a *finite duration*. It is assumed that $y(t)$ is real valued and that $y(t - iT)$ is a time-shifted version of $y(t - jT)$, at least where $|i - j|$ does not exceed the number of received signal elements effectively involved in the detection of any one data symbol. White Gaussian noise with zero mean and a two-sided power

DECISION-FEEDBACK EQUALIZERS 231

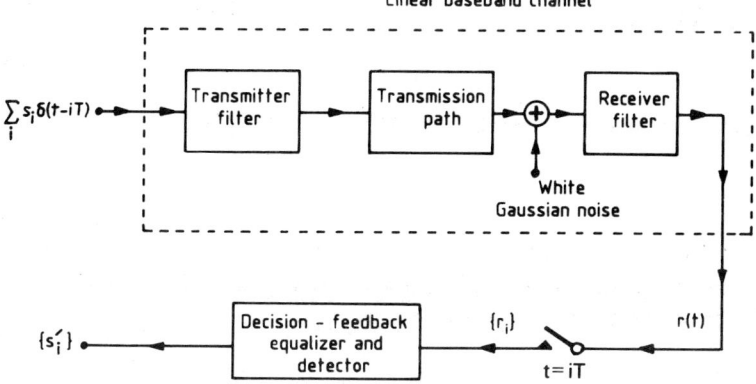

Fig. 5.1   Data-transmission system

spectral density of $\tfrac{1}{2}N_0$ is added to the data signal at the output of the transmission path, giving the Gaussian waveform $w(t)$ at the output of the receiver filter, that is, at the output of the baseband channel in Fig. 5.1. Thus, the resultant received signal waveform here is

$$r(t) = \sum_i s_i y(t - iT) + w(t) \tag{5.2}$$

and this is sampled once per data-symbol s_i, at the time instants $\{iT\}$ for $i = 0, 1, 2, \ldots$. The received signal at time $t = iT$, at the output of the sampler, is now

$$r_i = \sum_{j=0}^{g} s_{i-j} y_j + w_i \tag{5.3}$$

where $r_i = r(iT)$, $y_j = y(jT)$ and $w_i = w(iT)$. Each of these signals is *real valued*. As before, the *sampled impulse-response* of the linear baseband channel is given by the $(g+1)$-component vector (sequence)

$$Y = [y_0 \quad y_1 \quad \ldots \quad y_g] \tag{5.4}$$

whose z-transform is

$$Y(z) = y_0 + y_1 z^{-1} + \cdots + y_g z^{-g} \tag{5.5}$$

The delay in transmission, other than that involved in the time dispersion of the transmitted signal, is neglected here, so that $y_0 \neq 0$ and $y_i = 0$ for $i < 0$ and $i > g$.

As is shown in Section 1.2, the transmitter filter in Fig. 5.1 is such that the average transmitted energy per signal element is the mean-

232 DECISION-FEEDBACK EQUALIZERS

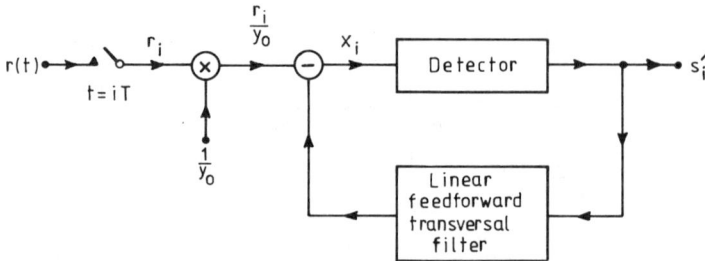

Fig. 5.2 Receiver using nonlinear equalization by decision-directed cancellation of intersymbol interference

square value of s_i, which is

$$\mathscr{E}[s_i^2] = k^2 \tag{5.6}$$

from Equation 1.22, and, from Equation 1.24, the receiver filter is such that the Gaussian noise components $\{w_i\}$ are statistically independent Gaussian random variables with zero mean and variance

$$\sigma^2 = \tfrac{1}{2} N_0 \tag{5.7}$$

where $\tfrac{1}{2} N_0$ is the two-sided power spectral density of the white Gaussian noise at the input to the receiver filter. Thus, the *signal/noise ratio*, measured in dB as the average transmitted energy per bit at the *input* to the transmission path, divided by the two-sided noise power spectral density at the *output* of the transmission path, is

$$\psi = 10 \log_{10} (k^2/\sigma^2) \tag{5.8}$$

The decision-feedback equalizer and detector in the receiver of Fig. 5.1 is now the nonlinear filter shown in Fig. 5.2. The signals given here are those at the time instant $t = iT$, and s_i' is the detected value of s_i. The decision-feedback equalizer is implemented as a linear feedforward transversal filter fed from the output of the detector, as shown in Fig. 5.3, where

$$v_j = y_j/y_0 \tag{5.9}$$

for $j = 1, 2, \ldots, g$. The output signal from the transversal filter in Figs. 5.2 and 5.3 is an estimate of the intersymbol interference in the signal r_i/y_0 at the output of the multiplier, and is subtracted from this signal to give the equalized signal x_i at the detector input. The equalizer therefore operates by *decision-feedback* (or quantized-feedback) correction, removing the intersymbol interference from the detector input signal, as will now be explained in more detail.

DECISION-FEEDBACK EQUALIZERS   233

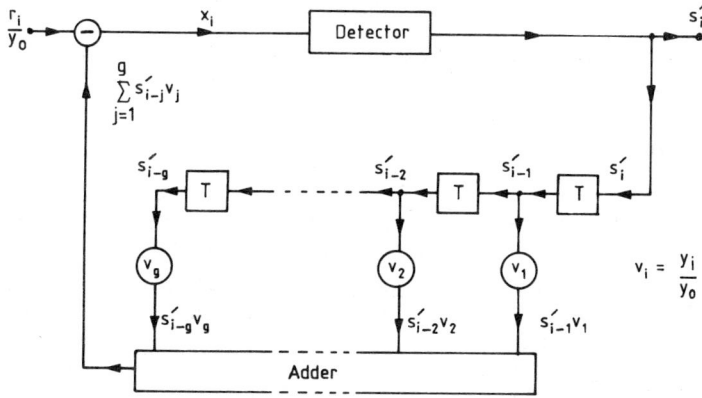

Fig. 5.3   Detector and pure nonlinear equalizer

It is first interesting to observe that, if the detector is removed from Fig. 5.3, the *linear* feedback transversal equalizer of Fig. 4.10 is obtained. The arrangement in Fig. 5.3 is therefore a *nonlinear feedback transversal equalizer*. The equalizer is nonlinear because the detector, which is a nonlinear device, is included in the feedback path of the equalizer. In order to distinguish the equalizer shown in Figs. 5.2 and 5.3 from the more general nonlinear equalizers referred to as *decision-feedback equalizers* (see Sections 5.5–5.8), the particular equalizers considered here will be referred to as *pure nonlinear equalizers*. This emphasizes the fact that the equalization process is achieved *entirely* by means of a *nonlinear* filter. The equalizer is, of course, a special case of a decision-feedback equalizer.

The received signal at the input to the multiplier in Fig. 5.2, at time $t = iT$, is

$$r_i = s_i y_0 + \sum_{j=1}^{g} s_{i-j} y_j + w_i \qquad (5.10)$$

where $s_i y_0$ is the wanted signal. The corresponding signal at the output of the multiplier is

$$\frac{r_i}{y_0} = s_i + \sum_{j=1}^{g} s_{i-j} \frac{y_j}{y_0} + \frac{w_i}{y_0}$$

$$= s_i + \sum_{j=1}^{g} s_{i-j} v_j + \frac{w_i}{y_0} \qquad (5.11)$$

where $v_j = y_j/y_0$. In Equation 5.11, s_i is the *wanted signal*, $\sum_{j=1}^{g} s_{i-j} v_j$ is the *intersymbol interference*, and w_i/y_0 is the *noise*.

The sampled impulse-response of the baseband channel, sampler and multiplier is

$$V = \frac{1}{y_0} Y = [1 \quad v_1 \quad v_2 \quad \ldots \quad v_g] \qquad (5.12)$$

whose z-transform is

$$V(z) = 1 + v_1 z^{-1} + v_2 z^{-2} + \cdots + v_g z^{-g} \qquad (5.13)$$

and both vectors (sequences) Y and V are assumed to be known at the receiver, as are the two possible values $\pm k$ of the data-symbol s_i.

The signals shown in Fig. 5.3 are those present at the time-instant $t = iT$. A square marked T is here a store that effectively introduces a delay of one sampling interval (T seconds), and a circle marked v_j is a multiplier that multiplies its input signal by v_j. Thus the detected data-symbols $\{s_i'\}$ are fed to a linear transversal filter that operates in a similar manner to the filter in Fig. 4.3. Associated with each store in Fig. 5.3 is a multiplier that multiplies the output signal s_{i-j}' from the store by v_j, and the resulting products are then added so that the output signal from the transversal filter, at time $t = iT$, is

$$\sum_{j=1}^{g} s_{i-j}' v_j \qquad (5.14)$$

Thus, the input signal to the detector, at time $t = iT$, is

$$x_i = \frac{r_i}{y_0} - \sum_{j=1}^{g} s_{i-j}' v_j$$

$$= s_i + \sum_{j=1}^{g} s_{i-j} v_j + \frac{w_i}{y_0} - \sum_{j=1}^{g} s_{i-j}' v_j \qquad (5.15)$$

and, with the correct detection of each s_{i-j} such that

$$s_{i-j}' = s_{i-j} \qquad (5.16)$$

for $j = 1, 2, \ldots, g$, this becomes

$$x_i = s_i + \frac{w_i}{y_0} \qquad (5.17)$$

In the correctly equalized signal x_i (Equation 5.17) s_i is the *wanted signal* and w_i/y_0 is the *noise*.

In the optimum detection process for s_i from x_i, the detected value of s_i is taken as its *possible* value *closest* to x_i. The data-symbol s_i is here detected from x_i by comparing the latter with a decision threshold of zero. Thus, when $x_i > 0$, $s_i' = k$, and when $x_i < 0$, $s_i' = -k$.

It can be seen from Equations 5.11 and 5.17 that both r_i/y_0 and x_i have the *same* wanted-signal component s_i and noise component w_i/y_0, so that the nonlinear equalizer removes the intersymbol interference *without changing* the signal/noise ratio. This is in contrast to the action of a linear equalizer, which, in the presence of any amplitude distortion, tends to *reduce* the signal/noise ratio (see Section 4.10).

If s_i is correctly detected from x_i such that $s_i'=s_i$, and given the correct detection of $s_{i-1}, s_{i-2}, \ldots, s_{i-g+1}$, then, on the receipt of r_{i+1} at time $t=(i+1)T$, the signal at the input to the pure nonlinear equalizer is

$$\frac{r_{i+1}}{y_0} = s_{i+1} + \sum_{j=1}^{g} s_{i+1-j} v_j + \frac{w_{i+1}}{y_0} \qquad (5.18)$$

From r_{i+1}/y_0 is subtracted the signal

$$\sum_{j=1}^{g} s_{i+1-j} v_j \qquad (5.19)$$

at the output of the transversal filter, to give the equalized signal

$$x_{i+1} = s_{i+1} + \frac{w_{i+1}}{y_0} \qquad (5.20)$$

at the detector input. The process continues in this way.

It can be seen that the pure nonlinear equalizer uses the detected data-symbols $\{s_i'\}$ to *synthesize* the intersymbol-interference component in a received signal r_i and it then removes the intersymbol interference by *subtraction*. This is a process of decision-directed cancellation of intersymbol interference.

It is clear that, so long as the data-symbols $\{s_i\}$ are correctly detected, their intersymbol interference is removed (cancelled) from the following equalized signals by the nonlinear equalizer, and the channel continues to be accurately equalized. To start this process of nonlinear equalization, a *known* sequence of more than g data-symbols $\{s_i\}$ is transmitted, and the intersymbol interference introduced by these data symbols is removed automatically in the nonlinear equalizer, without requiring the detection of the corresponding $\{s_i\}$. The channel is now correctly equalized, and the following received data-symbols are detected and cancelled as just described. It is interesting to observe that *no* assumptions have been made here concerning the zeros (roots) of $Y(z)$. There may or may not be zeros of $Y(z)$ that lie *on* the unit circle in the z plane, and, of course, there may be zeros of $Y(z)$ both inside and outside the unit circle or else *all* zeros of $Y(z)$ may lie either inside or outside the unit circle.

It can be seen that, when a received data-symbol is correctly detected and cancelled, the following equalized signals become as though the data symbol had *never been transmitted*. Thus when a received data-symbol s_i is being detected, following the *correct* detection and cancellation of the preceding data-symbols, the corresponding equalized signal x_i is as though the data-symbol being detected is the *first* of the received sequence of data symbols.

5.2 EXAMPLE

The $(i+1)$th sample of a received binary baseband signal is

$$r_i = s_i + as_{i-1} + bs_{i-2} + w_i \qquad (5.21)$$

where the $\{s_i\}$ are statistically independent and equally likely to be either ± 1. a and b are constants, and w_i is a Gaussian random variable with zero mean and fixed variance.

Describe, with the aid of a diagram, the pure nonlinear equalizer that operates on r_i so that s_i is detected without intersymbol interference.

Describe the corresponding arrangement for the case where

$$r_i = s_i + as_{i-1} + bs_{i-1}s_{i-2} + w_i \qquad (5.22)$$

How would you detect s_i from r_i for the case where

$$r_i = s_i + \tfrac{1}{2}s_i s_{i-1} + w_i \qquad (5.23)$$

and for the case where

$$r_i = s_i s_{i-1} + s_{i-1} w_i \qquad (5.24)$$

Solution

The pure nonlinear equalizer for the case where

$$r_i = s_i + as_{i-1} + bs_{i-2} + w_i \qquad (5.25)$$

is as shown in Fig. 5.4. The output signal from the 2-tap transversal filter, with

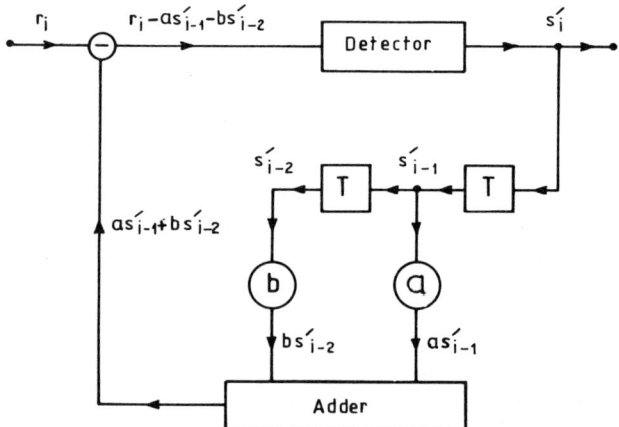

Fig. 5.4 Pure nonlinear equalizer for the case where $r_i = s_i + as_{i-1} + bs_{i-2} + w_i$

tap gains a and b, is

$$as'_{i-1} + bs'_{i-2}$$

With the correct detection of s_{i-1} and s_{i-2}, the output signal becomes

$$as_{i-1} + bs_{i-2}$$

and the input signal to the detector is now

$$x_i = r_i - as_{i-1} - bs_{i-2} = s_i + w_i \qquad (5.26)$$

Hence the intersymbol interference is eliminated from x_i, and s_i is detected by comparing x_i with a decision-threshold of zero. Thus, when $x_i > 0$, $s'_i = 1$, and when $x_i < 0$, $s'_i = -1$.

The pure nonlinear equalizer for the case where

$$r_i = s_i + as_{i-1} + bs_{i-1}s_{i-2} + w_i \qquad (5.27)$$

is as shown in Fig. 5.5. This operates in the same way as the previous equalizer, except that a third multiplier is now included to multiply the signal s'_{i-2} by s'_{i-1}, as shown, so that the output signal from the modified transversal filter is

$$as'_{i-1} + bs'_{i-1}s'_{i-2}$$

With the correct detection of s_{i-1} and s_{i-2}, the output signal becomes

$$as_{i-1} + bs_{i-1}s_{i-2}$$

and the input signal to the detector is now

$$x_i = r_i - as_{i-1} - bs_{i-1}s_{i-2} = s_i + w_i \qquad (5.28)$$

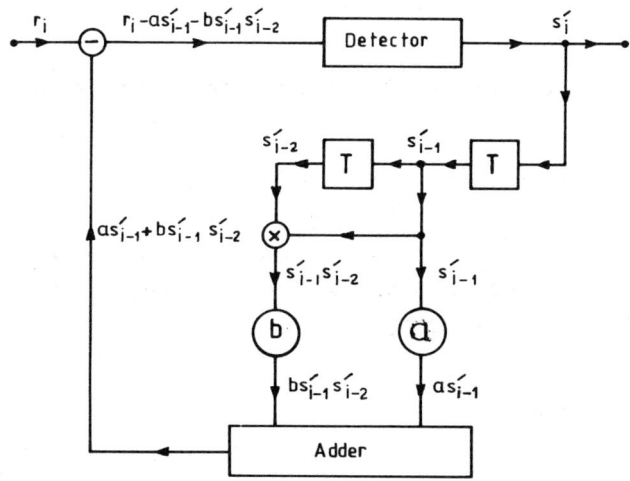

Fig. 5.5 Pure nonlinear equalizer for the case where
$r_i = s_i + as_{i-1} + bs_{i-1}s_{i-2} + w_i$

so that, again, all intersymbol interference is eliminated at the detector input. s_i is detected from x_i as before.

When
$$r_i = s_i + \tfrac{1}{2} s_i s_{i-1} + w_i \qquad (5.29)$$

the intersymbol interference term $\tfrac{1}{2} s_i s_{i-1}$ is a function of s_i and so it cannot be determined *prior* to the detection of s_i. Decision-directed cancellation of intersymbol interference cannot therefore be used here. However, the received sample can be rewritten as

$$r_i = s_i(1 + \tfrac{1}{2} s_{i-1}) + w_i \qquad (5.30)$$

and, following the detection of s_{i-1}, the receiver can determine $1 + \tfrac{1}{2} s_{i-1}$ as $1 + \tfrac{1}{2} s'_{i-1}$ and then form

$$\frac{r_i}{1 + \tfrac{1}{2} s'_{i-1}} = \frac{s_i(1 + \tfrac{1}{2} s_{i-1})}{1 + \tfrac{1}{2} s'_{i-1}} + \frac{w_i}{1 + \tfrac{1}{2} s'_{i-1}} \qquad (5.31)$$

which, with the correct detection of s_{i-1}, becomes

$$\frac{r_i}{1 + \tfrac{1}{2} s_{i-1}} = s_i + \frac{w_i}{1 + \tfrac{1}{2} s_{i-1}} \qquad (5.32)$$

The data-symbol s_i is now detected as 1 or -1, depending upon whether $r_i / (1 + \tfrac{1}{2} s'_{i-1})$ is positive or negative, respectively.

A simpler but equally effective detection process operates as follows. Since $\tfrac{1}{2} s_i s_{i-1}$ has the possible values $\pm \tfrac{1}{2}$ and s_i has the possible values ± 1, the intersymbol interference introduced by $\tfrac{1}{2} s_i s_{i-1}$ does not of itself prevent the correct detection of s_i from r_i. Clearly, s_i can be detected directly from r_i, such that, when $r_i > 0$, $s'_i = 1$, and when $r_i < 0$, $s'_i = -1$. The latter detection process is, in fact, effectively the *same* as that previously described.

When
$$r_i = s_i s_{i-1} + s_{i-1} w_i \qquad (5.33)$$

it can be seen that
$$s_{i-1}^{-1} r_i = s_i + w_i \qquad (5.34)$$

But, since $s_{i-1} = \pm 1$,
$$s_{i-1} r_i = s_i + w_i \qquad (5.35)$$

so that the simple arrangement in Fig. 5.6 can now be used, to give the signal

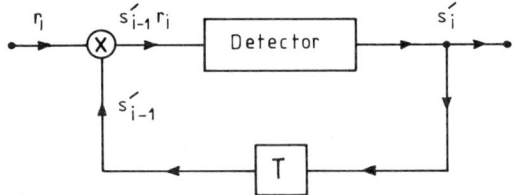

Fig. 5.6 *Pure nonlinear equalizer for the case where* $r_i = s_i s_{i-1} + s_{i-1} w_i$

$s'_{i-1}r_i$ at the detector input. With the correct detection of s_{i-1}, the signal at the detector input becomes $s_{i-1}r_i$ in Equation 5.35, which means that all intersymbol interference has been removed. The data-symbol s_i is detected from $s'_{i-1}r_i$, such that, when $s'_{i-1}r_i > 0$, $s'_i = 1$, and when $s'_{i-1}r_i < 0$, $s'_i = -1$.

5.3 TOLERANCE TO ADDITIVE GAUSSIAN NOISE

Consider the pure nonlinear equalizer of Figs. 5.2 and 5.3 operating with a channel whose z-transform $Y(z)$ is given by Equation 5.5, and assume the correct detection of the data-symbols $s_{i-1}, s_{i-2}, \ldots, s_{i-g}$. The equalized signal at the detector input is now

$$x_i = s_i + \frac{w_i}{y_0} \tag{5.36}$$

where
$$s_i = \pm k \tag{5.37}$$

as shown in Equations 5.1 and 5.17. In the detector, s_i is detected as k or $-k$, depending upon whether x_i is positive or negative, respectively. An error now occurs in s'_i (the detected value of s_i) when x_i has the opposite sign to s_i, and this occurs when the noise component w_i/y_0 has a magnitude greater than k and the opposite sign to s_i. But w_i/y_0 is a Gaussian random variable with zero mean and variance σ^2/y_0^2, so that the probability density function of w_i/y_0, at the sample-value w, is

$$p(w) = \frac{1}{\sqrt{(2\pi y_0^{-2}\sigma^2)}} \exp\left(-\frac{w^2}{2y_0^{-2}\sigma^2}\right) \tag{5.38}$$

and the probability of error in the dection of a received data-symbol, following the correct detection of the preceding g symbols, is

$$P_e = \int_k^\infty p(w)\,dw = \int_k^\infty \frac{1}{\sqrt{(2\pi y_0^{-2}\sigma^2)}} \exp\left(-\frac{w^2}{2y_0^{-2}\sigma^2}\right) dw$$

$$= \int_{k|y_0|/\sigma}^\infty \frac{1}{\sqrt{2\pi}} \exp(-\tfrac{1}{2}w^2)\,dw = Q\left(\frac{k|y_0|}{\sigma}\right) \tag{5.39}$$

as can be seen from Equations 4.187–4.190[78–80].

When a received data-symbol is incorrectly detected, its intersymbol interference in the following equalized signals, instead of being eliminated, is doubled, and this greatly increases the probability of error in the detection of the following data-symbols. Errors therefore tend to occur in bursts, and the system suffers from error-extension effects.

At error rates of less than 1 in 10, the error rate in the $\{s'_i\}$ can be taken to be bP_e, where b is the average number of errors in an error burst. Provided that $|y_0|$ is not too small, b is typically in the range 1–100, and the corresponding reduction in tolerance to noise, at error rates of around 1 in 10^5 or 1 in 10^6, lies in the range 0–2 dB. When y_0 is one of the larger components of Y and there are not too many components of significant magnitude, b is typically in the range 1–10, giving a reduction in tolerance to noise in the range 0–1 dB. Thus, under the more favourable conditions, no very serious inaccuracy is introduced by taking the error rate in the $\{s'_i\}$ to be P_e in Equation 5.39.

The above result relies heavily on the assumption that the $\{s_i\}$ are statistically independent and equally likely to have either binary value, which means that the $\{s_i\}$ must be well scrambled to avoid as far as possible the transmission of repetitive short sequences. Again, when the data-symbols $\{s_i\}$ are multilevel, such as 4- or 16-level, the value of b can become unacceptably large for the more severe signal distortion, such as occurs when there are several of the larger components of Y (Equation 5.4) with similar magnitudes. Under these conditions due account must be taken of b in evaluating the tolerance to noise. For our purposes, however, it is assumed that the data symbols are binary and statistically independent, with zero mean.

An important property of the nonlinear equalizer just described is that it operates correctly even when $Y(z)$, the z-transform of the channel, has all its zeros *outside* the unit circle in the z plane. However, the value of b may now become so large as to make the system unusable in practice.

When all the zeros of $Y(z)$ lie *inside* the unit circle, the $(m+1)$- tap *linear* feedforward transversal equalizer for the channel (Fig. 4.3) introduces *no delay* and has a z-transform $C(z)$ approximately equal to $Y^{-1}(z)$. Now

$$Y(z) = y_0 + y_1 z^{-1} + \cdots + y_g z^{-g} \qquad (5.40)$$

as in Equation 5.5, and, from Equation 4.24,

$$C(z) = c_0 + c_1 z^{-1} + \cdots + c_m z^{-m} \simeq Y^{-1}(z) \qquad (5.41)$$

so that

$$Y(z)C(z) \simeq 1 \qquad (5.42)$$

from which it follows that

$$c_0 = 1/y_0 \qquad (5.43)$$

Furthermore, it is evident that one or more of the remaining tap gains of the equalizer must be nonzero, which means that

$$|C|^2 > 1/y_0^2 \tag{5.44}$$

where $|C|^2$ is the sum of the squares of the equalizer tap gains. But the noise variance at the output of the linear equalizer is now

$$\eta^2 = \sigma^2 |C|^2 > \sigma^2/y_0^2 \tag{5.45}$$

as can be seen from Equations 4.185 and 5.44, so that the corresponding probability of error in the detection of s_i is

$$Q\left(\frac{k}{\eta}\right) = Q\left(\frac{k}{\sigma|C|}\right) > Q\left(\frac{k|y_0|}{\sigma}\right) \tag{5.46}$$

as can be seen from Equations 4.191 and 5.45. Thus, whenever all the zeros of the z-transform of the channel lie *inside* the unit circle in the z plane and the signal/noise ratio is high so that error-extension effects may be neglected, the nonlinear equalizer just described gives a *better* tolerance to additive white Gaussian noise than a linear equalizer.

In practice, $Y(z)$ frequently has zeros outside the unit circle, and the nonlinear equalizer now often gives a lower tolerance to additive white Gaussian noise than a linear equalizer. Under these conditions, a better performance is sometimes obtained by detecting s_i, not from r_i (after cancellation of the intersymbol interference) but from r_{i+1} or r_{i+2}.

Consider next the receiver shown in Fig. 5.7. The multiplier (at the input) is omitted here and s_i is detected from the signal x_{i+h} at the output of the nonlinear equalizer, at time $t = (i+h)T$, where h is an integer in the range 0 to g. The received sample at time $t = (i+h)T$ is

$$r_{i+h} = \sum_{j=0}^{h-1} s_{i+h-j} y_j + s_i y_h + \sum_{j=h+1}^{g} s_{i+h-j} y_j + w_{i+h} \tag{5.47}$$

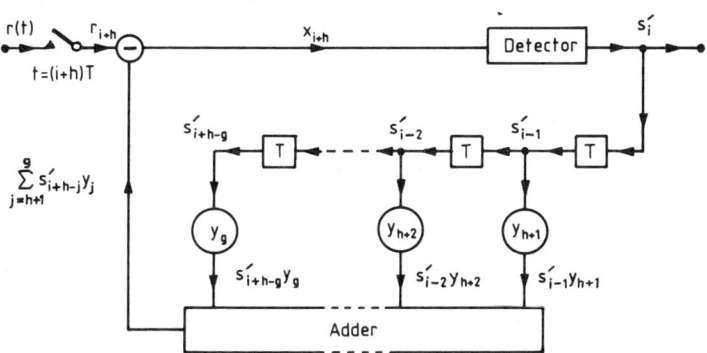

Fig. 5.7 Receiver using a pure nonlinear equalizer that achieves only partial equalization of the channel

where $s_i y_h$ is the *wanted signal*,

$$\sum_{j=0}^{h-1} s_{i+h-j} y_j$$

is the non-removable intersymbol interference, and

$$\sum_{j=h+1}^{g} s_{i+h-j} y_j$$

is the removable intersymbol interference.

The former of the two intersymbol-interference terms *cannot* be removed by decision-directed cancellation, since the data symbols involved have not yet been detected and are therefore not known at the receiver. The output signal from the linear feedforward transversal filter in Fig. 5.7 is

$$\sum_{j=1}^{g-h} s'_{i-j} y_{h+j} = \sum_{j=h+1}^{g} s'_{i+h-j} y_j \qquad (5.48)$$

so that the signal at the dector input, at time $t = (i+h)T$, is

$$x_{i+h} = r_{i+h} - \sum_{j=h+1}^{g} s'_{i+h-j} y_j \qquad (5.49)$$

Thus, with the correct detection of the data-symbols $s_{i-1}, s_{i-2}, \ldots, s_{i+h-g}$, the signal at the dector input becomes

$$\begin{aligned} x_{i+h} &= r_{i+h} - \sum_{j=h+1}^{g} s_{i+h-j} y_j \\ &= \sum_{j=0}^{h-1} s_{i+h-j} y_j + s_i y_h + w_{i+h} \end{aligned} \qquad (5.50)$$

The channel is here only *partially* equalized, such that $s_i y_h$ is the *wanted signal*, $\sum_{j=0}^{h-1} s_{i+h-j} y_j$ is the remaining *intersymbol interference*, and w_{i+h} is the *noise*.

The data-symbol s_i is now detected as k or $-k$, depending upon whether x_{i+h} is closer to ky_h or $-ky_h$, respectively. Thus, if y_h is positive, s_i is detected as k or $-k$, depending upon whether x_{i+h} is positive or negative, respectively, and if y_h is negative, s_i is detected as k or $-k$, depending upon whether x_{i+h} is negative or positive, respectively.

When $|y_h| \gg |y_j|$, for $j = 0, 1, \ldots, h-1$, the arrangement just described often gives a better tolerance to noise than the pure nonlinear equalizer where s_i is detected from x_i, with the complete elimination of *all* intersymbol interference. In the latter case, $h = 0$ in

Fig. 5.7 and s_i is detected, at time $t = iT$, from the equalized signal x_i at the detector input. With the correct detection of $s_{i-1}, s_{i-2}, \ldots, s_{i-g}$, the equalized signal becomes

$$x_i = r_i - \sum_{j=1}^{g} s_{i-j} y_j = s_i y_0 + w_i \tag{5.51}$$

An error now occurs in the detection of s_i when the noise component w_i has the opposite sign to $s_i y_0$ and a magnitude greater than $k|y_0|$. As before, w_i is a Gaussian random variable with zero mean and variance σ^2. Thus the probability of error in the detection of s_i, given the correct detection of the g preceding $\{s_j\}$, is

$$P_e = Q\left(\frac{k|y_0|}{\sigma}\right) \tag{5.52}$$

As can be seen from Equation 5.39, this is the same as the error probability of the receiver in Fig. 5.2, so that, when $h = 0$, the receivers in Figs. 5.2 and 5.7 are equivalent. The reason for describing the arrangement in Fig. 5.2 is to emphasize the relationship between this and some of the equalizers to be considered later.

An important property of the equalizers studied here is that they can be used to equalize a channel whose z-transform has one or more zeros *on* the unit circle in the z plane. Such a channel cannot be equalized *linearly*.

All equalizers studied in Sections 5.1–5.3 are *pure nonlinear equalizers* and are often referred to more simply as *nonlinear equalizers*. They form a special class of the more general type of equalizer, known as a *decision-feedback equalizer*, which is considered in Sections 5.5 to 5.8.

Consider the case where the z-transform of the baseband channel in Fig. 5.1 is

$$Y(z) = 1 + 2.5z^{-1} + z^{-2} \tag{5.53}$$

so that the z-transform of the $(i+1)$th received signal-element, at the output of the sampler, is

$$s_i z^{-i} Y(z) = s_i z^{-i} + 2.5 s_i z^{-i-1} + s_i z^{-i-2} \tag{5.54}$$

Assume that the receiver is as in Fig. 5.7. The received sample at time $t = iT$, at the input to the equalizer, is now

$$r_i = s_i + 2.5 s_{i-1} + s_{i-2} + w_i \tag{5.55}$$

where w_i is a Gaussian random variable with zero mean and variance σ^2, as before.

If $h = 0$ in Fig. 5.7, s_i is detected from the signal x_i at the input to the detector, at time $t=iT$. Now

$$x_i = r_i - 2.5s'_{i-1} - s'_{i-2} \tag{5.56}$$

and, with the correct detection of s_{i-1} and s_{i-2},

$$x_i = s_i + w_i \tag{5.57}$$

so that all intersymbol interference is eliminated from x_i.

When $x_i > 0$, s_i is detected as k, and when $x_i < 0$, s_i is detected as $-k$. An error occurs in the detection of s_i from x_i, when w_i has a magnitude greater than k and the opposite sign to s_i. Thus the probability of error in the detection of s_i from x_i, given the correct detection of s_{i-1} and s_{i-2}, is

$$P_1 = Q\left(\frac{k}{\sigma}\right) \tag{5.58}$$

The incorrect detection of a data symbol greatly increases the probability of error in the detection of the immediately following data symbols, so that errors tend to occur in bursts. In the arrangement considered here there are on average some four errors in a burst ($b = 4$), so that the actual error rate is of the order of $4P_1$ and the actual tolerance to additive Gaussian noise, at error rates around 10^{-5} or 10^{-6}, is about $\frac{1}{2}$ dB less than that given by P_1.

Suppose next that $h=1$ in Fig. 5.7, so that s_i is detected from the signal x_{i+1} at the input to the detector, at time $t=(i+1)T$. The received sample, at time $t=(i+1)T$, is

$$r_{i+1} = s_{i+1} + 2.5s_i + s_{i-1} + w_{i+1} \tag{5.59}$$

and the corresponding signal at the detector input is

$$x_{i+1} = r_{i+1} - s'_{i-1} \tag{5.60}$$

With the correct detection of s_{i-1},

$$x_{i+1} = s_{i+1} + 2.5s_i + w_{i+1} \tag{5.61}$$

When $x_{i+1} > 0$, s_i is detected as k, and when $x_{i+1} < 0$, s_i is detected as $-k$. At high signal/noise ratios and with the correct detection of s_{i-1}, practically all errors in s'_i occur when

$$s_{i+1} = -s_i \tag{5.62}$$

and now

$$x_{i+1} = 1.5s_i + w_{i+1} \tag{5.63}$$

Under these conditions, an error occurs in s'_i when w_{i+1} has a magnitude greater than $1.5k$ and the opposite sign to s_i. Thus the error probability, given that $s'_{i-1} = s_{i-1}$ and $s_{i+1} = -s_i$, is

$$P_2 = Q\left(\frac{1.5k}{\sigma}\right) \tag{5.64}$$

Since there is a probability of $\frac{1}{2}$ that $s_{i+1} = -s_i$, the average probability of error in the detection of s_i from x_{i+1}, given that $s'_{i-1} = s_{i-1}$, is effectively $\frac{1}{2} P_2$. But, at high signal/noise ratios with additive Gaussian noise (giving error rates around 10^{-5} or 10^{-6}), a change in the error probability by a factor of $\frac{1}{2}$ or 2 corresponds to a change of a little under $\frac{1}{3}$ dB in the signal/noise ratio, so that the probability of an error in s'_i, given that $s'_{i-1} = s_{i-1}$, can be taken to be P_2 in Equation 5.64. The average number of errors in an error burst, b, in the corresponding practical system here, is around 2, so that the actual error rate allowing for error-extension effects (error bursts) can be taken to be P_2 in Equation 5.64, at high signal/noise ratios. Now, when $P_1 = P_2$ and k has a given value, the value of σ in Equation 5.64 is 3.5 dB greater than that in Equation 5.58, implying an advantage of 3.5 dB in tolerance to noise for the pure nonlinear equalizer with $h=1$ over the pure nonlinear equalizer with $h=0$. When making an allowance for the approximations that have been made in Equations 5.58 and 5.64, the nonlinear equalizer with $h=1$ in fact gains an advantage over the nonlinear equalizer with $h=0$ of about 4 dB in tolerance to noise, at error rates between 10^{-5} and 10^{-6}. Table 5.1 compares the performances of these two equalizers with that of a *linear* equalizer, when operating with the given channel (Equation 5.53) at very high signal/noise ratios. Thus, whereas the nonlinear equalizer with $h=0$ has an inferior performance to that of the linear equalizer, the nonlinear equalizer with $h=1$ gains an advantage of a

TABLE 5.1 RELATIVE PERFORMANCES OF THREE DIFFERENT EQUALIZERS WITH A CHANNEL WHOSE Z-TRANSFORM IS $Y(z) = 1 + 2.5z^{-1} + z^{-2}$

Equalizer	Error probability	Tolerance to noise relative to that of the linear equalizer (dB)
Linear	$Q\left(\frac{1.164k}{\sigma}\right)$	0
Pure nonlinear ($h=0$)	$Q\left(\frac{k}{\sigma}\right)$	-1.3
Pure nonlinear ($h=1$)	$Q\left(\frac{1.5k}{\sigma}\right)$	2.2

little more than 2 dB, in tolerance to additive white Gaussian noise, over the linear equalizer.

The important results previously obtained in connection with the relative performances of linear and pure nonlinear equalizers are now, for convenience, summarized as follows.

(1) When all zeros of $Y(z)$ lie inside the unit circle in the z plane, the pure nonlinear equalizer with $h=0$ gains an advantage in tolerance to additive white Gaussian noise, at very high signal/noise ratios, over the linear equalizer. The nonlinear equalizer is now usually simpler to implement than the linear equalizer, since it requires only g taps compared with the somewhat larger number normally required by the linear equalizer.

(2) When $Y(z)$ represents nearly pure phase distortion, the linear equalizer effectively becomes a *matched filter* and the receiver approximates to a maximum-likelihood (optimum) detector (see Sections 1.5 and 4.12). Under these conditions the linear equalizer usually gives a much better tolerance to additive white Gaussian noise than does a pure nonlinear equalizer. The two equalizers here are of similar complexity.

5.4 EXAMPLES

Problem 1
The $(i+1)$th sample of a received signal is

$$r_i = s_i + 4s_{i-1} + 7s_{i-2} + w_i \tag{5.65}$$

where the data-symbols $\{s_i\}$ are statistically independent and equally likely to have either value $\pm k$, and the noise components $\{w_i\}$ are statistically independent Gaussian random variables with zero mean and variance σ^2. A high signal/noise ratio is assumed.

Derive an expression for the probability of error when s_{i-2} is detected directly from r_i in the presence of intersymbol interference from s_i and s_{i-1}.

Describe briefly the pure nonlinear equalizer that operates on r_i, for each of the following two cases: (a) such that s_i is detected without intersymbol interference, and (b) such that s_{i-1} is detected with intersymbol interference from s_i. Derive an expression for the probability of error in each case.

Which of the three data symbols is detected from r_i with the lowest probability of error?

Solution
When s_{i-2} is detected from r_i, a simple threshold-level detector is used to compare r_i with a threshold of zero, to give s'_{i-2}, the detected value of s_{i-2}. When $r_i > 0$, $s'_{i-2} = k$, and when $r_i < 0$, $s'_{i-2} = -k$.

The data-symbol s_{i-2} is here detected from r_i in the presence of intersymbol interference from s_{i-1} and s_i. When the signal/noise ratio is high so that there is a low probability of error in the detection of s_{i-2}, practically all errors in the detected values of s_{i-2} occur when

$$s_i = s_{i-1} = -s_{i-2} \tag{5.66}$$

Now
$$r_i = 2s_{i-2} + w_i \tag{5.67}$$

and an error occurs in the detection of s_{i-2} from r_i, when

$$|w_i| > 2k \tag{5.68}$$

and when w_i has the opposite sign to s_{i-2}.

The probability density function of w_i, at a sample value w, is

$$p(w) = \frac{1}{\sqrt{(2\pi\sigma^2)}} \exp\left(-\frac{w^2}{2\sigma^2}\right) \tag{5.69}$$

so that, when $s_i = s_{i-1} = -s_{i-2}$, the probability of error in the detection of s_{i-2} from r_i, is

$$\int_{2k}^{\infty} p(w)\, dw = \int_{2k}^{\infty} \frac{1}{\sqrt{(2\pi\sigma^2)}} \exp\left(-\frac{w^2}{2\sigma^2}\right) dw$$

$$= \int_{2/k\sigma}^{\infty} \frac{1}{\sqrt{2\pi}} \exp(-\tfrac{1}{2}w^2)\, dw = Q\left(\frac{2k}{\sigma}\right) \tag{5.70}$$

regardless of the binary value of s_{i-2}. But there is a probability of $\tfrac{1}{4}$ that $s_i = s_{i-1} = -s_{i-2}$ so that the average or actual probability of error in the detection of s_{i-2} from r_i is effectively

$$\tfrac{1}{4}Q\left(\frac{2k}{\sigma}\right) \tag{5.71}$$

At very low error rates, the actual probability of error can be taken to be

$$P_1 = Q\left(\frac{2k}{\sigma}\right) \tag{5.72}$$

(a) The pure nonlinear equalizer, that operates on r_i so that s_i is detected without intersymbol interference, is as shown in Fig. 5.8. The signals here are those obtained after the receipt of r_i. The transversal filter has two taps, with gains of 4 and 7, and its output signal

$$4s'_{i-1} + 7s'_{i-2}$$

is subtracted from r_i to give the equalized signal

$$x_i = r_i - 4s'_{i-1} - 7s'_{i-2} \tag{5.73}$$

which is fed to the detector.

With the correct detection of s_{i-1} and s_{i-2},

$$x_i = r_i - 4s_{i-1} - 7s_{i-2} = s_i + w_i \tag{5.74}$$

so that all intersymbol interference is eliminated from x_i.

The detector now compares x_i with a threshold of zero. When $x_i > 0$, $s'_i = k$, and when $x_i < 0$, $s'_i = -k$.

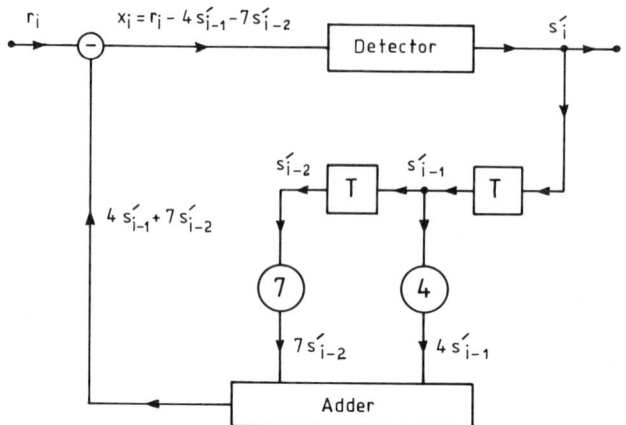

Fig. 5.8 Pure nonlinear equalizer with $h = 0$

With the correct detection of s_{i-1} and s_{i-2}, so that

$$x_i = s_i + w_i \tag{5.75}$$

an error occurs in the detection of s_i when $|w_i| > k$ and w_i has the opposite sign to s_i. Now, for either value of s_i, the probability of error in s'_i is

$$P_2 = \int_k^\infty p(w)\,dw = Q\left(\frac{k}{\sigma}\right) \tag{5.76}$$

The incorrect detection of s_i leads to severe intersymbol interference in the detection of s_{i+1} and s_{i+2}, thus very greatly increasing the probability of error in their detection, so that errors tend to occur in bursts. However, at very high signal/noise ratios, the increase in the error rate resulting from these error extension effects can be neglected, so that the actual error probability in s'_i can be taken to be P_2.

(b) The pure nonlinear equalizer, that operates on r_i so that s_{i-1} is detected with intersymbol interference from s_i, is as shown in Fig. 5.9. The signals here are those obtained after the receipt of r_i. The transversal filter now has only one tap and this has a gain of 7. The output signal $7s'_{i-2}$ from the transversal filter is subtracted from r_i to give the equalized signal

$$x_i = r_i - 7s'_{i-2} \tag{5.77}$$

and, with the correct detection of s_{i-2},

$$x_i = s_i + 4s_{i-1} + w_i \tag{5.78}$$

The detector ignores the intersymbol interference term s_i in x_i. When $x_i > 0$, $s'_{i-1} = k$, and when $x_i < 0$, $s'_{i-1} = -k$.

With the correct detection of s_{i-2}, so that Equation 5.78 is satisfied, practically all the errors in the $\{s'_{i-1}\}$ occur when

$$s_i = -s_{i-1} \tag{5.79}$$

DECISION-FEEDBACK EQUALIZERS 249

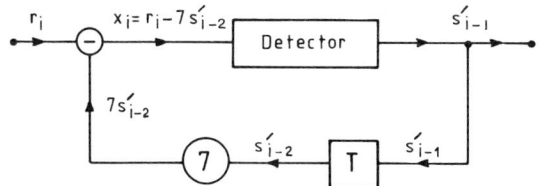

Fig. 5.9 Pure nonlinear equalizer with h=1

such that

$$x_i = 3s_{i-1} + w_i \tag{5.80}$$

If $s_i = -s_{i-1}$, an error occurs in the detection of s_{i-1} from x_i, when

$$|w_i| > 3k \tag{5.81}$$

and when w_i has the opposite sign to s_{i-1}, so that the error probability is now

$$\int_{3k}^{\infty} p(w)\,dw = Q\left(\frac{3k}{\sigma}\right) \tag{5.82}$$

But the probability that $s_i = -s_{i-1}$ is $\frac{1}{2}$, so that, for either value of s_{i-1} and with the correct detection of s_{i-2}, the probability of error in the detection of s_{i-1} from x_i is now effectively

$$\tfrac{1}{2}Q\left(\frac{3k}{\sigma}\right) \tag{5.83}$$

At very high signal/noise ratios, the error extension effects together with the above factor $\frac{1}{2}$ can be neglected, so that the actual probability of error in s'_{i-1} can be taken to be

$$P_3 = Q\left(\frac{3k}{\sigma}\right) \tag{5.84}$$

To compare the three systems, bear in mind that

$$P_1 = Q\left(\frac{2k}{\sigma}\right), \quad P_2 = Q\left(\frac{k}{\sigma}\right), \quad P_3 = Q\left(\frac{3k}{\sigma}\right)$$

At a given very low error probability, such that

$$P_1 = P_2 = P_3 \tag{5.85}$$

and at a given value of k, the corresponding values of σ effectively determine the relative tolerances to noise of the three systems. Thus, the first system (the simple threshold-level detector) gains an advantage of

$$20 \log_{10} 2 = 6 \tag{5.86}$$

dB over the second system (the pure nonlinear equalizer that achieves the exact equalization of the channel), but the third system gains an advantage of

$$20 \log_{10} 3 = 9.5 \tag{5.87}$$

dB over the second system, and so is the best of the three systems.
In the third system, of course, s_{i-1} is detected from r_i and the error probability is P_3.

Problem 2

The $(i+1)$th received sample, at time $t=iT$, of a serial digital baseband signal is

$$r_i = 0.2s_i + 1.1s_{i-1} + 0.5s_{i-2} + w_i \qquad (5.88)$$

where the data-symbols $\{s_i\}$ are statistically independent and equally likely to have either value ± 1. The noise components $\{w_i\}$ are statistically independent Gaussian random variables with zero mean and variance σ^2. A high signal/noise ratio is assumed.

Determine the tap gains of a 10-tap linear feedforward transversal equalizer for this channel, that introduces a peak distortion of less than 0.01 into the equalized signal. This is achieved by using (in the design process) 7 and 4 taps, respectively, in equalizing the factors of the z-transform of the channel with zeros (roots) inside and outside the unit circle in the z plane.

Describe the pure nonlinear equalizer for the channel, in which s_{i-1} is detected from r_i.

Derive an expression for the probability of error in the detection of a data symbol, for each equalizer.

Which equalizer gives the better tolerance to noise?

Solution

The sampled impulse-response of the channel is the sequence (vector)

$$Y = [0.2 \quad 1.1 \quad 0.5] \qquad (5.89)$$

with z-transform

$$Y(z) = 0.2 + 1.1z^{-1} + 0.5z^{-2} = (0.2 + z^{-1})(1 + 0.5z^{-1}) \qquad (5.90)$$

having zeros at $z = -0.2$ and $z = -2$.

To design the linear equalizer for the given channel, the linear equalizers for the two factors $0.2 + z^{-1}$ and $1 + 0.5z^{-1}$ of $Y(z)$ must be determined separately. Consider first the factor $1 + 0.5z^{-1}$.

$$(1 + 0.5z^{-1})^{-1} = 1 - 0.5z^{-1} + 0.25z^{-2} - 0.125z^{-3} + 0.0625z^{-4}$$

$$- 0.03125z^{-5} + 0.01563z^{-6} - \cdots \qquad (5.91)$$

so that the z-transform of the 7-tap linear equalizer for $1 + 0.5z^{-1}$ is

$$A(z) = 1 - 0.5z^{-1} + 0.25z^{-2} - 0.125z^{-3} + 0.0625z^{-4}$$

$$- 0.03125z^{-5} + 0.01563z^{-6} \qquad (5.92)$$

To determine the equalizer for the factor $0.2 + z^{-1}$, let

$$M(z) = 1 + 0.2z^{-1} \qquad (5.93)$$

Now $\quad M^{-1}(z) = (1 + 0.2z^{-1})^{-1}$

$$= 1 - 0.2z^{-1} + 0.04z^{-2} - 0.008z^{-3} + \cdots \qquad (5.94)$$

so that the z-transform of the required 4-tap linear equalizer for $0.2+z^{-1}$ is

$$B(z) = -0.008 + 0.04z^{-1} - 0.2z^{-2} + z^{-3} \quad (5.95)$$

The z-transform of the 10-tap linear equalizer for the channel with z-transform $Y(z)$ is now

$$A(z)B(z) = -0.008 + 0.044z^{-1} - 0.222z^{-2}$$
$$+ 1.111z^{-3} - 0.5555z^{-4} + 0.2778z^{-5} - 0.1389z^{-6}$$
$$+ 0.0694z^{-7} - 0.0344z^{-8} + 0.0156z^{-9} \quad (5.96)$$

and the z-transform of the channel and linear equalizer is

$$Y(z)A(z)B(z) \simeq -0.0016 + z^{-4} + 0.0078z^{-11} \simeq z^{-4} \quad (5.97)$$

Clearly, the peak distortion in the equalized signal is

$$(|-0.0016| + |0.0078|)/1 = 0.0094 \quad (5.98)$$

which can be neglected.

The tap gains of the linear feedforward transversal equalizer are

$-0.008, 0.044, -0.222, 1.111, -0.5555,$

$0.2778, -0.1389, 0.0694, -0.0344, 0.0156$

The signal at the output of the linear equalizer, at time $t = (i+4)T$, is

$$x_{i+4} \simeq s_i + u_{i+4} \quad (5.99)$$

where u_{i+4} is a Gaussian random variable with zero mean and variance

$$\eta^2 = \sigma^2(0.008^2 + 0.044^2 + 0.222^2 + 1.111^2 + 0.5555^2$$
$$+ 0.2778^2 + 0.1389^2 + 0.0694^2 + 0.0344^2 + 0.0156^2)$$
$$= 1.6969\, \sigma^2 \quad (5.100)$$

Thus the probability density function of u_{i+4}, at a sample value u, is

$$p_1(u) = \frac{1}{\sqrt{(2\pi\eta^2)}} \exp\left(-\frac{u^2}{2\eta^2}\right) \quad (5.101)$$

The detected value s'_i of the data-symbol s_i is determined as follows. When $x_{i+4} > 0$, $s'_i = 1$, and when $x_{i+4} < 0$, $s'_i = -1$.

An error occurs in s'_i when $|u_{i+4}| > 1$ and u_{i+4} has the opposite sign to s_i. Thus the probability of an error in s'_i, regardless of the binary value of s_i, is

$$P_1 = \int_1^\infty p_1(u)\, du = Q\left(\frac{1}{\eta}\right) = Q\left(\frac{1}{1.3026\sigma}\right) = Q\left(\frac{0.7677}{\sigma}\right) \quad (5.102)$$

The pure nonlinear equalizer in which s_{i-1} is detected from r_i is as shown in Fig. 5.10, the signals here being those after the receipt of r_i, at time $t = iT$. The partially equalized signal at the detector input is

$$x_i = r_i - 0.5 s'_{i-2} \quad (5.103)$$

With the correct detection of s_{i-2},

$$x_i = r_i - 0.5 s_{i-2} = 0.2 s_i + 1.1 s_{i-1} + w_i \quad (5.104)$$

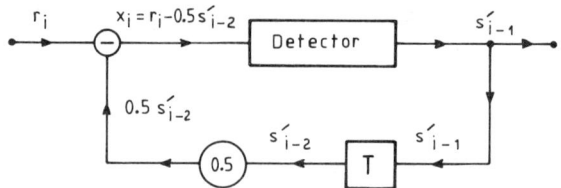

Fig. 5.10 Pure nonlinear equalizer with $h=1$

The detector ignores the intersymbol interference term $0.2s_i$ in x_i. When $x_i > 0$, $s'_{i-1} = 1$, and when $x_i < 0$, $s'_{i-1} = -1$.

With the correct detection of s_{i-2}, practically all errors in the $\{s'_{i-1}\}$ occur when

$$s_i = -s_{i-1} \tag{5.105}$$

such that

$$x_i = 0.9 s_{i-1} + w_i \tag{5.106}$$

The probability that $s_i = -s_{i-1}$ is $\frac{1}{2}$. For either value of s_{i-1} and with both $s_i = -s_{i-1}$ and the correct detection of s_{i-2}, an error occurs in the detection of s_{i-1} from x_i, when $|w_i| > 0.9$ and w_i has the opposite sign to s_{i-1}. But the probability density function of w_i, at a sample value w, is

$$p_2(w) = \frac{1}{\sqrt{(2\pi\sigma^2)}} \exp\left(-\frac{w^2}{2\sigma^2}\right) \tag{5.107}$$

so that the probability of an error in s'_{i-1} can be taken to be

$$\tfrac{1}{2} \int_{0.9}^{\infty} p_2(w)\, dw = \tfrac{1}{2} Q\left(\frac{0.9}{\sigma}\right) \tag{5.108}$$

At high signal/noise ratios, the error extension effects, resulting from the failure to cancel the intersymbol interference of an incorrectly detected data-symbol, can be neglected, together with the factor $\frac{1}{2}$, so that the actual probability of error in s'_{i-1} can be taken to be

$$P_2 = Q\left(\frac{0.9}{\sigma}\right) \tag{5.109}$$

Thus, at a given low error probability, such that $P_1 = P_2$ or

$$Q\left(\frac{0.7677}{\sigma_1}\right) = Q\left(\frac{0.9}{\sigma_2}\right) \tag{5.110}$$

where σ_1 and σ_2 are the corresponding values of σ for the linear and nonlinear equalizers, respectively, it is evident that

$$\frac{\sigma_2}{\sigma_1} = \frac{0.9}{0.7677} \tag{5.111}$$

Hence the value of σ in the case of the pure nonlinear equalizer is

$$20 \log_{10}(0.9/0.7677) = 1.4 \qquad (5.112)$$

dB above that for the linear equalizer, so that the nonlinear equalizer gains an advantage of 1.4 dB in tolerance to noise over the linear equalizer.

5.5 ZERO FORCING EQUALIZER

The improved tolerance to noise obtained by detecting s_i from x_{i+1} rather than from x_i, in the examples considered, suggests that a further improvement in tolerance to noise may be obtained with such arrangements by adding a linear filter at the input to the nonlinear equalizer, to remove the intersymbol-interference component in x_{i+1} that is not eliminated by the nonlinear equalizer. A linear filter is therefore added at the input to the equalizer in Fig. 5.7 to give the arrangement shown in Fig. 5.11, which is a combination of linear and nonlinear filters. The signals shown here are those at the time instant $t = (i+h)T$.

An important assumption that is made throughout the remainder of Chapter 5, except where specifically stated to the contrary, is that the z-transform of the sampled impulse-response of the linear baseband channel in Fig. 5.1 has *no* zeros (roots) that lie *on* the unit circle in the z plane.

Equalizers of the type shown in Fig. 5.11 are known as *decision-feedback equalizers*. Although *all* equalizers considered in this chapter are both *decision-feedback* and *nonlinear* equalizers, the term *decision-feedback* is applied specifically to equalizers containing both linear and nonlinear filters, as in Fig. 5.11, and the term *pure nonlinear* or *nonlinear* is applied to equalizers containing only a nonlinear filter. This enables a clear distinction to be made between the two different

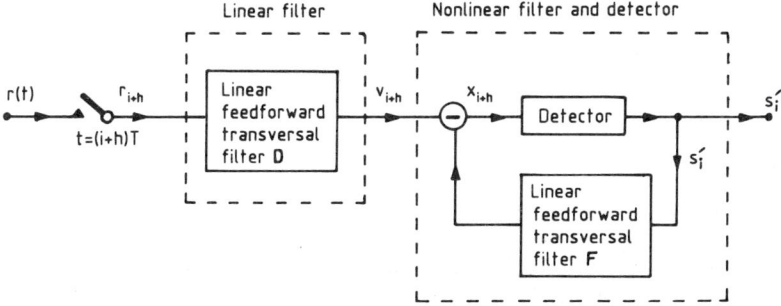

Fig. 5.11 Decision-feedback equalizer containing both a linear and a nonlinear filter

types of equalizer. The particular version of a decision-feedback equalizer under consideration here is known as a *zero-forcing* decision-feedback equalizer, which operates as follows.

The linear filter in Fig. 5.11 partially equalizes the channel by setting to zero all components of the channel sampled impulse-response preceding that of the largest magnitude, without however changing the relative values of the remaining components. The nonlinear filter then completes the equalization process, to give the accurate equalization of the linear baseband channel in Fig. 5.1. The nonlinear filter operates on the sampled impulse-response of the channel and linear filter in exactly the same way as that in which the nonlinear equalizer in Figs. 5.2 and 5.3 operates on the sampled impulse-response of the channel and multiplier. Furthermore, the basic structure of the nonlinear filter in Fig. 5.11 is the same as that in Fig. 5.3, the first tap of the linear feedforward transversal filter F occurring after a delay of T seconds.

As before, let the sampled impulse-response of the channel be

$$Y = [y_0 \quad y_1 \quad \cdots \quad y_g] \tag{5.113}$$

with z-transform

$$Y(z) = y_0 + y_1 z^{-1} + \cdots + y_g z^{-g} \tag{5.114}$$

and suppose that y_l is the component of Y having the greatest magnitude. From Equation 4.24, the z-transform of the $(m+1)$-tap linear equalizer for the channel is

$$C(z) = c_0 + c_1 z^{-1} + \cdots + c_m z^{-m} \tag{5.115}$$

which is such that

$$Y(z)C(z) \simeq z^{-h} \tag{5.116}$$

where h is a nonnegative integer in the range 0 to $m+g$. Use is, of course, made here of the assumption that $Y(z)$ has *no* zeros that lie *on* the unit circle in the z plane.

Let the sampled impulse-response (sequence of tap gains) of the $(n+1)$-tap linear feedforward transversal filter D in Fig. 5.11 be

$$D = [d_0 \quad d_1 \quad \cdots \quad d_n] \tag{5.117}$$

with z-transform

$$D(z) = d_0 + d_1 z^{-1} + \cdots + d_n z^{-n} \tag{5.118}$$

The required sampled impulse-response of the channel and linear filter is given by the $(g-l+1)$-component row vector (sequence)

$$E = \left[1 \quad \frac{y_{l+1}}{y_l} \quad \frac{y_{l+2}}{y_l} \quad \cdots \quad \frac{y_g}{y_l} \right] \tag{5.119}$$

with z-transform

$$E(z) = 1 + \frac{y_{l+1}}{y_l} z^{-1} + \frac{y_{l+2}}{y_l} z^{-2} + \cdots + \frac{y_g}{y_l} z^{-g+l} \quad (5.120)$$

where the delay introduced by the linear filter is *neglected*. Clearly, E is obtained from Y by removing the first l components and dividing each of the remaining components by y_l. From the assumption that

$$|y_l| > |y_i| \quad (5.121)$$

for $i = 0, 1, \ldots, g$ and $i \neq l$, it is evident that

$$\left| \frac{y_{l+i}}{y_l} \right| < 1 \quad (5.122)$$

for $i = 1, 2, \ldots, g-l$, which means that the first component of E (of value unity) is also the component of the greatest magnitude.

The z-transform $D(z)$ of the linear feedforward transversal filter D in Fig. 5.11 is now taken to be such that

$$D(z) = C(z) E(z) \quad (5.123)$$

so that the z-transform of the channel and linear filter is

$$Y(z) D(z) = Y(z) C(z) E(z) \simeq z^{-h} E(z) \quad (5.124)$$

where

$$n = m + g - l \quad (5.125)$$

The z-transform of the $(i+1)$th transmitted signal-element in Fig. 5.1 is $s_i z^{-i}$, so that, from Equation 5.124, the z-transform of the $(i+1)$th received signal-element at the output of the linear filter D in Fig. 5.11 is

$$s_i z^{-i-h} E(z) = s_i z^{-i-h} \left(1 + \frac{y_{l+1}}{y_l} z^{-1} + \cdots + \frac{y_g}{y_l} z^{-g+l} \right) \quad (5.126)$$

Whenever $E(z)$ has more than one nonzero term, as is usually the case, there is intersymbol interference between the signal elements at the output of the linear filter. This intersymbol interference is removed in the nonlinear filter, which operates exactly as the nonlinear equalizer in Fig. 5.2, except that it equalizes a response with z-transform $z^{-h} E(z)$ instead of the response with z-transform $y_0^{-1} Y(z)$. Clearly, the nonlinear filter is here as shown in Fig. 5.3, except that there are now $g - l$ taps, with tap gains

$$\frac{y_{l+1}}{y_l}, \frac{y_{l+2}}{y_l}, \ldots, \frac{y_g}{y_l}$$

instead of g taps with tap gains

$$v_1, v_2, \ldots, v_g$$

The linear and nonlinear filters in Fig. 5.11 together achieve the accurate equalization of the linear baseband channel in Fig. 5.1.

The signal at the input to the subtractor in Fig. 5.11, at time $t = (i+h)T$, is

$$v_{i+h} \simeq s_i + \frac{y_{l+1}}{y_l} s_{i-1} + \cdots + \frac{y_g}{y_l} s_{i-g+l} + u_{i+h} \quad (5.127)$$

where u_{i+h} is the noise component at time $t = (i+h)T$ at the output of the linear filter D. The equalized signal at the input to the detector is now

$$x_{i+h} = v_{i+h} - \frac{y_{l+1}}{y_l} s'_{i-1} - \frac{y_{l+2}}{y_l} s'_{i-2} - \cdots - \frac{y_g}{y_l} s'_{i-g+l} \quad (5.128)$$

where, as before, s'_{i-j} is the detected value of s_{i-j}. With the correct detection of $s_{i-1}, s_{i-2}, \ldots, s_{i-g+l}$, the signal at the detector input becomes

$$x_{i+h} \simeq s_i + u_{i+h} \quad (5.129)$$

When $x_{i+h} > 0$, $s'_i = k$, and when $x_{i+h} < 0$, $s'_i = -k$.

The noise component u_{i+h} in Equation 5.129 is given by

$$u_{i+h} = \sum_{j=0}^{n} w_{i+h-j} d_j \quad (5.130)$$

where the $\{w_{i+h-j}\}$ are statistically independent Gaussian random variables with zero mean and variance σ^2 (Equation 5.7). It follows from Equations 4.184 and 4.185 that u_{i+h} is a Gaussian random variable with zero mean and variance

$$\eta^2 = \sigma^2 \sum_{j=0}^{n} d_j^2 = \sigma^2 |D|^2 \quad (5.131)$$

where $|D|$ is the Euclidean length of the vector D. Thus the probability of error in the detection of s_i, given the correct detection of $s_{i-1}, s_{i-2}, \ldots, s_{i-g+l}$, is approximately

$$P_e = \int_k^\infty \frac{1}{\sqrt{2\pi\eta^2}} \exp\left(-\frac{u^2}{2\eta^2}\right) du = Q\left(\frac{k}{\eta}\right) = Q\left(\frac{k}{\sigma|D|}\right) \quad (5.132)$$

The error extension effects are not normally serious here, with an average of typically only a few errors in an error burst, so that, at very high signal/noise ratios, no significant inaccuracy is introduced by taking the average probability of error as P_e.

The decision-feedback equalizer just described is *not optimum* in any sense, but the study of this equalizer introduces some important techniques that are employed in the design of the optimum equalizer, and the relationship between this and other arrangements of the decision-feedback equalizer are quite illuminating.

When the z-transform of the linear baseband channel in Fig. 5.1 is

$$Y(z) = 1 + 2.5z^{-1} + z^{-2} \qquad (5.133)$$

the decision-feedback equalizer gains an advantage in tolerance to additive white Gaussian noise over the linear equalizer of about 2.9 dB, at high signal/noise ratios. The equalizer often gives a better performance than both the linear and nonlinear equalizers previously described, but it is sometimes inferior to one or the other of these, not being itself an optimum equalizer.

5.6 LINEAR AND NONLINEAR EQUALIZATION OF THE TWO FACTORS OF THE CHANNEL RESPONSE

When the linear filter D in Fig. 5.11 sets to zero the components preceding the largest in the sampled impulse-response of the channel, it is not necessary that the remaining components should be changed only by a constant factor. An alternative technique is as follows.

The z-transform of the sampled impulse-response of the channel is

$$Y(z) = Y_1(z) Y_2(z) \qquad (5.134)$$

where all the zeros (roots) of $Y_1(z)$ lie *inside* the unit circle in the z plane, and all the zeros of $Y_2(z)$ lie *outside* the unit circle. Let

$$Y_1(z) = 1 + p_1 z^{-1} + p_2 z^{-2} + \cdots + p_{g-f} z^{-g+f} \qquad (5.135)$$

and

$$Y_2(z) = q_0 + q_1 z^{-1} + q_2 z^{-2} + \cdots + q_f z^{-f} \qquad (5.136)$$

where f is an integer in the range 0 to g. All coefficients $\{p_i\}$ and $\{q_i\}$ here are *real valued*.

The $(n+1)$-tap linear feedforward transversal filter D in Fig. 5.11 now equalizes the factor $Y_2(z)$ of the channel z-transform, and the nonlinear filter equalizes the factor $Y_1(z)$. The z-transform $D(z)$ of the linear filter therefore satisfies the equation

$$Y_2(z) D(z) \simeq z^{-h} \qquad (5.137)$$

where

$$h = f + n \qquad (5.138)$$

so that
$$D(z) \simeq z^{-h} Y_2^{-1}(z) \qquad (5.139)$$

The z-transform of the channel and linear filter is
$$Y(z)D(z) \simeq z^{-h} Y_1(z) \qquad (5.140)$$

This is equalized by the nonlinear filter. From Equation 5.134, the sampled impulse-response of the channel and linear filter is given approximately by the $(g-f+1)$-component row vector (sequence)
$$E = [1 \quad p_1 \quad p_2 \quad \cdots \quad p_{g-f}] \qquad (5.141)$$

with z-transform
$$E(z) = Y_1(z) \qquad (5.142)$$

when *ignoring* the delay of hT seconds. Of course, $p_i = 0$ for $i < 0$ and $i > g - f$. Thus the first nonzero component in the sampled impulse-response of the channel and linear filter is unity and is one of the larger components, although it need not now be the largest as in Equation 5.119.

The linear feedforward transversal filter F, that is part of the nonlinear filter in Fig. 5.11, is here as shown in Fig. 5.3, except that there are $g - f$ taps instead of g taps and the tap gains are $p_1, p_2, \ldots, p_{g-f}$, instead of $v_1, v_2, \ldots, v_g$. As before, the nonlinear filter in Fig. 5.11 equalizes the resultant channel formed by the channel and linear filter, whose z-transform now approximates to $z^{-h} E(z)$, where $E(z)$ is given by Equation 5.142.

The signal at the input to the subtractor in Fig. 5.11, at time $t = (i+h)T$, is
$$v_{i+h} \simeq s_i + p_1 s_{i-1} + p_2 s_{i-2} + \cdots + p_{g-f} s_{i-g+f} + u_{i+h} \qquad (5.143)$$

where u_{i+h} is the noise component at time $t = (i+h)T$ at the output of the linear filter. The equalized signal at the input to the detector is now
$$x_{i+h} = v_{i+h} - p_1 s'_{i-1} - p_2 s'_{i-2} - \cdots - p_{g-f} s'_{i-g+f} \qquad (5.144)$$

and, with the correct detection of $s_{i-1}, s_{i-2}, \ldots, s_{i-g+f}$, the signal at the detector input becomes
$$x_{i+h} \simeq s_i + u_{i+h} \qquad (5.145)$$

As before (Equation 5.129), the data-symbol s_i is detected from the *sign* of x_{i+h}, and u_{i+h} is a Gaussian random variable with zero mean and variance $\sigma^2 |D|^2$ (Equation 5.131), so that the probability of error in the detection of s_i, given the correct detection of $s_{i-1}, s_{i-2}, \ldots,$

s_{i-g+f}, is approximately

$$P_e = Q\left(\frac{k}{\sigma|D|}\right) \qquad (5.146)$$

as in Equation 5.132. The error-extension effects are again not normally very serious, so that, at very high signal/noise ratios, no significant inaccuracy is introduced by taking the average error probability (and hence the average error rate) at P_e.

The linear filter here normally has fewer taps that that in the arrangement of zero forcing (Section 5.5), and the tolerance of the system to additive white Gaussian noise is more often than not a little better than that of the other arrangement. This technique is therefore preferable to that of zero forcing, but is again *not optimum*.

If $Y_1(z)$ and $Y_2(z)$ are equalized by the linear and nonlinear filters, respectively, the arrangement always has a *lower* tolerance to additive white Gaussian noise than that just described. This is consistent with the fact that, if $Y(z) = Y_1(z)$, the nonlinear equalizer in Fig. 5.2 gives a better tolerance to Gaussian noise than a linear equalizer, whereas, if $Y(z) = Y_2(z)$, a linear equalizer generally gives a better tolerance to noise.

Consider the case where the z-transform of the baseband channel in Fig. 5.1 is

$$Y(z) = 1 + 2\tfrac{1}{2}z^{-1} + z^{-2} = (1 + \tfrac{1}{2}z^{-1})(1 + 2z^{-1}) \qquad (5.147)$$

In the preferred arrangement of pure nonlinear equalization by decision-directed cancellation of intersymbol interference for the given channel, s_i is detected from the signal x_{i+1} (Equation 5.60) at the detector input, at time $t = (i+1)T$. The intersymbol interference of s_{i-1} has here been eliminated from x_{i+1} but not the intersymbol interference of s_{i+1}. The uncancelled intersymbol interference may, if required, be removed by a linear filter, as in the arrangement of zero forcing described in Section 5.5. The resultant z-transform of the channel and linear filter (Fig. 5.11) is now $z^{-h}(2\tfrac{1}{2} + z^{-1})$, and this is equalized by the nonlinear filter. Alternatively, the factor $1 + 2z^{-1}$ of $Y(z)$ may be equalized linearly and the factor $(1 + \tfrac{1}{2}z^{-1})$ may be equalized nonlinearly. This is the linear and nonlinear equalization of the two factors of the channel response, and will now be considered in some detail.

If the linear filter has 7 taps, its z-transform is

$$D(z) = \tfrac{1}{128} - \tfrac{1}{64}z^{-1} + \tfrac{1}{32}z^{-2} - \tfrac{1}{16}z^{-3} + \tfrac{1}{8}z^{-4} - \tfrac{1}{4}z^{-5} + \tfrac{1}{2}z^{-6} \qquad (5.148)$$

and the z-transform of the channel and linear filter is

$$Y(z)D(z) = (1 + 2\tfrac{1}{2}z^{-1} + z^{-2})$$
$$\times(\tfrac{1}{128} - \tfrac{1}{64}z^{-1} + \tfrac{1}{32}z^{-2} - \tfrac{1}{16}z^{-3} + \tfrac{1}{8}z^{-4} - \tfrac{1}{4}z^{-5} + \tfrac{1}{2}z^{-6})$$
$$= \tfrac{1}{128} + \tfrac{1}{256}z^{-1} + z^{-7} + \tfrac{1}{2}z^{-8}$$
$$\simeq z^{-7} + \tfrac{1}{2}z^{-8} = z^{-7}(1 + \tfrac{1}{2}z^{-1}) \tag{5.149}$$

Thus the z-transform of the channel and linear filter is

$$Y(z)D(z) \simeq z^{-7}E(z) \tag{5.150}$$

where
$$E(z) = 1 + \tfrac{1}{2}z^{-1} \tag{5.151}$$

and the sampled impulse-response of the channel and linear filter is given approximately by the two-component vector (sequence)

$$E = [1 \quad \tfrac{1}{2}] \tag{5.152}$$

when ignoring the delay introduced by the linear filter. The linear feedforward transversal filter F in Fig. 5.11 now has just one tap, with a gain of $\tfrac{1}{2}$.

The z-transform of the $(i+1)$th received signal-element at the output of the linear filter is approximately

$$s_i z^{-i-7}(1 + \tfrac{1}{2}z^{-1}) = s_i z^{-i-7} + \tfrac{1}{2}s_i z^{-i-8} \tag{5.153}$$

and the signal at the output of the linear filter, at time $t = (i+7)T$, is

$$v_{i+7} \simeq s_i + \tfrac{1}{2}s_{i-1} + u_{i+7} \tag{5.154}$$

where u_{i+7} is a Gaussian random variable with zero mean and variance

$$\eta^2 = \sigma^2[(\tfrac{1}{128})^2 + (\tfrac{1}{64})^2 + (\tfrac{1}{32})^2 + (\tfrac{1}{16})^2 + (\tfrac{1}{8})^2 + (\tfrac{1}{4})^2 + (\tfrac{1}{2})^2]$$
$$= 0.333\sigma^2 \tag{5.155}$$

so that

$$\eta = 0.577\sigma \tag{5.156}$$

With the correct detection of s_{i-1}, the equalized signal at the input to the detector, at time $t = (i+7)T$, is

$$x_{i+7} = v_{i+7} - \tfrac{1}{2}s_{i-1} \simeq s_i + u_{i+7} \tag{5.157}$$

It can be seen that the nonlinear filter has removed the intersymbol interference term in v_{i+7} to give x_{i+7}, without changing either the wanted-signal component s_i or the noise component u_{i+7}. Thus the nonlinear filter does not change the signal/noise ratio.

The probability of error in the detection of s_i, at high signal/noise ratios, can be taken to be

$$P_e = Q\left(\frac{k}{\eta}\right) = Q\left(\frac{k}{0.577\sigma}\right) = Q\left(\frac{1.732k}{\sigma}\right) \quad (5.158)$$

the error-extension effects being negligible here.

From Equation 4.287, the advantage in tolerance to additive white Gaussian noise, gained by the above decision-feedback equalizer over a linear equalizer, for the given channel at high signal/noise ratios, is approximately 3.5 dB. This compares with an advantage of 2.9 dB gained by the corresponding nonlinear equalizer using zero forcing, and an advantage of 2.2 dB gained by the better of the two arrangements of pure nonlinear equalization, where there is no linear filter.

Suppose now that, in the arrangement of the linear and nonlinear equalization of the two factors of the channel response, the factor $1 + \frac{1}{2}z^{-1}$ of the channel z-transform is equalized linearly and the factor $1 + 2z^{-1}$ is equalized nonlinearly.

If the linear filter has 7 taps, its z-transform is

$$D(z) = 1 - \tfrac{1}{2}z^{-1} + \tfrac{1}{4}z^{-2} - \tfrac{1}{8}z^{-3} + \tfrac{1}{16}z^{-4} - \tfrac{1}{32}z^{-5} + \tfrac{1}{64}z^{-6} \quad (5.159)$$

The z-transform of the channel and linear filter is

$$Y(z)D(z) = (1 + 2\tfrac{1}{2}z^{-1} + z^{-2})$$
$$\times (1 - \tfrac{1}{2}z^{-1} + \tfrac{1}{4}z^{-2} - \tfrac{1}{8}z^{-3} + \tfrac{1}{16}z^{-4} - \tfrac{1}{32}z^{-5} + \tfrac{1}{64}z^{-6})$$
$$\simeq 1 + 2z^{-1} \quad (5.160)$$

Thus the z-transform of the channel and linear filter is approximately

$$E(z) = 1 + 2z^{-1} \quad (5.161)$$

and the sampled impulse-response of the channel and linear filter is given approximately by the two-component vector (sequence)

$$E = \begin{bmatrix} 1 & 2 \end{bmatrix} \quad (5.162)$$

The linear feedforward transversal filter F in Fig. 5.11 again has just one tap, but now with a gain of 2.

The z-transform of the $(i+1)$th signal-element at the output of the linear filter is approximately

$$s_i z^{-i}(1 + 2z^{-1}) = s_i z^{-i} + 2s_i z^{-i-1} \quad (5.163)$$

and the signal at the output of the linear filter, at time $t = iT$, is

$$v_i \simeq s_i + 2s_{i-1} + u_i \quad (5.164)$$

where u_i is a Gaussian random variable with zero mean and variance

$$\eta^2 = \sigma^2(1^2 + (\tfrac{1}{2})^2 + (\tfrac{1}{4})^2 + (\tfrac{1}{8})^2 + (\tfrac{1}{16})^2 + (\tfrac{1}{32})^2 + (\tfrac{1}{64})^2)$$
$$= 1.333\sigma^2 \tag{5.165}$$

so that

$$\eta = 1.155\sigma \tag{5.166}$$

With the correct detection of s_{i-1}, the equalized signal at the input to the detector, at time $t = iT$, is

$$x_i = v_i - 2s_{i-1} \simeq s_i + u_i \tag{5.167}$$

The probability of error in the detection of s_i, at high signal/noise ratios, can be taken to be

$$P_e = Q\left(\frac{k}{\eta}\right) = Q\left(\frac{k}{1.155\sigma}\right) = Q\left(\frac{0.866k}{\sigma}\right) \tag{5.168}$$

The error-extension effects here lead to a doubling of the error rate, there being an average of two errors in a burst, and these do not seriously affect the signal/noise ratio at a given low error rate. The tolerance to noise is thus 6 dB below that of the previous decision-feedback equalizer, where the factor $1 + 2z^{-1}$ is equalized linearly and the factor $1 + \tfrac{1}{2}z^{-1}$ is equalized nonlinearly. It is also some 2.6 dB *below* that of a linear equalizer. Clearly, the previous decision-feedback equalizer is much to be preferred.

5.7 EQUALIZER THAT MINIMIZES THE MEAN-SQUARE ERROR IN THE EQUALIZED SIGNAL

It is clear from the previous analysis that when the channel is equalized by a *linear* filter, the filter is effectively determined by the channel. This, of course, follows from the fact that, if the channel has the z-transform $Y(z)$, the corresponding linear equalizer has a z-transform approximately equal to $z^{-h}Y^{-1}(z)$, where h is an appropriate integar, and, for the given value of h, $z^{-h}Y^{-1}(z)$ is uniquely determined by $Y(z)$. With the accurate equalization of the channel, the tolerance to noise of the resultant system is determined by the noise variance at the equalizer output. This, in turn, is determined by the input noise variance, together with the sum of the squares of the tap gains of the linear equalizer, and so by the channel itself. Thus the tolerance to noise is not significantly affected by the particular design technique applied to the linear equalizer.

However, when the channel is equalized by a *decision-feedback* equalizer, containing both a linear and a nonlinear filter, the channel

Sampled impulse-response
of channel and linear
equalizer :

Sampled impulse-response
of channel and linear filter
in decision - feedback
equalizer :

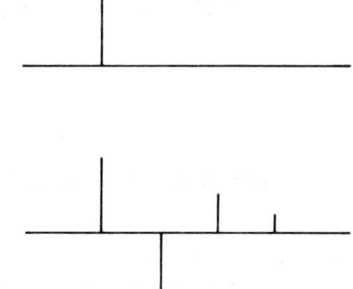

Fig. 5.12   Relationship between linear and decision-feedback equalizers

is equalized partly by the linear filter and partly by the nonlinear filter, the two filters together achieving the accurate equalization of the channel. Under these conditions the equalization of the channel may be achieved by an infinite number of different combinations of the linear and nonlinear filters. Thus there must be one or more particular combinations of the two filters that give the best tolerance to additive white Gaussian noise. The situation here is illustrated in Fig. 5.12.

With a decision-feedback equalizer, the tolerance of the system to additive white Gaussian noise at high signal/noise ratios is essentially determined by the noise variance at the output of the linear filter, for a given input noise variance, and therefore by the sum of the squares of the tap gains of this filter. The noise variance is not affected by the nonlinear filter. Furthermore, the signal component at the detector input is the first nonzero component of the corresponding signal-element at the output of the linear filter. Thus the wanted signal component and the noise component, at the output of the linear filter, are passed *unchanged* to the detector input and are therefore unaffected by the nonlinear filter. This merely removes the unwanted signal components that form the residual intersymbol interference at the output of the linear filter. In other words, the nonlinear filter removes the residual intersymbol interference at the output of the linear filter without however changing the ratio of the level of the wanted signal component to the level of the noise. To maximize the signal/noise ratio at the detector input, it is necessary to maximize the ratio of the magnitude of the first nonzero component in the sampled impulse-response of the channel and linear filter to the sum of the squares of the linear-filter tap gains. The linear filter can clearly be selected to maximize this ratio, and the nonlinear filter is now completely determined by the sampled impulse-response of the channel and linear filter, as can be seen in Fig. 5.12.

The error-extension effects, caused by the incorrect cancellation of wrongly detected data symbols, are dependent entirely on the nonlinear filter and therefore on the sampled impulse-response of the channel and linear filter. They may cause the error rate to be typically up to about ten times that with the corresponding idealized system, where there is always the correct cancellation of intersymbol interference, regardless of the errors in detection. However, at high signal/noise ratios, in the presence of additive Gaussian noise, this increase in error rate represents a reduction in tolerance to noise of up to only about 1 dB, which is not very serious and for our purposes may be neglected. Thus, if the signal/noise ratio at the detector input is maximized, then, for practical purposes, the error rate in the detection of the received signal-elements is minimized, to give the optimum decision-feedback equalizer.

At high signal/noise ratios and with statistically independent $\{s_i\}$ having zero mean, error-extension effects only become important when there is very severe intersymbol interference at the output of the linear filter (the first part of the decision-feedback equalizer) or else when multilevel data symbols are transmitted. Under either or both of these two conditions, due account must be taken of the error-extension effects, that is, of the average number of errors in an error burst, since these may well increase the average error rate by as much as 100 or even 1000 times. We are however only concerned here with binary signals and with moderate levels of intersymbol interference at the output of the linear filter, for which the average error rate is normally increased by less than 10 times.

It is important to notice that the optimization process just outlined is subject (constrained) to the *accurate equalization* of the channel. To minimize (without constraint) the error rate in the detected signal, at very low signal/noise ratios, a greater reduction in level must be applied to the noise at the detector input than to the intersymbol interference, these having a similarly *high* level here. Thus a significant level of intersymbol interference must now be accepted in the equalized signal, and the channel is not accurately equalized. On the other hand, to minimize (without constraint) the error rate in the detected signal, at very high signal/noise ratios, a greater reduction in level must be applied to the intersymbol interference at the detector input than to the noise, these having a similarly *low* level here, so that the channel is now quite accurately equalized. A sufficient number of taps is, of course, assumed for the equalizer. Thus, at very high signal/noise ratios, the tolerance to additive white Gaussian noise of the equalizer that minimizes the error rate, subject to the accurate equalization of the channel, is unlikely to be noticeably different from that of the optimum equalizer, that has not been constrained to the accurate equalization of the channel.

It will first be assumed that the decision-feedback equalizer of Fig. 5.11 is adjusted to minimize the mean-square error in the equalized signal x_{i+h}. With reasonably accurate equalization,

$$x_{i+h} \simeq s_i + u_{i+h} \tag{5.169}$$

where $s_i = \pm k$ and u_{i+h} is the additive Gaussian noise component at time $t = (i+h)T$ at the output of the filter D. No constraint is however applied here to ensure the accurate equalization of the channel, so that the error in the equalized signal is due partly to noise and partly also to intersymbol interference. The data-symbol s_i is detected from x_{i+h} at time $t = (i+h)T$, which means that the delay in detection is h sampling intervals. As before, the linear feedforward transversal filter D has $n+1$ taps and a sampled impulse-response

$$D = [d_0 \quad d_1 \quad \ldots \quad d_n] \tag{5.170}$$

with z-transform

$$D(z) = d_0 + d_1 z^{-1} + d_2 z^{-2} + \cdots + d_n z^{-n} \tag{5.171}$$

and now the sampled impulse-response of the channel and filter is given by the $(n+g+1)$-component vector

$$E = [e_0 \quad e_1 \quad e_2 \quad \ldots \quad e_{n+g}] \tag{5.172}$$

with z-transform

$$E(z) = e_0 + e_1 z^{-1} + e_2 z^{-2} + \cdots + e_{n+g} z^{-n-g} \tag{5.173}$$

such that

$$E(z) = Y(z)D(z) \tag{5.174}$$

where $Y(z)$ is the z-transform of the sampled impulse-response of the channel (Equation 5.114). Notice that in Equation 5.174 *no* approximations are made nor does the equation ignore any delay introduced by the linear filter, contrary to the case of Equations 5.120 and 5.142. The linear feedforward transversal filter F, that forms part of the nonlinear filter in Fig. 5.11, has μ taps and a sampled impulse-response

$$F = [f_1 \quad f_2 \quad \ldots \quad f_\mu] \tag{5.175}$$

Clearly, the tap gains of the filter D are given by $d_0, d_1, \ldots, d_n$ and the tap gains of the filter F are given by $f_1, f_2, \ldots, f_\mu$.

Now let[49]

$$S_i = [s_i \quad s_{i-1} \quad s_{i-2} \quad \ldots \quad s_{i-n-g}] \tag{5.176}$$

and
$$F_0 = [\overbrace{0 \quad 0 \quad \ldots \quad 0}^{h+1} \quad f_1 \quad f_2 \quad \ldots \quad f_\mu] \tag{5.177}$$
where both S_i and F_0 are $(n+g+1)$-component row vectors and
$$n+g = h+\mu \tag{5.178}$$
As before, following the correct detection of the data-symbols s_{i-1}, $s_{i-2}, \ldots, s_{i-\mu}$, the equalized signal at the detector input, at time $t = (i+h)T$, is given by Equation 5.169. The filters D and F in the decision-feedback equalizer of Fig. 5.11 are here assumed to be adjusted to minimize the mean-square error ε in the equalized signal, where
$$\varepsilon = \mathscr{E}[(x_{i+h} - s_i)^2] \tag{5.179}$$
given that the data-symbols $s_{i-1}, s_{i-2}, \ldots, s_{i-\mu}$ have been *correctly detected*. The term $\mathscr{E}[x]$ means the *expected value* of x.

The signal at the input to the detector in Fig. 5.11, at time $t = (i+h)T$ and with the correct detection of $s_{i-1}, s_{i-2}, \ldots, s_{i-\mu}$, is
$$x_{i+h} = \sum_{j=0}^{n+g} s_{i+h-j} e_j - \sum_{j=1}^{\mu} s_{i-j} f_j + u_{i+h}$$
$$= S_{i+h} E^T - S_{i+h} F_0^T + u_{i+h} \tag{5.180}$$
Let E_h be the $(n+g+1)$-component row vector
$$E_h = [\overbrace{0 \quad 0 \quad \ldots \quad 0}^{h} \quad 1 \quad 0 \quad \ldots \quad 0] \tag{5.181}$$
Then, from Equations 5.179 and 5.180, the mean-square error in the equalized signal x_{i+h} is
$$\varepsilon = \mathscr{E}[(S_{i+h} E^T - S_{i+h} F_0^T - S_{i+h} E_h^T + u_{i+h})^2]$$
$$= \mathscr{E}[(S_{i+h}(E^T - F_0^T - E_h^T) + u_{i+h})^2] \tag{5.182}$$
where u_{i+h} is a Gaussian random variable with zero mean and variance
$$\eta^2 = \sigma^2 |D|^2 \tag{5.183}$$
as can be seen from Equation 5.131. Since the noise component u_{i+h} and any data-symbol forming a component of S_{i+h} are *statistically independent*, and since u_{i+h} has zero mean, it follows that u_{i+h} and s_{i+h-j}, for any integer j, are *statistically orthogonal*[78-80], so that
$$\mathscr{E}[u_{i+h} s_{i+h-j}] = 0 \tag{5.184}$$
and also
$$\mathscr{E}[u_{i+h}^2] = \eta^2 \tag{5.185}$$

DECISION-FEEDBACK EQUALIZERS

Thus
$$\varepsilon = \mathscr{E}[(S_{i+h}(E^T - F_0^T - E_h^T))^2 + u_{i+h}^2]$$
$$= \mathscr{E}[(S_{i+h}(E^T - F_0^T - E_h^T))^2] + \sigma^2|D|^2 \qquad (5.186)$$

Let
$$E - F_0 - E_h = B = [b_0 \quad b_1 \quad \ldots \quad b_{n+g}] \qquad (5.187)$$

Now, since the data-symbols $\{s_i\}$ are statistically independent with zero mean and variance k^2, it follows that the $\{s_i\}$ are *statistically orthogonal*[78-80], so that

$$\mathscr{E}[s_i s_j] = 0 \qquad (5.188)$$

for any $i \neq j$, and also

$$\mathscr{E}[s_i^2] = k^2 \qquad (5.189)$$

Thus, from Equations 5.184–5.189,

$$\varepsilon = \mathscr{E}[(S_{i+h}B^T)^2] + \sigma^2|D|^2 = \mathscr{E}[(\sum_{j=0}^{n+g} s_{i+h-j}b_j)^2] + \sigma^2|D|^2$$

$$= \sum_{j=0}^{n+g} \mathscr{E}[(s_{i+h-j}b_j)^2] + \sigma^2|D|^2 = \sum_{j=0}^{n+g} k^2 b_j^2 + \sigma^2|D|^2$$

$$= k^2|B|^2 + \sigma^2|D|^2 \qquad (5.190)$$

It can be seen from Equations 5.187 and 5.190 that, for any *given* value of D, $k^2|B|^2$ and ε are minimized by setting $f_j = e_{h+j}$, for $j = 1, 2, \ldots, \mu$, such that, from Equation 5.175,

$$F = [e_{h+1} \quad e_{h+2} \quad \ldots \quad e_{h+\mu}] \qquad (5.191)$$

The term $\sigma^2|D|^2$ is, of course, *independent* of F. Clearly, when Equation 5.191 is satisfied, the last μ components of B (Equation 5.187) are set to zero. Under these conditions and with the correct detection of $s_{i-1}, s_{i-2}, \ldots, s_{i-\mu}$, exact cancellation of the last μ components of E (Equation 5.172) is achieved by means of the nonlinear filter in Fig. 5.11. Thus Equation 5.191 represents a *necessary* (but not sufficient) condition that must be satisfied to obtain the minimum value of ε.

The decision-feedback equalizer is assumed to have μ taps in the filter F of Fig. 5.11, where μ satisfies Equation 5.178. This is the smallest number of taps whereby exact decision–directed cancellation can be achieved of *all* intersymbol interference components in x_{i+h}, involving the data-symbols $s_{i-1}, s_{i-2}, s_{i-3}, \ldots$ that have been detected. It has also been shown that, for the minimum mean-square error in the equalized signal, *exact* cancellation must be applied here. No advantage is gained if more than μ taps are used in the filter F, and the performance of the equalizer is in general *degraded*, if fewer than μ

taps are used. Indeed, it is evident from Equation 5.190 that, under the assumed conditions, any *reduction* below μ in the number of taps of the filter F, without changing the filter D, must *increase* the intersymbol interference in the equalized signal without affecting the level of the noise, whenever any components of B that are no longer cancelled (set to zero) by the nonlinear filter become nonzero. Thus, in Equation 5.187,

$$B = [e_0 \quad e_1 \quad \ldots \quad e_{h-1} \quad (e_h - 1) \quad \overbrace{0 \quad 0 \quad \ldots \quad 0}^{\mu}] \quad (5.192)$$

bearing in mind that correct detection of the data-symbols $s_{i-1}, s_{i-2}, \ldots, s_{i-\mu}$ is assumed here.

Let Y_c be the $(n+1) \times (n+g+1)$ matrix whose $(i+1)$th row is

$$Y_i = [\overbrace{0 \quad 0 \quad \ldots \quad 0}^{i} \quad y_0 \quad y_1 \quad \ldots \quad y_g \quad 0 \quad \ldots \quad 0] \quad (5.193)$$

such that the *first* component of Y_0 is y_0 and the *last* component of Y_n is y_g. Y_c is, of course, the *convolution matrix* for the sampled impulse-response of the channel. Now let Z be the $(n+1) \times (n+g+1)$ matrix obtained from Y_c by setting all components of its last μ columns to zero. Since Z has $n+g+1$ columns and $\mu = n+g-h$ (from Equation 5.178), it is clear that the first $h+1$ columns of Z are the *same* as the corresponding columns of Y_c, the remaining columns being, of course, given by *zero* column vectors. It follows from Equation 5.174 that $E = DY_c$, and hence from Equation 5.192 that

$$B = DZ - E_h \quad (5.194)$$

so that (from Equation 5.190) the mean-square error in the equalized signal x_{i+h} is

$$\varepsilon = k^2 |DZ - E_h|^2 + \sigma^2 |D|^2 \quad (5.195)$$

The *correct detection* of the data-symbols $s_{i-1}, s_{i-2}, \ldots, s_{i-\mu}$ is assumed here, as before.

It can be seen from Equations 4.254, 4.256, 4.265 and 4.266 that the $(m+1)$-tap *linear equalizer* which minimizes the mean-square error

$$\varepsilon_0 = k^2 |CY_c - E_h|^2 + \sigma^2 |C|^2 \quad (5.196)$$

in the equalized signal has tap gains given by

$$C = E_h Y_c^T \left(Y_c Y_c^T + \frac{\sigma^2}{k^2} I \right)^{-1} \quad (5.197)$$

and the resultant mean-square error in the equalized signal is

$$\varepsilon_0 = k^2 \left[1 - E_h Y_c^T \left(Y_c Y_c^T + \frac{\sigma^2}{k^2} I \right)^{-1} Y_c E_h^T \right] \quad (5.198)$$

C is here an $(m+1)$-component row vector giving the equalizer tap gains (Equation 4.246), Y_c is the $(m+1) \times (m+g+1)$ convolution matrix for the channel sampled impulse-response (Equation 4.255), E_h is the $(m+g+1)$-component row vector all of whose components are zero except for the $(h+1)$th component which has the value unity (Equation 4.248), and I is the $(m+1) \times (m+1)$ identity matrix.

In Equation 5.195, D and E_h are, respectively, $(n+1)$- and $(n+g+1)$-component row vectors (Equations 5.170 and 5.181), and Z is an $(n+1) \times (n+g+1)$ matrix. It is evident now, from the close similarity between Equations 5.195 and 5.196, that exactly the same procedure as that employed in the derivation of Equations 5.197 and 5.198 from Equation 5.196, can be applied also to Equation 5.195, provided only that the $(n+1) \times (n+1)$ matrix $ZZ^T + (\sigma^2/k^2)I$ is *nonsingular*. Indeed, it is clear that, if Equation 5.195 is used in place of Equation 5.196 in this derivation, then the following results must be obtained. From Equation 5.197, the $(n+1)$-tap linear feedforward transversal filter D of the decision-feedback equalizer in Fig. 5.11, that minimizes the mean-square error in the equalized signal, has tap gains given by the $(n+1)$-component row vector

$$D = E_h Z^T \left(ZZ^T + \frac{\sigma^2}{k^2} I \right)^{-1} \quad (5.199)$$

where ZZ^T and $ZZ^T + (\sigma^2/k^2)I$ are $(n+1) \times (n+1)$ real symmetric matrices, and I is the $(n+1) \times (n+1)$ identity matrix. Also, from Equation 5.198, the mean-square error in the equalized signal is now

$$\varepsilon = k^2 \left[1 - E_h Z^T \left(ZZ^T + \frac{\sigma^2}{k^2} I \right)^{-1} Z E_h^T \right] \quad (5.200)$$

The *correct detection* of the data-symbols $s_{i-1}, s_{i-2}, \ldots, s_{i-\mu}$ is, of course, assumed in Equations 5.199 and 5.200, and the tap gains of the filter F in Figure 5.11 must satisfy Equations 5.175 and 5.191. It is clear from Equation 5.199 that D is given by the $(h+1)$th row of the $(n+g+1) \times (n+1)$ matrix

$$Z^T \left(ZZ^T + \frac{\sigma^2}{k^2} I \right)^{-1} \quad (5.201)$$

Since the last μ columns of Z are $(n+1)$-component *zero* column-vectors, the last μ rows of Z^T and therefore also the last μ rows of $Z^T(ZZ^T + (\sigma^2/k^2)I)^{-1}$ are $(n+1)$-component *zero* row-vectors. Also, since $h+\mu=n+g$ (Equation 5.178), D is given by the *last nonzero row* of $Z^T(ZZ^T+(\sigma^2/k^2)I)^{-1}$. If Z_h is the $(n+1)\times(h+1)$ matrix obtained from Z by omitting the last μ columns that are all zero vectors, then D is given by the *last row* of the $(h+1)\times(n+1)$ matrix $Z_h^T(Z_hZ_h^T+(\sigma^2/k^2)I)^{-1}$.

If now μ is *constrained* to some given small value, say 2 or 3, then a reduced value of the mean-square error in the equalized signal (for the given constraint) may well be obtained through the use of a *smaller* value of h than that given by Equation 5.178, the equation, therefore, being no longer satisfied. Under these conditions, the intersymbol interference involving *detected* data symbols can no longer be completely eliminated from the equalized signal by means of the nonlinear filter. The above analysis nevertheless still applies, although the matrix Z in Equations 5.199 and 5.200 is now derived from the corresponding matrix Y_c by setting its $(h+2)$th to $(h+\mu+1)$th columns to zero, leaving its first $h+1$ and last $n+g-\mu-h$ columns unchanged. This is, in fact, the most general form of the matrix Z. However, we are primarily concerned here with an equalizer having the most suitable number of taps, μ, of the filter F, or equivalently the best value of h, *given* that Equation 5.178 holds, since, for particular values of n and h, and given the correct detection of s_{i-1}, $s_{i-2}, \ldots$, the mean-square error in the equalized signal is *minimized* by selecting μ to satisfy this equation. The optimum equalizer, for a particular value of n and a particular signal/noise ratio, is therefore finally determined by selecting the value of h that minimizes the mean-square error in the equalized signal, under the assumed conditions (which included Equation 5.178).

In a decision-feedback equalizer, where the value of μ is restricted to a value *below* that satisfying Equation 5.178 and where there are suitable values of n and h together with a high signal/noise ratio, the sampled impulse-response of the channel and linear-filter D (Fig. 5.11) becomes, for practical purposes, the $(\mu+1)$-component sequence (vector)

$$[1 \quad f_1 \quad f_2 \quad \cdots \quad f_\mu]$$

neglecting here the delay in transmission. Furthermore, the μ tap gains of the filter F are given by the components of the vector

$$[f_1 \quad f_2 \quad \cdots \quad f_\mu]$$

The values of the $\{f_i\}$ here are such as to minimize the mean-square error in the equalized signal and are dependent on Y (Equation 5.113), μ, h and n. The sampled impulse-response of the channel and filter D is now *restricted* to a sequence of just $\mu+1$ components. Clearly, if μ is reduced to zero, for *given* values of n and h not satisfying Equation

5.178, the system degenerates into a conventional linear equalizer that minimizes the mean-square error in the equalized signal. Thus, as μ is increased from 0 to $n+g-h$ (for a given value of h) so the equalizer changes from a linear equalizer to the ideal decision-feedback equalizer. *All* intersymbol interference, involving data symbols that have been detected, is here eliminated from the equalized signal, given the correct detection of the data symbols. Obviously, μ could be chosen to be 1 or 2, together with appropriately large values of n and h (Equation 5.178 *not* now being satisfied) so that, with severe amplitude and phase distortions of the received signal, *most* of the equalization is carried out by the *linear* filter. Furthermore, if the μ tap gains of the filter F are *fixed* at some arbitrary set of values, then, under the assumed conditions, the filter D adjusts the sampled impulse-response of the channel and filter D to be, effectively,

$$[1 \quad f_1 \quad f_2 \quad \ldots \quad f_\mu]$$

where the values of the $\{f_i\}$ are now determined by the corresponding *fixed* tap gains of the filter F. In every case, of course, it is assumed the tap gains of the decision-feedback equalizer, that are *not* fixed in value, are adjusted to minimize the mean-square error in the equalized signal. It is evident that, if the tap gains of the equalizer can be adjusted *adaptively* to minimize the mean-square error in the equalized signal, then the equalizer can be used to adjust the tap gains of the linear filter in such a way as to *reduce the number of components* of the sampled impulse-response of the channel and linear filter to some desired value, or even to adjust the sampled impulse-response *itself* to some desired sequence. Such techniques have been studied for use with sophisticated detectors[81-87], and are not considered further here. Techniques for *adaptively* adjusting the tap gains of an equalizer are considered in Section 6.1. Our concern now is with the *ideal* adjustment of such an equalizer, under any particular set of conditions, so that Equations 5.178 and 5.199 are satisfied and the mean-square error in the equalized signal is minimized for the given values of n and h.

In the presence of noise, when $\sigma^2 \neq 0$, the $(n+1) \times (n+1)$ matrix

$$ZZ^T + \frac{\sigma^2}{k^2} I \quad (5.202)$$

is real, symmetric and positive definite, so that it must be *nonsingular*. This is because, for $i=1, 2, \ldots, n+1$, the ith eigenvalue of $ZZ^T + (\sigma^2/k^2)I$ is equal to the *sum* of the corresponding eigenvalues of ZZ^T and $(\sigma^2/k^2)I$, the eigenvalues of ZZ^T being real-valued and *non-negative* and all eigenvalues of $(\sigma^2/k^2)I$ being equal to σ^2/k^2. Thus, when $\sigma^2 \neq 0$, *all* eigenvalues of $ZZ^T + (\sigma^2/k^2)I$ are real valued and *positive*, and the matrix is nonsingular[88–90].

In the absence of noise, when $\sigma^2 = 0$, the matrix $ZZ^T + (\sigma^2/k^2)I$ becomes ZZ^T, which could well be a positive semi-definite matrix and may therefore be singular. The rank of Z cannot exceed the number of its nonzero columns, which is

$$n + g + 1 - \mu = h + 1$$

from Equation 5.178. Nor, of course, can it exceed the number of its rows, which is $n+1$. Thus, if $\mu > g$, Z has fewer than $n+1$ nonzero columns and so has rank *less* than $n+1$. The $(n+1) \times (n+1)$ matrix ZZ^T must now have rank less than $n+1$ and is therefore *singular*. Again, if the first few components of the sampled impulse-response of the channel are very small, the matrix ZZ^T becomes very nearly singular (ill conditioned) even when μ is a little less than g. However, for smaller values of μ, ZZ^T becomes nonsingular. When $\mu = 0$ and $n = m$ (Equation 5.178 *not* being satisfied) then $Z = Y_c$ and $D = C$ in Equations 5.195, 5.199 and 5.200. The decision-feedback equalizer that minimizes the mean-square error in the equalized signal here becomes the *same* as the corresponding linear equalizer, bearing in mind that h may take on *any* appropriate value.

When the matrix $ZZ^T + (\sigma^2/k^2)I$ becomes singular or very nearly singular, there is no longer a unique or well-defined vector D that minimizes the mean-square error in the equalized signal. Indeed, there are now an infinite number of such vectors $\{D\}$ that, for practical purposes, minimize the mean-square error. If ZZ^T is singular, the matrix $ZZ^T + (\sigma^2/k^2)I$ becomes nearly singular at very high signal/noise ratios, when $\sigma^2 \ll k^2$ (assuming that $|Y| \simeq 1$), which suggests that μ (Equation 5.178) should be chosen such that ZZ^T is *nonsingular* and therefore such that $\mu \leqslant g$. Furthermore, if $\mu = g+1$, the last $g+1$ components of the vector E (Equation 5.172) are effectively set to zero by the nonlinear filter (Fig. 5.11). Since E is the convolution of Y and D (Equations 5.113 and 5.170), this means that the last tap of the filter D, with gain d_n, no longer influences the vector B in Equation 5.192 and could therefore be omitted without affecting the performance of the equalizer. In other words, if $\mu > g$, the effective number of taps in the filter D is *reduced*, thus *degrading* the performance of the equalizer. Clearly, μ must *not* exceed g. On the other hand, increasing μ from zero towards g, with a sufficiently large value of n, correspondingly increases the permitted number of significant components in the sampled impulse-response of the channel and filter D (see Fig. 5.12) and thereby correspondingly increases the range of possible vectors $\{D\}$ from which the optimum vector is selected. In other words, it increases the number of degrees of freedom available in the optimization of the equalizer, probably leading therefore to a reduction in the resulting mean-square error of the equalized signal. This suggests the arrangement of the equalizer in which

$$\mu = g \tag{5.203}$$

and
$$h = n \tag{5.204}$$

bearing in mind that Equation 5.178 must be satisfied. The delay in detection is here constrained to be n sampling intervals.

Suppose now that, under these conditions, there is no noise, that is, an infinitely high signal/noise ratio, such that the error in the equalized signal is caused entirely by intersymbol interference. Now $\sigma^2 = 0$ and the matrix $ZZ^T + (\sigma^2/k^2)I$ in Equation 5.199 becomes ZZ^T, where

$$Z = \begin{bmatrix} y_0 & y_1 & y_2 & \cdots & 0 & 0 & 0 & 0 & \cdots & 0 \\ 0 & y_0 & y_1 & \cdots & 0 & 0 & 0 & 0 & \cdots & 0 \\ 0 & 0 & y_0 & \cdots & 0 & 0 & 0 & 0 & \cdots & 0 \\ \vdots & \vdots & \vdots & \vdots\vdots\vdots & \vdots & \vdots & \vdots & \vdots & \vdots\vdots\vdots & \vdots \\ 0 & 0 & 0 & \cdots & y_0 & y_1 & y_2 & 0 & \cdots & 0 \\ 0 & 0 & 0 & \cdots & 0 & y_0 & y_1 & 0 & \cdots & 0 \\ 0 & 0 & 0 & \cdots & 0 & 0 & y_0 & 0 & \cdots & 0 \end{bmatrix} \tag{5.205}$$

assuming that $n > g$. The first $n+1$ columns of Z here form an upper triangular matrix, with each component along the main diagonal being equal to y_0, and the remaining g columns of Z have all their components equal to zero. So long as $y_0 \neq 0$, the matrix Z has rank $n+1$, so that the $(n+1) \times (n+1)$ matrix ZZ^T is nonsingular. Thus, from Equations 5.199 and 5.204,

$$D = E_n Z^T (ZZ^T)^{-1} = E'_n Z_n^T (Z_n Z_n^T)^{-1} \tag{5.206}$$

where Z_n is the $(n+1) \times (n+1)$ matrix formed by the first $n+1$ columns of Z, and E'_n is the $(n+1)$-component vector formed by the first $n+1$ components of E_n. Hence

$$D = E'_n Z_n^T (Z_n^T)^{-1} Z_n^{-1} = E'_n Z_n^{-1} \tag{5.207}$$

It can be shown that Z_n^{-1} is an upper triangular matrix all of whose components along the main diagonal are y_0^{-1} [38,88-90]. Also, from Equation 5.207, D is formed by the *last row* of Z_n^{-1} so that

$$D = [0 \quad 0 \quad \cdots \quad 0 \quad y_0^{-1}] \tag{5.208}$$

The filter D (Fig. 5.11) here simply adjusts the level and delays the received signal by n sampling intervals. Filter F has g taps, with gains of $y_1/y_0, y_2/y_0, \ldots, y_g/y_0$. Thus, with the correct detection of $s_{i-1}, s_{i-2}, \ldots, s_{i-g}$, the equalized signal at the detector input, at time $t = (i+n)T$, is

$$x_{i+n} = s_i + y_0^{-1} w_{i+n} \qquad (5.209)$$

Clearly there is *no* intersymbol interference in the equalized signal. With a finite number of taps in the filter D, any other arrangement of the equalizer, that is, any other method of *sharing* the equalization process between the filters D and F, must lead to *some* intersymbol interference in the equalized signal and so to a *greater* mean-square error in the latter.

At extremely low signal/noise ratios, such that the error caused by noise in the equalized signal greatly exceeds the level of the equalized signal itself, $\sigma^2 \gg k^2$ (where again $|Y| \simeq 1$) and D is now given by

$$D \simeq E_n Z^T \left(\frac{\sigma^2}{k^2} I \right)^{-1} = \frac{k^2}{\sigma^2} E_n Z^T$$

$$= \frac{k^2}{\sigma^2} [0 \quad 0 \quad \ldots \quad 0 \quad y_g \quad y_{g-1} \quad \ldots \quad y_0] \qquad (5.210)$$

assuming that $n > g = \mu$. As before, it is assumed that the data-symbols $s_{i-1}, s_{i-2}, \ldots, s_{i-g}$ have been correctly detected. The filter D is here *matched* to each received signal-element, just as for the corresponding *linear* equalizer (Equation 4.268), the tap gains of filter F being, of course, appropriately adjusted. It is interesting to note that any *reduction* in μ, such that $\mu < g$ and therefore $h > n$, means that the filter D is no longer matched to each received signal-element but only to a *portion* of it, so that the performance is degraded. Thus the *only* value of μ, that has no potential disadvantages under the various conditions considered, is $\mu = g$, as in Equation 5.203. At very low signal/noise ratios, when Equations 5.203 and 5.210 are satisfied and there is significant *amplitude distortion* in the received signal, the components $\{e_j\}$ of the sampled impulse-response E of the channel and linear filter (Equation 5.172) do *not* now approximate to zero for $n - g \leqslant j < n$, but become quite large for some of these values of j, leading to serious intersymbol interference in the equalized signal. The assumption made in Equation 5.210 that, in the detection of s_i, the data-symbols $s_{i-1}, s_{i-2}, \ldots, s_{i-g}$ have been correctly detected, is clearly not practical here, since the error rate would, in fact, now be very high. Thus the preceding analysis, although illuminating, is perhaps somewhat academic.

The case of greatest interest is that where there are a sufficient number of taps in the equalizer to enable the accurate equalization of the channel to be achieved and where the signal/noise ratio is sufficiently high to permit this equalization to be achieved in an equalizer that minimizes the mean-square error in the equalized signal. If, in fact, the equalizer equalizes the channel *exactly* and, as before, $h = n$, then the sampled impulse-response of the channel and filter D is given by the $(n+g+1)$-component row vector

$$E = [\overbrace{0 \quad 0 \quad \ldots \quad 0}^{n} \quad 1 \quad e_{n+1} \quad e_{n+2} \quad \ldots \quad e_{n+g}] \quad (5.211)$$

and the tap gains of the filter F are given by the g-component row vector

$$F = [e_{n+1} \quad e_{n+2} \quad \ldots \quad e_{n+g}] \quad (5.212)$$

The adjustment of the decision-feedback equalizer is now that assumed in Fig. 5.12, where it can be seen that Equations 5.211 and 5.212 are both satisfied. Thus, in Equation 5.190, the component $k^2|B|^2$ that represents the mean-square error caused by intersymbol interference in the equalized signal, is set to *zero*, leaving only the component $\sigma^2|D|^2$, that represents the mean-square error caused by noise. Clearly, $|D|$ must now be minimized, subject to $|B|$ being held at zero. The constraint that the first n components of E are all zero and the $(n+1)$th component is unity, is not of itself sufficient to determine *uniquely* the tap gains of the equalizer, as has been mentioned before. However, when this constraint is combined with the requirement that the mean-square error caused by noise in the equalized signal (and hence $|D|$) be minimized, it leads to an elegant and practically very useful design for the decision-feedback equalizer. It does not, of course, *necessarily* follow that the assumed value of μ given by Equation 5.203 is now optimum, so that values of μ *less* than g (and hence values of h *greater* than n) must also be considered when optimizing the system. Ideally, $n \to \infty$ here. The decision-feedback equalizer that is constrained to achieve the *exact* equalization of the channel, will next be studied in some detail.

5.8 EQUALIZER THAT MAXIMIZES THE SIGNAL/NOISE RATIO IN THE EQUALIZED SIGNAL, SUBJECT TO THE EXACT EQUALIZATION OF THE CHANNEL

5.8.1 Derivation of equalizer via matrix algebra

A study will now be made of the decision-feedback equalizer that *maximizes* the signal/noise ratio at the detector input, subject to exact equalization. Under the latter constraint and at very high signal/noise ratios, the equalizer furthermore minimizes, at least for practical purposes, the error rate in the detected data signals *for the given equalizer structure*, as will presently be shown. The equalizer is therefore of great theoretical interest. It is, for convenience, referred to as the *optimum* decision-feedback equalizer.

Consider the data-transmission system of Fig. 5.1 and the decision-feedback equalizer of Fig. 5.11. Suppose that the baseband channel in Fig. 5.1 has the sampled impulse-response

$$Y = [y_0 \quad y_1 \quad \cdots \quad y_g] \tag{5.213}$$

with z-transform

$$Y(z) = y_0 + y_1 z^{-1} + \cdots + y_g z^{-g} \tag{5.214}$$

as before, and let the $(m+1)$-tap *linear* feedforward transversal equalizer for the channel to have the z-transform

$$C(z) = c_0 + c_1 z^{-1} + \cdots + c_m z^{-m} \tag{5.215}$$

as in Equation 4.24. It is assumed that $Y(z)$ has no zeros (roots) *on* the unit circle in the z plane. Thus

$$Y(z)C(z) \simeq z^{-h} \tag{5.216}$$

or

$$C(z) \simeq z^{-h} Y^{-1}(z) \tag{5.217}$$

for some appropriate integer h in the range 0 to $m+g$.

The linear feedforward transversal filter D of the decision-feedback equalizer in Fig. 5.11 is assumed to have $n+1$ taps and a sampled impulse-response.

$$D = [d_0 \quad d_1 \quad \cdots \quad d_n] \tag{5.218}$$

with z-transform

$$D(z) = d_0 + d_1 z^{-1} + \cdots + d_n z^{-n} \tag{5.219}$$

where $n > m$. Now let

$$D(z) = C(z)E(z) \tag{5.220}$$

where

$$E(z) = 1 + e_1 z^{-1} + e_2 z^{-2} + \cdots + e_{n-m} z^{-n+m} \tag{5.221}$$

so that the z-transform of the channel and linear filter is

$$Y(z)D(z) \simeq z^{-h} E(z) \tag{5.222}$$

Thus the sampled impulse-response of the channel and linear filter D in Fig. 5.11 is given approximately by the $(n-m+1)$-component row vector (sequence)

$$E = [1 \quad e_1 \quad e_2 \quad \cdots \quad e_{n-m}] \tag{5.223}$$

neglecting the delay of h sampling intervals. The components $e_1, e_2, \ldots, e_{n-m}$ in Equation 5.223 are not to be confused with the corresponding components in Equation 5.172. They are however related to each other in that the ith component of E in Equation 5.223 is equal to the $(h+i)$th component of E in Equation 5.172, for $i=1, 2, \ldots, n-m+1$, the remaining components of the latter vector being approximately zero and therefore ignored in the former vector (Equation 5.223).

The nonlinear filter in Fig. 5.11 equalizes $E(z)$, so that the linear feedforward transversal filter F has $n-m$ taps with gains equal, respectively, to the components $e_1, e_2, \ldots, e_{n-m}$ of the vector

$$F = [e_1 \quad e_2 \quad \cdots \quad e_{n-m}] \tag{5.224}$$

The signal at the output of the linear filter D, at time $t = (i+h)T$, is

$$v_{i+h} \simeq s_i + s_{i-1}e_1 + s_{i-2}e_2 + \cdots + s_{i-n+m}e_{n-m} + u_{i+h} \tag{5.225}$$

where u_{i+h} is the noise component, and the equalized signal at the input to the detector is

$$x_{i+h} = v_{i+h} - s'_{i-1}e_1 - s'_{i-2}e_2 - \cdots - s'_{i-n+m}e_{n-m} \tag{5.226}$$

With the correct detection of $s_{i-1}, s_{i-2}, \ldots, s_{i-n+m}$, the signal at the input to the detector becomes

$$x_{i+h} \simeq s_i + u_{i+h} \tag{5.227}$$

When $x_{i+h} > 0$, $s'_i = k$, and when $x_{i+h} < 0$, $s'_i = -k$.

The noise component u_{i+h} is a Gaussian random variable with zero mean and variance

$$\eta^2 = \sigma^2 \sum_{j=0}^{n} d_j^2 = \sigma^2 |D|^2 \tag{5.228}$$

where $|D|$ is the Euclidean length of the vector D[78-80].

Thus the probability of error in the detection of s_i, at high signal/noise ratios and given the correct detection of $s_{i-1}, s_{i-2}, \ldots, s_{i-n+m}$, can be taken to be

$$P_e = \int_k^\infty \frac{1}{\sqrt{2\pi\eta^2}} \exp\left(-\frac{u^2}{2\eta^2}\right) du = Q\left(\frac{k}{\eta}\right) = Q\left(\frac{k}{\sigma|D|}\right) \tag{5.229}$$

The error extension effects are not normally excessive here, at least not with the binary signals assumed, so that no serious inaccuracy is introduced by taking the average probability of error or average error rate, actually obtained, as P_e.

It is clear from Equations 5.227 and 5.228 that, to minimize the error rate in the detected data-symbols $\{s'_i\}$, it is necessary to minimize $|D|$. The minimization is, of course, subject to the constraints imposed by Equations 5.220, 5.221 and 5.224, which imply the *accurate* equalization of the channel by the decision-feedback equalizer, together with the appropriate adjustment in signal level, so that *exact* equality holds in

Equation 5.227. At low signal/noise ratios, a better performance can be obtained by relaxing the constraint of accurate equalization. However, our main interest is in applications at high signal/noise ratios, were a minimum mean-square error equalizer with a sufficient number of taps (Equation 5.199) in any case achieves a reasonably accurate equalization of the channel.

It is evident that, when exact equality holds in Equation 5.227,

$$\mathscr{E}[x_{i+h}] = \mathscr{E}[s_i + u_{i+h}] = \mathscr{E}[s_i] \quad (5.230)$$

since $E[u_{i+h}] = 0$. This means that x_{i+h} is now an *unbiased estimate* of s_i, and the mean-square error in the equalized signal becomes[78-80]

$$\mathscr{E}[(x_{i+h} - s_i)^2] = \mathscr{E}[u_{i+h}^2] = \sigma^2 |D|^2 \quad (5.231)$$

as has, in fact, been mentioned previously (see Equation 5.190). It follows immediately that if $|D|$ is minimized, subject to the *exact* equalization of the channel, this both *minimizes* the mean-square error in the equalized signal and *maximizes* the signal/noise ratio here, subject, of course, to exact equality being maintained in Equation 5.227. Throughout Section 5.8, the *exact* or *accurate* equalization of the channel is assumed, so that *all intersymbol interference* is eliminated (exactly or for practical purposes) in the equalized signal. Except where otherwise stated, it is also assumed that the data signal here is properly scaled (having the correct magnitude), such that the equalized signal x_{i+h} is an *unbiased estimate* of s_i.

It is clear now that the required decision-feedback equalizer must be adjusted to minimize $|D|$, subject to Equation 5.222 and with the first component of E being held at unity, as in Equation 5.223.

Let G be the $(n-m+1) \times (n+1)$ convolution matrix whose $(i+1)$th row is given by the $(n+1)$-component row vector

$$G_i = [\overbrace{0 \quad \ldots \quad 0}^{i} \quad c_0 \quad c_1 \quad \ldots \quad c_m \quad 0 \quad \ldots \quad 0] \quad (5.232)$$

so that

$$G = \begin{bmatrix} c_0 & c_1 & c_2 & \ldots & 0 & 0 & 0 \\ 0 & c_0 & c_1 & \ldots & 0 & 0 & 0 \\ 0 & 0 & c_0 & \ldots & 0 & 0 & 0 \\ \vdots & \vdots & \vdots & \ddots & \vdots & \vdots & \vdots \\ 0 & 0 & 0 & \ldots & c_m & 0 & 0 \\ 0 & 0 & 0 & \ldots & c_{m-1} & c_m & 0 \\ 0 & 0 & 0 & \ldots & c_{m-2} & c_{m-1} & c_m \end{bmatrix} \quad (5.233)$$

where c_i is the coefficient of z^{-i} in $C(z)$ (Equation 5.215). Then, from Equations 5.220 and 5.223,

$$D = EG = G_0 - LM \qquad (5.234)$$

where
$$L = [-e_1 \quad -e_2 \quad \ldots \quad -e_{n-m}] \qquad (5.235)$$

and M is the $(n-m) \times (n+1)$ matrix whose ith row is G_i. Clearly, the removal of the first row from G gives the matrix M.

It is now required to minimize

$$|D| = |G_0 - LM| \qquad (5.236)$$

But, for given values of $C(z)$ and n, G_0 and M are *fixed*, leaving L as the only variable in $|G_0 - LM|$. Thus L must be chosen to minimize this quantity, no restrictions being placed on any components of L.

G_0 and LM are $(n+1)$-component vectors and so can be represented as points in an $(n+1)$-dimensional Euclidean vector space[88,90], as illustrated in Fig. 5.13. Only three of the $n+1$ orthogonal axes are, of course, shown in this diagram. Now G_0 is *fixed* but LM can be moved by changing L. Thus L must be chosen to minimize $|G_0 - LM|$, which is the Euclidean distance between the vectors G_0 and LM. But

$$LM = \sum_{i=1}^{n-m} -e_i G_i \qquad (5.237)$$

and the $n-m$ vectors $\{G_i\}$ are *linearly independent*, since no one of them can be formed by a linear combination of any of the others. It follows that the $n-m$ vectors $\{G_i\}$ in Equation 5.237 span an $(n-m)$-dimensional subspace of the $(n+1)$-dimensional Euclidean vector space. But Equation 5.237 shows that LM is a linear combination of the $n-m$ vectors $\{G_i\}$, so that LM must lie in the $(n-m)$-dimensional subspace spanned by these vectors. Thus $|G_0 - LM|$ is the distance between the *fixed* point G_0 in Fig. 5.13 and the *moveable* point LM, which is confined to the given subspace. Since there are no constraints on L, the vector L may be selected such that LM is *any* required point in the $(n-m)$-dimensional subspace. It follows that $|G_0 - LM|$ *is minimum* when LM is the point in the subspace at the *minimum distance* from G_0. By the

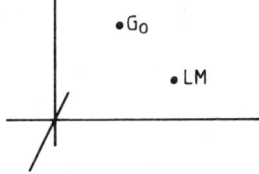

Fig. 5.13   Vectors G_0 and LM in the $(n+1)$-dimensional Euclidean vector space

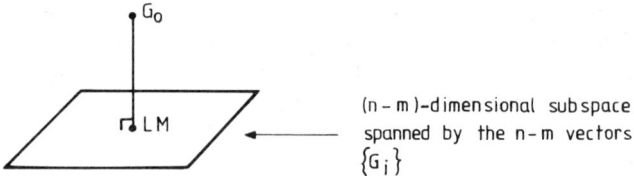

Fig. 5.14 Orthogonal projection of G_0 on to the $(n-m)$-dimensional subspace, to give LM

Projection theorem (Equation 4.269), LM is now the *orthogonal projection* of G_0 on to the subspace, as illustrated in Fig. 5.14, so that $G_0 - LM$ is *orthogonal* to the subspace and therefore also to every vector that lies in the subspace.

The subspace contains each G_i, for $i = 1, 2, \ldots, n-m$, and it also contains the origin of the vector space, that is, the zero vector. Thus the vector $G_0 - LM$ is *orthogonal* to each vector G_i, for $i = 1, 2, \ldots, n-m$, so that

$$(G_0 - LM)G_i^T = 0 \tag{5.238}$$

for $i = 1, 2, \ldots, n-m$. Hence, if the $(n+1)$-component zero row vector is designated 0,

$$(G_0 - LM)M^T = 0 \tag{5.239}$$

or

$$LMM^T = G_0 M^T \tag{5.240}$$

But the $(n-m) \times (n+1)$ matrix M is of rank $n-m$, since its $n-m$ rows, given by the $n-m$ vectors $\{G_i\}$, are linearly independent. Thus the $(n-m) \times (n-m)$ matrix MM^T is *nonsingular*. Now, from Equation 5.240,

$$L = G_0 M^T (MM^T)^{-1} \tag{5.241}$$

and, from Equation 5.234,

$$D = G_0 - LM = G_0 - G_0 M^T (MM^T)^{-1} M$$
$$= G_0 (I - M^T (MM^T)^{-1} M) \tag{5.242}$$

where I is the $(n+1) \times (n+1)$ identity matrix. The $n+1$ tap gains of the required linear feedforward transversal filter D, in Fig. 5.11, are therefore given by the $(n+1)$-component vector D in Equation 5.242. G_0 is here an $(n+1)$-component row vector, and $M^T(MM^T)^{-1}M$ and $I - M^T(MM^T)^{-1}M$ are both $(n+1) \times (n+1)$ matrices.

As before, the filter F in Fig. 5.11 has $n-m$ taps with tap gains equal, respectively, to the components $e_1, e_2, \ldots, e_{n-m}$ of the vector F in Equation 5.224, the values of the $\{e_i\}$ being now given by the vector L in Equations 5.235 and 5.241. Thus, with a sufficient number of taps in the

DECISION-FEEDBACK EQUALIZERS   281

filter D and an appropriate value of h in Equation 5.227, the decision-feedback equalizer achieves the accurate equalization of the channel.

It has been found that, for a sufficiently large value of n, L undergoes no further change as n is increased, having in fact no more than g consecutive nonzero components. Also, the number of taps with significantly nonzero gains required by the linear filter D is frequently smaller than $m+1$ (the number of tap gains of the corresponding *linear equalizer* for the channel) and the tap gains themselves are essentially independent of n. Assume now, for convenience, that $n = 2m$, so that, in Equation 5.234, L has m components and M is an $m \times (2m+1)$ matrix. Let D_i be the $(n+1)$-component vector obtained from D (Equation 5.242) by moving its nonzero components i places to the right, where $i = 1, 2, \ldots, m-g$, and m is sufficiently large such that the last $m-g$ of the $2m+1$ components of D are all effectively zero. Now, from Equation 5.234,

$$D_i = G_i - L_i M_i \qquad (5.243)$$

where L_i is the $(m-i)$-component row vector, whose components, other than the first g which are as in L, are all zero, and where M_i is the $(m-i) \times (2m+1)$ matrix obtained from the $m \times (2m+1)$ matrix M by removing its first i rows. The vector D_i is a linear combination of the vectors $G_i, G_{i+1}, \ldots, G_m$ and it therefore lies in the $(m-i+1)$-dimensional subspace spanned by these vectors. But, from Equations 5.234 and 5.238, D is orthogonal to *all* vectors $G_1, G_2, \ldots, G_m$, so that D is orthogonal to D_i, for $i = 1, 2, \ldots, m-g$. By further increasing the integer m, the nonzero components of D are confined to its first $m-g$ components, the number of its nonzero components no longer increasing with m. But the first $m-g$ components of the vector D_{m-g+i} are *zero* for any nonnegative integer i, which implies that D and D_{m-g+i} are orthogonal, *regardless* of the component values of D. It is evident now that D becomes *orthogonal* to *all* $\{D_i\}$, for $i > 0$, such that

$$DD_i^T = 0 \qquad (5.244)$$

for all positive integers $\{i\}$. It follows from Equation 5.244, that the linear feedforward transversal filter with sampled impulse-response D is an *allpass network*. This does not change any amplitude distortion in the received signal, affecting only the phase distortion (see Section 3.2).

5.8.2 Derivation of equalizer via z-transforms

An alternative derivation of the optimum decision-feedback equalizer just described will now be presented. This leads to the same result but expressed in a quite different and perhaps more useful form.

Consider the decision-feedback equalizer shown in Fig. 5.11. As before, let the z-transform of the baseband channel and sampler in Fig. 5.1 be

$$Y(z) = y_0 + y_1 z^{-1} + \cdots + y_g z^{-g} \qquad (5.245)$$

with no zeros (roots) *on* the unit circle in the z plane, and where $y_0 \neq 0$. Let the z-transform on the $(n+1)$-tap linear feedforward transversal filter D in Fig. 5.11 be

$$D(z) = d_0 + d_1 z^{-1} + \cdots + d_n z^{-n} \qquad (5.246)$$

Suppose that the z-transform of the baseband channel and linear filter is

$$Y(z)D(z) \simeq z^{-h} V_1(z) V_2(z) \qquad (5.247)$$

where *all* coefficients in $V_1(z)$ and $V_2(z)$ are *real valued*, and the first term of each factor contains z^0. $V_2(z)$ may be any factor of $Y(z)D(z)$ such that *all* zeros of $V_2(z)$ lie *outside* the unit circle. No assumptions are however made about the zeros of $V_1(z)$, except, of course, that any complex-valued zeros occur in complex-conjugate pairs. Zeros at the origin or infinity in the z plane are ignored throughout this discussion. Let

$$V_2(z) = v_0 + v_1 z^{-1} + \cdots + v_j z^{-j} \qquad (5.248)$$

where $v_0 \neq 0$ and j is a positive integer equal to the number of zeros of $V_2(z)$.

Suppose now that there is added, at the input of the linear filter D in Fig. 5.11, another linear filter with z-transform

$$A(z) \simeq z^{-l} V_2^{-1}(z) V_3(z) \qquad (5.249)$$

where
$$V_3(z) = v_j + v_{j-1} z^{-1} + \cdots + v_0 z^{-j} \qquad (5.250)$$

and l is the positive integer needed to make $A(z)$ physically realizable. $V_3(z)$ is obtained from $V_2(z)$ by reversing the order of its coefficients, so that the zeros of $V_3(z)$ are the *reciprocals* of those of $V_2(z)$ and therefore all lie *inside* the unit circle. $V_2^{-1}(z)$ has an infinite number of terms with positive powers of z, and in $A(z)$ this is truncated into an appropriately large finite number, for the required degree of accuracy. The integer l is now made large enough to remove all positive powers of z from $A(z)$. As is shown in Section 3.3, $V_2^{-1}(z)V_3(z)$ is the z-transform of the ideal linear equalizer for the pure phase distortion $V_2(z)V_3^{-1}(z)$. It is phase distortion because $V_2^{-1}(z)V_3(z)$ is the *inverse* of $V_2(z)V_3^{-1}(z)$ and is also obtained from the latter by *reversing* the order of its coefficients, the reversal being pivoted about the coefficient of z^0. Thus $V_2^{-1}(z)V_3(z)$ is both *matched* to $V_2(z)V_3^{-1}(z)$ and is its *inverse*. This means that $A(z)$ represents an *orthogonal transformation*, which does not change the level of the data signal, nor does it change the level or other statistics of the additive

Gaussian noise. The linear filter with z-transform $A(z)$ is therefore an *allpass network*.

The z-transform of the channel and two linear filters is

$$Y(z)A(z)D(z) \simeq z^{-h-l}V_1(z)V_3(z) \tag{5.251}$$

as can be seen from Equations 5.247 and 5.249. Let the first term of $Y(z)D(z)$ (Equation 5.247) be $e_0 z^{-h}$, and let the first term of $Y(z)A(z)D(z)$ (Equation 5.251) be $f_0 z^{-h-l}$. Also, let

$$V_1(z) = p_0 + p_1 z^{-1} + \cdots + p_m z^{-m} \tag{5.252}$$

Then, from Equations 5.248 and 5.250,

$$|e_0| = |p_0||v_0| \tag{5.253}$$

and

$$|f_0| = |p_0||v_j| \tag{5.254}$$

Now

$$V_2(z) = v_0(1 - \alpha_1 z^{-1})(1 - \alpha_2 z^{-1}) \ldots (1 - \alpha_j z^{-1}) \tag{5.255}$$

where $\alpha_1, \alpha_2, \ldots, \alpha_j$ are the j zeros of $V_2(z)$, such that

$$|\alpha_i| > 1 \tag{5.256}$$

for $i = 1, 2, \ldots, j$. Thus

$$|v_j| = |v_0||\alpha_1||\alpha_2| \ldots |\alpha_j| > |v_0| \tag{5.257}$$

and, from Equations 5.253 and 5.254,

$$|f_0| > |e_0| \tag{5.258}$$

It is assumed that the decision-feedback equalizer in Fig. 5.11, with the additional linear filter at its input, achieves the accurate equalization of the baseband channel, but now *without* appropriately adjusting the signal level such that x_{i+h+l} is an unbiased estimate of s_i. The z-transform of the channel and two linear filters is $Y(z)A(z)D(z)$ (Equation 5.251) whose first term is $f_0 z^{-h-l}$. Thus the nonlinear filter in Fig. 5.11 sets to zero all but the first of the components of the sampled impulse-response of the channel and two linear filters (by decision-directed cancellation of intersymbol interference, see Section 5.1) to give a resultant sampled impulse-response containing only the one nonzero component f_0. Furthermore, the nonlinear filter does not change the noise variance. It follows that, with the correct detection of the data-symbols $s_{i-1}, s_{i-2}, \ldots,$ the equalized signal at the detector input, at time $t = (i+h+l)T$, is

$$x_{i+h+l} \simeq s_i f_0 + u_{i+h+l} \tag{5.259}$$

where u_{i+h+l} is a Gaussian random variable with zero mean and variance dependent on the two filters. The data-symbol s_i is now detected from the sign of x_{i+h+l}.

It is clear from Section 4.12 that the noise components at the output of the linear filter with z-transform $A(z)$, which is connected between the sampler and the linear filter with z-transform $D(z)$, are statistically independent Gaussian random variables with zero mean and variance σ^2, just like the noise components at the input of this filter. It follows that the variance of u_{i+h+l} is

$$\eta^2 = \sigma^2 |D|^2 \tag{5.260}$$

where $|D|^2$ is the sum of the squares of the tap gains of the linear filter with z-transform $D(z)$. Thus the probability of error in the detection of s_i, at high signal/noise ratios, can be taken to be

$$P_1 = Q\left(\frac{k|f_0|}{\eta}\right) \tag{5.261}$$

Consider next the case where the decision-feedback equalizer in Fig. 5.11 contains just the single linear filter with z-transform $D(z)$, such that the z-transform of the channel and linear filter is $Y(z)D(z)$ (Equation 5.247), whose first term is $e_0 z^{-h}$, and where the nonlinear filter is appropriately modified to achieve the accurate equalization of the channel and linear filter. With the correct detection of the data-symbols $s_{i-1}, s_{i-2}, \ldots$, the equalized signal at the detector input, at time $t = (i+h)T$, is

$$x_{i+h} \simeq s_i e_0 + u_{i+h} \tag{5.262}$$

and the data-symbol s_i is detected from the sign of x_{i+h}. The noise component u_{i+h} is a Gaussian random variable with zero mean and variance $\eta^2 = \sigma^2 |D|^2$, as before. Thus the probability of error in the detection of s_i, at high signal/noise ratios, can be taken to be

$$P_2 = Q\left(\frac{k|e_0|}{\eta}\right) \tag{5.263}$$

and this is always *greater* than the error probability P_1 in Equation 5.261, since $|e_0| < |f_0|$. Clearly, the *signal/noise ratio* in the equalized signal given by Equation 5.259 must be *greater* than that given by Equation 5.262.

In conclusion, whenever the z-transform $Y(z)D(z)$ of the channel and linear filter, for the decision-feedback equalizer in Fig. 5.11, contains one or more zeros outside the unit circle in the z plane, an improved tolerance to additive white Gaussian noise at high signal/noise ratios is always achieved by adding, at the input to the linear filter, an allpass network that replaces the zeros outside the unit circle by their reciprocals, the nonlinear filter being appropriately modified to maintain the accurate equalization of the channel.

Suppose now that the linear filter in Fig. 5.11 comprises two linear feedforward transversal filters in cascade, the first filter being an allpass network with z-transform $A(z)$ and the second filter having a z-transform $B(z)$. Thus the resultant z-transform of the linear filter is

$$D(z) = A(z)B(z) \tag{5.264}$$

The decision-feedback equalizer in Fig. 5.11 is otherwise exactly as shown. As in Equations 4.94–4.96, the z-transform of the baseband channel in Fig. 5.1 is assumed to be

$$Y(z) = Y_1(z)Y_2(z) \tag{5.265}$$

where
$$Y_1(z) = 1 + p_1 z^{-1} + p_2 z^{-2} + \cdots + p_{g-f} z^{-g+f} \tag{5.266}$$

with all its zeros *inside* the unit circle in the z plane, and

$$Y_2(z) = q_0 + q_1 z^{-1} + q_2 z^{-2} + \cdots + q_f z^{-f} \tag{5.267}$$

with all its zeros *outside* the unit circle. The integer f lies in the range 0 to g. Now let

$$Y_3(z) = q_f + q_{f-1} z^{-1} + q_{f-2} z^{-2} + \cdots + q_0 z^{-f} \tag{5.268}$$

so that $Y_3(z)$ is obtained from $Y_2(z)$ by reversing the order of its coefficients. The zeros of $Y_3(z)$ are the reciprocals of those of $Y_2(z)$ and therefore lie *inside* the unit circle.

It is assumed that

$$A(z) \simeq z^{-h} Y_2^{-1}(z) Y_3(z) \tag{5.269}$$

so that the z-transform of the channel and allpass network is

$$Y(z)A(z) \simeq z^{-h} Y_1(z) Y_3(z) \tag{5.270}$$

all of whose zeros lie *inside* the unit circle. The allpass network with z-transform $A(z)$ can be considered to equalize the factor $Y_2(z)Y_3^{-1}(z)$ of

$$Y(z) = Y_1(z)Y_3(z) \cdot Y_2(z)Y_3^{-1}(z) \tag{5.271}$$

since
$$Y_2(z)Y_3^{-1}(z)A(z) \simeq z^{-h} \tag{5.272}$$

$Y_1(z)Y_3(z)$ is the *minimum phase* component of $Y(z)$ (see Section 3.6) and $Y_2(z)Y_3^{-1}(z)$ represents *pure phase distortion* (see Section 3.3), so that the latter is removed by the allpass network to leave only the minimum phase component.

The zeros of $Y(z)A(z)$ comprise the zeros of $Y(z)$ that lie inside the unit circle together with the reciprocals of the zeros of $Y(z)$ that lie outside the unit circle. Thus, in $Y(z)A(z)$, the allpass network replaces the zeros of $Y(z)$ that lie outside the unit circle by their reciprocals and leaves the remaining zeros unchanged. When all the zeros of $Y(z)$ lie inside the unit circle, $A(z) = 1$, and when $Y(z)$ represents pure phase distortion,

$A(z) \simeq z^{-h} Y^{-1}(z)$ so that $Y(z)A(z) \simeq z^{-h}$. For a general channel, the first term in $Y(z)A(z)$ is $q_f z^{-h}$.

Suppose that the second linear filter has the z-transform

$$B(z) = q_f^{-1} + b_1 z^{-1} + b_2 z^{-2} + \cdots + b_j z^{-j} \tag{5.273}$$

where j is a nonnegative integer, $B(z)$ being such that the resultant z-transform of the channel and two linear filters is

$$Y(z)A(z)B(z) \simeq z^{-h}(1 + a_1 z^{-1} + a_2 z^{-2} + \cdots + a_l z^{-l}) \tag{5.274}$$

with all its zeros *inside* the unit circle, but otherwise with no restrictions on the values of the $\{a_i\}$ which are, of course, all *real valued*, as are the $\{b_i\}$. The symbol l may be any nonnegative integer. If $Y(z)A(z)B(z)$ has any zeros *outside* the unit circle (contrary to the assumption just made) these must be introduced by $B(z)$. $A(z)$ can now be modified to replace these zeros by their reciprocals, without changing the level or other statistical properties of the noise at the output of the filter. Since this change must increase the level of the data signal at the detector input without changing the level of the noise, for the reasons previously explained, it must *increase the signal/noise ratio* here and so reduce the probability of error in the detection of s_i, given the correct detection of $s_{i-1}, s_{i-2}, \ldots$. Thus it will be assumed that all the zeros of $B(z)$ lie *inside* the unit circle. The second linear filter is therefore a minimum-phase (minimum delay) network whose main tap is the first tap. Since the first term in $Y(z)A(z)$ is $q_f z^{-h}$, it can be seen from Equation 5.274 that the gain of the *first* tap of the second filter must be q_f^{-1}, as in Equation 5.273, such that the first term in $Y(z)A(z)B(z)$ is z^{-h}. This means that the first component of the sampled impulse-response of the channel and two filters is unity, with a delay of h sampling intervals. The nonlinear filter in Fig. 5.11 is assumed to be appropriately adjusted to maintain the *accurate equalization* of the channel for any value of $B(z)$. With the correct detection of the data-symbols $s_{i-1}, s_{i-2}, \ldots, s_{i-l}$, the equalized signal at the input to the detector, at time $t = (i+h)T$, is

$$x_{i+h} \simeq s_i + u_{i+h} \tag{5.275}$$

where u_{i+h} is the noise component at the output of the second linear filter, at time $t = (i+h)T$. Clearly, x_{i+h} is now an *unbiased* estimate of s_i. Since the noise components at the output of the allpass network, with z-transform $A(z)$, have the *same* statistical properties as those at its input (Section 4.12), they are statistically independent Gaussian random variables with zero mean and variance σ^2. Thus the noise component u_{i+h} at the output of the second linear filter, with z-transform $B(z)$, is a Gaussian random variable with zero mean and variance[78-80]

$$\eta^2 = \sigma^2 (q_f^{-2} + b_1^2 + b_2^2 + \cdots + b_j^2) \tag{5.276}$$

The resultant probability of error in the detection of s_i, given the correct detection of $s_{i-1}, s_{i-2}, \ldots, s_{i-l}$, is approximately

$$P_e = Q\left(\frac{k}{\eta}\right) \tag{5.277}$$

which can also be taken to be the average error probability or average error rate actually obtained, at high signal/noise ratios, when the error-extension effects become unimportant.

To maximize the signal/noise ratio in the equalized signal x_{i+h}, and hence to minimize the value of P_e in Equation 5.277, it is necessary to minimize η^2 in Equation 5.276. The adjustable variables here are the tap gains $b_1, b_2, \ldots, b_j$ of the second linear filter (with z-transform $B(z)$) and no constraints are placed on the values of any of the $\{b_i\}$. The gain of the *first* tap of the second linear filter is, of course, held *fixed* at q_f^{-1}, for the reasons previously explained. It is clear now that the minimum value of η is

$$\eta = \frac{\sigma}{|q_f|} \tag{5.278}$$

and this is obtained when

$$b_i = 0 \tag{5.279}$$

for $i = 1, 2, \ldots, j$. Thus, for the minimum value of P_e,

$$B(z) = q_f^{-1} \tag{5.280}$$

The second linear filter here becomes a simple *multiplier* that multiplies each signal at its input by q_f^{-1}. In practice, the multiplier is combined with the *first* linear filter, having the z-transform $A(z)$ (Equation 5.269), by multiplying each tap gain of this filter by q_f^{-1}. The probability of error in the detection of s_i from the equalized signal x_{i+h}, given the correct detection of $s_{i-1}, s_{-2}, \ldots$, is now approximately

$$P_e = Q\left(\frac{k}{\eta}\right) = Q\left(\frac{k|q_f|}{\sigma}\right) \tag{5.281}$$

as can be seen from Equations 5.277 and 5.278. The *accurate equalization* of the channel is, of course, assumed throughout this analysis.

Suppose next that the z-transform of the linear feedforward transversal filter D in Fig. 5.11 is

$$D(z) = A(z)B(z) \tag{5.282}$$

as in Equation 5.264, where $A(z)$ is the z-transform of an allpass network, that replaces all zeros of $Y(z)$ *outside* the unit circle by their reciprocals, and $B(z)$ has all its zeros *inside* the unit circle. However, suppose now

that
$$D(z) \simeq z^{-h} Y_2^{-1}(z) \tag{5.283}$$
where $Y_2(z)$ is given by Equation 5.267. Thus
$$A(z)B(z) \simeq z^{-h} Y_2^{-1}(z) \tag{5.284}$$
so that
$$A(z) \simeq z^{-h} Y_2^{-1}(z) Y_3(z) \tag{5.285}$$
as in Equation 5.269, but now
$$B(z) \simeq Y_3^{-1}(z) \tag{5.286}$$
which differs from that in Equation 5.280. Furthermore, the z-transform of the channel and linear filter becomes
$$Y(z)D(z) \simeq Y_1(z) Y_2(z) z^{-h} Y_2^{-1}(z) = z^{-h} Y_1(z) \tag{5.287}$$
and the arrangement just described is, of course, the linear and nonlinear equalization of the two factors of the channel response, studied in Section 5.6. Consider now $B(z)$ in Equation 5.286. $Y_3(z)$ is given by Equation 5.268 and all the zeros of $Y_3(z)$ and poles of $Y_3^{-1}(z)$ lie *inside* the unit circle. But, since
$$Y_3(z) Y_3^{-1}(z) = 1 \tag{5.288}$$
and the first term of $Y_3(z)$ is q_f, the first term of $Y_3^{-1}(z)$ must be q_f^{-1} and $Y_3^{-1}(z)$ must also contain *other* terms with *nonzero* coefficients. The same therefore holds for $B(z)$. However, it has just been shown that, for the optimum performance of the decision-feedback equalizer, the coefficients of all terms of $B(z)$, other than the first, must be set to zero, as in Equation 5.280. It follows that the decision-feedback equalizer, where $D(z)$ is given by Equation 5.283, is a *suboptimum* system. Indeed, if the z-transform of the linear feedforward transversal filter D in Fig. 5.11 contains the factor $V_2^{-1}(z)$ without the factor $V_3(z)$, where $V_2(z)$ is any factor of $Y_2(z)$ and $V_3(z)$ is the corresponding factor of $Y_3(z)$, then $B(z)$ must contain the factor $V_3^{-1}(z)$ together with an appropriate constant multiplier, leading therefore to a *suboptimum* system.

It has now been shown that, in the optimum decision-feedback equalizer, the z-transform of the linear filter D that forms the first part of the equalizer (Fig. 5.11) is
$$D(z) \simeq q_f^{-1} z^{-h} Y_2^{-1}(z) Y_3(z) \tag{5.289}$$
as can be seen from Equations 5.264, 5.269 and 5.280. It is of interest next to determine the optimum value of h, given that the linear filter has $n+1$ taps.

Consider first the $(n+1)$-tap *linear* feedforward transversal equalizer for a channel whose z-transform is

$$Y_2(z) = q_0 + q_1 z^{-1} + q_2 z^{-2} + \cdots + q_f z^{-f} \quad (5.290)$$

as in Equation 5.267. $Y_2(z)$ is, of course, the factor of $Y(z)$ (Equation 5.265) containing all zeros of the latter that lie *outside* the unit circle in the z plane. It follows immediately that the linear equalizer for $Y_2(z)$ is a maximum-phase (maximum-delay) network whose main tap (that passes the wanted signal component through to the output) is the *last* tap. Furthermore, the wanted (main) component of the $(i+1)$th received signal-element, with z-transform

$$s_i z^{-i} Y_2(z) = s_i q_0 z^{-i} + s_i q_1 z^{-i-1} + \cdots + s_i q_f z^{-i-f} \quad (5.291)$$

is its *last* component $s_i q_f$ that occurs at time $t = (i+f)T$. Thus the channel delays a signal element effectively by f sampling intervals and the $(n+1)$-tap linear feedforward transversal equalizer for this channel delays a signal element by n sampling intervals, to give a total delay of $f+n$ sampling intervals. This can be seen also from Equation 4.98, which assumes an (m_2+1)-tap linear feedforward transversal equalizer for $Y_2(z)$. Thus, if the z-transform of the given $(n+1)$-tap linear equalizer is $C_2(z)$, the z-transform of the channel $Y_2(z)$ and linear equalizer is

$$Y_2(z) C_2(z) \simeq z^{-f-n} \quad (5.292)$$

Clearly, $\quad\quad C_2(z) \simeq z^{-f-n} Y_2^{-1}(z) \quad (5.293)$

and the z-transform of the channel $Y(z)$ and linear equalizer is

$$Y(z) C_2(z) \simeq z^{-f-n} Y_1(z) \quad (5.294)$$

as follows from Equation 5.265. Now the linear filter with z-transform $D(z)$ (Equation 5.289) not only *removes* the factor $Y_2(z)$ from the resulting z-transform of the channel $Y(z)$ and filter but it also *inserts* the factor $Y_3(z)$. This could be achieved by adding, at the output of the linear filter with z-transform $C_2(z)$, a linear feedforward transversal filter with $f+1$ taps and the z-transform

$$Y_3(z) = q_f + q_{f-1} z^{-1} + q_{f-2} z^{-2} + \cdots + q_0 z^{-f} \quad (5.295)$$

as in Equation 5.268. However, rather than have two separate filters, the two can be combined into a single linear feedforward transversal filter having $n+f+1$ taps and giving exactly the same performance (see, for example, Equation 4.82). The z-transform of this filter is $C_2(z) Y_3(z)$ and the resultant z-transform of the channel $Y(z)$ and the filter is

$$Y(z) C_2(z) Y_3(z) \simeq z^{-f-n} Y_1(z) Y_3(z) \quad (5.296)$$

Since the required linear filter has only $n+1$ taps and since the larger tap gains occur towards the *output end* of the filter, the first few tap gains

290 DECISION-FEEDBACK EQUALIZERS

being normally quite small or negligible, the reduction in the number of taps, that usually results in the smallest increase in intersymbol interference in the equalized signal, is achieved by omitting the *first f* multipliers together with the associated stores (delay circuits) from the *input* end of the filter, thus *reducing* the delay introduced by the filter by f sampling intervals, without otherwise significantly affecting the z-transform of the filter. Finally, to obtain the required output signal level from the filter, each of its $n+1$ tap gains must be multiplied by q_f^{-f}, to give the resultant z-transform

$$D(z) \simeq q_f^{-1} z^f C_2(z) Y_3(z) \simeq q_f^{-1} z^{-n} Y_2^{-1}(z) Y_3(z) \qquad (5.297)$$

Comparing Equations 5.289 and 5.297 it is clear that the preferred value of h, with an $(n+1)$-tap linear feedforward transversal filter, is

$$h = n \qquad (5.298)$$

It is interesting to note firstly that h is *not* a function of $Y(z)$, which is most encouraging for the use of such a decision-feedback equalizer with a *time-varying* channel, and secondly that the preferred value of h here is the *same* as its suggested value in a decision-feedback equalizer that minimizes the mean-square error in the equalized signal, but *without* the constraint of the exact equalization of the channel (see Equation 5.204).

It can be seen from Equations 5.289 and 5.298, that in the optimum decision-feedback equalizer of Fig. 5.11, that minimizes the mean-square error in the equalized signal, subject to the accurate equalization of the channel, the $(n+1)$-tap linear feedforward transversal filter D is an allpass network with the z-transform

$$D(z) \simeq q_f^{-1} z^{-n} Y_2^{-1}(z) Y_3(z) \qquad (5.299)$$

and the z-transform of the channel and linear filter is

$$Y(z) D(z) \simeq q_f^{-1} z^{-n} Y_1(z) Y_3(z) \qquad (5.300)$$

From Equations 5.266 and 5.268,

$$Y(z) D(z) \simeq z^{-n} E(z) \qquad (5.301)$$

where $\qquad E(z) = 1 + e_1 z^{-1} + e_2 z^{-2} + \cdots + e_g z^{-g} \qquad (5.302)$

so that the sampled impulse-response of the channel and linear filter is given by the $(g+1)$-component row vector

$$E = [1 \quad e_1 \quad e_2 \quad \ldots \quad e_g] \qquad (5.303)$$

neglecting the delay of n sampling intervals.

The signal at the output of the linear feedforward transversal filter D in Fig. 5.11, at time $t = (i+n)T$, is

$$v_{i+n} \simeq s_i + s_{i-1} e_1 + s_{i-2} e_2 + \cdots + s_{i-g} e_g + u_{i+n} \qquad (5.304)$$

The linear feedforward transversal filter F in Fig. 5.11 has g taps whose gains are given, respectively, by the g components of the vector

$$F = [e_1 \quad e_2 \quad \ldots \quad e_g] \tag{5.305}$$

With the correct detection of the data-symbols $s_{i-1}, s_{i-2}, \ldots, s_{i-g}$, the equalized signal at the detector input, at time $t = (i+n)T$, is

$$x_{i+n} = v_{i+n} - s_{i-1}e_1 - s_{i-2}e_2 - \cdots - s_{i-g}e_g \simeq s_i + u_{i+n} \tag{5.306}$$

The presence or absence of the component q_f^{-1} in $D(z)$ (Equation 5.299) does not affect the signal-noise ratio at the detector input and hence does not affect P_e (Equation 5.281), assuming, as always, that the nonlinear filter in Fig. 5.11 is adjusted for the accurate equalization of the channel. The presence of q_f^{-1} in $D(z)$, of course, ensures that the first nonzero component of the sampled impulse-response E of the channel and linear filter has the value unity, so that Equation 5.306 is satisfied.

It is evident from Equations 5.299–5.303 that, if the decision-feedback equalizer is replaced by the corresponding *linear equalizer*, the latter must have a z-transform that approximates to

$$D(z)E^{-1}(z)$$

Considering the linear equalizer as the feedforward transversal filters with z-transforms $D(z)$ and $E^{-1}(z)$ connected in cascade, it can be seen from the previous discussion that the noise components at the input to the second filter (with z-transform $E^{-1}(z)$) are statistically independent Gaussian random variables with zero mean and variance σ^2/q_f^2. Since $E(z)$ is minimum phase, so also is $E^{-1}(z)$, which means that the main tap of the transversal filter with z-transform $E^{-1}(z)$ is its *first* tap, whose gain is *unity*. Except when $E(z) = 1$, which is the case where the channel introduces pure phase distortion and the decision-feedback equalizer degenerates into a linear equalizer, $E^{-1}(z)$ must have one or more *nonzero* terms in addition to its first term (that is fixed at unity). Thus the second transversal filter must have one or more taps with nonzero gains *in addition* to its first tap, which means that the noise variance at its output must *exceed* σ^2/q_f^2. This follows from the discussion involving Equations 5.269–5.280. Hence, in the presence of any *amplitude distortion*, the signal/noise ratio at the detector input is always *greater* for a decision-feedback equalizer than for the corresponding linear equalizer, giving the former a better tolerance to noise at high signal/noise ratios (when error-extension effects can be neglected).

It may be helpful now to summarize some of the more important properties of the linear filter with z-transform $D(z)$ in Equation 5.299. If the sequence with z-transform $z^{-n}Y_2^{-1}(z)Y_3(z)$ is *reversed* in time, the reversal being pivoted about its component at time $t = 0$, the sequence with z-transform $z^n Y_3^{-1}(z)Y_2(z)$ is obtained which is clearly the *inverse* of the original sequence. It follows that the filter with z-transform $q_f^{-1}z^{-n}Y_2^{-1}(z)Y_3(z)$ is an *allpass* network that does not change any amplitude distortion in the received signal, but affects only the phase distortion. Furthermore, the filter is such that, in the z-transform of the channel and filter, $q_f^{-1}z^{-n}Y_1(z)Y_3(z)$, all zeros (roots) cf $Y(z)$ that lie outside the unit circle are replaced by their reciprocals, the remaining zeros being left unchanged. Clearly, $q_f^{-1}z^{-n}Y_1(z)Y_3(z)$ has *no* zeros *outside* the unit circle in the z plane. As in Chapter 3, any zeros or poles at the origin or infinity in the z plane are ignored, since they do not either represent or contribute to signal distortion, nor do they affect the signal level. Thus, for our purposes, $q_f^{-1}z^{-n}Y_1(z)Y_3(z)$ has *no* zeros *outside* the unit circle, and the sampled impulse-response of the channel and filter is *minimum phase*. Strictly speaking, it is minimum phase for the given delay of n sampling intervals represented by z^{-n}. An important property of a minimum-phase sequence is that the energy of the sequence is concentrated into the earliest components. It follows from these considerations that, if the level change caused by q_f^{-1} is removed to give the z-transform $z^{-n}Y_2^{-1}(z)Y_3(z)$, the filter introduces an *orthogonal* transformation into the received signal and so does not change the inner products, lengths, distances or angles between the different possible received signal vectors (where the latter correspond to the different possible received messages). The filter does not change the signal or noise levels nor any of the noise statistics, and subject to the constraint of being an allpass network with unit gain, it *maximizes* the magnitude of the *first* nonzero component of the sampled impulse-response of the channel and filter. Of course, the various properties just described for the filter only hold exactly if the filter has an *infinite* number of taps. However, by using an appropriately large number of taps (which is usually not excessive) the properties of the filter can be made to approach as closely as required to those described.

The most important single property of the linear filter with z-transform $D(z)$ in Equation 5.299 is that, under the assumed conditions, it *maximizes* the ratio of the magnitude of the *first* nonzero component of the sampled impulse-response of the channel and filter to the noise variance at the output of the filter.

It is interesting to observe that in the optimum decision-feedback equalizer, both the linear and nonlinear filters perform the particular operations for which each of them always gives a greater output signal/noise ratio than can be achieved by the other, leading to the optimum sharing of the equalization of the channel between the linear and nonlinear filters. Thus the linear filter equalizes pure phase distortion and the nonlinear filter equalizes a z-transform all of whose zeros lie inside the unit circle in the z plane.

With the linear and nonlinear equalization of the two factors of the channel response, the linear filter merely equalizes the factor $Y_2(z)$ of the channel z-transform, as shown by Equation 5.140. In the optimum decision-feedback equalizer, the linear filter not only equalizes $Y_2(z)$ but also introduces the factor $q_f^{-1}Y_3(z)$, as shown by Equation 5.300.

It may readily be shown that if $Y(z)$ contains one or more zeros *on* the unit circle in the z plane, the optimum decision-feedback equalizer is obtained by taking $Y_1(z)$ as the factor of $Y(z)$ that includes all the zeros *on or inside* the unit circle[16].

When the baseband channel introduces pure phase distortion, the optimum decision-feedback equalizer degenerates into a pure linear equalizer which, of course, now achieves the best available tolerance to additive white Gaussian noise. Again, when all the zeros of $Y(z)$ lie on or inside the unit circle, the optimum decision-feedback equalizer becomes a pure nonlinear equalizer as in Fig. 5.2, that removes all the signal distortion by decision-directed cancellation of intersymbol interference.

Suppose that the z-transform of the baseband channel in Fig. 5.1 is

$$Y(z) = 1 + 2.5z^{-1} + z^{-2} \tag{5.307}$$

so that the z-transform of the $(i+1)$th received signal-element, at the output of the sampler, is

$$s_i z^{-i} Y(z) = s_i z^{-i} + 2.5 s_i z^{-i-1} + s_i z^{-i-2} \tag{5.308}$$

and the received sample, at time $t = iT$, at the input to the equalizer is

$$r_i = s_i + 2.5 s_{i-1} + s_{i-2} + w_i \tag{5.309}$$

where w_i is a Gaussian random variable with zero mean and variance σ^2, as before.

Consider the decision-feedback equalizer of Fig. 5.11 that maximizes the signal/noise ratio in the equalized signal subject to the accurate equalization of the channel. From Equation 4.281, $Y(z)$ is equalized *linearly* by an 11-tap feedforward transversal filter with the z-transform

$$C(z) = -0.012 + 0.035z^{-1} - 0.079z^{-2} + 0.164z^{-3}$$
$$- 0.331z^{-4} + 0.665z^{-5} - 0.331z^{-6} + 0.164z^{-7}$$
$$- 0.079z^{-8} + 0.035z^{-9} - 0.012z^{-10} \tag{5.310}$$

such that the z-transform of the channel and linear equalizer is

$$Y(z)C(z) \simeq z^{-6} \tag{5.311}$$

Let G_0 be the 16-component row vector

$$G_0 = [-0.012 \quad 0.035 \quad -0.079 \quad 0.164 \quad -0.331$$
$$0.665 \quad -0.331 \quad 0.164 \quad -0.079 \quad 0.035$$
$$-0.012 \quad 0.000 \quad 0.000 \quad 0.000 \quad 0.000 \quad 0.000] \tag{5.312}$$

294 DECISION-FEEDBACK EQUALIZERS

and let M be the 5×16 matrix whose ith row, for $i = 1, 2, 3, 4, 5$, is

$$G_i = [\overbrace{0 \ \ldots \ 0}^{i} \ \overbrace{-0.012 \ \ 0.035 \ \ldots \ -0.012}^{11} \ 0 \ \ldots \ 0] \qquad (5.313)$$

the 11 nonzero components being as in G_0.

The 16 tap gains of the required linear feedforward transversal filter D in Fig. 5.11 are given by the 16 components of the row vector

$$D = G_0(I - M^T(MM^T)^{-1}M) \qquad (5.314)$$

where I is the 16×16 identity matrix. Thus

$$D = [\ -0.012 \quad 0.023 \quad -0.047 \quad 0.094 \quad -0.188$$
$$0.376 \quad 0.249 \quad 0.001 \quad -0.002 \quad 0.003$$
$$-0.003 \quad 0.000 \quad -0.005 \quad 0.001 \quad 0.000 \quad 0.000] \qquad (5.315)$$

It is clear that only the first seven taps of the linear filter need in fact be used, the remaining taps having only a small effect on the sampled impulse-response of the channel and linear filter. Thus the linear filter in the decision-feedback equalizer is *smaller* than the corresponding linear equalizer for the channel, which requires some eleven taps, as can be seen from Equation 5.310. The design of the linear filter does not appear to be significantly affected by the number of the components of the vector D, so long as this number is not less than $m + 3$ (that is, $m + g + 1$), where $m + 1$ is the number of nonzero components of G_0 and hence the number of taps of the linear equalizer that achieves the reasonably accurate equalization of the channel.

From Equation 5.241,

$$L = G_0 M^T (MM^T)^{-1} = [\ -1.00 \quad -0.25 \quad 0.00 \quad 0.00 \quad 0.00] \qquad (5.316)$$

so that, from Equations 5.223 and 5.235, the sampled impulse-response of the channel and linear filter is given approximately by the 6-component vector

$$E = [1.00 \quad 1.00 \quad 0.25 \quad 0.00 \quad 0.00 \quad 0.00] \qquad (5.317)$$

neglecting the delay in transmission. The corresponding z-transform is

$$E(z) = 1 + z^{-1} + 0.25z^{-2} \qquad (5.318)$$

The *linear equalizer* for the channel, given by Equation 5.310, has the z-transform

$$C(z) \simeq z^{-6} Y^{-1}(z) \qquad (5.319)$$

The z-transform of the linear feedforward transversal filter D in Fig. 5.11 is

$$D(z) = C(z)E(z) \simeq z^{-6} Y^{-1}(z) E(z) \qquad (5.320)$$

so that the z-transform of the channel and linear filter is

$$Y(z)D(z) \simeq z^{-6}E(z) \tag{5.321}$$

and the z-transform of the $(i+1)$th received signal-element at the output of the linear filter is approximately

$$s_i z^{-i-6} E(z) = s_i z^{-i-6}(1+z^{-1}+0.25z^{-2}) \tag{5.322}$$

The linear feedforward transversal filter F, in Fig. 5.11, has two taps, with gains 1.00 and 0.25.

The signal at the output of the linear filter D, at time $t = (i+6)T$, is

$$v_{i+6} \simeq s_i + s_{i-1} + 0.25 s_{i-2} + u_{i+6} \tag{5.323}$$

where u_{i+6} is the noise component. With the correct detection of s_{i-1} and s_{i-2}, the signal at the input to the detector is

$$x_{i+6} = v_{i+6} - s_{i-1} - 0.25 s_{i-2} \simeq s_i + u_{i+6} \tag{5.324}$$

where u_{i+6} is a Gaussian random variable with zero mean and variance

$$\eta^2 = \sigma^2 |D|^2 = 0.25\sigma^2 \tag{5.325}$$

When $x_{i+6} > 0$, $s_i' = k$, and when $x_{i+6} < 0$, $s_i' = -k$, s_i' being the detected value of s_i.

The probability of error in the detection of s_i, given the correct detection of s_{i-1} and s_{i-2}, is approximately

$$P_e = Q\left(\frac{k}{\eta}\right) = Q\left(\frac{2k}{\sigma}\right) \tag{5.326}$$

Since error-extension effects are not serious here, there being an average of some two errors in an error burst at low error rates, the average error probability (or average error rate) actually obtained, at high signal/noise ratios, can be taken to be P_e in Equation 5.326.

The optimum decision-feedback equalizer for the given channel may alternatively be derived as follows. The z-transform of the channel is

$$Y(z) = (1 + \tfrac{1}{2}z^{-1})(1 + 2z^{-1}) \tag{5.327}$$

so that, in Equation 5.265, $Y_1(z) = 1 + \tfrac{1}{2}z^{-1}$ and $Y_2(z) = 1 + 2z^{-1}$. From Equation 5.299, the $(n+1)$-tap linear feedforward transversal filter D, in Fig. 5.11, has the z-transform

$$D(z) \simeq \tfrac{1}{2} z^{-n}(1+2z^{-1})^{-1}(2+z^{-1}) \tag{5.328}$$

and the z-transform of the channel and linear filter is

$$Y(z)D(z) \simeq \tfrac{1}{2} z^{-n}(1+\tfrac{1}{2}z^{-1})(2+z^{-1}) \tag{5.329}$$

in which the zero of $Y(z)$ outside the unit circle in the z plane has been replaced by its reciprocal.

Suppose that the linear filter has seven taps. Then

$$D(z) = \tfrac{1}{2}(-\tfrac{1}{64} + \tfrac{1}{32}z^{-1} - \tfrac{1}{16}z^{-2} + \tfrac{1}{8}z^{-3} - \tfrac{1}{4}z^{-4} + \tfrac{1}{2}z^{-5})(2 + z^{-1})$$
$$= -\tfrac{1}{64} + \tfrac{3}{128}z^{-1} - \tfrac{3}{64}z^{-2} + \tfrac{3}{32}z^{-3} - \tfrac{3}{16}z^{-4} + \tfrac{3}{8}z^{-5} + \tfrac{1}{4}z^{-6} \quad (5.330)$$

so that

$$D = [-0.016 \quad 0.023 \quad -0.047 \quad 0.094 \quad -0.188 \quad 0.375 \quad 0.250]$$
$$(5.331)$$

which approximates quite well to D in Equation 5.315.

The z-transform of the channel and linear filter is now

$$Y(z)D(z) \simeq z^{-6}(1 + z^{-1} + 0.25z^{-2}) \quad (5.332)$$

which is the *same* as that in Equation 5.321. Thus the linear feedforward transversal filter F in Fig. 5.11 again has two taps, with gains 1.00 and 0.25, and s_i is detected from the equalized signal x_{i+6}, which, with the correct detection of s_{i-1} and s_{i-2}, is given by Equation 5.324. It is evident from Equations 5.315 and 5.331 that the sum of the squares of the tap gains of the linear filter D is essentially the same as before, which means that the variance of the noise component in the equalized signal (Equation 5.324) is again given by η^2 in Equation 5.325, so that the probability of error in the detection of s_i can be taken to be P_e in Equation 5.326.

The probability of error in the detection of s_i can alternatively (and more simply) be derived from Equation 5.281, which shows that, with the correct detection of s_{i-1} and s_{i-2}, the error probability is approximately

$$P_e = Q\left(\frac{k|q_f|}{\sigma}\right) \quad (5.333)$$

where q_f is the coefficient of z^{-f} in $Y_2(z)$ (Equation 5.267). In this application, $f = 1$ and $q_f = 2$, so that P_e is given by Equation 5.326, as before.

The second method of evaluating the decision-feedback equalizer is clearly the simpler of the two, and the method also gives the more accurate optimization of the equalizer, when only a limited number of taps can be used for the linear filter.

It can be seen from Equations 4.286 and 5.326 that the optimum decision-feedback equalizer gains an advantage of 4.7 dB in tolerance to additive white Gaussian noise over a linear equalizer, for the given channel at very high signal/noise ratios.

The performances of the more important of the various equalizers studied in Chapters 4 and 5 are compared in Table 5.2 for the particular channel that has just been considered (Equation 5.307). The results here

Table 5.2 RELATIVE PERFORMANCES OF VARIOUS EQUALIZERS WITH A CHANNEL WHOSE Z-TRANSFORM IS $Y(z) = 1 + 2.5z^{-1} + z^{-2}$

Equalizer	Tolerance to noise relative to that of the linear equalizer (dB)
Linear	0
Pure nonlinear ($h=0$)	−1.3
Pure nonlinear ($h=1$)	2.2
Decision-feedback with zero forcing	2.9
Decision-feedback with linear and nonlinear equalization of the two factors of the channel response	3.5
Optimum decision-feedback	4.7

neglect error extension effects and therefore apply at very high signal/noise ratios. The pure nonlinear equalizers are, of course, particular arrangements of the more general class of decision-feedback equalizers.

5.8.3 Complex-valued signals

When a QAM signal is transmitted over a bandpass channel, the signals in the data-transmission system of Fig. 5.1 become *complex valued* (see Section 3.7). Thus the received sample, at time $t = iT$, is

$$r_i = \sum_{j=0}^{g} s_{i-j} y_j + w_i \qquad (5.334)$$

as in Equation 5.3, but r_i, s_{i-j}, y_j and w_i are now *complex valued*. Consider the simple case, as in Equations 4.288–4.301, where the $\{s_i\}$ are statistically independent and equally likely to have any of the four values $\pm k \pm jk$, where $j = \sqrt{-1}$. Thus the mean-square value of s_i is

$$\mathscr{E}[|s_i|^2] = 2k^2 \qquad (5.335)$$

Again, with a receiver filter corresponding to that for the real-valued signals in Fig. 5.1, the real and imaginary parts of the noise components $\{w_i\}$ are statistically independent Gaussian random variables with zero mean and variance σ^2, so that the $\{w_i\}$ themselves are statistically independent complex-valued Gaussian random variables with zero mean and variance $2\sigma^2$. The signal/noise ratio is k^2/σ^2 and is therefore again given by ψ in Equation 5.8.

Under these conditions, the mean-square error in the equalized signal

(Equation 5.190) becomes

$$\varepsilon = \mathscr{E}[|x_{i+h} - s_i|^2] = 2k^2|B|^2 + 2\sigma^2|D|^2 \tag{5.336}$$

where $|B|$ and $|D|$ are the *unitary lengths* of the vectors B and D, respectively, in the Equations 5.187 and 5.170.

The linear feedforward transversal filter D, that forms the first part of the decision-feedback equalizer in Fig. 5.11, where the equalizer is adjusted to minimize the mean-square error due to *both* intersymbol interference and noise in the equalized signal, may now be derived along exactly the same lines as before (Equations 5.190–5.199). The $n+1$ tap gains of the resulting filter are given by the $(n+1)$-component *complex* row vector

$$D = E_h Z^* \left(ZZ^* + \frac{\sigma^2}{k^2} I\right)^{-1} \tag{5.337}$$

in place of D in Equation 5.199, the tap gains of the filter F being given by Equations 5.175 and 5.191, as before. Z^* is here the *conjugate transpose* of Z. The mean-square error in the equalized signal is now

$$\varepsilon = 2k^2 \left[1 - E_h Z^* \left(ZZ^* + \frac{\sigma^2}{k^2} I\right)^{-1} Z E_h^T\right] \tag{5.338}$$

in place of ε in Equation 5.200. The matrices E_h, Z and I are as previously defined in Section 5.7, and the matrix

$$ZZ^* + \frac{\sigma^2}{k^2} I \tag{5.339}$$

is a positive-definite Hermitian matrix, being positive-definite so long as $\sigma^2 \neq 0$ or ZZ^* is nonsingular.

Consider next the linear feedforward transversal filter D (Figure 5.11), where the equalizer achieves the accurate equalization of the channel and, subject to this constraint, is adjusted to maximize the signal/noise ratio in the equalized signal. The design of the equalizer may be derived along the same lines as before (Section 5.8.2). Thus the z-transform of the $(n+1)$-tap linear feedforward transversal filter D is

$$D(z) \simeq (\overline{q_f})^{-1} z^{-n} Y_2^{-1}(z) \overline{Y_3}(z) \tag{5.340}$$

in place of $D(z)$ in Equation 5.299, the tap gains of the filter F being given by Equation 5.305, as before. $\overline{q_f}$ is the *complex conjugate* of q_f, and $\overline{Y_3}(z)$ is obtained from $Y_3(z)$ replacing each *coefficient* in $Y_3(z)$ by its *complex conjugate*. Thus the z-transform of the channel and linear filter D is

$$Y(z)D(z) \simeq (\overline{q_f})^{-1} z^{-n} Y_1(z) \overline{Y_3}(z) \tag{5.341}$$

DECISION-FEEDBACK EQUALIZERS 299

The zeros of $Y(z)$ do *not* here occur in complex-conjugate pairs, and the zeros of $\bar{Y}_3(z)$ are the *complex conjugates* of the *reciprocals* of the zeros of $Y_2(z)$. Hence in $Y(z)D(z)$ all zeros of $Y(z)$ that lie outside the unit circle are replaced by the *complex conjugates* of their *reciprocals*, the remaining zeros being left unchanged. As before, all zeros of $Y(z)D(z)$ lie *inside* the unit circle, so that, when neglecting the delay of n sampling intervals, the sampled impulse-response of the channel and filter is *minimum phase*. The noise variance in the equalized signal x_{i+n} (Equation 5.306) is now

$$\eta^2 = 2\sigma^2 |q_f|^{-2} \qquad (5.342)$$

in place of that given by Equation 5.278.

5.9 EXAMPLES

Problem 1

The $(i+1)$th received sample of a serial digital baseband signal is the real-valued signal

$$r_i = 0.4s_i + 1.2s_{i-1} + 0.5s_{i-2} + w_i \qquad (5.343)$$

where the data-symbols $\{s_i\}$ are statistically independent and equally likely to have either value ± 1. The noise components $\{w_i\}$ are statistically independent Gaussian random variables with zero mean and variance σ^2.

Determine the advantage in tolerance to noise gained here by the optimum decision-feedback equalizer over the suboptimum decision-feedback equalizer employing linear and nonlinear equalization of the two factors of the channel response. The accurate equalization of the channel and a high signal/noise ratio are assumed in each case.

Solution

The sampled impulse-response of the channel is the 3-component sequence (vector)

$$Y = [0.4 \quad 1.2 \quad 0.5] \qquad (5.344)$$

with z-transform

$$Y(z) = 0.4 + 1.2z^{-1} + 0.5z^{-2} = (0.4 + z^{-1})(1 + 0.5z^{-1}) \qquad (5.345)$$

In the ideal case, and neglecting the delay required to make the filter physically realisable, the linear filter D that forms the first part of the optimum decision-feedback equalizer (Fig. 5.11) has the z-transform

$$\begin{aligned}
D(z) &= (0.4 + z^{-1})^{-1}(1 + 0.4z^{-1}) = z(1+0.4z)^{-1}(1+0.4z^{-1}) \\
&= z(1+0.4z^{-1})(1 - 0.4z + 0.16z^2 - 0.064z^3 + 0.0256z^4 \\
&\quad - 0.01024z^5 + 0.004096z^6 - 0.0016384z^7 + \cdots) \\
&= z(0.4z^{-1} + 0.84 - 0.336z + 0.1344z^2 - 0.05376z^3 \\
&\quad + 0.02150z^4 - 0.00860z^5 + 0.00344z^6 - \cdots) \\
&= 0.4 + 0.84z - 0.336z^2 + 0.1344z^3 - 0.05376z^4 \\
&\quad + 0.02150z^5 - 0.00860z^6 + 0.00344z^7 - \cdots \qquad (5.346)
\end{aligned}$$

300 DECISION-FEEDBACK EQUALIZERS

Thus the tap gains of the linear filter (ending with the last tap) are

$$\ldots, 0.02150, -0.05376, 0.1344, -0.336, 0.84, 0.4$$

The noise component at the output of the linear filter, at time $t=iT$, is

$$u_i = 0.4w_i + 0.84w_{i+1} - 0.336w_{i+2} + 0.1344w_{i+3}$$
$$- 0.05376w_{i+4} + 0.02150w_{i+5} - \cdots \quad (5.347)$$

Since the $\{w_i\}$ are statistically independent Gaussian random variables with zero mean and variance σ^2, u_i is a Gaussian random variable with zero mean and variance

$$\sigma^2(0.4^2 + 0.84^2 + 0.336^2 + 0.1344^2 + 0.05376^2 + 0.02150^2 + \cdots) = \sigma^2 \quad (5.348)$$

so that the probability density function of u_i, at a sample value u, is

$$p_1(u) = \frac{1}{\sqrt{2\pi\sigma^2}} \exp\left(-\frac{v^2}{2\sigma^2}\right) \quad (5.349)$$

This result can in fact be obtained directly, without evaluating the tap gains of the linear filter, since its z-transform is

$$(0.4 + z^{-1})^{-1}(1 + 0.4z^{-1}) \quad (5.350)$$

which is a pure orthogonal transformation, that is, a process of pure phase equalization (although not for the channel itself). The linear filter therefore does not change the level or any other statistics of the noise, leaving its mean and variance unchanged together with the statistical independence of different samples.

The z-transform of the channel and linear filter is now

$$Y(z)D(z) = (0.4 + z^{-1})(1 + 0.5z^{-1})(0.4 + z^{-1})^{-1}(1 + 0.4z^{-1})$$
$$= (1 + 0.5z^{-1})(1 + 0.4z^{-1}) = 1 + 0.9z^{-1} + 0.2z^{-2} \quad (5.351)$$

with a sampled impulse-response

$$E = [1 \quad 0.9 \quad 0.2] \quad (5.352)$$

Thus the signal at the output of the linear filter, at time $t=iT$, is

$$v_i = s_i + 0.9s_{i-1} + 0.2s_{i-2} + u_i \quad (5.353)$$

The delay required to make the system physically realizable has, of course, been ignored here.

The intersymbol interference in v_i is removed by the feedback filter F (Fig. 5.11) that uses the detected data-symbols s'_{i-1} and s'_{i-2} to form the signal

$$0.9s'_{i-1} + 0.2s'_{i-2}$$

which is then subtracted from v_i to give the equalized signal

$$x_i = v_i - 0.9s'_{i-1} - 0.2s'_{i-2} \quad (5.354)$$

With the correct detection of s_{i-1} and s_{i-2},

$$x_i = s_i + u_i \quad (5.355)$$

x_i is compared with a threshold level of zero to give s'_i, the detected value of s_i. When $x_i > 0$, $s'_i = 1$, and when $x_i < 0$, $s'_i = -1$.

An error occurs here in the detection of s_i when u_i has a magnitude greater than 1 and the opposite sign to s_i. Thus the probability of an error in s'_i, given the correct detection of s_{i-1} and s_{i-2}, is

$$P_1 = \int_1^\infty p_1(u)\,du = Q\left(\frac{1}{\sigma}\right) \qquad (5.356)$$

regardless of the sign of s_i. The error-extension effects, resulting from the incorrect cancellation of the intersymbol interference in x_i, are only likely to increase the error rate by a factor of some two times, so that, at high signal/noise ratios, the actual probability of error in the detection of s_i can be taken to be P_1.

In the ideal case, and neglecting the delay required to make the filter physically realizable, the linear filter D that forms the first part of the suboptimum decision-feedback equalizer (Fig. 5.11), employing linear and nonlinear equalization of the two factors of the channel response, has the z-transform

$$B(z) = (0.4 + z^{-1})^{-1} = z(1 + 0.4z)^{-1}$$
$$= z(1 - 0.4z + 0.16z^2 - 0.064z^3 + 0.0256z^4$$
$$\quad - 0.01024z^5 + 0.004096z^6 - \cdots) \qquad (5.357)$$

so that the tap gains of the linear filter (ending with the last tap) are

$$\ldots, -0.01024, 0.0256, -0.064, 0.16, -0.4, 1$$

The noise component at the output of the linear filter, at time $t = iT$, is

$$u_i = w_{i+1} - 0.4w_{i+2} + 0.16w_{i+3} - 0.064w_{i+4}$$
$$\quad + 0.0256w_{i+5} - 0.01024w_{i+6} + \cdots \qquad (5.358)$$

Since the $\{w_i\}$ are statistically independent Gaussian random variables with zero mean and variance σ^2, u_i is a Gaussian random variable with zero mean and variance

$$\eta^2 = \sigma^2(1 + 0.4^2 + 0.16^2 + 0.064^2 + 0.0256^2 + 0.01024^2 + \cdots) = 1.190\sigma^2 \qquad (5.359)$$

so that the probability density function of u_i, at the sample value u, is

$$p_2(u) = \frac{1}{\sqrt{2\pi\eta^2}} \exp\left(-\frac{u^2}{2\eta^2}\right) \qquad (5.360)$$

The z-transform of the channel and linear filter is now

$$Y(z)B(z) = (0.4 + z^{-1})(1 + 0.5z^{-1})(0.4 + z^{-1})^{-1} = 1 + 0.5z^{-1} \qquad (5.361)$$

with a sampled impulse-response

$$E = [1 \quad 0.5] \qquad (5.362)$$

Thus the signal at the output of the linear filter, at time $t = iT$, is

$$v_i = s_i + 0.5s_{i-1} + u_i \qquad (5.363)$$

The feedback filter F (Fig. 5.11) uses the detected data-symbol s'_{i-1} to form $0.5s'_{i-1}$ which is subtracted from v_i to give the equalized signal

$$x_i = v_i - 0.5s'_{i-1} \qquad (5.364)$$

With the correct detection of s_{i-1},

$$x_i = s_i + u_i \qquad (5.365)$$

When $x_i > 0$, $s'_i = 1$, and when $x_i < 0$, $s'_i = -1$.

The probability of error in the detection of s_i, given the correct detection of s_{i-1}, is

$$P_2 = \int_1^\infty p_2(u)\,du = Q\left(\frac{1}{\eta}\right) = Q\left(\frac{1}{1.091\sigma}\right) \qquad (5.366)$$

since $\eta = 1.091\sigma$. The error-extension effects are unimportant here so that, at high signal/noise ratios, the actual probability of error in the detection of s_i can be taken to be P_2.

It can be seen from Equations 5.356 and 5.366 that, for a given low probability of error, the value of σ in the optimum equalizer is 1.091 times that in the suboptimum equalizer, so that, at very high signal/noise ratios, the optimum equalizer gains an advantage of

$$20\log_{10}\left(\frac{\eta}{\sigma}\right) = 20\log_{10}(1.091) = 0.76 \qquad (5.367)$$

dB in tolerance to noise over the other.

Problem 2

The $(i+1)$th sample of a received binary baseband signal, at time $t = iT$, is

$$r_i = 0.3s_i + s_{i-1} + w_i \qquad (5.368)$$

where the data-symbols $\{s_i\}$ are statistically independent and equally likely to have either value $\pm k$, and the $\{w_i\}$ are statistically independent Gaussian random variables with zero mean and variance σ^2.

Compare the tolerances to noise of the following detection processes.

(a) A linear equalizer.
(b) A simple threshold-level detector.
(c) A pure nonlinear equalizer.
(d) Linear and nonlinear equalization of the two factors of the channel response.
(e) The optimum decision-feedback equalizer.

Comment on the influence of the coefficient of s_i (in the received sample r_i) on the performances of the detection processes, as this coefficient is increased from 0.3 towards unity.

How do the performances of these detection processes compare, when

$$r_i = s_i + 0.3s_{i-1} + w_i \qquad (5.369)$$

Solution
The sampled impulse-response of the channel is given by the two-component

sequence (vector)
$$Y = [0.3 \quad 1] \tag{5.370}$$
so that the z-transform of the sampled impulse-response is
$$Y(z) = 0.3 + z^{-1} \tag{5.371}$$

(a) The z-transform of the sampled impulse-response of the 6-tap linear equalizer that minimizes the peak distortion in the equalized signal for the case where the z-transform of the channel is
$$M(z) = 1 + 0.3z^{-1} \tag{5.372}$$
is given by the first six terms in the expansion of $(1 + 0.3z^{-1})^{-1}$. This is best evaluated through the division of 1 by $1 + 0.3z^{-1}$, using a normal process of long division, to give
$$N(z) = 1 - 0.3z^{-1} + 0.09z^{-2} - 0.027z^{-3} + 0.0081z^{-4} - 0.00243z^{-5} \tag{5.373}$$
Since $0.3 + z^{-1}$ is obtained from $1 + 0.3z^{-1}$ simply by reversing the order of the coefficients, the z-transform $C(z)$ of the 6-tap linear equalizer that minimizes the peak distortion for $0.3 + z^{-1}$ is obtained from $N(z)$ again by reversing the order of the coefficients, to give
$$C(z) = -0.00243 + 0.0081z^{-1} - 0.027z^{-2} + 0.09z^{-3} - 0.3z^{-4} + z^{-5} \tag{5.374}$$
Thus the tap gains of the required 6-tap linear equalizer are
$$-0.00243, 0.0081, -0.027, 0.09, -0.3, 1$$
The z-transform of the channel and linear equalizer is
$$Y(z)C(z) = (0.3 + z^{-1})(-0.00243 + 0.0081z^{-1} - 0.027z^{-2} + 0.09z^{-3}$$
$$-0.3z^{-4} + z^{-5})$$
$$= -0.000729 + z^{-6} \simeq z^{-6} \tag{5.375}$$
This means that the equalizer achieves the accurate equalization of the channel (for practical purposes) and it also introduces a delay of six sampling intervals ($6T$ seconds). Thus, the equalized signal at time $t = (i+6)T$, at the output of the linear equalizer, is
$$x_{i+6} \simeq s_i + u_{i+6} \tag{5.376}$$
where
$$u_{i+6} = -0.00243w_{i+6} + 0.0081w_{i+5} - 0.027w_{i+4} + 0.09w_{i+3}$$
$$-0.3w_{i+2} + w_{i+1} \tag{5.377}$$
Since the $\{w_i\}$ are statistically independent Gaussian random variables with zero mean and variance σ^2, u_{i+6} is a Gaussian random variable with zero mean and variance
$$\eta^2 = 0.00243^2\sigma^2 + 0.0081^2\sigma^2 + 0.027^2\sigma^2 + 0.09^2\sigma^2 + 0.3^2\sigma^2 + \sigma^2$$
$$= \sigma^2(0.0000059 + 0.0000656 + 0.000729 + 0.0081 + 0.09 + 1.0)$$
$$= 1.0989\sigma^2 \tag{5.378}$$

The detected data-symbol s'_i is determined from the equalized signal x_{i+6} by comparing this with a decision threshold of zero. When $x_{i+6} > 0$, $s'_i = k$, and when $x_{i+6} < 0$, $s'_i = -k$.

An error occurs in s'_i when s_i and x_{i+6} have opposite signs, that is, when u_{i+6} has a magnitude greater than k and the opposite sign to s_i.

The probability density function of u_{i+6}, at a sample value u, is

$$p_1(u) = \frac{1}{\sqrt{2\pi\eta^2}} \exp\left(-\frac{u^2}{2\eta^2}\right) \tag{5.379}$$

so that, regardless of the sign of s_i, the probability of error in s'_i is approximately

$$P_1 = \int_k^\infty p_1(u) \, du = Q\left(\frac{k}{\eta}\right) = Q\left(\frac{k}{1.0483\sigma}\right) = Q\left(\frac{0.9539k}{\sigma}\right) \tag{5.380}$$

(b) A simple threshold-level detector detects s_i from the received sample

$$r_{i+1} = 0.3 s_{i+1} + s_i + w_{i+1} \tag{5.381}$$

by comparing r_{i+1} with a threshold level of zero. Thus, when $r_{i+1} > 0$, $s'_i = k$, and when $r_{i+1} < 0$, $s'_i = -k$. The detector here treats the intersymbol-interference component $0.3 s_{i+1}$ as noise and therefore simply ignores this component in the detection process.

At high signal/noise ratios, practically all errors in s'_i occur when

$$s_{i+1} = -s_i \tag{5.382}$$

that is, when

$$r_{i+1} = 0.7 s_i + w_{i+1} \tag{5.383}$$

Under these conditions an error occurs in s'_i when w_{i+1} has a magnitude greater than $0.7k$ and the opposite sign to s_i. The probability density function of w_{i+1} at a sample value w is

$$p_2(w) = \frac{1}{\sqrt{2\pi\sigma^2}} \exp\left(-\frac{w^2}{2\sigma^2}\right) \tag{5.384}$$

so that, under the given conditions and regardless of the sign of s_i, the probability of error in s'_i is

$$P_2 = \int_{0.7k}^\infty p_2(w) \, dw = Q\left(\frac{0.7k}{\sigma}\right) \tag{5.385}$$

There is a probability of $\frac{1}{2}$ that $s_{i+1} = -s_i$, so that the actual error probability in s'_i is effectively $\frac{1}{2} Q(0.7k/\sigma)$. At high signal/noise ratios the factor $\frac{1}{2}$ in the error probability can be ignored, so that the error probability can be taken to be P_2.

(c) On the receipt of r_i, a pure nonlinear equalizer (Fig. 5.7 with $h = 0$) has determined s'_{i-1} and forms the equalized signal

$$x_i = r_i - s'_{i-1} \tag{5.386}$$

With the correct detection of s_{i-1},

$$x_i = 0.3 s_i + w_i \tag{5.387}$$

and s_i is now detected from x_i, as follows. When $x_i > 0$, $s'_i = k$, and when $x_i < 0$, $s'_i = -k$.

An error occurs here in the detection of s_i when w_i has a magnitude greater than $0.3k$ and the opposite sign to s_i. Regardless of the sign of s_i, the probability of error is

$$P_3 = \int_{0.3k}^{\infty} p_2(w)\,dw = Q\left(\frac{0.3k}{\sigma}\right) \tag{5.388}$$

The incorrect detection of s_i greatly increases the probability of error in the detection of $s_{i+1}, s_{i+2}, \ldots$, so that errors tend to occur in bursts and the actual (or average) error probability is bP_3, where b is the average number of errors in a burst. In this particular case, $b \simeq 2$. At high signal/noise ratios the actual error probability can be taken to be P_3.

(d) The linear and nonlinear equalization of the two factors of the channel response here degenerates into a linear equalizer, since $Y(z)$ has no zeros inside the unit circle so that the corresponding factor of $Y(z)$ is missing. Thus the probability of error in the detection of s_i is now approximately P_1 (Equation 5.380).

(e) The linear feedforward transversal filter D (Fig. 5.11) that forms the first part of the optimum decision-feedback equalizer performs an orthogonal transformation on the received signal, such that the z-transform of the channel and filter becomes approximately

$$z^{-h}(1 + 0.3z^{-1}) \tag{5.389}$$

where h is an appropriate positive integer. The filter does not change the signal/noise ratio, the noise statistics or any amplitude distortion in the received signal, and its output signal, at time $t = (i+h)T$, is

$$v_{i+h} \simeq s_i + 0.3s_{i-1} + u_{i+h} \tag{5.390}$$

where u_{i+h} is a Gaussian random variable with zero mean and variance σ^2. Thus the probability density function of u_{i+h}, at a sample value u, is

$$p_2(u) = \frac{1}{\sqrt{2\pi\sigma^2}} \exp\left(-\frac{u^2}{2\sigma^2}\right) \tag{5.391}$$

which is clearly the same as that for w_{i+1} in Equation 5.384.

On the receipt of v_{i+h}, the nonlinear filter (in the optimum decision-feedback equalizer) has formed $0.3s'_{i-1}$ which is subtracted from v_{i+h} to give the equalized signal

$$x_{i+h} = v_{i+h} - 0.3s'_{i-1} \tag{5.392}$$

that is fed to the detector. With the correct detection of s_{i-1}, the equalized signal becomes

$$x_{i+h} \simeq s_i + u_{i+h} \tag{5.393}$$

When $x_{i+h} > 0$, $s'_i = k$, and when $x_{i+h} < 0$, $s'_i = -k$.

An error now occurs in the detection of s_i from x_{i+h} when u_{i+h} has a magnitude greater than k and the opposite sign to s_i, so that, regardless of the sign of s_i, the probability of error in the detection of s_i, given the correct detection of s_{i-1}, is approximately

$$P_4 = \int_k^\infty p_2(u)\,du = Q\left(\frac{k}{\sigma}\right) \tag{5.394}$$

The error-extension effects (tendency for errors to occur in bursts) are not very significant here, so that, at high signal/noise ratios, the actual probability of error can be taken to be P_4.

The error probabilities for the five detection processes, at high signal/noise ratios, are now:

(a) $P_1 = Q\left(\dfrac{0.9539k}{\sigma}\right)$, (b) $P_2 = Q\left(\dfrac{0.7k}{\sigma}\right)$, (c) $P_3 = Q\left(\dfrac{0.3k}{\sigma}\right)$,

(d) $P_1 = Q\left(\dfrac{0.9539k}{\sigma}\right)$, (e) $P_4 = Q\left(\dfrac{k}{\sigma}\right)$

Thus, measured relative to the linear equalizer, the tolerances to noise of the detection processes are:

(a) 0 dB, (b) −2.7 dB, (c) −10.0 dB, (d) 0 dB, and (e) 0.4 dB

The best detection process is the optimum decision-feedback equalizer and the poorest is the pure nonlinear equalizer.

It can be seen that, if

$$Y(z) = y + z^{-1} \tag{5.395}$$

where $0.3 < y < 1.0$, then, as y increases, so the tolerances to noise of the detection processes (a), (b) and (d) all *decrease* ((a) and (d) being, of course, the same linear equalizer), the tolerance to noise of the detection process (c) *increases*, whereas that of the detection process (e) remains *unchanged*. A high signal/noise ratio is assumed. Furthermore, when $0.5 < y < 1.0$, the tolerance to noise of the detection process (c) is better than that of the detection process (b), and, as y approaches 1.0, so the detection process (e) gains a very large advantage in tolerance to noise over the detection processes (a), (b) and (d). The performance of the detection process (c) here approaches that of (e).

Suppose next that

$$r_i = s_i + 0.3 s_{i-1} + w_i \tag{5.396}$$

so that the sampled impulse-response of the channel is given by the two-component sequence (vector)

$$Y = [1 \quad 0.3] \tag{5.397}$$

with z-transform

$$Y(z) = 1 + 0.3 z^{-1} \tag{5.398}$$

Clearly, $Y(z)$ here is derived from that in Equation 5.371, by reversing the order of the coefficients.

(a) The 6-tap linear feedforward transversal equalizer for the channel is derived from the previous linear equalizer by simply reversing the order of the tap gains. Thus the linear equalizer now introduces no delay in detection but has the same error probability P_1 (Equation 5.356) as that of the linear equalizer for the previous channel.

(b) The simple threshold-level detector now detects s_i from r_i, using a decision threshold of zero as before, and, at high signal/noise ratios, the probability of error can again be taken to be P_2 (Equation 5.385).

(c) On the receipt of r_i, the pure nonlinear equalizer has determined s'_{i-1} and forms the equalized signal

$$x_i = r_i - 0.3 s'_{i-1} \tag{5.399}$$

With the correct detection of s_{i-1},

$$x_i = s_i + w_i \tag{5.400}$$

and s_i is detected, as before, by comparing x_i with a decision threshold of zero.

It is evident that the probability of error in the detection of s_i from x_i in Equation 5.400 is the same as that in the detection of s_i from x_{i+h} in Equation 5.393. Furthermore, the error extension effects are the same in the two cases. Thus, at high signal/noise ratios, the actual probability of error in the case of the pure nonlinear equalizer can now be taken to be P_4 (Equation 5.394).

(d) The linear and nonlinear equalization of the two factors of the channel response here degenerates into a pure nonlinear equalizer, since $Y(z)$ has no zeros outside the unit circle so that the corresponding factor of $Y(z)$ is missing. Thus the probability of error in the detection of s_i, at high signal/noise ratios, can here be taken to be P_4.

(e) Since all zeros of $Y(z)$ now lie inside the unit circle, the optimum decision-feedback equalizer for the channel degenerates into a pure nonlinear equalizer, whose error probability, at high signal/noise ratios, can be taken to be P_4.

The error probabilities for the five detection processes, at high signal/noise ratios, are now:

(a) $P_1 = Q\left(\dfrac{0.9539 k}{\sigma}\right)$, (b) $P_2 = Q\left(\dfrac{0.7 k}{\sigma}\right)$, (c) $P_4 = Q\left(\dfrac{k}{\sigma}\right)$,

(d) $P_4 = Q\left(\dfrac{k}{\sigma}\right)$, (e) $P_4 = Q\left(\dfrac{k}{\sigma}\right)$

Thus, measured relative to the linear equalizer, the tolerances to noise of the detection processes are:

(a) 0 dB, (b) -2.7 dB, (c) 0.4 dB, (d) 0.4 dB, and (e) 0.4 dB

5.10 EQUALIZER FOR SEVERE AMPLITUDE DISTORTION

Consider the data-transmission system shown in Fig. 5.1 with the conventional decision-feedback equalizer of Fig. 5.11 at the receiver. The

308 DECISION-FEEDBACK EQUALIZERS

received sample, at time $t = iT$, is

$$r_i = \sum_{j=0}^{g} s_{i-j} y_j + w_i \qquad (5.401)$$

where
$$s_i = \pm 1 \qquad (5.402)$$

so that $k = 1$. As before, the $\{s_i\}$ are statistically independent and equally likely to have either binary value, and the $\{w_i\}$ are statistically independent Gaussian random variables with zero mean and variance σ^2. The z-transform of the baseband channel is $Y(z)$ (Equation 5.245) and the z-transform of the $(n+1)$-tap linear feedforward transversal filter D is $D(z)$ (Equation 5.246). The filter D is adjusted as described in Section 5.8.2 to maximize the signal/noise ratio in the equalized signal, subject to the *exact* equalization of the channel, but now with the filter constrained such that

$$|D|^2 = \sum_{j=0}^{n} d_j^2 = 1 \qquad (5.403)$$

The z-transform of the channel and filter D is

$$Y(z)D(z) \simeq z^{-n} E(z) \qquad (5.404)$$

where
$$E(z) = e_0 + e_1 z^{-1} + \cdots + e_g z^{-g} \qquad (5.405)$$

so that the sampled impulse-response of the channel and filter D is given, approximately, by the $(g+1)$-component vector (sequence)

$$E = [e_0 \quad e_1 \quad \ldots \quad e_g] \qquad (5.406)$$

neglecting the delay of n sampling intervals.

The signal at the output of the filter D, at time $t = (i+n)T$, is

$$v_{i+n} \simeq \sum_{j=0}^{g} s_{i-j} e_j + u_{i+n} \qquad (5.407)$$

where
$$u_{i+n} = \sum_{j=0}^{n} w_{i+n-j} d_j \qquad (5.408)$$

so that u_{i+n} is a Gaussian random variable with zero mean and variance

$$\eta^2 = \sigma^2 |D|^2 = \sigma^2 \qquad (5.409)$$

from Equation 5.403.

The detector and filter F of Fig. 5.11 are now as shown in Fig. 5.15, and the tap gains of the filter F are given by the g components of the vector

$$F = [e_1 \quad e_2 \quad \ldots \quad e_g] \qquad (5.410)$$

DECISION-FEEDBACK EQUALIZERS

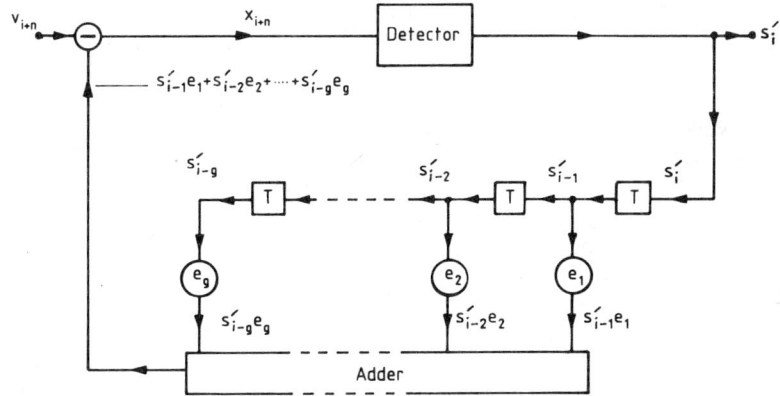

Fig. 5.15 Detector and transversal filter F in a conventional decision-feedback equalizer

Thus the output signal from the filter, at time $t = (i+n)T$, is

$$\sum_{j=1}^{g} s'_{i-j} e_j \tag{5.411}$$

and this is subtracted from v_{i+n} to give the equalized signal

$$x_{i+n} = v_{i+n} - \sum_{j=1}^{g} s'_{i-j} e_j \tag{5.412}$$

at the input to the detector. As before, s'_{i-j} is the detected value of s_{i-j}. With the correct detection of the data-symbols $s_{i-1}, s_{i-2}, \ldots, s_{i-g}$,

$$x_{i+n} = v_{i+n} - \sum_{j=1}^{g} s_{i-j} e_j \simeq s_i e_0 + u_{i+n} \tag{5.413}$$

It can be seen that the decision-feedback equalizer equalizes the channel accurately, in the sense that it removes *all* intersymbol interference, but the equalized signal is *not* here an *unbiased* estimate of s_i, as assumed in Section 5.8, since normally $e_0 \neq 1$.

The detected value s'_i of the data-symbol s_i is taken as its possible value (± 1) such that $s_i e_0$ is closest to x_{i+n}. Clearly, an error is now caused in s'_i by the noise component u_{i+n} when the latter has a magnitude greater than $|e_0|$ and the opposite sign to $s_i e_0$. From Equations 5.402, 5.409 and 5.413, the probability of error in the detection of s_i, given the correct detection of $s_{i-1}, s_{i-2}, \ldots, s_{i-g}$, is

$$P_1 = Q\left(\frac{|e_0|}{\sigma}\right) \tag{5.414}$$

310 DECISION-FEEDBACK EQUALIZERS

In practice, the incorrect detection of a data symbol leads usually to errors in one or more of the following detected data-symbols so that errors occur in bursts. If b_1 is the average number of errors in an error burst, then the average error probability or actual error rate can be taken to be $b_1 P_1$. Clearly, P_1 is a lower bound to the actual error rate, the discrepancy between P_1 and the actual error rate decreasing with the error rate, and generally becoming quite small at extremely low error rates. An error burst is defined here as a group of errors in the $\{s'_i\}$ in which no two adjacent errors are separated by g or more consecutive correctly detected data-symbols.

Bearing in mind that the linear filter D in Fig. 5.11 is an allpass network that makes the sampled impulse-response of the channel and filter minimum phase, and which therefore maximizes the magnitudes of the earliest components of the sampled impulse-response without changing the amplitude distortion in this response, it is most unlikely that

$$|e_0| \gg |e_1| \ll |e_i| \tag{5.415}$$

for any $i > 1$, where $|e|$ is the absolute value (modulus) of e. Indeed, with very low amplitude distortion,

$$|e_0| \gg |e_i| \tag{5.416}$$

for $i = 1, 2, \ldots, g$, and, as the amplitude distortion increases, so the value of i for which $|e_i|$ is maximum tends to increase, with usually now a fairly steady reduction in $|e_i|$ as i either decreases or increases from its value for which $|e_i|$ is maximum. The properties of the amplitude distortion of a sampled data signal are considered in Sections 3.4 and 3.5. Severe amplitude distortion is most often caused by the correspondingly severe band-limiting of the data signal.

When there is very severe amplitude distortion in the received signal, such that $|e_1| \gg |e_0|$, it is possible to achieve a useful advantage in tolerance to noise over the conventional equalizer by means of the following simple system.[59] The receiver now detects s_i not from x_{i+n} (Fig. 5.15) but from x_{i+n+1}, at time $t = (i+n+1)T$, using the arrangement shown in Fig. 5.16. The linear feedforward transversal filter F here has $g-1$ taps, holding the detected data-symbols $s'_{i-1}, s'_{i-2}, \ldots, s'_{i-g+1}$ and associated with the tap-gains $e_2, e_3, \ldots, e_g$, respectively. Thus the output signal from the filter F, at time $t = (i+n+1)T$, is

$$\sum_{j=2}^{g} s'_{i+1-j} e_j \tag{5.417}$$

and this is subtracted from v_{i+n+1} to give the signal

$$x_{i+n+1} = v_{i+n+1} - \sum_{j=2}^{g} s'_{i+1-j} e_j \tag{5.418}$$

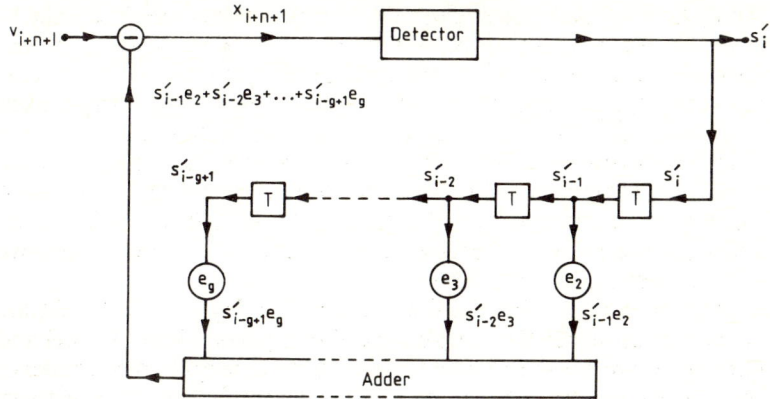

Fig. 5.16 Detector and transversal filter F in the modified decision-feedback equalizer

at the input to the detector. With the correct detection of $s_{i-1}, s_{i-2}, \ldots, s_{i-g+1}$,

$$x_{i+n+1} = s_{i+1}e_0 + s_i e_1 + u_{i+n+1} \tag{5.419}$$

where u_{i+n+1} is the noise component in v_{i+n+1}. Clearly, $s_{i+1}e_0$ is an intersymbol interference component so that the channel is here only *partially equalized*. The data-symbol s_i is detected as its possible value s'_i such that $s'_i e_0$ is closest to x_{i+n+1}. The intersymbol interference component $s_{i+1}e_0$ is ignored in the detection process and is therefore treated as noise.

At high signal/noise ratios practically all errors in the $\{s'_i\}$ occur when $s_{i+1}e_0 + s_i e_1$ has its *minimum* magnitude which is $|e_1| - |e_0|$, since, as is assumed throughout this analysis,

$$|e_1| > |e_0| \tag{5.420}$$

and $s_i = \pm 1$. An error now occurs in the detection of s_i from x_{i+n+1}, when u_{i+n+1} has a magnitude greater than $|e_1| - |e_0|$ and the opposite sign to $s_{i+1}e_0 + s_i e_1$ and therefore also the opposite sign to $s_i e_1$. Thus, under the various conditions just assumed, the probability of error in the detection of s_i from x_{i+n+1} is

$$P_2 = Q\left(\frac{|e_1| - |e_0|}{\sigma}\right) \tag{5.421}$$

Since there is a probability of $\frac{1}{2}$ that $s_{i+1}e_0 + s_i e_1$ has its minimum magnitude, the actual error probability in the detection of s_i, given the correct detection of $s_{i-1}, s_{i-2}, \ldots, s_{i-g+1}$, is approximately $\frac{1}{2}P_2$. As in the case of the conventional decision-feedback equalizer, errors tend to

312 DECISION-FEEDBACK EQUALIZERS

Table 5.3 SAMPLED IMPULSE-RESPONSE OF THE CHANNEL AND FILTER D FOR EACH OF THE SEVEN CHANNELS TESTED

Channel	Sampled impulse-response E						
1	0.408	0.816	0.408				
2	0.548	0.789	0.273	−0.044	0.027		
3	0.321	0.620	0.633	0.322	0.087		
4	0.167	0.500	0.667	0.500	0.167		
5	0.29	0.50	0.58	0.50	0.29		
6	0.085	0.289	0.493	0.577	0.493	0.289	0.085
7	0.19	0.35	0.46	0.50	0.46	0.35	0.19

occur in bursts. If b_2 is the average number of errors in an error burst here, the average error probability or actual error rate can be taken to be $\frac{1}{2}b_2 P_2$. P_2 is now an *approximate* lower bound to the actual error rate, being in fact a correct estimate of the error rate when $b_2 = 2$. Furthermore, for a given sampled impulse-response E (Equation 5.406) normally $b_2 < b_1$, so that P_2 is generally a *better* estimate of the error rate for the modified equalizer than P_1 is for the conventional equalizer. Again, the discrepancy between P_2 and the actual error rate decreases with the error rate and becomes quite small at extremely low error rates.

Computer-simulation tests have been carried out by R.S. Marshall to compare the performances of the modified and conventional equalizers over each of seven different channels[59]. The sampled impulse-responses of the channel and filter D, for the seven channels, are shown in Table 5.3. These sampled impulse-responses must, of course, be minimum phase, such that all zeros (roots) of the corresponding z-transforms lie inside or on the unit circle in the z plane. Each vector E here (Equation 5.406) has unit length so that no signal gain or attenuation is introduced by any resultant channel. Bearing in mind that the linear filter D introduces an *orthogonal transformation* into the received signal, being an allpass network that introduces no gain or loss (Equation 5.403), it is evident that the linear baseband channel itself (Fig. 5.1), with z-transform $Y(z)$, also introduces no signal gain or loss. The various channels have been selected to represent a wide range of different conditions, such that each channel introduces a significant level of amplitude distortion. Since the filter D removes the phase distortion introduced by the linear baseband channel in Fig. 5.1, through making the channel and filter minimum phase, it is evident that, whenever the channel introduces negligible amplitude distortion, the channel is approximately equalized by the filter D to give a sampled impulse-response E in which $e_0 \simeq 1$ and $e_i \simeq 0$ for $i = 1, 2, \ldots, g$. Channel 1 is a well known partial-response channel[91], and channel 2 was obtained with an actual telephone circuit in the transmission path[92]. Channel 3 has been widely studied as one introducing very severe amplitude distortion and yet having a sampled impulse-response whose z-transform has no zeros on or near the unit circle in the z plane[30,32,38,58,59,92]. Channels 4 and 6 are particularly unfavourable to decision-feedback equalizers, and the sampled impulse-responses of channels 5 and 7 are the five- and seven-component sampled responses, respectively, that most seriously degrade the tolerance to additive white Gaussian noise of a maximum-likelihood detector[93]. Whereas the z-transforms of the sampled impulse-responses of channels 2 and 3 have *no* zeros on the unit circle, *all* zeros of the z-transforms of the channels 1 and 4–7 lie on the unit circle. When the received signal is sampled at the Nyquist rate, as is assumed here (at least for practical purposes), not only does the filter D remove all phase distortion introduced into the data signal by the linear baseband channel (Fig. 5.1) but it also corrects any shift in phase of the sampling instants from the ideal or optimum value, since any such shift introduces the corresponding phase distortion into the sampled impulse-response of the channel, which is then removed by the filter D. A consequence of this is that the sampled impulse-response of the channel is not uniquely determined by the corresponding vector E in Table 5.3. However, in the particular case where the channel introduces no phase distortion into the data signal and where the ideal or optimum sampling phase is used at the receiver, the filter D involved in the channels 1 and 4–7 in Table 5.3 degenerates into a direct connection (introducing no change into the received signal), so that E is here also the sampled impulse-response of the linear baseband channel in Fig. 5.1.

Each individual measurement of the error rate in the $\{s_i'\}$ was made with the transmission of between 10,000 and 40,000 data-symbols at the given signal/noise ratio, the latter being expressed as ψ dB, where

$$\psi = 10 \log_{10}(1/\sigma^2) \tag{5.422}$$

and where the mean-square value of s_i has the value unity. The results of the tests are shown in Figs. 5.17–5.23. The 95% confidence limits of these results are generally better than ± 0.5 dB, except at the lowest error rates where they are somewhat wider.

It can be seen from the results of the computer-simulation tests that the modified equalizer achieves its best performance, relative to the conventional equalizer, over channel 6, where it gains an advantage of over 8 dB in tolerance to additive white Gaussian noise. The poorest relative performance of the modified equalizer is over channel 2, where its tolerance to noise is up to 7 dB inferior to that of the conventional equalizer. These results apply, of course, to *binary* signals (Equation 5.402). It has been confirmed by computer-simulation tests (whose detailed results are not shown here) that, with 4-level signals, the performance of the modified equalizer over each of the channels 1–7 is inferior to that of the conventional equalizer. The modified equalizer

314 DECISION-FEEDBACK EQUALIZERS

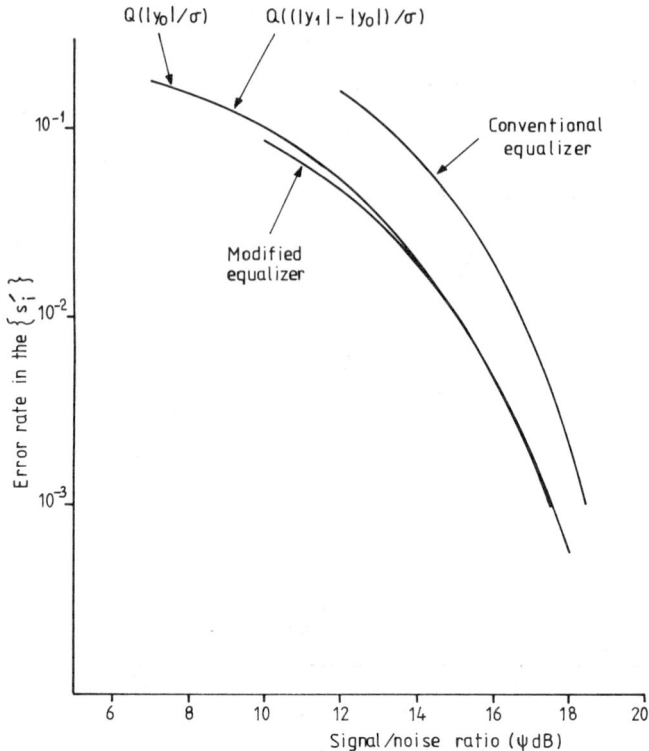

Fig. 5.17 Performance of modified and conventional equalizers with channel 1

does not therefore appear to be suitable for use with multilevel signals. In Figs. 5.17–5.23 the degradation in performance given by the computer-simulation results, for both the modified and conventional equalizer, relative to the theoretically predicted performance given by Equations 5.414 and 5.421, is due to the error-extension effects that cause the error bursts, these being neglected in the equations. The degradation in performance increases with the error rate, the degradation being in general greater and increasing at a faster rate in the case of the conventional equalizer than in the case of the modified equalizer. Thus, at the higher error rates around 1 in 10, the performance of the modified equalizer relative to the conventional equalizer is appreciably better than that predicted theoretically.

The useful feature of the modified equalizer is that, under the appropriate conditions, it enables a useful improvement in performance to be achieved together with a small reduction in complexity of the

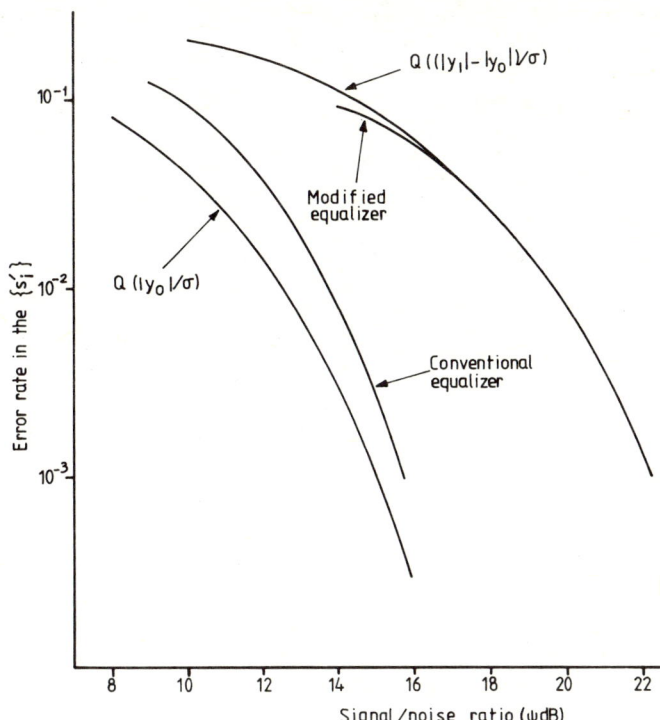

Fig. 5.18 Performance of modified and conventional equalizers with channel 2

equalizer. The improvement is achieved through the deliberate introduction of intersymbol interference into the equalized signal.

In any of the decision-feedback equalizers studied here a received data-symbol is detected from just *one* sample v_i at the output of the linear filter D in Fig. 5.11, which means that only *one* component of a received signal-element at the output of the filter D is used in the detection of the corresponding data-symbol. Any components of the signal element preceding that used in the detection process are ignored (treated as noise) by the detector, and the following (remaining) components are removed by subtraction in the nonlinear filter (Fig. 5.11). If now the detection of a data symbol is delayed until all $g+1$ components of the corresponding signal-element have been received by the detector, all components of the already *detected* data symbols having been removed from the $g+1$ samples carrying the required signal-element, then the appropriate data-symbol s_i can be detected from these samples. Since s_i is now detected from *all* components of the corresponding signal-element, an

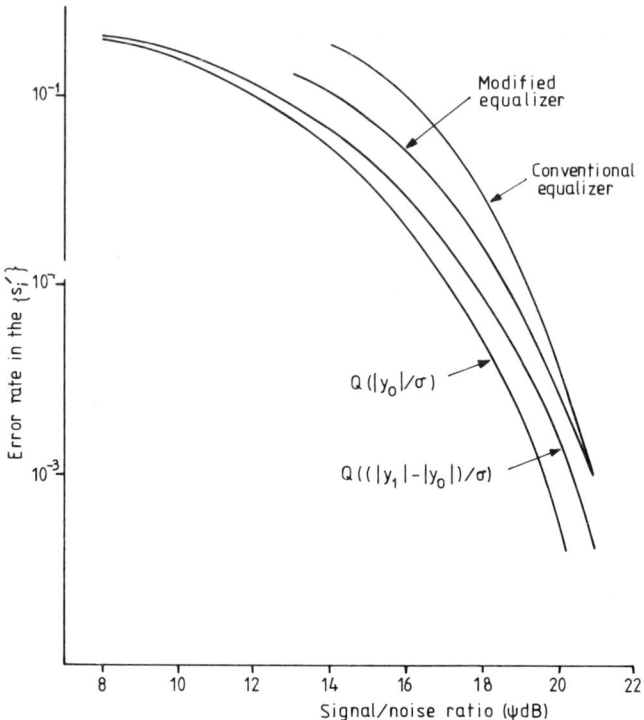

Fig. 5.19 Performance of modified and conventional equalizers with channel 3

improved tolerance to noise is normally achieved. This is intuitively obvious from the fact that more information on s_i, carried by the received samples, is now used in the detection process. Unfortunately this detection process involves also the temporary detection of the data-symbols $s_{i+1}, s_{i+2}, \ldots, s_{i+g}$, which leads to a *complex* detection process. Furthermore, in order to achieve the best available tolerance to noise with such a detection process, the delay in detection must often be increased to several times g, thus further complicating the detection process. Detection processes of this type are not strictly speaking *equalizers* but, for the sake of completeness, they are considered briefly in Section 7.8. They are described in detail elsewhere[38,50,59,92].

REFERENCES

1. Gorog, E. 'A new approach to time-domain equalization', *IBM J. Res. Develop.*, **9**, 228–232 (1965)

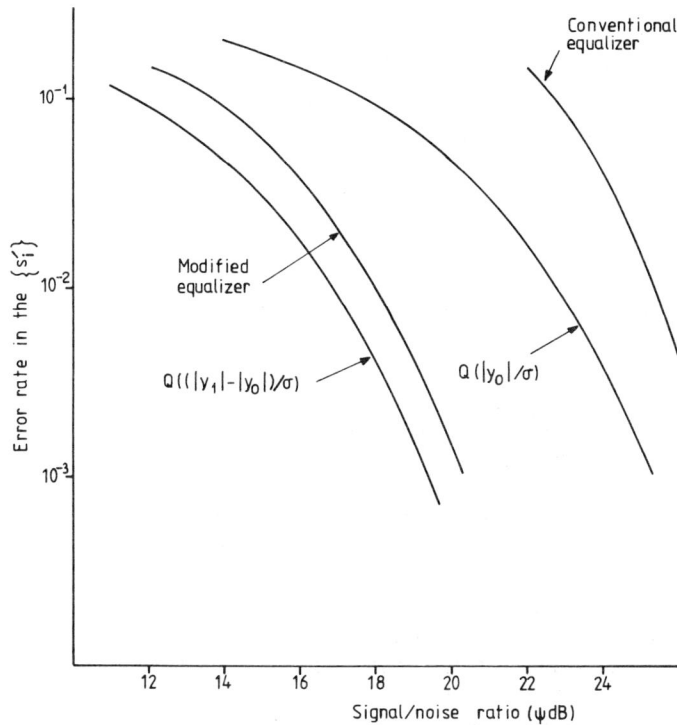

Fig. 5.20 Performance of modified and conventional equalizers with channel 4

2. Boyd, R.T. and Monds, F.C. 'Adaptive equaliser for multipath channels', *Electronics Letters*, **6**, 556–558 (1970)
3. Boyd, R.T. and Monds, F.C. 'Equaliser for digital communication', *Electronics Letters*, **7**, 58–60 (1971)
4. Monsen, P. 'Feedback equalization for fading dispersive channels', *IEEE Trans. Inform. Theory*, **IT-17**, 56–64 (1971)
5. Tomlinson, M. 'New automatic equalizer employing modulo arithmetic', *Electronics Letters*, **7**, 138–139 (1971)
6. Taylor, D.P. 'Nonlinear feedback equalizer employing a soft limiter', *Electronics Letters*, **7**, 265–267 (1971)
7. George, D.A., Bowen, R.R. and Storey, J.R. 'An adaptive decision feedback equalizer', *IEEE Trans. Commun. Technol.*, **COM-19**, 281–293 (1971)
8. Costello, P.J. and Patrick, E.A. 'Unsupervised estimation of signals with intersymbol interference', *IEEE Trans. Inform. Theory*, **IT-17**, 620–622 (1971)
9. Ungerboeck, G. 'Nonlinear equalization of binary signals in Gaussian noise', *IEEE Trans. Commun. Technol.*, **COM-19**, 1128–1137 (1971)
10. Bershad, N.J. and Vena, P.A. 'Eliminating intersymbol interference—a state space approach', *IEEE Trans. Inform. Theory*, **IT-18**, 275–281 (1972)
11. Brownlie, J.D. 'Effects of decision-errors on adaptive equalization', *IERE Conf. Proc.*, No. 23, 221–240 (1972)

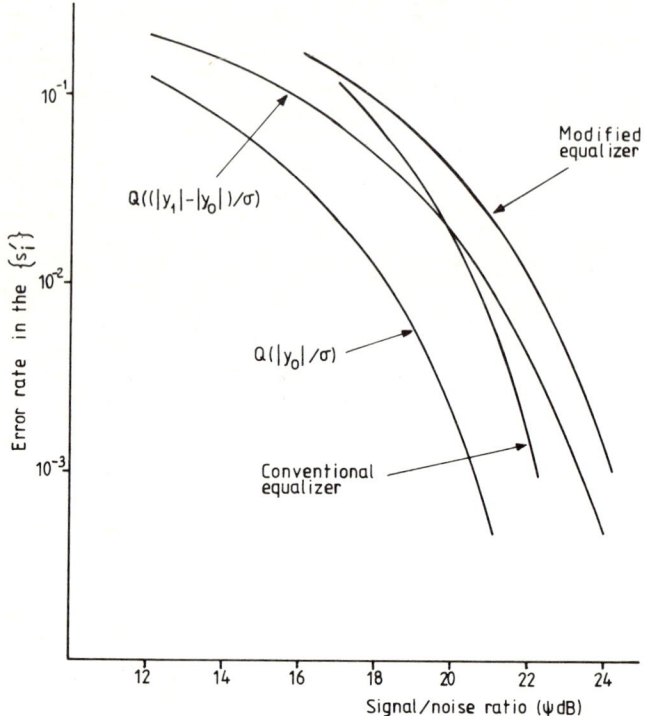

Fig. 5.21 *Performance of modified and conventional equalizers with channel 5*

12. Price, R. 'Nonlinearly feedback equalized PAM vs. capacity for noisy filter channels', *Conf. Rec. IEEE Int. Conf. Communications*, 22.12–22.17 (1972)
13. Harashima, H. and Miyakawa, H. 'Matched-transmission technique for channels with intersymbol interference', *IEEE Trans. Commun.*, **COM-20**, 774–780 (1972)
14. Westcott, R.J. 'An experimental adaptively equalised modem for data transmission over the switched telephone network', *Radio Electron. Eng.*, **42**, 499–507 (1972)
15. Monsen, P. 'Digital transmission performance on fading dispersive diversity channels', *IEEE Trans. Commun.*, **COM-21**, 33–39 (1973)
16. Clark, A.P. 'Design technique for nonlinear equalisers', *Proc. IEE*, **120**, 329–333 (1973)
17. Mark, J.W. 'A note on the modified Kalman filter for channel equalization', *Proc. IEEE*, **61**, 481–483 (1973)
18. Mark, J.W. and Budihardjo, P.S. 'Performance of jointly optimized prefilter-equalizer receivers', *IEEE Trans. Commun.*, **COM-21**, 941–945 (1973)
19. Kobayashi, H. and Tang, D.T. 'A decision-feedback receiver for channels with strong intersymbol interference', *IBM J. Res. Develop.*, **17**, 413–419 (1973)
20. Taylor, D.P. 'The estimate feedback equalizer: A suboptimum non-linear receiver', *IEEE Trans. Commun.*, **COM-21**, 979–990 (1973)
21. Salz, J. 'Optimum mean-square decision feedback equalization', *Bell Syst. Tech. J.*, **52**, 1341–1373 (1973)
22. Lucky, R.W. 'A survey of the communication theory literature: 1968–1973', *IEEE Trans. Inform. Theory*, **IT-19**, 725–739 (1973)

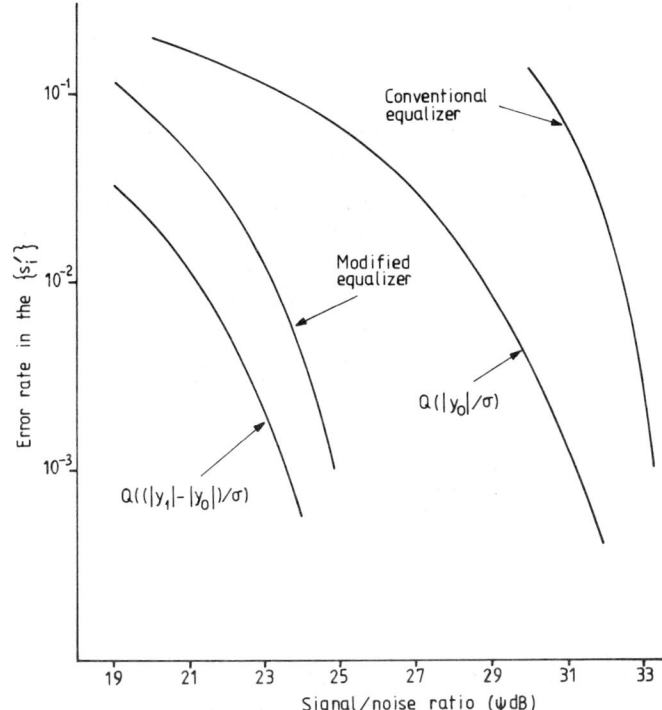

Fig. 5.22 Performance of modified and conventional equalizers with channel 6

23. Messerschmitt, D.G. 'A geometric theory of intersymbol interference. Part 1: Zero-forcing and decision feedback equalization', *Bell Syst. Tech. J.*, **52**, 1483–1519 (1973)
24. Falconer, D.D. and Foschini, G.J. 'Theory of minimum mean square error QAM systems employing decision feedback equalization', *Bell Syst. Tech. J.*, **52**, 1821–1849 (1973)
25. Salazar, A.C. 'Design of transmitter and receiver filters for decision feedback equalization', *Bell Syst. Tech. J.*, **53**, 503–523 (1974)
26. Koeth, H. and Schollmeier, G. 'An adaptive equalizer for partial response signals with improved convergence properties', *IEEE Trans. Commun.*, **COM-22**, 884–885 (1974)
27. Shamash, E. and Yao, K. 'On the structure and performance of a linear decision feedback equalizer based on the minimum error probability criterion', *IEEE Int. Conf. Commun.*, Minneapolis, MN, USA (1974)
28. Duttweiler, D.L., Mazo, J.E. and Messerschmitt, D.G. 'An upper bound on the error probability in decision feedback equalization', *IEEE Trans. Inform. Theory*, **IT-20**, 490–497 (1974)
29. Monsen, P. 'Adaptive equalization of the slow fading channel', *IEEE Trans. Commun.*, **COM-22**, 1064–1075 (1974)
30. Clark, A.P. and Tint, U.S. 'Linear and non-linear transversal equalizers for baseband channels', *Radio Electron. Eng.*, **45**, 271–283 (1975)
31. Macleod, C.J., Ciapala, E. and Jelonek, Z.J. 'Study of recursive equalisers for data

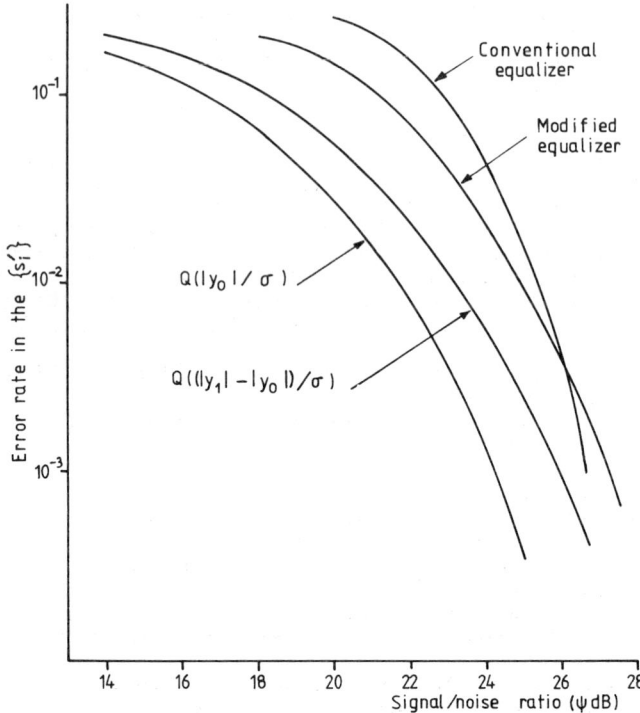

Fig. 5.23 Performance of modified and conventional equalizers with channel 7

transmission with a comparison of the performance of six systems', *Proc. IEE*, **122**, 1097–1104 (1975)
32. Clark, A.P. and Serinken, M.N. 'Nonlinear equalizers using modulo arithmetic', *Proc. IEE*, **123**, 32–38 (1976)
33. Makarov, S.B. and Tsikin, I.A. 'Noise immunity of an algorithm for element-by-element reception with decision feedback in the presence of intersymbol interference', *Radiotekhnika Moskya Telecomm. and Radio Eng.*, Pt. 2, **31**, 53–57 (1976)
34. Tamburelli, G. 'Decision feedback and feedforward receiver (for rates faster than Nyquist's), *Alta Frequenza*, **45**, 587–594 (1976)
35. Falconer, D.D. 'Application of passband decision feedback equalization in two-dimensional data communication systems', *IEEE Trans. Commun.*, **COM-24**, 1159–1166 (1976)
36. Falconer, D.D. and Magee, F.R. 'Evaluation of decision feedback equalization and Viterbi algorithm detection for voiceband data transmission—Part I', *IEEE Trans. Commun.*, **COM-24**, 1130–1138 (1976)
37. Falconer, D.D. and Magee, F.R. 'Evaluation of decision feedback equalization and Viterbi algorithm detection for voiceband data transmission—Part II', *IEEE Trans. Commun.*, **COM-24**, 1238–1245 (1976)
38. Clark, A.P. *Advanced Data-Transmission Systems*, Pentech Press, London (1977)
39. Monsen, P. 'Theoretical and measured performance of a DFE modem on a fading multipath channel', *IEEE Trans. Commun.*, **COM-25**, 1144–1153 (1977)

40. Park, J.H. and Anderson, T.A. 'Decision feedback equalization simulations', *MIDCON Conv. Rec.*, 14.2/1–8 (1977)
41. Salz, J. 'On mean-square decision feedback equalization and timing phase', *IEEE Trans. Commun.*, **COM-25**, 1471–1476 (1977)
42. Roza, E. 'A practical design approach to decision feedback receivers with conventional filters', *IEEE Trans. Commun.*, **COM-26**, 679–689 (1978)
43. Gibson, B. 'Equalization design for a 600 MBd quantized feedback PCM repeater', *IEEE Trans. Commun.*, **COM-27**, 134–142 (1979)
44. Luvison, A., Sacchi, L. and Tamburelli, G. 'Theory and implementation of adaptive equalizers', *IEEE Int. Conf. on Communications*, Boston, MA, USA, 45.6/1–6 (1979)
45. Dalle Mese, E. and Giuli, D. 'Generalised decision feedback equalisation for digital communication channels', *Electronic Circuits and Systems*, **3**, 145–152 (1979)
46. Kawas-Kaleh, G. 'Double-decision feedback equalizer', *Frequenz*, **33**, 146–149 (1979)
47. Belfiore, C.A. and Park, J.H. 'Decision feedback equalization', *Proc. IEEE*, **67**, 1143–1156 (1979)
48. Shensa, M.J. 'A least-squares lattice decision feedback equalizer', *IEEE Int. Conf. on Communications*, Seattle, USA, Pt. 3, 57.6/1–5 (1980)
49. Harvey, J.D. 'Synchronisation of a synchronous modem', SERC report GR/A/1200.7, SERC, Swindon, UK (1980)
50. Clark, A.P. 'Equalisation and detection techniques', IEE Colloquium on 'A Review of Modem Techniques', London (1981)
51. Speidel, J. 'An automatic decision feedback equalizer with variable optimum step size', *IEEE Int. Symp. on Circuits and Systems*, Chicago, USA, **2**, 653–657 (1981)
52. Monsen, P. 'MMSE equalization of interference on fading diversity channels', *IEEE Int. Conf. on Communications*, Denver, CO, USA, **1**, 12.2/1–5 (1981)
53. Tronca, G.P. and Zenezini, G.G. 'On the optimum decision feedback receiver for PCM high frequency cable transmission', *IEEE Int. Conf. on Communications*, Denver, CO, USA, **1**, 12.7/1–6 (1981)
54. Kurzweil, J. and Bingham, A.C. 'Adaptive equalizer design considerations', *IEEE Int. Conf. on Communications*, Denver, CO, USA, **3**, 56.5/1–7 (1981)
55. Gersho, A. and Lim, T.L. 'Adaptive cancellation of intersymbol interference for data transmission', *Bell Syst. Tech. J.*, **60**, 1997–2021 (1981)
56. Pennington, J. 'Comparative measurements of parallel and serial 2.4 kbps modems', *IEE Conf. on HF Commun. Systems and Techniques*, London, 141–144 (1982)
57. Sano, A. and Yamazaki, J.-I. 'Generalized decision-feedback equalizer for approximate maximum-likelihood bit detection of digital signals', *Int. J. Electron.*, **52**, 167–176 (1982)
58. Clark, A.P. and Hussein, B.A. 'Non-linear equalizers having an improved tolerance to noise', *Radio Electron. Eng.*, **52**, 145–153 (1982)
59. Clark, A.P., Lee, L.H. and Marshall, R.S. 'Developments of the conventional nonlinear equaliser', *IEE Proc.*, Pt. F, 129, 85–94 (1982)
60. Ling, F. and Proakis, J.G. 'Generalized least squares lattice algorithm and its application to decision feedback equalization', *IEEE Int. Conf. on Acoustics, Speech and Signal Processing*, Paris, France, **3**, 1764–1769 (1982)
61. Bello, P.A. and Pahlavan, K. 'Performance of adaptive equalization for staggered QPSK and QPR over frequency-selective LOS microwave channels', *IEEE Int. Conf. on Communications*, Philadelphia, PA, USA, **2**, 3H.1/1–6 (1982)
62. Monsen, P. 'Decision feedback equalizer performance on mixed diffraction/scatter paths', *IEEE Int. Conf. on Communications*, Philadelphia, PA, USA, **2**, 3H.2/1–4 (1982)
63. Hsu, F.M. 'Square root Kalman filtering for high-speed data received over fading dispersive HF channels', *IEEE Trans. Inform. Theory*, **IT-28**, 753–763 (1982)
64. Anderson, P.H., Hsu, F.M. and Sandler, M.N. 'A new adaptive modem for long haul HF digital communications at data rates greater than 1 bps/Hz', *IEEE Military Commun. Conf.*, Boston, MA, USA, **2**, 29.2/1–7 (1982)

65. Sari, H. 'A comparison of equalization techniques on 16 QAM digital radio systems during selective fading', *IEEE Global Telecommun. Conf.*, Miami, FL, USA, 3, F3.5/1–6 (1982)
66. Tamburelli, G. 'The parallel decision feedback and feedforward equalizer', *IEEE Trans. Commun.*, **COM-31**, 224–231 (1983)
67. Hodgkiss, W. and Turner, L.F. 'Practical equalization and synchronization strategies for use in serial data transmission over h.f. channels', *Radio Electron. Eng.*, **53**, 141–146 (1983)
68. Bogush, R.L., Guigliano, F.W. and Knepp, D.L. 'Frequency selective scintillation effects and decision-feedback equalization in high data-rate satellite links', *Proc. IEEE*, **71**, 754–767 (1983)
69. El-Hefnawi. 'Performance evaluation of decision feedback equalizer under mismatch', *Proc. IEEE Nat. Aerospace and Electronics Conf.*, Dayton, Ohio, USA, 528–532 (1983)
70. Clark, A.P., Slater, M. and Parama Raj, K. 'Simple nonlinear equalisers for binary baseband signals', *IEE Proc.*, Pt. F, **130**, 495–505 (1983)
71. Araki, M. 'Time domain equalizer performance in the non-minimum phase shift fading channel', *Trans. IECE Japan*, **E66**, 671–677 (1983)
72. Hodgkiss, W., Turner, L.F. and Pennington, J. 'Serial data transmission over HF radio links', *IEE Proc.*, Pt. F, **131**, 107–116 (1984)
73. Clark, A.P. and Hau, S.F. 'Adaptive adjustment of receiver for distorted digital signals', *IEE Proc.*, Pt. F, **131**, 526–536 (1984)
74. Savage, J.E. 'Some simple self-synchronizing digital data scramblers', *Bell Syst. Tech. J.*, **46**, 449–487 (1967)
75. Nakamura, K. and Iwadare, Y. 'Data scramblers for multilevel pulse sequences', *Electron. Commun. Japan*, **55-A**, No. 6, 8–16 (1972)
76. Leeper, D.G. 'A universal digital data scrambler', *Bell Syst. Tech. J.*, **52**, 1851–1865 (1973)
77. Kasai, H., Senmoto, S. and Matsushita, M. 'PCM jitter suppression by scrambling', *IEEE Trans. Commun.*, **COM-22**, 1114–1122 (1974)
78. Papoulis, A. *Probability, Random Variables and Stochastic Processes*, McGraw-Hill, New York (1965)
79. Thomas, J.B. *An Introduction to Statistical Communication Theory*, Wiley, New York (1968)
80. Davenport, W.B. *Probability and Random Processes*, McGraw-Hill, New York (1970)
81. Qureshi, S.U.H. and Newhall, E.E. 'An adaptive receiver for data transmission over time-dispersive channels', *IEEE Trans. Inform. Theory*, **IT-19**, 448–457 (1973)
82. Falconer, D.D. and Magee, F.R. 'Adaptive channel memory truncation for maximum likelihood sequence estimation', *Bell Syst. Tech. J.*, **52**, 1541–1562 (1973)
83. Messerschmitt, D. 'Design of a finite impulse response for the viterbi algorithm and decision feedback equalizer', *IEEE Int. Conf. on Communications*, Minneapolis, MN, USA (1974)
84. Cantoni, A. and Kwong, K. 'Further results on the Viterbi algorithm equalizer', *IEEE Trans. Inform. Theory*, **IT-20**, 764–767 (1974)
85. Fredricsson, S.A. 'Joint optimization of transmitter and receiver filters in digital PAM systems with a Viterbi detector', *IEEE Trans. Inform. Theory*, **IT-22**, 200–210 (1976)
86. Desblache, A.E. 'Optimal short desired impulse response for maximum likelihood sequence estimation', *IEEE Trans. Commun.*, **COM-25**, 735–738 (1977)
87. Beare, C.T. 'The choice of the desired impulse response in combined linear-Viterbi algorithm equalizer', *IEEE Trans. Commun.*, **COM-26**, 1301–1307 (1978)
88. Browne, E.T. *Introduction to the Theory of Determinants and Matrices*, Chapel Hill, University of North Carolina Press (1958)
89. Paige, L.J. and Swift, J.D. *Elements of Linear Algebra*, Blaisdell Publishing Co., New York (1961)
90. Ayres, F. *Matrices*, McGraw-Hill, New York (1962)

91. Clark, A.P. *Principles of Digital Data Transmission*, second edition, Pentech Press, London (1983)
92. Clark, A.P., Harvey, J.D. and Driscoll, J.P. 'Near-maximum-likelihood detection processes for distorted digital signals', *Radio Electron Eng.*, **48**, 301–309 (1978)
93. Magee, F.R. and Proakis, J.G. 'An estimate of the upper bound on error probability for maximum likelihood sequence estimation on channels having a finite-duration pulse response', *IEEE Trans. Inform. Theory*, **IT-19**, 699–702 (1973)

Chapter 6

Various topics on linear and decision-feedback equalizers

6.1 ADAPTIVE EQUALIZERS

When the sampled impulse-response of the channel varies with time, it is necessary to keep readjusting the tap gains of a transversal equalizer in order to hold it correctly set for the channel. The most cost-effective method of doing this is to derive the information needed for the correct setting of the tap gains from the received data signal itself[1-4]. A brief outline will now be given of the basic principles of a simple and widely used technique, that works quite well so long as the data-symbols $\{s_i\}$ are reasonably uncorrelated. The technique is, in fact, the well-known *gradient algorithm* which has, for many applications, been found to be highly cost-effective[1-4]. Details of a wide range of different techniques for the adaptive adjustment of an equalizer are given in the references listed at the end of each of Chapters 4 and 5.

Consider first the *linear* feedforward transversal equalizer that minimizes the mean-square error in its output signal[2], as shown in Fig. 6.1. The signals shown here are those present at the time instant $t = iT$. As before, the sampled impulse-response of the channel (Figs. 4.1 and 5.1) is given by the $(g+1)$-component vector (sequence)

$$Y = [y_0 \quad y_1 \quad \ldots \quad y_g] \quad (6.1)$$

with z-transform

$$Y(z) = y_0 + y_1 z^{-1} + \cdots + y_g z^{-g} \quad (6.2)$$

and the sampled impulse-response of the linear equalizer is given by the $(m+1)$-component row vector

$$C = [c_0 \quad c_1 \quad \ldots \quad c_m] \quad (6.3)$$

with z-transform

$$C(z) = c_0 + c_1 z^{-1} + \cdots + c_m z^{-m} \quad (6.4)$$

The equalizer is ideally adjusted such that

$$Y(z)C(z) \simeq z^{-h} \quad (6.5)$$

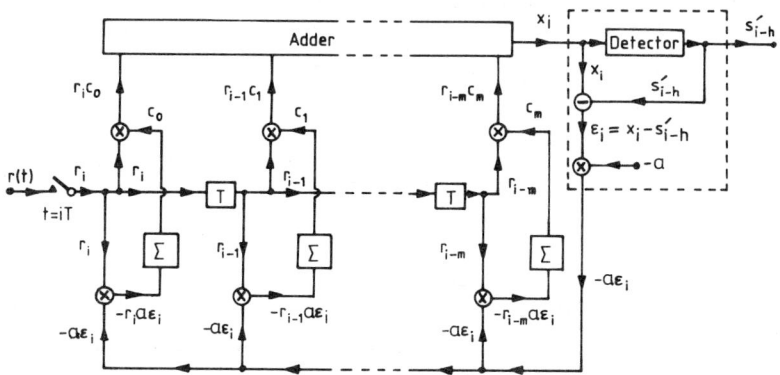

Fig. 6.1 *Adaptive linear equalizer*

which means that the channel and equalizer together introduce a delay of h sampling intervals. Clearly, with a sufficiently large value of m,

$$C(z) \simeq z^{-h} Y^{-1}(z) \tag{6.6}$$

assuming, as before, that $Y(z)$ has no zeros (roots) *on* the unit circle in the z plane and that h is suitably valued to permit the above approximation to be achieved to the required accuracy.

With the reasonably accurate equalization of the channel, the output signal from the equalizer, at time $t = iT$, is

$$x_i \simeq s_{i-h} + u_i \tag{6.7}$$

and s'_{i-h}, the detected value of s_{i-h}, is taken as k or $-k$ depending upon whether $x_i > 0$ or $x_i < 0$, respectively. The integer h is the number of sampling intervals delay in detection, and u_i is the noise component. The error in x_i is $x_i - s_{i-h}$, and at low error rates, when usually $s'_{i-h} = s_{i-h}$, the error in x_i can be taken to be

$$\varepsilon_i = x_i - s'_{i-h} \tag{6.8}$$

When $s'_{i-h} = s_{i-h}$, ε_i is composed partly of noise and partly of intersymbol interference.

For a given vector C, the mean-square error in x_i can be taken to be

$$e(C) = \mathscr{E}[\varepsilon_i^2] = \mathscr{E}[(x_i - s'_{i-h})^2] \tag{6.9}$$

assuming again that $s'_{i-h} = s_{i-h}$. But

$$x_i = \sum_{l=0}^{m} r_{i-l} c_l \tag{6.10}$$

and

$$\frac{\partial \varepsilon_i}{\partial c_j} = \frac{\partial}{\partial c_j}\left(\sum_{l=0}^{m} r_{i-l}c_l - s'_{i-h}\right) = r_{i-j} \quad (6.11)$$

so that

$$\frac{\partial e(C)}{\partial c_j} = \mathscr{E}\left[\frac{\partial \varepsilon_i^2}{\partial \varepsilon_i}\frac{\partial \varepsilon_i}{\partial c_j}\right] = \mathscr{E}[2\varepsilon_i r_{i-j}] \quad (6.12)$$

The *gradient* of $e(C)$ with respect to C is defined to be the $(m+1)$-component vector

$$\nabla e(C) = \left[\frac{\partial e(C)}{\partial c_0} \quad \frac{\partial e(C)}{\partial c_1} \quad \cdots \quad \frac{\partial e(C)}{\partial c_m}\right] \quad (6.13)$$

Thus the *maximum increase* in $e(C)$ for a given small magnitude of change in C (that is, for a given small distance moved by C in the $(m+1)$-dimensional Euclidean vector space containing this vector) is achieved when the vector moves in the direction given by $\nabla e(C)$ and so along the *gradient* of $e(C)$ with respect to C. But it can be shown that $e(C)$ is a *convex* function of C with *one* global minimum at which

$$\frac{\partial e(C)}{\partial c_j} = 0 \quad (6.14)$$

for $j = 0, 1, \ldots, m$. The equalizer is now correctly adjusted. Hence, for any arbitrary value of C, a more accurate adjustment of the equalizer and hence a better value of C is given by

$$C - \frac{a}{2}\nabla e(C) \quad (6.15)$$

where a is a small positive real-valued constant. The reason for using $a/2$ in place of a here will become clear presently. From Equation 6.13 and Expression 6.15 it can be seen that c_j, the $(j+1)$th component of C, is here changed to

$$c_j - \frac{a}{2}\frac{\partial e(C)}{\partial c_j} \quad (6.16)$$

for $j = 0, 1, \ldots, m$. But Equation 6.12 shows that an *estimate* of $\frac{1}{2}\partial e(C)/\partial c_j$ is given by $\varepsilon_i r_{i-j}$. Thus, after each time interval of T seconds when a new signal r_i is received, the gain c_j of the $(j+1)$th tap of the $(m+1)$-tap transversal equalizer is incremented by

$$\delta c_j = -a\varepsilon_i r_{i-j} \quad (6.17)$$

for $j = 0, 1, \ldots, m$. Over a number of l received signal-elements, this arrangement effectively measures the *cross-correlation* between the corresponding l error signals $\{\varepsilon_i\}$ and the l received samples $\{r_{i-j}\}$, for any given value of j. The resultant change in the tap gain c_j is now

$$\Delta c_j = -la\mathscr{E}_s[\varepsilon_i r_{i-j}] \tag{6.18}$$

where $\mathscr{E}_s[\varepsilon_i r_{i-j}]$ is the *short-term average* of $\varepsilon_i r_{i-j}$, determined over the l values of i^{1-3}. The values of the $\{c_j\}$ now converge towards the values at which the mean-square error in x_i is minimum, and as the channel varies with time the tap gains are appropriately adjusted to hold the equalizer correctly set for the channel. The conditions for correct convergence and reliable adaptive operation have been widely studied (see the references at the end of Chapter 4). The smaller the value of a in Equation 6.18, the greater the value of l for a given magnitude of Δc_j. Hence the slower the response of the tap gains and the greater the effective averaging or integration period of the system.

A necessary condition for the correct operation of the arrangement is that the $\{s_i\}$ are uncorrelated, but even now a small value of a is required to obtain a sufficiently accurate measure of the true average value of $\varepsilon_i r_{i-j}$. A small value of a is also necessary to reduce the effects of noise on the settings of the $\{c_j\}$.

Since the arrangement minimizes $\mathscr{E}[\varepsilon_i^2]$ in Equation 6.9, it minimizes the mean-square error due to *both* intersymbol interference and noise. Thus, even with an extremely small value of a (and so the effective averaging over a very large number of received samples), the settings of the $\{c_j\}$ are affected by the noise variance σ^2. For instance, an increase in σ^2 causes a general reduction in the magnitudes of the $\{c_j\}$, regardless of the value of a. However, at high signal/noise ratios, where the predominant source of error in the $\{x_i\}$ is the intersymbol interference, the tap gains are effectively adjusted to minimize the mean-square error caused by intersymbol interference alone.

The adaptive equalizer may be modified in various different ways. For instance, either or both ε_i and r_{i-j} may be quantized in Equation 6.17. Again, the changes in the $\{c_j\}$ may only be applied after every lT seconds or alternatively only when the magnitudes of the changes exceed a given threshold level[1-4]. However, the method of operation is still basically similar to that just described.

Consider next the *decision-feedback* equalizer that minimizes the mean-square error in its output signal (see Fig. 5.11). The nonlinear filter of this equalizer is shown in Fig. 6.2. The linear filter has $n+1$ taps with gains given by the $n+1$ components of the row vector

$$D = [d_0 \quad d_1 \quad \ldots \quad d_n] \tag{6.19}$$

This filter is very similar to the portion of Fig. 6.1 outside the dotted rectangle, but with c_j replaced by d_j and with m replaced by n. The μ tap gains of the feedback transversal filter are given by the μ components of

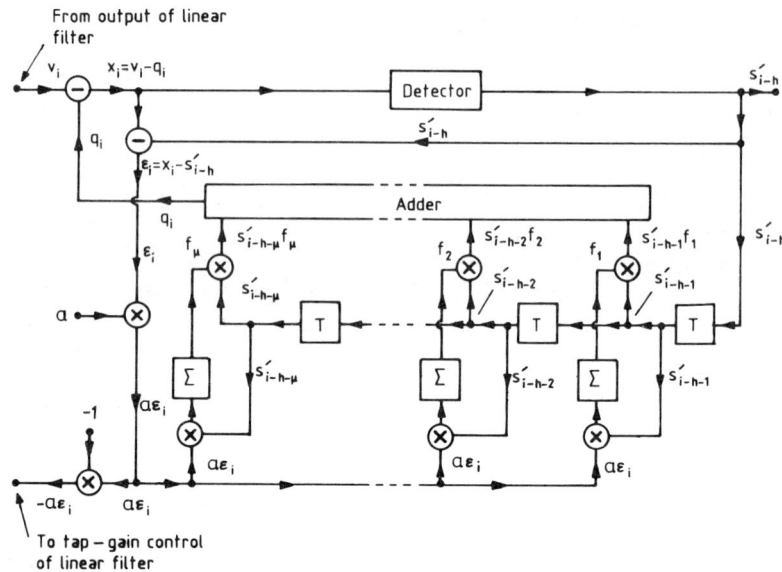

Fig. 6.2 Nonlinear filter of adaptive decision-feedback equalizer

the row vector

$$F = [f_1 \quad f_2 \quad \cdots \quad f_\mu] \qquad (6.20)$$

The signals shown in Fig. 6.2 are those present at the time instant $t = iT$. With the reasonably accurate equalization of the channel, the output signal from the decision-feedback equalizer at this time instant is

$$x_i \simeq s_{i-h} + u_i \qquad (6.21)$$

where the delay in detection is again h sampling intervals and u_i is the noise component at the output of the $(n+1)$-tap linear feedforward transversal filter. As before, the error in the equalized signal x_i can be taken to be

$$\varepsilon_i = x_i - s'_{i-h} \qquad (6.22)$$

assuming that $s'_{i-h} = s_{i-h}$.

The sampled impulse-response of the channel and $(n+1)$-tap linear feedforward transversal filter is now given by the $(n+g+1)$-component row vector

$$E = [e_0 \quad e_1 \quad \cdots \quad e_{n+g}] \qquad (6.23)$$

and *ideally* (as is shown in Section 5.7)

$$f_i = e_{h+i} \tag{6.24}$$

for $i = 1, 2, \ldots, \mu$, where

$$h + \mu = n + g \tag{6.25}$$

so that the nonlinear filter has the effect of setting to zero the *last* μ components of E, where μ is an appropriately large integer.

For the *given* vectors D and F (Equations 6.19 and 6.20) the mean-square error in the equalized signal x_i (Fig. 6.2) can be taken to be

$$e(D, F) = \mathscr{E}[\varepsilon_i^2] = \mathscr{E}[(x_i - s'_{i-h})^2] \tag{6.26}$$

where now

$$x_i = \sum_{j=0}^{n} r_{i-j} d_j - \sum_{j=1}^{\mu} s'_{i-h-j} f_j \tag{6.27}$$

As in Equation 6.11,

$$\frac{\partial \varepsilon_i}{\partial d_j} = \frac{\partial}{\partial d_j}\left(\sum_{l=0}^{n} r_{i-l} d_l - \sum_{l=1}^{\mu} s'_{i-h-l} f_l - s'_{i-h}\right) = r_{i-j} \tag{6.28}$$

and

$$\frac{\partial \varepsilon_i}{\partial f_j} = \frac{\partial}{\partial f_j}\left(\sum_{l=0}^{n} r_{i-l} d_l - \sum_{l=1}^{\mu} s'_{i-h-l} f_l - s'_{i-h}\right) = -s'_{i-h-j} \tag{6.29}$$

so that

$$\frac{\partial e(D,F)}{\partial d_j} = \mathscr{E}\left[\frac{\partial \varepsilon_i^2}{\partial \varepsilon_i} \cdot \frac{\partial \varepsilon_i}{\partial d_j}\right] = \mathscr{E}[2\varepsilon_i r_{i-j}] \tag{6.30}$$

and

$$\frac{\partial e(D,F)}{\partial f_j} = \mathscr{E}\left[\frac{\partial \varepsilon_i^2}{\partial \varepsilon_i} \cdot \frac{\partial \varepsilon_i}{\partial f_j}\right] = \mathscr{E}[-2\varepsilon_i s'_{i-h-j}] \tag{6.31}$$

When the equalizer is correctly adjusted, to minimize the mean-square error in the equalized signal, both $\partial e(D, F)/\partial d_j$ and $\partial e(D, F)/\partial f_j$ are zero for each value of j.

Let A be the $(n + \mu + 1)$-component row vector given by

$$A = [d_0 \quad d_1 \quad \ldots \quad d_n \quad f_1 \quad f_2 \quad \ldots \quad f_\mu] \tag{6.32}$$

the first segment of A being formed by the vector D and the remaining segment by the vector F. Then the gradient of $e(D, F)$ with respect to A is

$$\nabla e(D, F) = \left[\frac{\partial e(D, F)}{\partial d_0} \quad \ldots \quad \frac{\partial e(D, F)}{\partial d_n} \quad \frac{\partial e(D, F)}{\partial f_1} \quad \ldots \quad \frac{\partial e(D, F)}{\partial f_\mu}\right] \tag{6.33}$$

For any arbitrary value of A, a more accurate adjustment of the equalizer and hence a better value of A is given by

$$A - \frac{a}{2}\nabla e(D,F) \tag{6.34}$$

where a is a small positive real-valued constant. Now, d_j, the $(j+1)$th component of D is changed to

$$d_j - \frac{a}{2}\frac{\partial e(D,F)}{\partial d_j} \tag{6.35}$$

for $j=0, 1, \ldots, n$, and f_j, the jth component of F, is changed to

$$f_j - \frac{a}{2}\frac{\partial e(D,F)}{\partial f_j} \tag{6.36}$$

for $j=1, 2, \ldots, \mu$. It can be seen from Equations 6.30 and 6.31, that an estimate of $\frac{1}{2}\partial e(D,F)/\partial d_j$ is given by $\varepsilon_i r_{i-j}$ and an estimate of $\frac{1}{2}\partial e(D,F)/\partial f_j$ is given by $-\varepsilon_i s'_{i-h-j}$. Thus, after each time interval of T seconds when a new signal r_i is received, the tap gain d_j is incremented by

$$\delta d_j = -a\varepsilon_i r_{i-j} \tag{6.37}$$

for $j=0, 1, \ldots, n$, and the tap gain f_j is incremented by

$$\delta f_j = a\varepsilon_i s'_{i-h-j} \tag{6.38}$$

for $j=1, 2, \ldots, \mu$.

Over l received signals $\{r_i\}$, the tap gain d_j changes by

$$\Delta d_j = -la\mathscr{E}_s[\varepsilon_i r_{i-j}] \tag{6.39}$$

where $\mathscr{E}_s[\varepsilon_i r_{i-j}]$ is the short-term average of $\varepsilon_i r_{i-j}$ determined over the l values of i, and the tap gain f_j changes by

$$\Delta f_j = la\mathscr{E}_s[\varepsilon_i s'_{i-h-j}] \tag{6.40}$$

where $\mathscr{E}_s[\varepsilon_i s'_{i-h-j}]$ is the short-term average of $\varepsilon_i s'_{i-h-j}$, determined over the l values of i. As for the linear equalizer, $\mathscr{E}_s[\varepsilon_i r_{i-j}]$ is a measure of the cross-correlation between the error signals $\{\varepsilon_i\}$ and the corresponding l received signals $\{r_{i-j}\}$, whereas $\mathscr{E}_s[\varepsilon_i s'_{i-h-j}]$ is a measure of the cross-correlation between the error signals $\{\varepsilon_i\}$ and the corresponding l detected data-symbols $\{s'_{i-h-j}\}$. It can be seen that the tap gains of the equalizer are adjusted so as to set to zero each of the cross-correlations just mentioned, over the appropriate range of values of j, which means that, when the equalizer is correctly adjusted, the error signal ε_i tends to be *uncorrelated* both with the received samples $\{r_{i-j}\}$ and with the detected data-symbols $\{s'_{i-h-j}\}$.

Since the algorithm for adjusting the tap gains of the decision-feedback equalizer is such as to minimize the mean-square error in the equalized signal, the accuracy of the adjustment improving as the magnitude of a is reduced, it is evident that with a sufficiently small value of a, the adjustment of the equalizer, for the given values of Y, n, μ, h and σ^2, is as given by Equation 5.199 in Section 5.7. At high signal/noise ratios and with a sufficient number of taps in the equalizer (together with an appropriate value of h) the channel is equalized quite accurately. Now, in the sampled impulse-response of the channel and $(n+1)$-tap linear feedforward transversal filter, given by the $(n+g+1)$-component vector E in Equation 6.23, $e_j \simeq 0$ for $j = 0, 1, \ldots, h-1$, and $e_h \simeq 1$. The remaining $\{e_j\}$ take on appropriate values, which may or may not be nonzero. From Equation 6.25, there are

$$\mu = n + g - h \tag{6.41}$$

of these components, and they are subsequently set to zero by the nonlinear filter (Fig. 6.2). This has μ taps with gains ideally equal to the components $e_{h+1}, e_{h+2}, \ldots, e_{n+g}$ of E. As the signal/noise ratio increases and with an appropriately large number of taps in the equalizer, the latter approaches the decision-feedback equalizer that minimizes the mean-square error in the equalized signal, subject to the *exact* equalization of the channel (Section 5.8). For this equalizer

$$\mu = g \tag{6.42}$$

and
$$h = n \tag{6.43}$$

as can be seen from Equations 5.298 and 5.305. The given values of μ nd h also appear to be suitable for the much wider range of conditions considered in Section 5.7, so that they should normally be used under the general conditions assumed here.

Equation 6.43 is a most important result because it shows that the delay in detection of h sampling intervals is *independent* of the component *values* of Y (Equation 6.1). Thus, when Y varies with time, there is no need to *change* the delay in detection in order to maintain the optimum adjustment of the equalizer. The precise reasons why Equation 6.43 holds are illuminated here by the simple example considered in Table 6.1. This shows the gains $d_{n-6}, d_{n-5}, \ldots, d_n$ of the last seven taps of the linear feedforward transversal filter D in Fig. 5.11, together with the last eight components $e_{n+g-7}, e_{n+g-6}, \ldots, e_{n+g}$ of the resulting sampled impulse-response E for the channel and filter, for each of six different values of the sampled impulse-response Y of the channel, where the latter has just two components. The decision-feedback equalizer is here adjusted to minimize the mean-square error in the equalized signal, subject to the *exact* equalization of the channel. A very large (ideally infinite) value of n is assumed. The z-transform of the sampled impulse-

Table 6.1 VECTORS D AND E FOR VARIOUS VECTORS Y

(1)	Y:	1.100	1.000						
	D:	0.000	0.000	0.000	0.000	0.000	0.000	0.909	
	E:	0.000	0.000	0.000	0.000	0.000	0.000	1.000	0.909
(2)	Y:	1.000	1.000						
	D:	0.000	0.000	0.000	0.000	0.000	0.000	1.000	
	E:	0.000	0.000	0.000	0.000	0.000	0.000	1.000	1.000
(3)	Y:	0.990	1.000						
	D:	−0.019	0.019	−0.019	0.020	−0.020	0.020	0.990	
	E:	0.000	0.000	0.000	0.000	0.000	0.000	1.000	0.990
(4)	Y:	0.900	1.000						
	D:	−0.112	0.125	−0.139	0.154	−0.171	0.190	0.900	
	E:	0.000	0.000	0.000	0.000	0.000	0.000	1.000	0.900
(5)	Y:	0.500	1.000						
	D:	−0.024	0.047	−0.094	0.188	−0.375	0.750	0.500	
	E:	0.000	0.000	0.000	0.000	0.000	0.000	1.000	0.500
(6)	Y:	0.100	1.000						
	D:	0.000	0.000	−0.001	0.010	−0.099	0.990	0.100	
	E:	0.000	0.000	0.000	0.000	0.000	0.000	1.000	0.100

Only the last seven components of D and the last eight components of E are shown here.

response of the channel is

$$Y(z) = y_0 + y_1 z^{-1} = y_0(1 - \alpha z^{-1}) \tag{6.44}$$

where α is the zero (root) of $Y(z)$. Clearly,

$$\alpha = -y_1/y_0 \tag{6.45}$$

When the larger of the two components of Y is its *first* component y_0, so that the zero of $Y(z)$ lies *inside* the unit circle in the z plane, the $(n+1)$-tap linear feedforward transversal filter D (Fig. 5.11) simply introduces a delay of n sampling intervals together with the appropriate level change, and the equalization of the channel is carried out entirely by the nonlinear filter (Fig. 5.11) whose transversal filter F has just *one* tap with a gain of y_1/y_0. When the larger of the two components of Y is its *second* component y_1, so that the zero of $Y(z)$ lies *outside* the unit circle, the larger of the two nonzero components of E (Equation 6.23 and Table 6.1) is the *first* of these, as before, so that the delay measured between the two *larger* components is now $n-1$ sampling intervals, as can be seen from Table 6.1. However, when the delay is measured between the *first nonzero* components of Y and E it remains equal to n sampling intervals, as for the case where the zero of $Y(z)$ lies inside the unit circle. It appears therefore that a delay of one sampling interval in the position of the larger component of Y is balanced by the corresponding reduction in delay between this component and the first nonzero component of E, leaving the delay between the respective *first* components *unchanged*. The same principle applies in the general case where Y has $g+1$ components and $Y(z)$ has f zeros outside the unit circle. The presence of these f zeros implies that the main component of Y is its $(f+1)$th component y_f (which need not however be its largest component). More precisely, the presence of f zeros outside the unit circle implies a *delay* of f sampling intervals, such that, if the first component y_0 of Y occurs at time $t=0$, the sequence is effectively *centred* on the main component y_f at time $t=fT$. The $(n+1)$-tap filter D must now introduce f zeros at the reciprocals of those of $Y(z)$ outside the unit circle, which means that to maximize the delay introduced by the filter and so to optimize its performance, it must introduce a delay of $n-f$ sampling intervals measured between the main components of Y and E. But the main component of E is always its *first nonzero* component, so that the main component of E is always its $(n+1)$th component e_n at time $t=nT$, the preceding $\{e_j\}$ being ideally all set to zero. It follows that, with a time-varying channel such that the zeros of $Y(z)$ keep crossing the unit circle, the optimum adjustment of the linear filter D (Fig. 5.11) can, at least in principle, be maintained at all times, without needing to change the delay in detection. An alternative and rather more rigorous derivation of this result is given in Section 5.8.2.

In moving from (1) to (6) in Table 6.1, the zero of $Y(z)$ moves from just inside the unit circle to well outside, being exactly *on* the unit circle for (2). It can be seen here that when the zero of $Y(z)$ is just outside the unit circle, a small change in Y leads to a correspondingly *much larger* change in D, which suggests that there may be some difficulty in holding the linear filter D correctly adjusted for the channel under these conditions, the difficulty increasing with the number of zeros that cross the unit circle. Unfortunately, when operating over an HF radio link, there may well be several zeros of $Y(z)$ simultaneously crossing the unit circle, which implies the need for a relatively rapid adjustment of the tap gains of the linear filter. Indeed, it may well now be better to adjust the linear filter by locating the *zeros* of $Y(z)$ that lie outside the unit circle and using the knowledge of these zeros to determine the vectors D and E (from Equations 5.299 and 5.300) and hence to achieve the required adaptive adjustment of the receiver. Such a system has been developed and is described elsewhere[5]. Further details of the system are, however, beyond the scope of this book.

The rate of convergence (speed of approach towards the correct setting) in the adjustment of the tap gains by the *gradient method* under consideration here (Figs. 6.1 and 6.2), in the case of a linear or decision-feedback equalizer, is dependent on the number of tap gains and on the level of amplitude distortion introduced by the channel. It also tends to increase with a, but the system becomes unstable when a exceeds a certain value. In the presence of nearly pure phase distortion there is relatively little *coupling* between the tap gains, which means that the adjustment of any one tap gain does not unduly influence the adjustment of any other, and a rapid rate of convergence therefore tends to occur. Most of the equalization in the decision-feedback equalizer is now carried out by the linear filter D (Fig. 5.11), with only small tap gains in the filter F. In general, the more taps in the transversal filter D and the greater the level of the amplitude distortion introduced by the channel, the *slower* the rate of convergence. Furthermore, it has been found that a decision-feedback equalizer tends to have a *slower* rate of convergence than the corresponding *linear* equalizer. This may be due to the coupling between the tap gains of the linear and nonlinear filters. It is interesting to observe that the filter F (Fig. 5.11) acts as an *estimator* for the last μ components of the sampled impulse-response E of the channel and filter D. For a *given* channel and filter D, the rate of convergence of the filter F is usually much faster than is the rate of convergence of the filters D and F for the given channel or even than the rate of convergence of the corresponding linear equalizer for the channel.

6.2 DOUBLE-SAMPLING EQUALIZERS

6.2.1 Introduction

Consider the synchronous serial data-transmission system of Fig. 6.3, which, for convenience, is shown as employing a *linear* equalizer at the receiver. Where a *decision-feedback* equalizer is used, the receiver is as

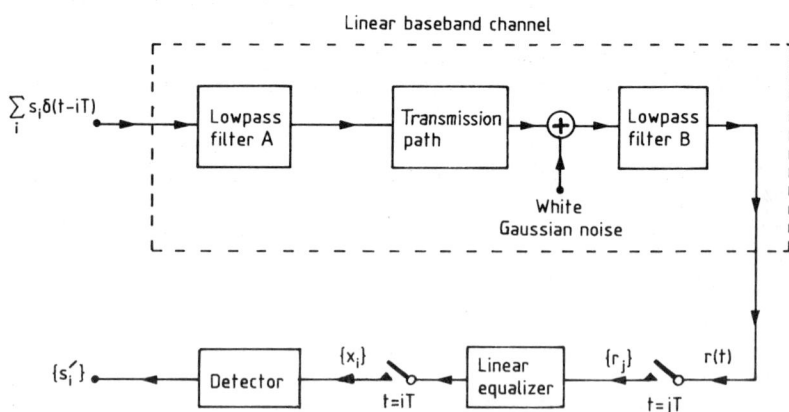

Fig. 6.3 Data-transmission system with a double-sampling linear equalizer at the receiver

shown in Fig. 5.11. The transmission path is a *linear baseband channel* whose output waveform $r(t)$ is sampled at the time instants $\{jT\}$, where, for the present, it is assumed that j takes on all integer values. Thus $j = i$, and $r(t)$ is sampled once per data symbol and therefore once per received signal-element. It is assumed that all signals here are *real-valued*.

In some applications, where linear or decision-feedback equalizers are used, the transmitted signal-element rate is well below the Nyquist rate for the given channel. If the received waveform here is sampled once per signal element, the sampling rate is well below the Nyquist rate for the receiver filter (lowpass filter B in Fig. 6.3), so that, if the bandwidth of the latter is B Hz, the sampling rate is well below $2B$ samples/second. Under these conditions there may be an appreciable loss of information in the samples of the received waveform. This loss of information appears in the form of *aliasing* which can cause one or more zeros (roots) of the z-transform of the sampled impulse-response of the channel to approach close to the unit circle. Thus, with an unfavourable choice of sampling phase at the receiver, the amplitude distortion of the sampled data symbol can become unduly high, with the corresponding severe noise enhancement and serious degradation in tolerance to noise, if a linear equalizer is used, and with an appreciable reduction in tolerance to noise even with a decision-feedback equalizer.

Suppose that the impulse-response of the transmission path is known and that the receiver input filter (Fig. 6.3) can be arranged to be *matched* to each received signal-element, that is, to the resultant impulse-response of the transmitter filter and transmission path in cascade, which means that the impulse-response of the receiver input filter (assumed here to be real-valued) is the *time reverse* of that of the transmitter filter and transmission path. Now, provided only that the waveform at the output of the receiver input filter is sampled once per signal element and in the correct phase, *all* the useful information in the received waveform is contained in the samples. Indeed, this is an ideal and optimum arrangement at the receiver, enabling, at least in principle, the optimum detection of the received data signal to be achieved from the given samples (see Sections 1.4–1.6). Unfortunately, there are certain problems with an arrangement of this type. Firstly, the resultant impulse-response of the transmitter filter and transmission path may not be known at the receiver, requiring therefore an additional estimation process for the correct adjustment of the receiver filter and hence leading to a more complex system[6,7]. Secondly, in the presence of additive white Gaussian noise (as is assumed throughout this analysis) it is highly desirable that the noise components in different samples of the received signal be as near as possible uncorrelated. Unfortunately, when the transmitter filter and transmission path together introduce appreciable amplitude distortion, the *same* amplitude distortion is introduced by the receiver filter that is matched to each received signal-element, so that, with additive white Gaussian noise at the input to the receiver filter, there is significant *correlation* between neighbouring noise samples at the output of the filter. A noise-whitening network can, of course, now be employed at the output of the sampler, but this further complicates the design of the receiver, particularly when it needs to be adaptively adjusted[8-10]. The best that can normally be done in practice is to design the receiver filter to be matched to each received signal-element in the *absence* of any signal distortion introduced by the transmission path, and this is, in fact, the arrangement assumed throughout the book (see Section 1.2). Under these conditions, when signal distortion is introduced by the transmission path, a satisfactory performance can often be obtained by suitably adjusting the phase of the sampling instants at the receiver. However, the need to optimize the phase of the sampling instants increases the complexity of the system[4].

The problem just described may be avoided through bandlimiting the data signal by means of the transmitter and receiver filters (filters A and B in Fig. 6.3), so that, as far as the latter are concerned, the signal is transmitted effectively at the Nyquist rate. Unfortunately, there may now be quite an inefficient use of the available bandwidth of the transmission path in that only, say, one half of this is used by the data signal, with a corresponding reduction in tolerance to effects such as frequency offset, fading, sudden changes in signal level or carrier phase, and so on. Under the above conditions, a better approach is to transmit a signal using the *whole* of the available frequency band of the transmission path and to sample the received data signal at *twice* the signal-element rate[11-15]. This is the arrangement now to be studied. The receiver filter is here assumed to be an ideal lowpass filter with a constant value of its transfer function over the frequency band 0 to $1/T$ Hz, with a sharp cut-off at $1/T$ Hz and a very high attenuation over all higher frequencies, the phase characteristic of the filter being linear over the passband.

6.2.2 Model of system

The model of the digital data-transmission system assumed here is shown in Fig. 6.3. The information to be transmitted is carried by the binary data-symbols $\{s_i\}$, where

$$s_i = \pm k \tag{6.46}$$

The $\{s_i\}$ are assumed to be statistically independent and equally likely to have either binary value. They are transmitted at $1/T$ symbols/second in the form of the impulses $\{s_i \delta(t - iT)\}$ at the input of the lowpass filter A.

The lowpass filter A has an impulse response $a(t)$, and it appropriately shapes the transmitted signal-elements $\{s_i a(t - iT)\}$. The transmission path is a linear baseband channel that could be a cable or else could comprise a modulator, bandpass channel and demodulator, all of which are linear. Stationary white Gaussian noise with zero mean and a two-sided power spectral density of $\frac{1}{2}N_0$ is added to the data signal at the output of the transmission path, and the resulting noise waveform at the output of the lowpass filter B is the band-limited zero-mean Gaussian waveform $w(t)$. The lowpass filter B has the transfer function

$$B(f) = \begin{cases} \sqrt{\frac{T}{2}}, & -\frac{1}{T} < f < \frac{1}{T} \\ 0, & \text{elsewhere} \end{cases} \quad (6.47)$$

and the sampling interval used at the receiver is $\frac{1}{2}T$ seconds. As before, frequency and time are measured in units of Hz and seconds, respectively. For practical purposes, each of the lowpass filters A and B can be made physically realizable by introducing a sufficient delay into the corresponding impulse-response. However, to simplify the analysis, the appropriate nonphysically realizable filters A and B are assumed, where necessary. It can be shown (Section 1.2) that the samples of $w(t)$, spaced at regular intervals of $\frac{1}{2}T$ seconds, are statistically independent Gaussian random variables with zero mean and variance

$$\sigma^2 = \frac{1}{2}N_0 \quad (6.48)$$

The lowpass filter A, transmission path and lowpass filter B together form a linear baseband channel with a real-valued impulse-response $y(t)$. For practical purposes, $y(t)$ has a finite duration and is also time invariant, such that $y(t - iT)$ is a time-shifted version of $y(t - jT)$, for any integers $i \neq j$. Thus the waveform at the output of the lowpass filter B is

$$r(t) = \sum_i s_i y(t - iT) + w(t) \quad (6.49)$$

and this is fed to the equalizer (Fig. 6.3). All signals here are *real-valued*.

6.2.3 Linear equalizer

Consider the double-sampling linear equalizer shown in Fig. 6.4. The received signal is the real-valued waveform $r(t)$ given by Equation 6.49. The waveform $r(t)$ is sampled *twice* per received signal-element $s_i y(t - iT)$ and therefore *twice* per data-symbol s_i, at the time instants $\{iT\}$ and $\{(i - \frac{1}{2})T\}$ to give the samples $\{r_i\}$ and $\{r_{i-1/2}\}$, where $r_i = r(iT)$ and $r_{i-1/2} = r((i - \frac{1}{2})T)$. i takes on all integer values, and the delay in transmission is neglected, as before. Thus two signals r_i and $r_{i-1/2}$ are received for the evaluation of each detected data-symbol s'_i at the detector output (Fig. 6.4). The subscript $i - 1/2$ here represents $i - (1/2)$.

Regardless of the impulse-response $y(t)$ of the linear baseband channel

LINEAR AND DECISION-FEEDBACK EQUALIZERS

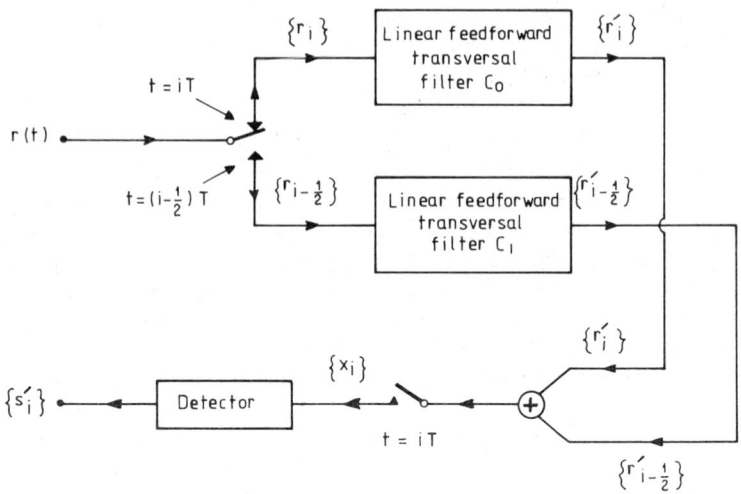

Fig. 6.4 Double-sampling linear equalizer

(Fig. 6.3), it follows that, with the appropriate adjustment of the delay in transmission, which will now be assumed, and with the appropriate integer g,

$$r_i = \sum_{j=0}^{g} s_{i-j} y_{0,j} + w_i \qquad (6.50)$$

and

$$r_{i-1/2} = \sum_{j=0}^{g} s_{i-j} y_{1,j} + w_{i-1/2} \qquad (6.51)$$

where

$$y_{0,j} = y(jT) \qquad (6.52)$$

$$y_{1,j} = y((j-\tfrac{1}{2})T) \qquad (6.53)$$

$$w_i = w(iT) \qquad (6.54)$$

$$w_{i-1/2} = w((i-\tfrac{1}{2})T) \qquad (6.55)$$

and the noise components $\{w_i\}$ and $\{w_{i-1/2}\}$ are statistically independent Gaussian random variables with zero-mean and variance $\sigma^2 = \tfrac{1}{2}N_0$. It is assumed that $y_{0,j}$ and $y_{1,j}$ are both zero (at least, approximately) when $j < 0$ and $j > g$. Thus the sampled impulse-responses of the linear baseband channel (Fig. 6.3), at the time instants $\{iT\}$ and $\{(i-\tfrac{1}{2})T\}$, are given by the $(g+1)$-component vectors (sequences)

$$Y_0 = [y_{0,0} \quad y_{0,1} \quad \cdots \quad y_{0,g}] \qquad (6.56)$$

and

$$Y_1 = [y_{1,0} \quad y_{1,1} \quad \cdots \quad y_{1,g}] \qquad (6.57)$$

respectively, Y_0 and Y_1 being usually *quite different*. The z-transforms of these sampled impulse-responses are

$$Y_0(z) = y_{0,0} + y_{0,1}z^{-1} + \cdots + y_{0,g}z^{-g} \quad (6.58)$$

and
$$Y_1(z) = y_{1,0} + y_{1,1}z^{-1} + \cdots + y_{1,g}z^{-g} \quad (6.59)$$

the delays of $y_{0,0}$ and $y_{1,0}$ being, for convenience, ignored here. In other words, each received sample $r_{i-1/2}$ is treated as though it arrives at time $t = iT$, and so is coincident in time with r_i. This is achieved, in practice, through *delaying* by $\tfrac{1}{2}T$ each received sample $r_{i-1/2}$, relative to the corresponding sample r_i, which is equivalent to delaying each component of Y_1 by $\tfrac{1}{2}T$, so that it coincides in time with the corresponding component of Y_0. The delay in $y_{0,0}$ is set to zero. Thus the received sampled signal is equivalent to that where, for each i, the data-symbol s_i is fed simultaneously, at the time instant iT, to two different linear baseband channels connected *in parallel*, and the received signals from the two channels are sampled *simultaneously*, once per data symbol and again at the time instants $\{iT\}$. The given model of the system simplifies the following analysis but does not affect the results obtained. Of course, in each z-transform $Y_0(z)$ and $Y_1(z)$, z^{-1} corresponds to a delay of T seconds.

In practice, as will be considered in more detail later, a single $2(m+1)$-tap linear feedforward transversal filter is used. The effective time delay here between adjacent samples stored in the transversal filter is $T/2$ seconds, this being *half* the corresponding time delay in a single-sampling equalizer. Thus *two* received samples are fed into the filter for every *one* output sample fed to the detector, such that only every *second* output sample is in fact employed in the detection process. Under these conditions the single transversal filter is equivalent to the two separate filters shown in Fig. 6.4.

The transversal filters C_0 and C_1 (Fig. 6.4) are $(m+1)$-tap linear feedforward transversal filters having tap gains, and hence sampled impulse-responses, given by the two $(m+1)$-component vectors

$$C_0 = [c_{0,0} \quad c_{0,1} \quad \cdots \quad c_{0,m}] \quad (6.60)$$

and
$$C_1 = [c_{1,0} \quad c_{1,1} \quad \cdots \quad c_{1,m}] \quad (6.61)$$

with z-transforms

$$C_0(z) = c_{0,0} + c_{0,1}z^{-1} + \cdots + c_{0,m}z^{-m} \quad (6.62)$$

and
$$C_1(z) = c_{1,0} + c_{1,1}z^{-1} + \cdots + c_{1,m}z^{-m} \quad (6.63)$$

respectively. The corresponding *single-sampling* linear equalizer is obtained simply by omitting the filter C_1 (together with the following adder and sampler) so that the output signal from the filter C_0 is fed directly to the detector. The filter C_0 is now adjusted as described in Section 4.11.

Let the sampled impulse-response of the channel and filter C_0 be given by the row vector (sequence)

$$E_0 = [e_{0,0} \quad e_{0,1} \quad \ldots \quad e_{0,m+g}] \quad (6.64)$$

which is the *convolution* of Y_0 and C_0, and let the sampled impulse-response of the channel and filter C_1 be given by the $(m+g+1)$-component row vector

$$E_1 = [e_{1,0} \quad e_{1,1} \quad \ldots \quad e_{1,m+g}] \quad (6.65)$$

which is the *convolution* of Y_1 and C_1. The resultant sampled impulse-response of the linear baseband channel and double-sampling equalizer, as obtained at the detector input in Fig. 6.4, is now given by the $(m+g+1)$-component row vector

$$E = E_0 + E_1 \quad (6.66)$$

Assuming that the linear equalizer is adjusted to introduce a delay of hT seconds into the detected signal, the ideal or desired sampled impulse-response of the channel and linear equalizer is given by the $(m+g+1)$-component row vector

$$E_h = [\overbrace{0 \quad 0 \quad \ldots \quad 0}^{h} \quad 1 \quad 0 \quad 0 \quad \ldots \quad 0] \quad (6.67)$$

The symbol or term E_h is, of course, not to be confused with E_0 or E_1.

The smaller the Euclidean distance $|E - E_h|$ between the vectors E and E_h, in the $(m+g+1)$-dimensional vector space containing these vectors, the more accurate is the equalization of the channel (see Section 4.11). It is, however, interesting to observe that quite accurate equalization of the channel can be achieved, under the appropriate conditions, even when $|E_0 - \frac{1}{2}E_h|$ and $|E_1 - \frac{1}{2}E_h|$ are *both* quite large. Thus, if

$$e_{0,h} + e_{1,h} \simeq 1 \quad (6.68)$$

and

$$e_{0,j} + e_{1,j} \simeq 0 \quad (6.69)$$

for $j = 0, 1, \ldots, m+g$, and $j \neq h$, then $E \simeq E_h$ as required. Hence it is not, in fact, always necessary for the two individual parallel channels (Y_0 and Y_1) to be equalized accurately, in order to achieve the accurate equalization of the resultant channel. This will now be considered in more detail.

From Equations 6.56, 6.60 and 6.64,

$$E_0 = C_0 Y_{c,0} \quad (6.70)$$

where $Y_{c,0}$ is the $(m+1) \times (m+g+1)$ convolution matrix whose $(i+1)$th

LINEAR AND DECISION-FEEDBACK EQUALIZERS

row is given by the $(m+g+1)$-component vector

$$Y_{0,i} = [\overbrace{0 \ \ldots \ 0}^{i} \ y_{0,0} \ y_{0,1} \ \ldots \ y_{0,g} \ 0 \ \ldots \ 0] \qquad (6.71)$$

Since the vectors $\{Y_{0,i}\}$ forming the $m+1$ rows of $Y_{c,0}$ are *linearly independent*, they span an $(m+1)$-dimensional subspace of the $(m+g+1)$-dimensional vector space containing these vectors. But, from Equations 6.70 and 6.71,

$$E_0 = \sum_{i=0}^{m} c_{0,i} Y_{0,i} \qquad (6.72)$$

Thus E_0 is a *linear combination* of the $m+1$ vectors $\{Y_{0,i}\}$ and so must lie in the $(m+1)$-dimensional subspace spanned by these (see Section 1.3).

Similarly, from Equations 6.57, 6.61 and 6.65,

$$E_1 = C_1 Y_{c,1} \qquad (6.73)$$

where $Y_{c,1}$ is the $(m+1) \times (m+g+1)$ convolution matrix whose $(i+1)$th row is given by the $(m+g+1)$-component vector

$$Y_{1,i} = [\overbrace{0 \ \ldots \ 0}^{i} \ y_{1,0} \ y_{1,1} \ \ldots \ y_{1,g} \ 0 \ \ldots \ 0] \qquad (6.74)$$

Since the vectors $\{Y_{1,i}\}$ forming the $m+1$ rows of $Y_{c,1}$ are *linearly independent*, they span an $(m+1)$-dimensional subspace of the $(m+g+1)$-dimensional vector space containing these vectors. But, from Equations 6.73 and 6.74,

$$E_1 = \sum_{i=0}^{m} c_{1,i} Y_{1,i} \qquad (6.75)$$

Thus E_1 is a linear combination of the $m+1$ vectors $\{Y_{1,i}\}$ and so must lie in the $(m+1)$-dimensional subspace spanned by these.

The two $(m+1)$-dimensional subspaces could, in principle, range from being the *same subspace* (such that any vector in either of the two subspaces must lie also in the other) to being *disjoint* or *linearly independent* subspaces, in the sense that *no* nonzero vector in either subspace lies also in the other. The two subspaces may *share* a lower dimensional subspace (the *intersection space*) *all* of whose vectors lie in both $(m+1)$-dimensional subspaces, the remaining vectors in either subspace not lying in the other. It now follows from Equation 6.66 that the vector E must lie in one or other or both of the two subspaces, and is therefore confined to a subspace given by the *sum* of the two $(m+1)$-dimensional subspaces. The *sum space* containing E is consequently an l-dimensional vector space, where

$$m+1 \leqslant l \leqslant 2(m+1) \qquad (6.76)$$

If now

$$l = 2(m+1)$$

so that the two subspaces are *disjoint*, the most accurate equalization of the resultant channel (such that $E \simeq E_h$) corresponds to a *unique* pair of vectors E_0 and E_1 and normally involves also the reasonably accurate equalization of each of the two constituent parallel channels (such that $E_0 \propto E_1 \propto E_h$). Of course, under these conditions, Equation 6.68 holds and usually neither $e_{0,h}$ nor $e_{1,h}$ approximates to unity. In all other cases, that is, whenever the *intersection space* (for E_0 and E_1) has dimension greater than zero, so that the $(m+1)$-dimensional subspaces containing E_0 and E_1 *share* a common set of vectors, the above requirement no longer holds. Under these conditions, the adjustment of the double-sampling equalizer to minimize the mean-square error due to intersymbol interference alone (in the absence of noise) is *not* unique. There are, in fact, now an infinite number of different combinations of the two filters C_0 and C_1 (Fig. 6.4) that together achieve the accurate equalization of the channel, according to Equations 6.68 and 6.69. However, in the presence of noise, there is a *unique* combination of the two filters C_0 and C_1 that minimize the mean-square error due to intersymbol interference and noise in the equalized signal, although again this need not involve the most accurate equalization of either of the two individual parallel channels.

When the linear baseband channel in Fig. 6.3 is strictly bandlimited and sampled at or above the Nyquist rate, such that the highest frequency component passed by the channel is less than $1/T$ Hz, then g in Equations 6.56 and 6.57 tends to infinity, so that the effective *finite* value of g is, for practical purposes, such that

$$g \gg m \qquad (6.77)$$

Now, the number of components of E are

$$m + g + 1 \gg 2(m+1)$$

which suggests that the two $(m+1)$-dimensional subspaces containing E_0 and E_1 may well be disjoint, such that

$$l = 2(m+1)$$

It has, in fact, been shown that this is so[15]. Practical channels, of course, are *not* strictly bandlimited, but, so long as $g \gg m$, this need not invalidate the above conclusion. On the other hand, when $m \geqslant g$,

$$m + g + 1 < 2(m+1)$$

which implies that the two $(m+1)$-dimensional subspaces containing E_0

LINEAR AND DECISION-FEEDBACK EQUALIZERS

and E_1 *cannot* now be *disjoint*, since there cannot be *more* than $m+g+1$ linearly independent vectors having $m+g+1$ components.

Consider next the general case where the linear equalizer is adjusted to minimize the mean-square error due to *both* intersymbol interference and noise. With a delay in detection of hT seconds, the equalized signal at the detector input (Fig. 6.4), at time $t=(i+h)T$, is

$$x_{i+h} = \sum_{j=0}^{m+g} s_{i+h-j}(e_{0,j}+e_{1,j}) + u_{i+h}$$

$$\simeq s_i + u_{i+h} \tag{6.78}$$

where

$$u_{i+h} = \sum_{j=0}^{m+g} (w_{i+h-j}c_{0,j} + w_{i+h-1/2-j}c_{1,j}) \tag{6.79}$$

Since the noise components $\{w_{i+h-j}\}$ and $\{w_{i+h-1/2-j}\}$ are statistically independent Gaussian random variables with zero mean and variance σ^2, the noise component u_{i+h} is a Gaussian random variable with zero mean and variance

$$\eta^2 = \sum_{j=0}^{m+g} (\sigma^2 c_{0,j}^2 + \sigma^2 c_{1,j}^2) = \sigma^2|C_0|^2 + \sigma^2|C_1|^2 \tag{6.80}$$

The mean-square error in the equalized signal x_{i+h} is

$$\mathscr{E}[(x_{i+h}-s_i)^2] = \mathscr{E}\left[\left(\sum_{j=0}^{m+g} s_{i+h-j}(e_{0,j}+e_{1,j}) + u_{i+h} - s_i\right)^2\right]$$

$$= k^2|E_0 + E_1 - E_h|^2 + \sigma^2(|C_0|^2 + |C_1|^2) \tag{6.81}$$

as is evident from Equations 6.78 and 4.254, where E is now replaced by $E_0 + E_1$ and $|C|^2$ is replaced by $|C_0|^2 + |C_1|^2$ (see Equations 4.185 and 6.80).

Let C be the $2(m+1)$-component row vector whose $2(i+1)$th component is $c_{0,i}$ for $i=0, 1, \ldots, m$, and whose $(2i+1)$th component is $c_{1,i}$ for $i=0, 1, \ldots, m$, so that

$$C = [c_{1,0} \quad c_{0,0} \quad c_{1,1} \quad c_{0,1} \quad \cdots \quad c_{1,m} \quad c_{0,m}] \tag{6.82}$$

Also let Y_s be the $2(m+1) \times (m+g+1)$ matrix whose $2(i+1)$th row is $Y_{0,i}$ (Equation 6.71) for $i=0, 1, \ldots, m$, and whose $(2i+1)$th row is $Y_{1,i}$ (Equation 6.74) for $i=0, 1, \ldots, m$. Thus, from Equations 6.66 and 6.70–6.75, the resultant sampled impulse-response of the channel and double-sampling equalizer (for the arrangement in Fig. 6.4, at the sampling instants $\{iT\}$) is

$$E = E_0 + E_1$$

$$= C_0 Y_{c,0} + C_1 Y_{c,1} = CY_s \tag{6.83}$$

and it is clear from the definition of C that

$$|C_0|^2 + |C_1|^2 = |C|^2 \tag{6.84}$$

and, from Equation 6.81, the mean-square error in the equalized signal x_{i+h} is

$$\mathscr{E}[(x_{i+h} - s_i)^2] = k^2|CY_s - E_h|^2 + \sigma^2|C|^2 \tag{6.85}$$

$k^2|CY_s - E_h|^2$ is here the mean-square error due to *intersymbol interference* and $\sigma^2|C|^2$ is the mean-square error due to *noise*. From the close similarity between this and the first line of Equation 4.257, it follows from Equations 4.257–4.265 that, for the minimum mean-square error in x_{i+h},

$$C = E_h Y_s^T \left(Y_s Y_s^T + \frac{\sigma^2}{k^2} I \right)^{-1} \tag{6.86}$$

where I is the $2(m+1) \times 2(m+1)$ identity matrix, E_h is the $(m+g+1)$-component row vector ($1 \times (m+g+1)$ matrix) given by Equation 6.67,

$$Y_s^T \left(Y_s Y_s^T + \frac{\sigma^2}{k^2} I \right)^{-1} \tag{6.87}$$

is an $(m+g+1) \times 2(m+1)$ matrix, and $Y_s Y_s^T$ is a $2(m+1) \times 2(m+1)$ real symmetric positive-definite or -semidefinite matrix. When $\sigma^2 \neq 0$, the $2(m+1) \times 2(m+1)$ matrix

$$Y_s Y_s^T + \frac{\sigma^2}{k^2} I \tag{6.88}$$

is a real symmetric positive-definite matrix, as can be seen from Expression 5.202. The matrix is therefore nonsingular and Equation 6.86 now holds, giving a unique vector C that minimizes the mean-square error in the equalized signal, for the assumed delay in detection of hT seconds. Clearly, C is given by the $(h+1)$th row of the matrix in Expression 6.87. Again, from Equation 4.266, the resulting value of the mean-square error in the equalized signal is

$$k^2 \left[1 - E_h Y_s^T \left(Y_s Y_s^T + \frac{\sigma^2}{k^2} I \right)^{-1} Y_s E_h^T \right] \tag{6.89}$$

As before, the delay in detection hT can be chosen to minimize the mean-square error given by Expression 6.89.

In the absence of noise (such that $\sigma^2 = 0$) and when the matrix $Y_s Y_s^T$ is *nonsingular*, such as is likely to be the case when $g \gg m$, the $2(m+1)$ tap gains of the double-sampling linear equalizer, that minimizes the mean-square error due to intersymbol-interference alone, are given uniquely by the vector

$$C = E_h Y_s^T (Y_s Y_s^T)^{-1} \tag{6.90}$$

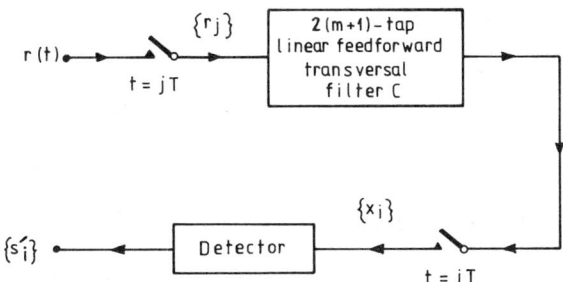

Fig. 6.5 Practical implementation of a double-sampling linear equalizer

This corresponds to the single-sampling linear equalizer given by Equation 4.274.

From the definition of the vector C, it is evident that this gives the tap gains of a double-sampling $2(m+1)$-tap linear feedforward transversal filter, fed with the received samples $r_i, r_{i+1/2}, r_{i+1}, r_{i+3/2}, \ldots$, in sequence and at intervals of $\frac{1}{2}T$, the filter equalizing the channel and introducing a delay of hT seconds. Thus the practical implementation of the double-sampling equalizer is as shown in Fig. 6.5. The symbol i here takes on all integer values, whereas the symbol j takes on the values of all integer multiples of $\frac{1}{2}$. Thus, *two* received samples $r_{i-1/2}$ and r_i are fed sequentially to the filter for every *one* output signal fed to the detector, which means that only every *second* output signal from the equalizer is fed to the detector, the other output signals being ignored. In a digital implementation of the equalizer, of course, only the required output signals are, in fact, generated.

It can now be seen from Figs. 6.1 and 6.5 together with Section 6.1, that the double-sampling equalizer can be adjusted *adaptively* in a similar manner to that described for the single-sampling linear equalizer in Section 6.1. The adjustment of the tap gains, of course, again takes place *once* for each received data-symbol s_i, and therefore at intervals of T. The performance of such systems is described elsewhere[11-15].

6.2.4 Decision-feedback equalizer that achieves the exact equalization of the channel

The decision-feedback equalizer considered here is, in practice, likely to be optimum or somewhere close to optimum, where the *optimum* equalizer is taken as that which maximizes the signal/noise ratio in the equalized signal, subject to the *exact* equalization of the channel and given the *correct* detection of the preceding g data-symbols. A further development of the equalizer, considered in Section 7.5, leads to a very simple and potentially cost-effective system.

346 LINEAR AND DECISION-FEEDBACK EQUALIZERS

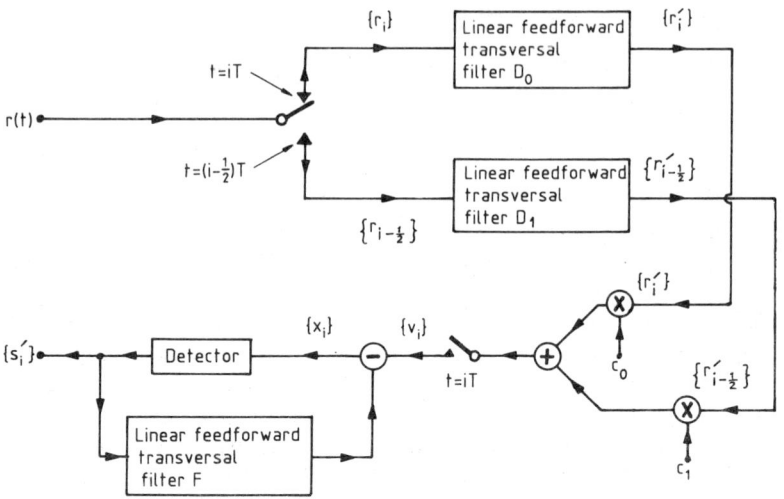

Fig. 6.6 Double-sampling decision-feedback equalizer

Consider the double-sampling decision-feedback equalizer shown in Fig. 6.6. The received signal is the real-valued waveform $r(t)$ given by Equation 6.49. The waveform $r(t)$ is sampled *twice* per data-symbol s_i, at the time instants $\{iT\}$ and $\{(i-\tfrac{1}{2})T\}$, to give the samples $\{r_i\}$ and $\{r_{i-1/2}\}$, respectively, as in Equations 6.50 and 6.51. The symbol i here takes on all integer values, and the noise components $\{w_i\}$ and $\{w_{i-1/2}\}$ are statistically independent Gaussian random variables with zero mean and variance $\sigma^2 = \tfrac{1}{2}N_0$. The sampled impulse-responses of the linear baseband channel in Fig. 6.3, at the time instants $\{iT\}$ and $\{(i-\tfrac{1}{2})T\}$, are given respectively by Y_0 and Y_1 in Equations 6.56 and 6.57. Thus Equations 6.50–6.59 all hold, as before, all signals being *real-valued*. The linear equalizer in Fig. 6.3 is however now replaced by the decision-feedback equalizer of Fig. 6.6.

The linear feedforward transversal filters D_0 and D_1 each have $n+1$ taps where n is assumed to be an appropriately large positive integer. The sampled impulse-response of the filter D_0 is given by the $(n+1)$-component vector (sequence)

$$D_0 = [d_{0,0} \quad d_{0,1} \quad \ldots \quad d_{0,n}] \qquad (6.91)$$

with z-transform

$$D_0(z) = d_{0,0} + d_{0,1}z^{-1} + \cdots + d_{0,n}z^{-n} \qquad (6.92)$$

and the sampled impulse-response of the filter D_1 is given by the $(n+1)$-

component vector
$$D_1 = [d_{1,0} \quad d_{1,1} \quad \cdots \quad d_{1,n}] \tag{6.93}$$
with z-transform
$$D_1(z) = d_{1,0} + d_{1,1}z^{-1} + \cdots + d_{1,n}z^{-n} \tag{6.94}$$
As before, z^{-1} corresponds to a delay of T seconds.

Suppose next that the z-transforms of Y_0 and Y_1 are
$$Y_0(z) = Y_{0,1}(z)Y_{0,2}(z) \tag{6.95}$$
and
$$Y_1(z) = Y_{1,1}(z)Y_{1,2}(z) \tag{6.96}$$
where all zeros (roots) of $Y_{0,1}(z)$ and $Y_{1,1}(z)$ lie *inside* or *on* the unit circle in the z plane and all zeros of $Y_{0,2}(z)$ and $Y_{1,2}(z)$ lie *outside* the unit circle. The delays of $y_{0,0}$ and $y_{1,0}$ are again ignored, as in Equations 6.58 and 6.59. This involves introducing a delay of $\frac{1}{2}T$ into the sequence Y_1 relative to the sequence Y_0, so that their respective ith components are coincident in time, for any i. Now let $Y_{0,3}(z)$ and $Y_{1,3}(z)$ be formed from $Y_{0,2}(z)$ and $Y_{1,2}(z)$ respectively, by reversing the order of the coefficients. This means, of course, that the zeros of $Y_{0,3}(z)$ and $Y_{1,3}(z)$ are the *reciprocals* of those of $Y_{0,2}(z)$ and $Y_{1,2}(z)$, respectively, and so lie *inside* the unit circle. As before, any zeros or poles at the origin or infinity in the z plane are ignored.

The transversal filters D_0 and D_1 are adjusted according to Equation 5.299, but with the scaling factor q_f^{-1} omitted, such that their z-transforms are
$$D_0(z) \simeq z^{-n} Y_{0,2}^{-1}(z) Y_{0,3}(z) \tag{6.97}$$
and
$$D_1(z) \simeq z^{-n} Y_{1,2}^{-1}(z) Y_{1,3}(z) \tag{6.98}$$
respectively. It is assumed here that n is sufficiently large for approximate equality to hold in Equations 6.97 and 6.98. Clearly, the z-transform of the channel and filter D_0 is $Y_0(z)D_0(z)$, and the z-transform of the channel and filter D_1 is $Y_1(z)D_1(z)$. Hence the sampled impulse-response of the channel and filter D_0 is given approximately by the $(g+1)$-component vector (sequence)
$$E_0 = [e_{0,0} \quad e_{0,1} \quad \cdots \quad e_{0,g}] \tag{6.99}$$
with z-transform
$$E_0(z) = z^{-n} Y_{0,1}(z) Y_{0,3}(z) \tag{6.100}$$
and the sampled impulse-response of the channel and filter D_1 is given approximately by the $(g+1)$-component vector
$$E_1 = [e_{1,0} \quad e_{1,1} \quad \cdots \quad e_{1,g}] \tag{6.101}$$

with z-transform

$$E_1(z) = z^{-n} Y_{1,1}(z) Y_{1,3}(z) \quad (6.102)$$

The first component of each sampled impulse-response (which is *not* usually equal to unity here) is delayed by n sampling intervals of T in the filter, the preceding n components having all been set to zero. It can be seen from Equations 6.91–6.102, that, for each case, in the z-transform of the channel and filter, all zeros of the z-transform of the channel that lie outside the unit circle are replaced by their reciprocals, the remaining zeros being left unchanged. Thus all zeros of the z-transform of the channel and filter lie *inside* or *on* the unit circle (see Equations 6.100 and 6.102). Furthermore, since the z-transform of each filter represents *pure phase distortion* (see Section 3.3), the filters do not change any *amplitude distortion* in the received signal, nor do they change the signal or noise level. Hence, each of the filters D_0 and D_1 is an *allpass* network that adjusts the sampled impulse-response of the channel and filter to be *minimum phase* (neglecting the delay of n sampling intervals of T). Each filter therefore introduces an *orthogonal* transformation into the received signal and so does not change the inner products, lengths, distances or angles between the different possible received signal vectors, corresponding to the different possible received messages. Furthermore, the filters do not change the signal/noise ratio nor any of the noise statistics. Perhaps the most important function performed by each filter is that it *maximizes* the ratio of the magnitude of the *first* nonzero component ($e_{0,0}$, or $e_{1,0}$) of the sampled impulse-response of the channel and filter to the noise variance at the output of the filter (see Section 5.8.2).

The signals at the outputs of the filters D_0 and D_1, at time $t = (i+n)T$, are

$$r'_{i+n} \simeq \sum_{j=0}^{g} s_{i-j} e_{0,j} + w'_{i+n} \quad (6.103)$$

and

$$r'_{i+n-1/2} \simeq \sum_{j=0}^{g} s_{i-j} e_{1,j} + w'_{i+n-1/2} \quad (6.104)$$

respectively, and the noise components w'_{i+n} and $w'_{i+n-1/2}$ are statistically independent Gaussian random variables with zero mean and variance σ^2. Strictly speaking, these conditions only hold *exactly* when $n \to \infty$, but it is assumed that n is sufficiently large to render negligible any errors resulting from the finite value of n.

The two multipliers and adder in Fig. 6.6 form the signal

$$v_{i+n} = c_0 r'_{i+n} + c_1 r'_{i+n-1/2} \quad (6.105)$$

at the input to the subtractor. Now, from Equations 6.103–6.105,

LINEAR AND DECISION-FEEDBACK EQUALIZERS

$$v_{i+n} \simeq s_i(c_0 e_{0,0} + c_1 e_{1,0}) + \sum_{j=1}^{g} s_{i-j}(c_0 e_{0,j} + c_1 e_{1,j})$$

$$+ w'_{i+n} c_0 + w'_{i+n-1/2} c_1 \qquad (6.106)$$

The linear feedforward transversal filter F (Fig. 6.6) has g taps, with gains $f_1, f_2, \ldots, f_g$, where

$$f_j = c_0 e_{0,j} + c_1 e_{1,j} \qquad (6.107)$$

for $j = 1, 2, \ldots, g$, the z-transform of the filter being

$$F(z) = f_1 z^{-1} + f_2 z^{-2} + \cdots + f_g z^{-g} \qquad (6.108)$$

Thus, with the correct detection of the data-symbols $s_{i-1}, s_{i-2}, \ldots, s_{i-g}$, the signal at the output of the filter F, at time $t = (i+n)T$, is

$$\sum_{j=1}^{g} s_{i-j} f_j = \sum_{j=1}^{g} s_{i-j}(c_0 e_{0,j} + c_1 e_{1,j}) \qquad (6.109)$$

and the corresponding signal at the input to the detector is the equalized signal

$$x_{i+n} = v_{i+n} - \sum_{j=1}^{g} s_{i-j}(c_0 e_{0,j} + c_1 e_{1,j})$$

$$\simeq s_i(c_0 e_{0,0} + c_1 e_{1,0}) + w'_{i+n} c_0 + w'_{i+n-1/2} c_1 \qquad (6.110)$$

With the correct adjustment of the equalizer, such that the first nonzero component of the *resultant* sampled impulse-response of the channel and two filters D_0 and D_1 has the value unity,

$$c_0 e_{0,0} + c_1 e_{1,0} = 1 \qquad (6.111)$$

so that

$$c_1 = \frac{1 - c_0 e_{0,0}}{e_{1,0}} \qquad (6.112)$$

To maximize the signal/noise ratio in the equalized signal, subject to exact equalization and given Equation 6.111, it is necessary to minimize the variance of the noise component $w'_{i+n} c_0 + w'_{i+n-1/2} c_1$ in the equalized signal x_{i+n}. Since w'_{i+n} and $w'_{i+n-1/2}$ are statistically independent Gaussian random variables with zero mean and variance σ^2, $w'_{i+n} c_0 + w'_{i+n-1/2} c_1$ is a Gaussian random variable with zero mean and variance

$$\eta^2 = \sigma^2(c_0^2 + c_1^2) \qquad (6.113)$$

From Equations 6.112 and 6.113,

$$\frac{\eta^2}{\sigma^2} = c_0^2 + \frac{(1-c_0 e_{0,0})^2}{e_{1,0}^2}$$

$$= c_0^2 + \frac{1}{e_{1,0}^2}(c_0^2 e_{0,0}^2 - 2c_0 e_{0,0} + 1)$$

$$= \left(1 + \frac{e_{0,0}^2}{e_{1,0}^2}\right)c_0^2 - 2\frac{e_{0,0}}{e_{1,0}^2}c_0 + \frac{1}{e_{1,0}^2} \quad (6.114)$$

η^2/σ^2 is a *convex* function of c_0 with a global minimum at

$$\frac{\partial}{\partial c_0}\left(\frac{\eta^2}{\sigma^2}\right) = 0 \quad (6.115)$$

Thus to minimize η^2/σ^2, and hence to minimize η^2, c_0 must be adjusted such that

$$\frac{\partial}{\partial c_0}\left(\frac{\eta^2}{\sigma^2}\right) = 2\left(1 + \frac{e_{0,0}^2}{e_{1,0}^2}\right)c_0 - 2\frac{e_{0,0}}{e_{1,0}^2} = 0 \quad (6.116)$$

or

$$\left(1 + \frac{e_{0,0}^2}{e_{1,0}^2}\right)c_0 = \frac{e_{0,0}}{e_{1,0}^2} \quad (6.117)$$

or

$$(e_{0,0}^2 + e_{1,0}^2)c_0 = e_{0,0} \quad (6.118)$$

or

$$c_0 = \frac{e_{0,0}}{e_{0,0}^2 + e_{1,0}^2} \quad (6.119)$$

and, from Equation 6.112,

$$c_1 = \frac{e_{1,0}}{e_{0,0}^2 + e_{1,0}^2} \quad (6.120)$$

Now, from Equations 6.110 and 6.111, the equalized signal, at time $t = (i+n)T$, becomes

$$x_{i+n} \simeq s_i + \frac{e_{0,0}w'_{i+n} + e_{1,0}w'_{i+n-1/2}}{e_{0,0}^2 + e_{1,0}^2} \quad (6.121)$$

But the noise components w'_{i+n} and $w'_{i+n-1/2}$ are *statistically orthogonal*[16-18] such that

$$\mathscr{E}[w'_{i+n}w'_{i+n-1/2}] = 0 \quad (6.122)$$

and

$$\mathscr{E}[(w'_{i+n})^2] = \mathscr{E}[(w'_{i+n-1/2})^2] = \sigma^2 \quad (6.123)$$

Thus the mean-square error in x_{i+n} is

$$\mathcal{E}[(x_{i+n}-s_i)^2] = \mathcal{E}\left[\left(\frac{e_{0,0}w'_{i+n}+e_{1,0}w'_{i+n-1/2}}{e_{0,0}^2+e_{1,0}^2}\right)^2\right]$$

$$= \mathcal{E}\left[\frac{e_{0,0}^2(w'_{i+n})^2+e_{1,0}^2(w'_{i+n-1/2})^2}{(e_{0,0}^2+e_{1,0}^2)^2}\right]$$

$$= \frac{\sigma^2(e_{0,0}^2+e_{1,0}^2)}{(e_{0,0}^2+e_{1,0}^2)^2} = \frac{\sigma^2}{e_{0,0}^2+e_{1,0}^2} \quad (6.124)$$

which is also the *variance* η^2 of the resultant zero-mean Gaussian noise component

$$(e_{0,0}w'_{i+n}+e_{1,0}w'_{i+n-1/2})/(e_{0,0}^2+e_{1,0}^2)$$

in Equation 6.121.

The detector in Fig. 6.6 detects s_i as k or $-k$ depending upon whether x_{i+n} is positive or negative, respectively. The probability of error in the detection of s_i, at high signal/noise ratios, can now be taken to be

$$Q\left(\frac{k}{\eta}\right) = Q\left(\frac{k\sqrt{e_{0,0}^2+e_{1,0}^2}}{\sigma}\right) \quad (6.125)$$

when neglecting error-extension effects.

6.2.5 Decision-feedback equalizer that minimizes the mean-square error in the equalized signal

The equalizer here minimizes the mean-square error due to both *intersymbol interference* and *noise* in the equalized signal. The equalizer is as shown in Fig. 6.6, where now

$$c_0 = c_1 = 1 \quad (6.126)$$

so that the corresponding two multipliers are, in fact, omitted. Equations 6.50–6.59 and 6.91–6.94 all hold, as before, together with the statistical properties of the noise components $\{w_i\}$ and $\{w_{i-1/2}\}$.

The sampled impulse-response of the channel and filter D_0 is given by the $(n+g+1)$-component row vector

$$E_0 = [e_{0,0} \quad e_{0,1} \quad \cdots \quad e_{0,n+g}] \quad (6.127)$$

with z-transform

$$E_0(z) = e_{0,0} + e_{0,1}z^{-1} + \cdots + e_{0,n+g}z^{-n-g} \quad (6.128)$$

and, similarly, the sampled impulse-response of the channel and filter D_1 is given by the $(n+g+1)$-component row vector

$$E_1 = [e_{1,0} \quad e_{1,1} \quad \cdots \quad e_{1,n+g}] \quad (6.129)$$

with z-transform

$$E_1(z) = e_{1,0} + e_{1,1}z^{-1} + \cdots + e_{1,n+g}z^{-n-g} \tag{6.130}$$

E_0 is the *convolution* of Y_0 and D_0 (Equations 6.56 and 6.91), and E_1 is the *convolution* of Y_1 and D_1 (Equations 6.57 and 6.93). As before, the received samples $\{r_{i-1/2}\}$ are assumed to be delayed by $\frac{1}{2}T$ so that, for each i, $r_{i-1/2}$ coincides in time with r_i. Furthermore, the delay in transmission, other than that involved in the time dispersion of the received signal, is neglected, so that the components $e_{0,0}$ and $e_{1,0}$ of E_0 and E_1, respectively, are taken to arrive with *no* delay.

The *resultant* sampled impulse-response of the linear baseband channel and linear filters D_0 and D_1, at the sampling instants $\{iT\}$, is given by the $(n+g+1)$-component row vector

$$E = E_0 + E_1 = [e_0 \quad e_1 \quad \cdots \quad e_{n+g}] \tag{6.131}$$

The double-sampling decision-feedback equalizer is assumed to introduce a delay in detection of hT seconds so that the data-symbol s_i is detected from the equalized signal x_{i+h}, at time $t = (i+h)T$. The integer h may be equal to or greater than n, as before (see Section 5.7), although again its most suitable value is likely to be $h = n$.

The linear feedforward transversal filter F in Fig. 6.6 has μ taps whose gains are given by the μ components of the vector

$$F = [f_1 \quad f_2 \quad \cdots \quad f_\mu] \tag{6.132}$$

with z-transform

$$F(z) = f_1 z^{-1} + f_2 z^{-2} + \cdots + f_\mu z^{-\mu} \tag{6.133}$$

As before (from Equation 5.178)

$$h + \mu = n + g \tag{6.134}$$

The ideal or optimum adjustment of the filter F is such that

$$F = [e_{h+1} \quad e_{h+2} \quad \cdots \quad e_{h+\mu}] \tag{6.135}$$

and, with the *correct detection* of $s_{i-1}, s_{i-2}, \ldots, s_{i-\mu}$, the output signal from the filter F, at time $t = (i+h)T$, is

$$\sum_{j=1}^{\mu} s_{i-j} e_{h+j}$$

The corresponding signal at the output of the sampler, following the adder in Fig. 6.6 and preceding the subtractor, is

$$v_{i+h} = \sum_{j=0}^{n+g} s_{i+h-j} e_j + u_{i+h} \tag{6.136}$$

where
$$u_{i+h} = \sum_{j=0}^{n} (w_{i+h-j} d_{0,j} + w_{i+h-1/2-j} d_{1,j}) \quad (6.137)$$

The output signal from the filter F is subtracted from v_{i+h} to give, at the detector input, the equalized signal

$$x_{i+h} = \sum_{j=0}^{h} s_{i+h-j} e_j + u_{i+h} \quad (6.138)$$

With a sufficient number of taps in the equalizer and at high signal/noise ratios, such that $e_j \simeq 0$, for $j = 0, 1, \ldots, h-1$, and with the correct detection of $s_{i-1}, s_{i-2}, \ldots, s_{i-\mu}$, such that Equation 6.138 holds, the equalized signal becomes

$$x_{i+h} \simeq s_i + u_{i+h} \quad (6.139)$$

Let D be the $2(n+1)$-component row vector whose $2(i+1)$th component is $d_{0,i}$, for $i = 0, 1, \ldots, n$, and whose $(2i+1)$th component is $d_{1,i}$, for $i = 0, 1, \ldots, n$, so that

$$D = [d_{1,0} \quad d_{0,0} \quad d_{1,1} \quad d_{0,1} \quad \ldots \quad d_{1,n} \quad d_{0,n}] \quad (6.140)$$

Also let Y_s be the $2(n+1) \times (n+g+1)$ convolution matrix whose $2(i+1)$th row is $Y_{0,i}$ (Equation 6.71) for $i = 0, 1, \ldots, n$, and whose $(2i+1)$th row is $Y_{1,i}$ (Equation 6.74) for $i = 0, 1, \ldots, n$. Then, from Equation 6.131, the *resultant* sampled impulse-response of the channel and two filters D_0 and D_1, at the sampling instants $\{iT\}$, is

$$E = DY_s \quad (6.141)$$

and it is clear from the definition of D that

$$|D_0|^2 + |D_1|^2 = |D|^2 \quad (6.142)$$

The detailed derivation of Equation 6.141 follows along exactly the same lines as that of Equation 6.83.

From the analysis in Equations 5.179–5.192, it can be seen that, with the ideal adjustment of the filter F (Equation 6.135) and with the *correct detection* of the data-symbols $s_{i-1}, s_{i-2}, \ldots, s_{i-\mu}$ (leading to Equation 6.139), the mean-square error in the equalized signal x_{i+h} is

$$\mathcal{E}[(x_{i+h} - s_i)^2] = k^2 |B|^2 + \sigma^2 |D|^2 \quad (6.143)$$

where B is the $(n+g+1)$-component row vector

$$B = [e_0 \quad e_1 \quad \ldots \quad e_{h-1} \quad (e_h - 1) \quad \overbrace{0 \quad 0 \quad \ldots \quad 0}^{\mu}] \quad (6.144)$$

Let Z_s be the $2(n+1) \times (n+g+1)$ matrix obtained from Y_s by setting its last μ columns to zero, the remaining columns being left unchanged.

Then, as in Equation 5.194,

$$B = DZ_s - E_h \tag{6.145}$$

where E_h is the $(n+g+1)$-component row vector

$$E_h = [\overbrace{0 \ 0 \ \ldots \ 0 \ 1}^{h} \ 0 \ 0 \ \ldots \ 0] \tag{6.146}$$

as in Equation 5.181. Thus the mean-square error in the equalized signal is

$$\mathscr{E}[(x_{i+h} - s_i)^2] = k^2 |DZ_s - E_h|^2 + \sigma^2 |D|^2 \tag{6.147}$$

It now follows from the analysis in Equations 5.195–5.200 that the $2(n+1)$ tap gains of the filters D_0 and D_1 in Fig. 6.6, that minimize the mean-square error in the equalized signal, are given by the corresponding $2(n+1)$ components of the vector

$$D = E_h Z_s^T \left(Z_s Z_s^T + \frac{\sigma^2}{k^2} I \right)^{-1} \tag{6.148}$$

where $Z_s Z_s^T$ and $Z_s Z_s^T + (\sigma^2/k^2)I$ are $2(m+1) \times 2(m+1)$ real symmetric matrices and I is the $2(m+1) \times 2(m+1)$ identity matrix. Also, the mean-square error in the equalized signal is now

$$k^2 \left[1 - E_h Z_s^T \left(Z_s Z_s^T + \frac{\sigma^2}{k^2} I \right)^{-1} Z_s E_h^T \right]$$

The *correct detection* of the data-symbols $s_{i-1}, s_{i-2}, \ldots, s_{i-\mu}$ is, of course, assumed here, and the tap gains of the filter F in Fig. 6.6 must satisfy Equation 6.135. The value of h can now be selected to minimize the mean-square error. It seems likely that, in general, the most suitable value of h is given by $h = n$, as for a single-sampling equalizer.

It can be seen from Equation 6.148 that D is given by the $(h+1)$th row of the $(n+g+1) \times 2(n+1)$ matrix

$$Z_s^T \left(Z_s Z_s^T + \frac{\sigma^2}{k^2} I \right)^{-1} \tag{6.149}$$

which is the *last* nonzero row, the remaining μ rows being all zero. In the presence of noise, when $\sigma^2 \neq 0$, the $2(n+1) \times 2(n+1)$ matrix

$$Z_s Z_s^T + \frac{\sigma^2}{k^2} I \tag{6.150}$$

is real, symmetric and positive definite, so that it is *nonsingular*. Equation 6.148 now holds and gives a *unique* vector D that minimizes the mean-square error in the equalized signal. However, when $\sigma^2 = 0$, the matrix (Expression 6.150) becomes $Z_s Z_s^T$ which is probably positive semi-

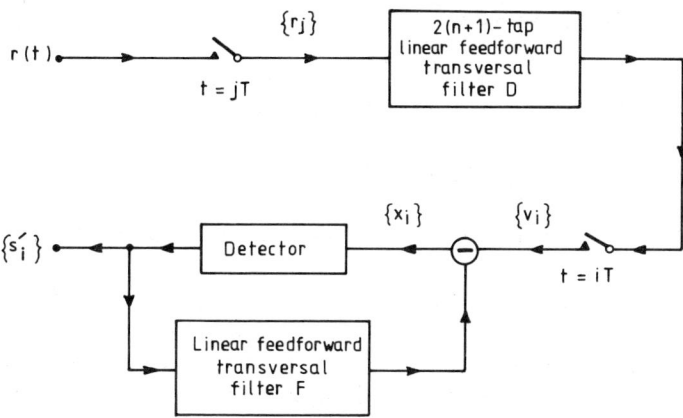

Fig. 6.7 Practical implementation of a double-sampling decision-feedback equalizer

definite and so is likely to be *singular*. Under these conditions, Equation 6.148 does not hold and there is no longer a unique vector D that minimizes the mean-square error.

From the definition of D, it is evident that this gives the tap gains of a double-sampling $2(n+1)$-tap linear feedforward transversal filter, fed with the received samples r_i, $r_{i+1/2}$, r_{i+1}, $r_{i+3/2}$, ..., in sequence and at intervals of $\frac{1}{2}T$ seconds. Thus the practical implementation of the double-sampling equalizer is as shown in Fig. 6.7. The symbol i here takes on all integer values, whereas the symbol j takes on the values of all integer multiples of $\frac{1}{2}$. Thus *two* received samples $r_{i-1/2}$ and r_i are fed sequentially to the filter for every *one* output signal fed to the subtractor, the other output signals being ignored.

It can be seen from Figs. 6.1, 6.2 and 6.7, together with Section 6.1, that the double-sampling equalizer can be adjusted *adaptively* in a similar manner to that described for the single-sampling decision-feedback equalizer in Section 6.1. The adjustment of the tap gains, of course, again takes place *once* for each received data-symbol s_i, and therefore at intervals of T seconds. It seems likely, however, that under unfavourable conditions, there may well be somewhat greater problems with the correct adjustment of the tap gains here than with the corresponding double-sampling linear equalizer.

As has previously been stated, a much greater degradation in tolerance to noise can result from the use of a non-optimum sampling phase in the case of a linear equalizer than in the case of a decision-feedback equalizer, when the received signal is sampled at *below* the Nyquist rate. For this reason, the technique of double-sampling is perhaps of most importance in its application to *linear* equalizers. When the received signal is sampled at or above the Nyquist rate, the performances of both linear and decision-feedback equalizers, that are adjusted to minimize the mean-square error in the equalized signal and have a sufficient number of taps, are *unaffected* by the sampling phase[19,20]. Thus, a useful advantage in tolerance to noise at a suboptimum timing phase is only likely to be achieved through the use of double-sampling in a decision-feedback equalizer, when the transmitted signal-element rate is significantly below the Nyquist rate for the given channel.

6.3 EQUALIZATION PROCESS SHARED BETWEEN TRANSMITTER AND RECEIVER

6.3.1 Data transmission over a time-invariant channel

In all the systems considered so far, the equalization of the channel is carried out at the receiver. This is generally the most cost-effective arrangement in those applications where the channel characteristics vary with time, since the information needed to hold the equalizer correctly set for the channel is derived from the received signal. If all or part of the equalizer is located at the transmitter, the information needed to hold the equalizer correctly adjusted must now be fed back from the receiver to the transmitter, thus introducing additional equipment complexity. On the other hand, with a time-invariant channel whose sampled impulse-response is *known*, both the transmitter and receiver have at all times the necessary prior knowledge of the channel characteristics, so that no additional equipment complexity need be involved in sharing the equalization process between the transmitter and receiver, and such an arrangement may in some cases be more cost-effective than that where the whole of the channel equalization is carried out at the receiver[21-28]. The typical magnitude of the improvement obtainable in this way was first evaluated by M.N.Y. Shum.

A *known time-invariant* baseband channel is assumed throughout the following discussion, and the data-transmission system is now as shown in Fig. 6.8. This data-transmission system is the same as that in Fig. 4.1 or 5.1, except that there are now *two* equalizers, one at the transmitter and the other at the receiver. Various combinations of transmitter and receiver equalizers will be considered, including of course the important case where all the equalization of the channel is carried out at the transmitter. The transmitter and receiver equalizers together achieve the accurate equalization of the channel, and a high signal/noise ratio is assumed, as before.

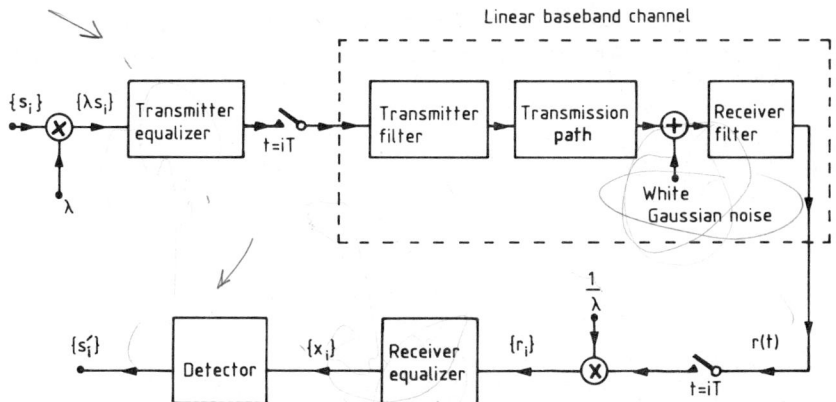

Fig. 6.8 Data-transmission system with the equalization of the channel shared between the transmitter and receiver

6.3.2 Linear equalization

Each binary data-symbol s_i (Equation 4.1) at the input to the transmitter in Fig. 6.8 is multiplied by the positive real-valued scalar quantity λ, before feeding it to the transmitter equalizer. The sequence of values at the output of the equalizer are sampled once for each data-symbol s_i, at the time instants $\{iT\}$, to give the corresponding sequence of impulses which are fed to the baseband channel.

The transmitter and receiver equalizers are linear feedforward transversal filters whose sampled impulse-responses are given by the n-component row vectors (sequences)

$$A = [a_0 \quad a_1 \quad \ldots \quad a_l \quad 0 \quad 0 \quad \ldots \quad 0] \tag{6.151}$$

and
$$B = [b_0 \quad b_1 \quad \ldots \quad b_l \quad 0 \quad 0 \quad \ldots \quad 0] \tag{6.152}$$

respectively, with z-transforms

$$A(z) = a_0 + a_1 z^{-1} + \cdots + a_l z^{-l} \tag{6.153}$$

and
$$B(z) = b_0 + b_1 z^{-1} + \cdots + b_l z^{-l} \tag{6.154}$$

where
$$n \geqslant 2l + 1 \tag{6.155}$$

The $\{a_i\}$ and $\{b_i\}$ are all *real valued* but they need not all be nonzero. In every case,

$$A(z)B(z) \simeq z^{-\theta} C(z) \tag{6.156}$$

where θ is an appropriate nonnegative integer and

$$C(z) = c_0 + c_1 z^{-1} + \cdots + c_m z^{-m} \tag{6.157}$$

is the z-transform of the $(m+1)$-tap linear equalizer for the channel, that minimizes the mean-square error due to intersymbol interference in the equalized signal. The integer m is, for convenience, taken to be even. As before, the z-transform of the baseband channel is

$$Y(z) = y_0 + y_1 z^{-1} + \cdots + y_g z^{-g} \qquad (6.158)$$

and the z-transform of the channel and single linear equalizer is

$$Y(z)C(z) \simeq z^{-h} \qquad (6.159)$$

where h is an appropriate positive integer in the range 0 to $m+g$. It is assumed, as before, that $Y(z)$ has *no* zeros (roots) *on* the unit circle in the z plane.

When $A(z)$ and $B(z)$ are simple factors of $C(z)$, Equation 6.156 holds exactly and $\theta = 0$. Now, in Equations 6.153 and 6.154, $l \geqslant \frac{1}{2}m$. When $A(z)$ and $B(z)$ are *not* simple factors of $C(z)$, then $\theta \geqslant 0$, $l \gg \frac{1}{2}m$, and Equation 6.156 holds only approximately, the approximation however improving as l increases.

The transfer function of the transmitter filter in Fig. 6.8 is assumed to be such that a unit impulse at its input gives a waveform of unit energy at its output, and such that *orthogonal* output waveforms are obtained from any two input impulses separated by a multiple of T seconds. These conditions are satisfied by the transmitter filter assumed for the data-transmission system in Fig. 4.1 and described in Section 1.2. The $(i+1)$th transmitted signal-element, at the input to the baseband channel in Fig. 6.8, is the sequence of $l+1$ impulses,

$$\lambda s_i \sum_{j=0}^{l} a_j \delta[t - (i+j)T] \qquad (6.160)$$

and the energy of this individual signal-element, at the input to the transmission path, is

$$\lambda^2 s_i^2 \sum_{j=0}^{l} a_j^2 = \lambda^2 k^2 |A|^2 \qquad (6.161)$$

since $s_i = \pm k$. $|A|$ is here the *Euclidean length* of the vector A. Since the data-symbols $\{s_i\}$ are statistically independent and have zero mean, they are *statistically orthogonal*, so that the expected energy of a sequence of n signal-elements, at the input to the transmission path, is equal to the sum of the expected energies of the individual elements[16-18]. Thus the average transmitted energy per signal element, at the input to the transmission path in Fig. 6.8, is $\lambda^2 k^2 |A|^2$. The value of λ is selected to set the average transmitted energy per signal element to k^2, as in Fig. 4.1 or 5.1, so that

$$\lambda = \frac{1}{|A|} \qquad (6.162)$$

The receiver filter in Fig. 6.8 has all the properties assumed for this filter in the data-transmission system of Fig. 4.1 or 5.1 and described in Section 1.2. Thus the noise components in the samples of the received waveform $r(t)$, at time instants $\{iT\}$, are statistically independent Gaussian random variables with zero mean and variance σ^2.

After sampling at the receiver, the received samples are multiplied by $1/\lambda$ to give the sequence of samples $\{r_i\}$ that are fed to the receiver equalizer.

Since, in Fig. 6.8 the signal is sampled at the time instants $\{iT\}$, both between the transmitter equalizer and baseband channel and between the baseband channel and receiver equalizer, the z-transform of the transmitter equalizer, baseband channel and receiver equalizer, in cascade, is

$$A(z)Y(z)B(z) \simeq z^{-h-\theta} \tag{6.163}$$

as can be seen from Equations 6.156 and 6.159. Thus the z-transform of the $(i+1)$th received signal-element, at the output of the receiver equalizer, is

$$s_i z^{-i} \lambda A(z) Y(z) \frac{1}{\lambda} B(z) \simeq s_i z^{-i-h-\theta} \tag{6.164}$$

From Equation 6.159, this is the same as that obtained with the corresponding single linear equalizer at the receiver, in the data-transmission system of Fig. 4.1, except for the additional delay of θT seconds.

6.3.3 Optimum combination of transmitter and receiver linear-equalizers

From Equation 6.164, the signal at the output of the receiver equalizer, at time $t = (i+h+\theta)T$, is

$$x_{i+h+\theta} \simeq s_i + u_{i+h+\theta} \tag{6.165}$$

where $u_{i+h+\theta}$ is a Gaussian random variable with zero mean and variance

$$\eta^2 = \frac{\sigma^2}{\lambda^2} \sum_{i=0}^{l} b_i^2 = \frac{\sigma^2}{\lambda^2} |B|^2 = \sigma^2 |A|^2 |B|^2 \tag{6.166}$$

as can be seen from Equations 6.152 and 6.162, bearing in mind that the noise components $\{w_i\}$, in the samples $\{r_i\}$ at the input to the receiver equalizer, are statistically independent Gaussian random variables with zero mean and variance σ^2/λ^2.

The detector detects s_i from the sign of $x_{i+h+\theta}$, so that, from Equations

6.165 and 6.166, the probability of error in the detection of s_i is

$$Q\left(\frac{k}{\eta}\right) = Q\left(\frac{k}{\sigma|A||B|}\right) \qquad (6.167)$$

bearing in mind that $s_i = \pm k$. Clearly, to minimize the probability of error in a detection process, it is necessary to minimize $|A||B|$.

Let the n-component discrete Fourier transforms (DFT's) of A and B be

$$E = [e_0 \quad e_1 \quad \ldots \quad e_{n-1}] \qquad (6.168)$$

and
$$F = [f_0 \quad f_1 \quad \ldots \quad f_{n-1}] \qquad (6.169)$$

respectively, where $n \geq 2l + 1$. The components of E and F are in general complex valued, whereas the components of A, B and C are always real valued. Furthermore, from Section 2.13,

$$|A|^2 = \frac{1}{n}\sum_{i=0}^{n-1}|e_i|^2 = \frac{1}{n}|E|^2 \qquad (6.170)$$

and
$$|B|^2 = \frac{1}{n}\sum_{i=0}^{n-1}|f_i|^2 = \frac{1}{n}|F|^2 \qquad (6.171)$$

so that

$$|A||B| = \frac{1}{n}|E||F| \qquad (6.172)$$

Thus to minimize the probability of error in a detection process, it is necessary to minimize $|E||F|$. $|E|$ and $|F|$ are here the *unitary lengths* of the vectors E and F, $|e_i|$ and $|f_i|$ being the *absolute values* of e_i and f_i.

From Equation 6.157, the sampled impulse-response of the $(m+1)$-tap linear feedforward transversal equalizer for the channel is given by the n-component row vector

$$C = [c_0 \quad c_1 \quad \ldots \quad c_m \quad 0 \quad 0 \quad \ldots \quad 0] \qquad (6.173)$$

But the sampled impulse-response of the transmitter and receiver equalizers in cascade (Fig. 6.8) is the *convolution* of the vectors (sequences) A and B, and is given (at least approximately) by the n-component row vector

$$C_\theta = [\overbrace{0 \quad 0 \quad \ldots \quad 0}^{\theta} \quad c_0 \quad c_1 \quad \ldots \quad c_m \quad 0 \quad 0 \quad \ldots \quad 0] \qquad (6.174)$$

as can be seen from Equation 6.156. Clearly, C_θ is derived from C by moving the nonzero components of the latter θ places to the right, n being always large enough to accommodate the nonzero components of

C_θ. Now let the n-component DFT of C_θ be

$$G = [g_0 \quad g_1 \quad \cdots \quad g_{n-1}] \tag{6.175}$$

so that, from Section 2.6 and Equations 6.168 and 6.169,

$$g_i = e_i f_i \tag{6.176}$$

for $i = 0, 1, \ldots, n-1$.

Let E_0 and F_0 be the nonnegative real n-component row vectors whose $(i+1)$th components are $|e_i|$ and $|f_i|$, respectively. Clearly,

$$|E||F| = \left(\sum_{i=0}^{n-1} |e_i|^2\right)^{1/2} \left(\sum_{i=0}^{n-1} |f_i|^2\right)^{1/2} = |E_0||F_0| \tag{6.177}$$

Also
$$E_0 F_0^T = \sum_{i=0}^{n-1} |e_i||f_i| = \sum_{i=0}^{n-1} |g_i| \tag{7.178}$$

since $|g_i| = |e_i||f_i|$, from Equation 6.176.

But C, and therefore also G, are *fixed* by the baseband channel, if we neglect the delay of θT seconds introduced by the transmitter and receiver equalizers relative to the corresponding single equalizer (Equation 6.156). It can be seen from Section 2.8 that this delay in fact affects the $\{g_i\}$ but not the $\{|g_i|\}$, so that it does not affect Equation 6.178 and need not therefore be considered for our purposes here. Thus $E_0 F_0^T$ has a *fixed* nonnegative value. Now, by the Schwarz inequality (Equation 1.37) or by using the fact that the angle ϕ between the vectors E_0 and F_0 satisfies the relationship

$$\cos \phi = \frac{1}{|E_0||F_0|} E_0 F_0^T \tag{6.179}$$

where
$$E_0 F_0^T \geq 0 \tag{6.180}$$

it is clear that

$$|E_0||F_0| \geq E_0 F_0^T \tag{6.181}$$

with equality when

$$E_0 = \varepsilon F_0 \tag{6.182}$$

where ε is a positive real scalar. Thus the minimum value of $|E_0||F_0|$ and hence of $|E||F|$ is given by $E_0 F_0^T$ in Equation 6.178, and this is obtained when

$$|e_i| = \varepsilon |f_i| = (\varepsilon |g_i|)^{1/2} \tag{6.183}$$

for $i = 0, 1, \ldots, n-1$.

It now follows from Equations 6.172 and 6.178 that the minimum value of $|A||B|$ is

$$v = \frac{1}{n}\sum_{i=0}^{n-1}|g_i| \qquad (6.184)$$

where $|g_i|$ is the absolute value (modulus) of the $(i+1)$th component of the n-component DFT of the sampled impulse-response of the single linear equalizer for the baseband channel, given by the vector C in Equation 6.173. Finally, from Equation 6.167, the minimum value of the probability of error in the detection of s_i is

$$P_e = Q\left(\frac{k}{\sigma v}\right) \qquad (6.185)$$

being, of course, obtained when Equation 6.183 is satisfied.

It is clear from Equation 6.183 that there is not in general a *unique* combination of transmitter and receiver equalizers that minimizes the probability of error. A particular combination of some interest is that where $E = F$ and $A = B$, so that the transmitter and receiver equalizers are the same. Now

$$A(z) = B(z) \simeq z^{-\tau}C^{1/2}(z) \qquad (6.186)$$

where 2τ is an appropriate nonnegative integer. From Equation 6.156, $\theta = 2\tau$. A possible method of evaluating $C^{1/2}(z)$ is described in Section 3.6.

It has been shown by U.S. Tint that if the conventional algebraic techniques for evaluating the square root of a polynomial are applied to $C(z)$[29], these fail when $C(z)$ has any zeros (roots) *outside* the unit circle in the z plane. The correct technique for evaluating $A(z)$ and $B(z)$ is therefore as follows. Let

$$C(z) = C_1(z)C_2(z) \qquad (6.187)$$

where the zeros of $C_1(z)$ lie inside the unit circle and the zeros of $C_2(z)$ lie outside. $C_1^{1/2}(z)$ can now readily be evaluated from $C_1(z)$ by conventional techniques[29], a sufficient number of terms of the infinite expansion of $C_1^{1/2}(z)$ being taken to obtain the required degree of accuracy (see also Section 3.6). To evaluate $C_2^{1/2}(z)$, let $C_3(z)$ be the polynomial obtained from $C_2(z)$ by reversing the order of its coefficients. The zeros of $C_3(z)$ are the reciprocals of the zeros of $C_2(z)$ and therefore all lie inside the unit circle. $C_3^{1/2}(z)$ can now readily be evaluated from $C_3(z)$[29], the evaluation being taken to a sufficient number of terms to give the required approximation to $C_3^{1/2}(z)$. The polynomial obtained from the approximation to $C_3^{1/2}(z)$, by reversing the order of its coefficients, is now an approximation to $z^{-\tau}C_2^{1/2}(z)$. Finally, $A(z)$ and $B(z)$ are each given by the product of the approximations to $C_1^{1/2}(z)$ and $z^{-\tau}C_2^{1/2}(z)$, so

that
$$A(z) = B(z) \simeq C_1^{1/2}(z) z^{-\tau} C_2^{1/2}(z) = z^{-\tau} C^{1/2}(z) \tag{6.188}$$

To ensure that the coefficients in $A(z)$ and $B(z)$ are real valued, it may be necessary to let $A(z) = -B(z)$.

Suppose next that $A(z) \neq \pm B(z)$, but now

$$A(z) = M_1(z) N_1(z) \tag{6.189}$$

and
$$B(z) = M_2(z) N_2(z) \tag{6.190}$$

where
$$M_1(z) = \mu_0 + \mu_1 z^{-1} + \cdots + \mu_j z^{-j} \tag{6.191}$$

and
$$M_2(z) = \mu_j + \mu_{j-1} z^{-1} + \cdots + \mu_0 z^{-j} \tag{6.192}$$

$A(z)$ and $B(z)$ also satisfy Equations 6.153 and 6.154, as before, and j is any nonnegative integer not greater than l. Let M_1 and M_2 be the n-component row vectors

$$M_1 = [\mu_0 \quad \mu_1 \quad \ldots \quad \mu_j \quad 0 \quad 0 \quad \ldots \quad 0] \tag{6.193}$$

and
$$M_2 = [\mu_j \quad \mu_{j-1} \quad \ldots \quad \mu_0 \quad 0 \quad 0 \quad \ldots \quad 0] \tag{6.194}$$

where
$$n \geq 2l + 1 \tag{6.195}$$

as before, and let N_1 and N_2 be the n-component row vectors corresponding to $N_1(z)$ and $N_2(z)$, respectively. M_1, M_2, N_1 and N_2 are all real vectors. Clearly, the vector (sequence) A in Equation 6.151 is the *convolution* of M_1 and N_1 and the vector B in Equation 6.152 is the *convolution* of M_2 and N_2. Finally, let the DFT's of M_1 and M_2 be the n-component row vectors whose $(i+1)$th components are α_i and β_i, respectively, and let the DFT's of N_1 and N_2 be the n-component row vectors whose $(i+1)$th components are γ_i and δ_i, respectively.

Since M_2 is obtained from M_1 by reversing the order of its nonzero components, it can be seen from the Reversal theorem (Section 2.10) that

$$|\alpha_i| = |\beta_i| \tag{6.196}$$

for $i = 0, 1, \ldots, n-1$.

Using the fact that the ith component of the DFT of the *convolution* of two vectors is simply the *product* of the ith components of the DFT's of the individual vectors, it follows from Equations 6.168, 6.169, 6.189 and 6.190 that

$$|E|^2 = \sum_{i=0}^{n-1} |e_i|^2 = \sum_{i=0}^{n-1} |\alpha_i|^2 |\gamma_i|^2 \tag{6.197}$$

and
$$|F|^2 = \sum_{i=0}^{n-1} |f_i|^2 = \sum_{i=0}^{n-1} |\beta_i|^2 |\delta_i|^2 \tag{6.198}$$

So that, if now $A(z)=M_2(z)N_1(z)$ and $B(z)=M_1(z)N_2(z)$, the symbols α_i and β_i in Equations 6.197 and 6.198 are interchanged, and, from Equations 6.196 and 6.177, the value of $|E\|F|$ remains unchanged. Hence, from Equations 6.167 and 6.172, the probability of error in the detection of a received data-symbol also remains unchanged.

It can be seen that the j zeros of $M_2(z)$ are the reciprocals of the j zeros of $M_1(z)$. Thus, if i of the zeros of $B(z)$ are the reciprocals of the corresponding i zeros of $A(z)$, $A(z)$ and $B(z)$ may be modified by interchanging one or more of these zeros of $A(z)$ with the zeros of $B(z)$ that are their reciprocals, and this will not affect the tolerance of the system to additive white Gaussian noise. Complex-valued zeros must here be interchanged in complex-conjugate pairs. If a factor

$$V_1(z)=v_0+v_1z^{-1}+\cdots+v_hz^{-h} \qquad (6.199)$$

of the z-transform $Y(z)$ of the baseband channel is equalized by a linear filter with z-transform $M_1(z)$, given by Equation 6.191, then a factor

$$V_2(z)=v_h+v_{h-1}z^{-1}+\cdots+v_0z^{-h} \qquad (6.200)$$

is equalized by a linear filter with z-transform $M_2(z)$, given by Equation 6.192. h is here an integer in the range 1 to g and all coefficients are real valued. Furthermore, the zeros of $V_2(z)$ and $M_2(z)$ are the reciprocals of the zeros of $V_1(z)$ and $M_1(z)$, respectively. It follows that, if the factor of $Y(z)$ equalized by $B(z)$ contains i zeros which are reciprocals of the corresponding i zeros of the factor of $Y(z)$ equalized by $A(z)$, then $A(z)$ and $B(z)$ may be modified to interchange one or more of these i zeros equalized by $A(z)$ with the corresponding zeros equalized by $B(z)$, without affecting the tolerance of the system to additive white Gaussian noise. In other words, the *same* interchange of factors containing reciprocal zeros may be carried out between the portions of $Y(z)$ that are *equalized* by $A(z)$ and $B(z)$, as may be carried out between $A(z)$ and $B(z)$ themselves, without affecting the tolerance of the system to noise. Again, complex-valued zeros must be interchanged in complex-conjugate pairs.

Consider now the case where the z-transform of the channel is

$$Y(z)=Y_1(z)Y_2(z)=y_0+y_1z^{-1}+\cdots+y_gz^{-g} \qquad (6.201)$$

and satisfies the following conditions. The zeros of $Y_1(z)$ lie inside the unit circle in the z plane, and the zeros of $Y_2(z)$ lie outside the unit circle, $Y(z)$ having, of course, no zeros *on* the unit circle. Furthermore, $Y(z)$ has an odd number of terms (so that g is even) and its $g+1$ coefficients are symmetrical in the sense that $y_i=y_{g-i}$, for $i=0, 1, \ldots, g$. Now $Y_1(z)$ and $Y_2(z)$ each have $\frac{1}{2}g+1$ terms, with real-valued coefficients, and the coefficients of $Y_2(z)$ are the same as those of $Y_1(z)$ but in the reverse order. Clearly, the zeros of $Y_2(z)$ are the reciprocals of the zeros of $Y_1(z)$. Let the zeros of $Y_1(z)$ be $\phi_1, \phi_2, \ldots, \phi_{g/2}$, and let the zeros of $Y_2(z)$ be $\psi_1, \psi_2, \ldots, \psi_{g/2}$. With the appropriate ordering of the zeros of $Y_1(z)$ and $Y_2(z)$, these must now satisfy

$$\phi_i = \frac{1}{\psi_i} \tag{6.202}$$

for each i. Evidently, if $A(z)$ equalizes $Y_1(z)$ and $B(z)$ equalizes $Y_2(z)$, $B(z)$ is obtained from $A(z)$ by reversing the order of its coefficients, so that the zeros of $B(z)$ are the reciprocals of the zeros of $A(z)$. From the Reversal theorem (Section 2.10) it follows that

$$|e_i| = |f_i| \tag{6.203}$$

for each of the corresponding pair of components of the n-component DFT's of the vectors A and B. From Equation 6.183, this satisfies the condition for the minimum probability of error in the detection of a received data-symbol.

If we now lift the restriction that the zeros of $Y_1(z)$ lie inside the unit circle in the z plane and the zeros of $Y_2(z)$ lie outside the unit circle, it can be seen that, so long as the coefficients in $Y_1(z)$ and $Y_2(z)$ remain real valued, any one or more pairs of reciprocal zeros of $Y_1(z)$ and $Y_2(z)$ may be interchanged, without affecting Equations 6.202 and 6.203 and therefore without affecting the tolerance of the system to additive white Gaussian noise. Strictly speaking, when one of the two interchanged factors is a constant times the reverse of the other, Equation 6.202 is still satisfied but Equation 6.203 is replaced by Equation 6.183. Again, in the more general case where the $\{y_i\}$ are *complex-valued*, any one or more pairs of *complex-conjugate* reciprocal zeros of $Y_1(z)$ and $Y_2(z)$ may be interchanged without affecting the tolerance to noise.

Thus, whenever the sampled impulse-response of the channel is real valued and symmetrical about the central sample, the tolerance of the system to additive white Gaussian noise is optimized by equalizing, at the transmitter, any real factor of $Y(z)$ whose zeros are the reciprocals of the zeros of the remaining factor. The latter is equalized at the receiver.

It is clear from Equations 6.156 and 6.167 that for *any* combination of transmitter and receiver equalizers in the data-transmission system of Fig. 6.8, which need *not* now be optimum (satisfying Equation 6.183), the error probability in the detection of a received data-symbol is *unaffected* by an interchange of transmitter and receiver equalizers. In the particular case where only a single equalizer is used, the error probability is the *same* whether the channel is equalized at the receiver or at the transmitter.

If in the vector G (Equation 6.175)

$$|g_i| = d \tag{6.204}$$

for $i = 0, 1, \ldots, n-1$, where d is any positive real-valued constant, then, for the optimum tolerance to noise, the channel may either be equalized entirely at the transmitter or receiver, or else by any combination of transmitter and receiver equalizers satisfying Equation 6.183. This follows because the *absence* of an equalizer at the transmitter, say, corresponds to setting $e_i = 1$, for $i = 0, 1, \ldots, n-1$, in Equation 6.168. But it is shown in Section 3.2 that Equation 6.204 represents pure phase distortion, with possibly some attenuation or gain. Thus the transmitter and receiver equalizers here combine to form a phase equalizer, which also appropriately adjusts the signal level. This means, of course, that the channel must introduce pure phase distortion, with possibly some attenuation or gain. But, under the conditions assumed here, the vectors E and F in Equations 6.168 and 6.169 are such that

$$|e_i| = d_1 \quad \text{and} \quad |f_i| = d_2 \qquad (6.205)$$

for $i = 0, 1, \ldots, n-1$, and

$$d_1 d_2 = d \qquad (6.206)$$

Thus, in any combination of transmitter and receiver equalizers that minimize the probability of error under these conditions, both the transmitter and receiver equalizers must *themselves* be *phase equalizers*, and the way in which the phase equalization of the channel is shared between the transmitter and receiver does not affect the performance of the system. In the general case, where the channel introduces both amplitude and phase distortions, the transmitter or receiver equalizer in an optimum combination of these equalizers may introduce any required amount of phase distortion, so long as the resultant phase distortion of the channel and this equalizer is corrected by the other equalizer.

Finally, it can be seen from Equation 6.183 and the definition of amplitude distortion (Section 3.4) that, in an optimum combination of linear equalizers, the equalization of the amplitude distortion must be shared *equally* between the transmitter and receiver equalizers, although these need not introduce the same attenuation or gain. This means that, in the particular case where the channel introduces pure amplitude distortion (together with the appropriate delay) and neither equalizer introduces any phase distortion, the tap gains of the transmitter equalizer are in a constant ratio to the corresponding tap gains of the receiver equalizer. In an arrangement of the optimum linear equalization of a channel introducing pure amplitude distortion and some delay, where the zeros of the factor of $Y(z)$ equalized by the transmitter equalizer are the reciprocals of the zeros of the factor equalized by the receiver equalizer, each equalizer introduces some phase distortion, the phase distortion introduced by one of these being such as to correct the phase distortion introduced by the other. One of the two equalizers is

LINEAR AND DECISION-FEEDBACK EQUALIZERS

now the *reverse* of the other, so that the two equalizers are not the same or in a constant ratio. Clearly, by permitting the transmitter and receiver equalizers to introduce the appropriate levels of phase distortion, the linear equalizers that minimize the error probability in the detected data-symbols can sometimes be implemented more simply than when the two equalizers are constrained to be the same or in a constant ratio.

It is reasonable to conclude from the preceding analysis that the smaller the fraction of the signal distortion corresponding to pure phase distortion, the greater the advantage in tolerance to additive white Gaussian noise likely to be achieved when a single equalizer at the receiver is replaced by an optimum combination of transmitter and receiver equalizers.

6.3.4 Optimum combination of a linear equalizer at the transmitter and a decision-feedback equalizer at the receiver

When a decision-feedback equalizer, containing both a linear and a nonlinear filter, is used at the receiver (Fig. 5.11), the linear filter with z-transform $D(z)$ partially equalizes the channel, the equalization process being completed by the nonlinear filter. For any given combination of linear and nonlinear filters, the linear filter may be shared between the transmitter and receiver, the nonlinear filter remaining at the receiver.

It can be seen from Equations 4.191 and 5.229 that where the whole of the equalization process is carried out at the receiver, the probability of error in the detection of a received data-symbol, under the assumed conditions, is the same function of the sampled impulse-response of the linear filter, whether the linear filter itself equalizes the channel or whether an additional nonlinear filter is used, that is, regardless of whether linear or decision-feedback equalization is used. It follows from Equations 6.156 and 6.167 that the optimum sharing of the linear filter between the transmitter and receiver is unaffected by whether or not a nonlinear filter is used at the receiver.

If now the z-transforms of the transmitter and receiver linear-filters are $A(z)$ and $B(z)$, respectively, in the case where a decision-feedback equalizer is used at the receiver, then

$$A(z)B(z) \simeq z^{-\theta}D(z) \quad (6.207)$$

where θ is an appropriate nonnegative integer. The condition for the minimum probability of error in the detection of a received data-symbol is given by Equation 6.183, as before, but now only at high signal/noise ratios, when error-extension effects can be neglected. The corresponding value of the error probability is given by Equation 6.185, if error-extension effects are again neglected, bearing in mind that $|g_i|$ in Equation 6.184 is now the absolute value (modulus) of the $(i+1)$th

6.3.5 Pure nonlinear equalizer at the transmitter

An interesting technique has been proposed whereby it is possible to equalize a channel nonlinearly at the transmitter[30-32]. Consider the arrangement where the channel is equalized entirely by a pure nonlinear equalizer at the transmitter. The data-transmission system is now as shown in Fig. 6.9, where the baseband channel is the same as that in Fig. 6.8.

The nonlinear equalizer converts the sequence of binary data-symbols $\{s_i\}$ into the sequence of values $\{f_i\}$ which are sampled and transmitted as the corresponding impulses $\{f_i \delta(t - iT)\}$. As before, $s_i = \pm k$, but the signal level at the input to the baseband channel is *not* now assumed to be such that the average transmitted energy per signal element, $\mathscr{E}[f_i^2]$, at the input to the transmission path, is necessarily equal to k^2. Again, the noise components $\{w_i\}$ at the output of the sampler in the receiver, are statistically independent Gaussian random variables with zero mean and variance σ^2. Thus the tolerance of the data-transmission system to additive white Gaussian noise is given by the signal/noise ratio $\mathscr{E}[f_i^2]/\sigma^2$ for a specified average error rate in the detected data-symbols $\{s_i'\}$. The z-transform of the baseband channel is $Y(z)$ in Equation 6.158, where $y_0 \neq 0$, and the signal at the output of the sampler in the receiver, at time $t = iT$, is

$$r_i = \sum_{j=0}^{g} f_{i-j} y_j + w_i \tag{6.208}$$

The corresponding signal at the output of the multiplier is r_i/y_0.

The square marked M in Fig. 6.9 is a nonlinear processor that

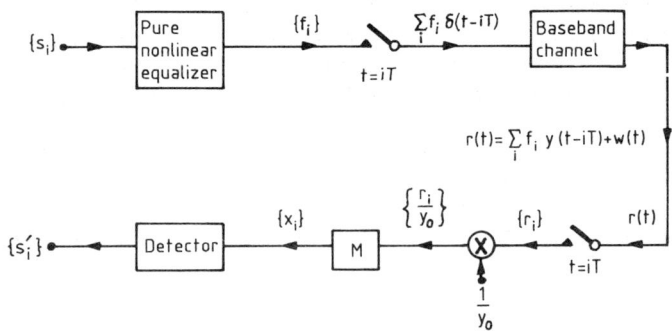

Fig. 6.9 Equalization of the channel by a pure nonlinear equalizer at the transmitter

performs a variant of the conventional *modulo* $-m$ operation on its input signal, that can be explained as follows. The conventional function x modulo $-m$ is defined to be such that

$$x \text{ modulo} - m = x - jm \tag{6.209}$$

where m need not be an integer and j is the appropriate integer, such that

$$0 \leqslant x \text{ modulo} - m < m \tag{6.210}$$

Suppose now that the input signal to the nonlinear processor has the value q, where q may be any positive or negative real number. Then the output signal from the processor is

$$M[q] = [(q + 2k) \text{modulo} - 4k] - 2k = q - 4jk \tag{6.211}$$

where k need not be an integer and j is the appropriate integer such that

$$-2k \leqslant M[q] < 2k \tag{6.212}$$

If $-2k \leqslant q < 2k$, $M[q] = q$. The nonlinear transformation M is implemented quite simply by means of a quantizer and subtraction circuit, or, in a digital system, by means of the appropriate look-up table that gives the value of $M[q]$ for each different possible value of q.

The data-transmission system of Fig. 6.9 can be represented in terms of the z-transforms of the signals, as in Fig. 6.10. The nonlinear processor M at the transmitter is here identical to that at the receiver, and the baseband channel includes the transmitter and receiver samplers. Every signal in Fig. 6.10 is a sequence of samples at the time instants $\{iT\}$.

The nonlinear equalizer in Fig. 6.10 is the same as the nonlinear equalizer in Fig. 5.3, except that in the former the detector is replaced by the nonlinear processor, M. The linear feedforward transversal filter in each of these nonlinear equalizers has a sampled impulse-response with z-transform $y_0^{-1} Y(z) - 1$. Thus the transversal filter in Fig. 6.10 has g taps, the ith of which has a gain $v_i = y_i/y_0$, as in Fig. 5.3. Let the signal at the output of the nonlinear equalizer be the sequence of samples $\{f_i\}$ with z-transform $F(z)$. Assuming the same transmitter filter as before (Sections 1.2, 4.2 and 5.1), the average transmitted energy per signal element at the input to the transmission path (Fig. 6.8) is equal to $\mathscr{E}[f_i^2]$, the mean-square value of f_i.

Suppose first that the nonlinear processor M is removed from the nonlinear equalizer in the transmitter of Fig. 6.10, to give the arrangement shown in Fig. 6.11. In $F(z)$ the coefficient of z^{-i} is f_i and in $S(z)$ it is s_i. Thus the z-transform of the signal at the input to the baseband channel is

$$\begin{aligned} F(z) &= S(z) - F(z)(y_0^{-1} Y(z) - 1) \\ &= S(z) - y_0^{-1} Y(z) F(z) + F(z) \end{aligned} \tag{6.213}$$

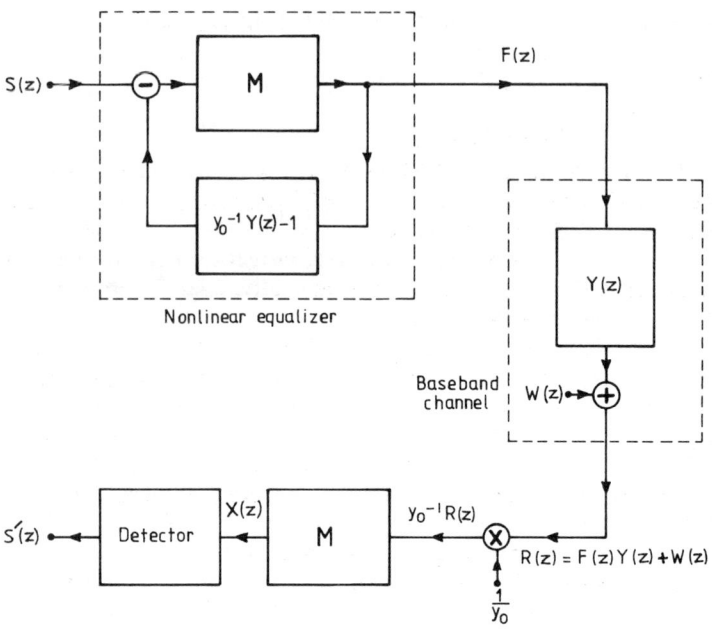

Fig. 6.10 Mathematical model of the data-transmission system in Fig. 6.9

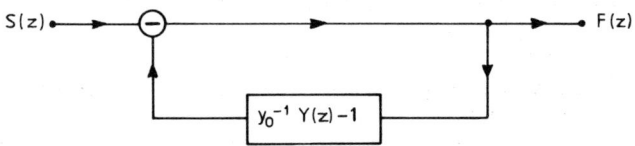

Fig. 6.11 Modified equalizer

so that

$$S(z) = y_0^{-1} Y(z) F(z) \qquad (6.214)$$

or

$$F(z) = \frac{y_0 S(z)}{Y(z)} \qquad (6.215)$$

where $S(z)/Y(z)$ signifies the division of $S(z)$ by $Y(z)$, using a normal process of long division.

It is shown in Section 4.7 that this arrangement is unstable whenever $Y(z)$ has any zeros (roots) outside the unit circle in the z plane. However, when all the zeros of $Y(z)$ lie inside the unit circle,

$$\frac{S(z)}{Y(z)} = S(z)Y^{-1}(z) \tag{6.216}$$

The arrangement is now stable and achieves the accurate linear equalization of the channel. The nonlinear processor M at the receiver is no longer needed in the data-transmission system of Fig. 6.10 and, as is shown in Section 6.3.3, the tolerance of the resultant system to additive white Gaussian noise is the same as that of the corresponding system with a linear equalizer at the receiver.

Consider next the data-transmission system in Fig. 6.10. The inclusion of the nonlinear processor M in the equalizer ensures that the value (area) f_i of a transmitted impulse $f_i \delta(t - iT)$ satisfies $-2k \leqslant f_i < 2k$. By this means it prevents instability in the equalizer, in the sense that it prevents a steady build-up in the magnitude of the output signal, regardless of $Y(z)$[30]. Now the nonlinear processor M operates, as described in Equations 6.211 and 6.212, separately on *each* signal (sample value) at its input, to give the corresponding output signal. The operation of the processor on a *sequence* of signals (samples), given by the vector A with z-transform $A(z)$, can conveniently be written $M[A(z)]$, which is the z-transform of the *output* sequence from the processor and signifies that the nonlinear transformation M (Equations 6.211 and 6.212) is applied separately to each *coefficient* in $A(z)$, and therefore, of course, to each input signal. Thus, the z-transform of the signal at the input to the baseband channel in Fig. 6.10 is now[32]

$$F(z) = M[S(z) - F(z)(y_0^{-1}Y(z) - 1)] \tag{6.217}$$

so that

$$M[F(z)] = M[S(z) + F(z) - y_0^{-1}Y(z)F(z)] \tag{6.218}$$

since, of course, $M[F(z)] = F(z)$. From Equations 6.211 and 6.218,

$$M[y_0^{-1}Y(z)F(z)] = M[S(z)] \tag{6.219}$$

The z-transform of the signal at the output of the baseband channel is

$$R(z) = F(z)Y(z) + W(z) \tag{6.220}$$

and the z-transform of the signal at the input to the nonlinear network M in the receiver is $y_0^{-1}R(z)$. In $R(z)$ here the coefficient of z^{-i} is the received sample r_i and in $W(z)$ it is the Gaussian noise component w_i with zero mean and variance σ^2.

The z-transform of the signal at the output of the nonlinear network

M in the receiver is

$$X(z) = M[y_0^{-1}R(z)] = M[y_0^{-1}Y(z)F(z) + y_0^{-1}W(z)]$$
$$= M[S(z) + y_0^{-1}W(z)] \quad (6.221)$$

as can be seen from Equations 6.220 and 6.219. Equation 6.221 can be expressed in terms of the signals (sample values) at time $t = iT$, to give[32]

$$x_i = M[y_0^{-1}r_i] = M[s_i + y_0^{-1}w_i] \quad (6.222)$$

where x_i is the signal at the detector input at this time instant. Clearly, all intersymbol interference has been eliminated in x_i, so that the channel is accurately equalized.

In the detector, s_i is detected as k or $-k$ depending upon whether x_i is positive or negative, respectively, to give s_i', the detected value of s_i. The z-transform of the sequence of the $\{s_i'\}$ is designated $S'(z)$.

It can be seen from Equations 6.211 and 6.222 that an error occurs in the detection of s_i whenever

$$(4j - 3)k < |y_0^{-1}w_i| < (4j - 1)k \quad (6.223)$$

for all positive integers $\{j\}$. $y_0^{-1}w_i$ is a Gaussian random variable with zero mean and variance $y_0^{-2}\sigma^2$, and $|y_0^{-1}w_i|$ is the absolute value (modulus) of $y_0^{-1}w_i$.

At practical signal/noise ratios, the probability of error in the detection of s_i is approximately equal to the probability that $|y_0^{-1}w_i| > k$ and is therefore approximately

$$2\int_k^\infty \frac{1}{\sqrt{2\pi y_0^{-2}\sigma^2}} \exp\left(-\frac{w^2}{2y_0^{-2}\sigma^2}\right) dw = 2Q\left(\frac{k|y_0|}{\sigma}\right) \quad (6.224)$$

There are no error-extension effects here. At high signal/noise ratios, with error probabilities around 1 in 10^5 or 1 in 10^6, the error probability in Equation 6.224 can be taken to be

$$P_e = Q\left(\frac{k|y_0|}{\sigma}\right) \quad (6.225)$$

with an inaccuracy of only a small fraction of 1 dB in the corresponding signal/noise ratio.

When the baseband channel in Fig. 6.10 introduces no signal distortion, so that $Y(z) = y_0$, bearing in mind that the transmission delay is neglected here, the average transmitted energy per signal element, at the input to the transmission path, is

$$\mathcal{E}[f_i^2] = k^2 \quad (6.226)$$

Suppose now that the baseband channel introduces signal distortion.

The value (area) of the impulse at the input to the baseband channel, at time $t = iT$, is now

$$f_i = M[s_i + a_i] \tag{6.227}$$

where a_i is a function of the $\{s_j\}$, for $j < i$. This is clear from the structure of the nonlinear equalizer. But the $\{s_i\}$ are statistically independent and equally likely to have either value k or $-k$. Thus, if $|f_i|$ can have the value $k - b$, where $0 < b < k$, then it is equally likely to have the value $k + b$. This necessarily implies that, when the channel introduces signal distortion, the average transmitted energy per signal element, at the input to the transmission path, must *exceed* k^2. Hence, for any channel,

$$\mathscr{E}[f_i^2] \geqslant k^2 \tag{6.228}$$

But $-2k \leqslant f_i < 2k$, and when $|f_i|$ can have its maximum value $2k$, it is equally likely to have the value 0. Thus

$$\mathscr{E}[f_i^2] \leqslant 2k^2 \tag{6.229}$$

which means that

$$k^2 \leqslant \mathscr{E}[f_i^2] \leqslant 2k^2 \tag{6.230}$$

Clearly, the average transmitted energy per signal element *must* lie in the range 0 to 3 dB above k^2. When the channel introduces severe signal distortion, a good estimate of $\mathscr{E}[f_i^2]$ is obtained by assuming that f_i is equally likely to have *any* value in the range $-2k$ to $2k$. Now $\mathscr{E}[f_i^2] = 1.3333k^2$, which is 1.25 dB above its minimum value of k^2. This analysis suggests that the average transmitted energy per signal element, $\mathscr{E}[f_i^2]$, should most often lie in the range 0 to 1.5 dB above k^2 and, in the presence of severe signal distortion, it is likely to be somewhere near 1.25 dB above k^2. Since the tolerance of the system to noise is measured by the value of the signal/noise ratio, $\mathscr{E}[f_i^2]/\sigma^2$, for a given error rate in the $\{s_i'\}$, it follows that the actual tolerance to noise of the system may be typically up to about 1.5 dB below that evaluated on the assumption that $\mathscr{E}[f_i^2] = k^2$.

Consider now the data-transmission system shown in Fig. 5.1, where the receiver is as shown in Fig. 5.2 and contains the pure nonlinear equalizer shown in Fig. 5.3 (see Section 5.1). The linear feedforward transversal filter here has the z-transform $y_0^{-1} Y(z) - 1$, and is therefore the same as the transversal filter in the nonlinear equalizer of Fig. 6.10. It is clear that the data-transmission system with the pure nonlinear equalizer at the receiver is obtained from the system in Fig. 6.10 by transferring the nonlinear equalizer from the transmitter to the receiver and removing the two nonlinear processors M. In the data-transmission system with the pure nonlinear equalizer at the receiver, the average transmitted energy per signal element is *exactly* k^2 and so the signal/noise ratio is exactly k^2/σ^2. Also, from Equation 5.39, the probability of error in the detection of s_i can be taken to be

$$P_e = Q\left(\frac{k|y_0|}{\sigma}\right) \qquad (6.231)$$

when error-extension effects are neglected. At high signal/noise ratios, the error-extension effects tend to reduce the tolerance to noise by typically up to 1 dB and sometimes by as much as 2 dB or even more.

From Equations 6.225 and 6.231, the two systems have theoretically the *same* tolerance to additive white Gaussian noise, when it is assumed that the average transmitted energy per signal element is k^2 in each case, and bearing in mind the other approximations that have been made in these equations. At high signal/noise ratios, the reduction in tolerance to additive white Gaussian noise caused by error-extension effects, in the data-transmission system with a pure nonlinear equalizer at the receiver, is of the same order as the reduction in tolerance to noise of the system with a pure nonlinear equalizer at the transmitter. This follows from the following facts. Firstly, the actual error probability of the latter system is *twice* that in Equation 6.225, and secondly, the transmitted signal level here generally exceeds the assumed value of k^2, often by a little over 1 dB. Thus, at high signal/noise ratios, the two systems have similar tolerances to additive white Gaussian noise, these being normally within about 1 dB of each other. The two systems can be considered as duals of each other, since they use the same equalization process, which is located in one case at the transmitter and in the other case at the receiver.

6.3.6. Optimum nonlinear equalizer at the transmitter

Just as the data-transmission system, with a pure nonlinear filter (equalizer) at the receiver, can be modified by the addition of a linear filter and the appropriate change in the nonlinear filter, to achieve the optimum decision-feedback equalization of the channel (Section 5.8.2), so also can the data-transmission system with a pure nonlinear filter (equalizer) at the transmitter, shown in Fig. 6.10. The resultant system is shown in Fig. 6.12 and operates as follows.

The linear filter added at the transmitter is an $(n+1)$-tap feedforward transversal filter with z-transform

$$D(z) = d_0 + d_1 z^{-1} + \cdots + d_n z^{-n} \qquad (6.232)$$

The output signal from this filter is fed to the baseband channel, which is that in Fig. 6.9 together with the two samplers.

The z-transform of the linear filter and baseband channel in Fig. 6.12 is

$$D(z)Y(z) \simeq z^{-h}E(z) \qquad (6.233)$$

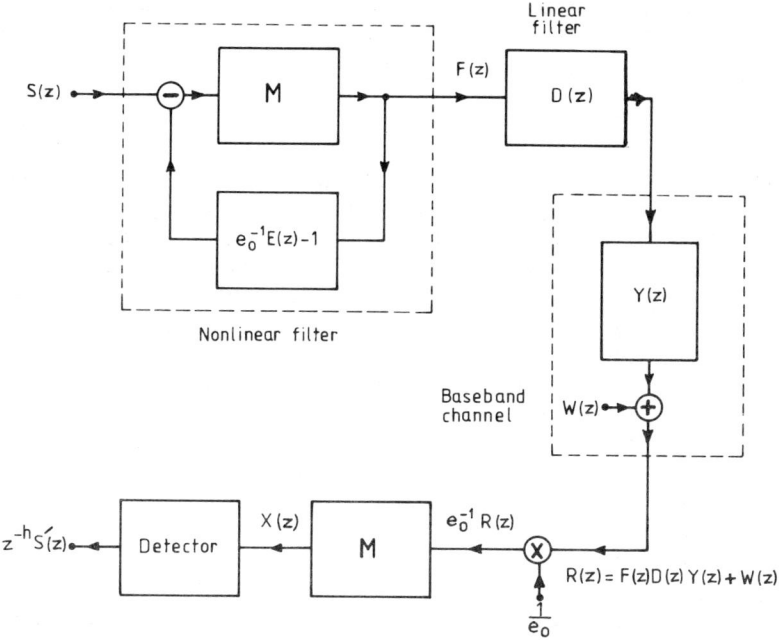

Fig. 6.12 Equalization of the channel by the optimum nonlinear equalizer at the transmitter

where
$$E(z) = e_0 + e_1 z^{-1} + \cdots + e_l z^{-l} \quad (6.234)$$

h and l are appropriate positive integers. Equation 6.233 holds for any $Y(z)$ and $D(z)$, provided only that g and n are finite integers, as is assumed here. The linear filter and baseband channel, with z-transform $z^{-h}E(z)$, are equalized by the nonlinear filter at the transmitter in Fig. 6.12, in a manner similar to that for the equalization of the baseband channel in Fig. 6.10. Thus the z-transform of the transversal filter, that forms part of the nonlinear filter, is $e_0^{-1}E(z) - 1$.

Let the signal at the output of the nonlinear filter be the sequence of samples $\{f_i\}$ with z-transform $F(z)$. It can now be seen from the previous analysis that $\mathscr{E}[f_i^2]$ usually lies in the range 0 to about 1.5 dB above k^2. Thus, to a first approximation, $\mathscr{E}[f_i^2] = k^2$. Suppose furthermore that

$$\sum_{i=0}^{n} d_i^2 = 1 \quad (6.235)$$

Now the signals $\{f_i\}$ have *zero mean* (so long as f_i never has the *exact* value $-2k$, as is usually the case) and the correlation introduced into these by the feedback filter with z-transform $e_0^{-1}E(z) - 1$ tends to be offset, or, at least, greatly reduced by the nonlinear processor M. The precise analysis of this effect is not easy and for our purposes here it will be assumed that the $\{f_i\}$ are effectively *uncorrelated*. This means that the $\{f_i\}$ are also *statistically orthogonal*[16-18], such that

$$\mathscr{E}[f_i f_j] = 0 \qquad (6.236)$$

for any $i \neq j$. Under these conditions, as can be seen from Equation 6.235, the mean-square value of the signal at the *output* of the linear filter with z-transform $D(z)$ is the *same* as that at its *input*. It follows from the previous analysis that the average transmitted energy per signal element, at the input to the transmission path is, to a first approximation, equal to k^2, as in the case of the data-transmission system with a pure nonlinear equalizer at the transmitter.

The z-transform of the signal at the output of the nonlinear filter, at the transmitter in Fig. 6.12, is[32]

$$F(z) = M[S(z) - F(z)(e_0^{-1} E(z) - 1)] \qquad (6.237)$$

so that

$$M[F(z)] = M[S(z) + F(z) - e_0^{-1} E(z) F(z)] \qquad (6.238)$$

or

$$M[e_0^{-1} E(z) F(z)] = M[S(z)] \qquad (6.239)$$

The z-transform of the signal at the input to the baseband channel is $F(z)D(z)$ and the z-transform of the signal at the receiver input is

$$R(z) = F(z)D(z)Y(z) + W(z) \simeq z^{-h} E(z) F(z) + W(z) \qquad (6.240)$$

from Equation 6.233. This signal is multiplied by e_0^{-1} and fed to the nonlinear processor M.

The z-transform of the signal at the input to the detector is

$$\begin{aligned} X(z) &= M[e_0^{-1} R(z)] \\ &\simeq M[e_0^{-1} z^{-h} E(z) F(z) + e_0^{-1} W(z)] \\ &= M[z^{-h} S(z) + e_0^{-1} W(z)] \end{aligned} \qquad (6.241)$$

from Equation 6.239. It follows that the signal at the detector input, at time $t = (i+h)T$, is[32]

$$x_{i+h} \simeq M[s_i + e_0^{-1} w_{i+h}] \qquad (6.242)$$

The data-symbol s_i is detected from the sign of x_{i+h}. At high signal/noise ratios, the probability of error in the detection of s_i can be taken to equal the probability that $e_0^{-1} w_{i+h} > k$, which is

$$P_e = Q\left(\frac{k|e_0|}{\sigma}\right) \qquad (6.243)$$

The derivation of Equation 6.243 follows along exactly the same lines as that given by Equations 6.223–6.225. Clearly, to minimize P_e, $|e_0|$ must be *maximized* and this can be achieved by the appropriate choice of $D(z)$. e_0 is, of course, the first term of $E(z)$ (Equation 6.234).

The value of $D(z)$ that maximizes $|e_0|$, given that $\sum_{i=0}^{n} d_i^2 = 1$, is the same as (or a constant times) the value of $D(z)$ that minimizes $\sum_{i=0}^{n} d_i^2$, given e_0, and this is determined in Section 5.8.2. The fact that e_0 is set to unity in Section 5.8.2 does not affect the minimization process other than by multiplying the terms of both $D(z)$ and $E(z)$ by the appropriate constant. Furthermore, it does not affect the tolerance of the system to noise, since *both* the signal and the noise are affected equally by the chosen value of e_0.

From Equations 5.265–5.268, the z transform of the baseband channel is

$$Y(z) = Y_1(z) Y_2(z) \tag{6.244}$$

where
$$Y_1(z) = 1 + p_1 z^{-1} + p_2 z^{-2} + \cdots + p_{g-f} z^{-g+f} \tag{6.245}$$

with all its zeros *inside* the unit circle in the z plane, and

$$Y_2(z) = q_0 + q_1 z^{-1} + q_2 z^{-2} + \cdots + q_f z^{-f} \tag{6.246}$$

with all its zeros *outside* the unit circle. The integer f lies in the range 0 to g. Also

$$Y_3(z) = q_f + q_{f-1} z^{-1} + q_{f-2} z^{-2} + \cdots + q_0 z^{-f} \tag{6.247}$$

so that $Y_3(z)$ is obtained from $Y_2(z)$ by reversing the order of its coefficients. The zeros of $Y_3(z)$ are the reciprocals of those of $Y_2(z)$ and therefore lie *inside* the unit circle.

From Section 5.8.2, the linear filter that minimizes $\sum_{i=0}^{n} d_i^2$, given that $e_0 = 1$, has the z-transform

$$D_0(z) \simeq q_f^{-1} z^{-h} Y_2^{-1}(z) Y_3(z) \tag{6.248}$$

Thus the linear filter that maximizes $|e_0|$, given that $\sum_{i=0}^{n} d_i^2 = 1$, has the z-transform

$$D(z) \simeq z^{-h} Y_2^{-1}(z) Y_3(z) \tag{6.249}$$

since $z^{-h} Y_2^{-1}(z) Y_3(z)$ represents an *orthogonal transformation* with *no* attenuation or gain. The linear filter equalizes a signal with z-transform $Y_2(z) Y_3^{-1}(z)$, which represents pure phase distortion, so that the linear filter is a pure phase equalizer and therefore an *allpass* network. It follows that the average energy per signal element at the output of this filter is the *same* as that at its input, regardless of any correlation between the symbols $\{f_i\}$ at its input. Thus no inaccuracy is introduced here by this assumption.

The z-transform of the linear filter and baseband channel is

$$D(z)Y(z) \simeq z^{-h}Y_1(z)Y_3(z) \qquad (6.250)$$

and, from Equation 6.233,

$$E(z) = Y_1(z)Y_3(z) \qquad (6.251)$$

The first term of $Y_1(z)$ is 1, the first term of $Y_3(z)$ is q_f, and $Y_1(z)Y_3(z)$ has $g+1$ terms. Thus $e_0 = q_f$ and $l = g$. Also $h = n$. From Equation 6.243 the probability of error in the detection of s_i can now be taken to be

$$P_e = Q\left(\frac{k|q_f|}{\sigma}\right) \qquad (6.252)$$

In practice, the average transmitted energy per signal element is likely to be up to about 1.5 dB above the value k^2 assumed (for the reasons explained in Section 6.3.5), giving the corresponding reduction in tolerance to noise.

The linear filter with z-transform $D(z)$ is a pure phase equalizer such that, in the z-transform of the channel and linear filter, all zeros of $Y(z)$ lying outside the unit circle in the z plane are replaced by their reciprocals. The linear filter partially equalizes the channel, the equalization process being completed by the nonlinear filter (at the transmitter) that comprises a feedback transversal filter and the nonlinear processor M.

If the nonlinear processor at the transmitter is removed, the baseband channel is now equalized *linearly* at the transmitter and the nonlinear processor M at the receiver should be removed. The linear equalizer is stable since all zeros of $E(z)$ lie inside the unit circle in the z plane. This system, of course, has the same tolerance to noise as that where the channel is equalized linearly at the receiver.

Consider now the data-transmission system shown in Fig. 5.1, where the receiver is as shown in Fig. 5.11, and uses the optimum decision-feedback equalizer described in Section 5.8.2. The linear feedforward transversal filter F, that forms part of the *nonlinear filter* here, has the z-transform $e_0^{-1}E(z) - 1$ and is therefore the same as the transversal filter in the nonlinear filter of Fig. 6.12. In Section 5.8.2, of course, the first term of $E(z)$ is 1, and not e_0 as it is here, so that $e_0^{-1}E(z) - 1$ becomes $E(z) - 1$. It is clear that the data-transmission system with the optimum decision-feedback equalizer at the receiver is obtained from the system in Fig. 6.12 by transferring the nonlinear equalizer (including both the linear and nonlinear filters) from the transmitter to the receiver and removing the two nonlinear processors M. Since the first term of $E(z)$ is now taken to be e_0, the multiplier in the receiver of Fig. 6.12 must be retained.

From Equations 5.281 and 6.252, the two systems have theoretically the same tolerance to additive white Gaussian noise, given the various approximations that have been made. In practice, the tolerance to noise of each system at high signal/noise ratios is reduced a little, for the same reasons as those given in Section 6.3.5 for the corresponding two systems using pure nonlinear equalizers, so that the former two systems should in fact have similar tolerances to additive white Gaussian noise. The two systems can be considered as duals of each other, since they use the same equalization process, which is located in one case at the transmitter and in the other at the receiver.

It may readily be shown that if $Y(z)$ contains one or more zeros *on* the unit circle in the z plane, the design of the optimum nonlinear equalizer, located at the transmitter, is obtained by taking $Y_1(z)$ in Equation 6.245 as the factor of $Y(z)$ that includes all zeros *on or inside* the unit circle.

For convenience, in the following discussion, the data-transmission system of Fig. 6.10 (involving the pure nonlinear equalizer at the transmitter) will now be referred to as system 1, and the data-transmission system of Fig. 6.12 (involving the optimum nonlinear equalizer at the transmitter) will be referred to as system 2.

The tolerances to additive white Gaussian noise of systems 1 and 2, when operating over 21 different channels, have been measured by M.N. Serinken, using computer-simulation tests[32]. The different channels used in the tests have been selected to represent a wide range of signal distortions, including many different combinations of amplitude and phase distortions. For every channel,

$$\sum_{i=0}^{g} y_i^2 = 1 \qquad (6.253)$$

so that no attenuation or gain is introduced in transmission, when the input signals $\{f_i\}$ are statistically orthogonal[16-18]. Again, $s_i = \pm 1$ so that $k=1$. In each test, the noise variance σ^2 was adjusted to give effectively 30 errors in the detection of 30 000 data symbols, this being an error rate of 1 in 10^3. The mean-square value of f_i was also measured to give the signal/noise ratio $\mathscr{E}[f_i^2]/\sigma^2$. From the assumptions made concerning the transmitter and receiver filters (Sections 1.2, 4.2 and 5.1), the signal/noise ratio is the ratio of the average transmitted energy per signal element at the input to the transmission path to the two-sided power spectral density of the white Gaussian noise at the receiver input (see Figs. 6.8 and 6.12). The tolerance to noise of each system, when operating over any channel, is now measured by the number of dB reduction in the signal/noise ratio $\mathscr{E}[f_i^2]/\sigma^2$ needed to maintain the error rate of 1 in 10^3, when the given channel is replaced by one introducing no distortion or attenuation. The 95% confidence limits for each of these measured values are of the order of ± 0.25 dB.

The results of the tests are given in Table 6.2. Also shown in Table 6.2

Table 6.2 NUMBER OF DECIBELS REDUCTION IN TOLERANCE TO ADDITIVE WHITE GAUSSIAN NOISE, AT HIGH SIGNAL/NOISE RATIOS, WHEN THE GIVEN CHANNEL REPLACES ONE THAT INTRODUCES NO DISTORTION OR ATTENUATION

Channel	Sampled impulse-response of channel					Results of computer-simulation tests		Results of theoretical analysis		
						System 1	System 2	System 1	System 2	System 3
1	0.161	0.974	0.161	0	0	17.6	1.5	15.9	0.5	1.0
2	0.262	0.929	0.262	0	0	13.3	3.5	11.6	1.4	3.1
3	0.348	0.870	0.348	0	0	10.6	4.8	9.2	3.1	7.9
4	0.378	0.845	0.378	0	0	10.3	6.0	8.5	4.3	12.0
5	0.219	0.750	0.625	0	0	15.9	5.8	13.2	4.1	11.4
6	0.625	0.750	0.219	0	0	5.9	5.8	4.1	4.1	11.4
7	0.575	0.776	0.259	0	0	6.6	6.6	4.8	4.8	14.5
8	0.262	0.928	−0.262	0	0	13.5	0.6	11.6	0.0	0.0
9	0.348	0.870	−0.348	0	0	10.8	0.7	9.2	0.1	0.1
10	0.575	0.776	−0.259	0	0	6.4	2.0	4.8	0.6	1.5
11	0.625	0.750	−0.219	0	0	5.7	2.6	4.1	0.9	2.5
12	0.196	0.392	0.785	0.392	0.196	15.9	6.4	14.2	4.9	9.0
13	0.167	0.471	0.707	0.471	0.167	17.4	11.7	15.5	9.9	20.4
14	0.203	0.339	0.749	0.406	0.343	15.7	8.0	13.9	6.2	13.4
15	0.137	0.457	0.684	0.479	0.274	19.1	10.1	17.3	8.4	15.4
16	0.265	−0.486	0.728	−0.368	0.169	13.3	8.3	11.5	7.1	14.0
17	0.152	−0.429	0.643	−0.597	0.152	18.2	9.5	16.4	7.7	15.0
18	0.182	−0.571	0.642	−0.428	0.214	16.6	10.8	14.8	8.9	18.6
19	0.273	0.447	0.682	−0.455	−0.161	13.1	1.3	11.3	0.5	0.8
20	0.152	0.597	0.643	−0.429	−0.152	18.2	2.6	16.4	0.9	2.1
21	0.307	0.510	0.702	−0.314	−0.234	11.9	2.6	10.3	1.0	2.1

are the theoretically calculated performances of systems 1 and 2, derived from Equations 6.225 and 6.252, respectively, together with the theoretically calculated performance of the corresponding *linear equalizer* at the *transmitter*. The latter will, for convenience, be referred to as system 3.

It can be seen from Table 6.2 that system 2 has in general a considerably better performance than systems 1 and 3, particularly when there is severe amplitude distortion. System 1 has on balance a poorer performance than system 3. It is also clear that the tolerances to additive white Gaussian noise of systems 1 and 2 are typically 1.7 dB below the theoretically calculated values. From theoretical considerations it appears that about 1.2 dB of this reduction is due to the fact that the transmitted signal level is greater than the assumed value, and the remaining 0.5 dB is due to the fact that the actual error probability is twice that assumed.

Computer simulation tests were also carried out on a pure nonlinear equalizer at the receiver, with channels 13, 15 and 17, these channels giving a very considerable reduction in tolerance to noise of this system, which, for convenience, will be referred to as system 4 (see Figs. 5.2 and 5.3). In each test a total of about 30 independent error bursts were counted, giving 95% confidence limits of about ± 0.25 dB in the measured signal/noise ratios, for a given error rate now of 4 in 10^3. System 1 was also tested over these channels at the same error rate and to the same confidence limits. In each case the measured tolerance to noise of system 4 was about 2.2 dB below the theoretical value, the latter being the same as that of system 1 in Table 6.2. However, the measured tolerance to noise of system 1 was about 2.0 dB below the theoretical value. Thus the measured tolerance to noise of system 4 was about 0.2 dB below that of system 1. The relatively large discrepancy between the theoretical and measured performances of system 4 is due to the fact that at error probabilities around 4 in 10^3, the error extension effects produce an appreciably greater increase in the signal/noise ratio for the given error probability, than the increase of about 1 dB to be expected at error probabilities around 1 in 10^5 or 1 in 10^6. At these lower error probabilities, system 4 would have a slightly better tolerance to Gaussian noise than system 1.

Some of the more important results obtained here are now summarized, as follows. A pure nonlinear equalizer at the transmitter, using modulo arithmetic (system 1), is the dual of the corresponding pure nonlinear equalizer using decision-directed cancellation of intersymbol interference at the receiver, and has a similar tolerance to additive white Gaussian noise. In the same way, the equalizer formed by the optimum combination of linear and nonlinear filters at the transmitter (system 2) is the dual of the corresponding decision-feedback equalizer at the receiver, and again has a similar tolerance to additive white Gaussian noise. The latter two equalizers have, in general, a much better performance than the former two, and achieve effectively the best tolerance to additive white Gaussian noise obtainable with a conventional transversal equalizer, at high signal/noise ratios and with the accurate equalization of the channel.

Systems 1 and 2 are the arrangements obtained when a pure nonlinear equalizer (as in Fig. 5.2) and an optimum decision-feedback equalizer (as in Fig. 5.11), respectively, are transferred from the receiver to the transmitter. The transfer of a pure-nonlinear or decision-feedback equalizer from the receiver to the transmitter is made possible by the use of two identical nonlinear processors, one at the transmitter and the other at the receiver, where the processors perform modulo $-m$ transformations on the respective input signals. Just as the feedback transversal filter in a pure-nonlinear or decision-feedback equalizer at the receiver removes the intersymbol interference by decision-directed signal cancellation, with no change in the noise level and therefore with no change in the signal/noise ratio, so also the feedback transversal filter and nonlinear processor in the transmitter of system 1 or 2 removes the intersymbol interference at the receiver, with essentially no change in the transmitted signal level and therefore again with no significant change in the signal/noise ratio.

6.4 COMPARISON OF DIFFERENT EQUALIZERS

The tolerances to additive white Gaussian noise of the different equalizers studied in Chapters 4–6 have been measured by U.S. Tint, using the results of the theoretical analysis presented here[33]. It is assumed that the data-transmission system is as shown in Figs. 4.1, 5.1 or 6.8 and as described in Section 4.2, 5.1 or 6.3, respectively. The systems have been tested over the 21 different channels listed in Table 6.2 and the results of these tests are given in Table 6.3[33]. The meanings of the system numbers in Table 6.3 are given in Table 6.4. In every case the channel is accurately equalized and there is a very high signal/noise ratio. Each entry in Table 6.3 gives the number of decibels increase in the noise level, which is necessary to maintain a given low error probability, when a channel with the corresponding sampled impulse-response is replaced by one introducing no distortion or attenuation, and the transmitted signal level is held unchanged. All systems have the same tolerance to noise in the presence of no distortion and no attenuation. The 5-component row vector, giving the sampled impulse-response of the channel, has unit length in every case, so that each channel introduces distortion but no attenuation or gain. The different values of the sampled impulse-

Table 6.3 NUMBER OF DECIBELS REDUCTION IN TOLERANCE TO ADDITIVE WHITE GAUSSIAN NOISE, AT VERY HIGH SIGNAL/NOISE RATIOS, WHEN THE GIVEN CHANNEL REPLACES ONE THAT INTRODUCES NO DISTORTION OR ATTENUATION

Channel	Sampled impulse-response of channel					System								
						1A	2A	3A	4A	5A	1B	3B	4B	5B
1	0.161	0.974	0.161	0	0	1.0	15.9	0.6	0.6	0.5	0.7	0.5	0.5	0.5
2	0.262	0.929	0.262	0	0	3.1	11.6	2.0	1.9	1.4	2.3	1.7	1.6	1.4
3	0.348	0.870	0.348	0	0	7.9	9.2	5.0	4.4	3.1	5.6	4.1	3.8	3.1
4	0.378	0.845	0.378	0	0	12.0	8.5	8.0	6.4	4.3	8.5	6.1	5.3	4.3
5	0.219	0.750	0.625	0	0	11.4	13.2	4.7	11.4	4.1	8.0	4.4	8.0	4.1
6	0.625	0.750	0.219	0	0	11.4	4.1	9.0	4.1	4.1	8.0	6.5	4.1	4.1
7	0.575	0.776	0.259	0	0	14.5	4.8	11.5	4.8	4.8	10.2	8.1	4.8	4.8
8	0.262	0.928	−0.262	0	0	0.0	11.6	0.4	0.3	0.0	0.0	0.2	0.2	0.0
9	0.348	0.870	−0.348	0	0	0.1	9.2	0.8	0.6	0.1	0.1	0.4	0.3	0.1
10	0.575	0.776	−0.259	0	0	1.5	4.8	2.9	2.6	0.6	1.0	1.7	1.6	0.6
11	0.625	0.750	−0.219	0	0	2.5	4.1	4.0	3.7	0.9	1.6	2.4	2.3	0.9
12	0.196	0.392	0.785	0.392	0.196	9.0	14.2	6.5	6.2	4.9	7.4	5.8	5.6	4.9
13	0.167	0.471	0.707	0.471	0.167	20.4	15.5	16.1	13.5	9.9	17.1	13.6	11.9	9.9
14	0.203	0.339	0.749	0.406	0.343	13.4	13.9	9.1	7.6	6.2	10.4	7.6	6.9	6.2
15	0.137	0.457	0.684	0.479	0.274	15.4	17.3	11.2	12.4	8.4	13.1	10.0	10.6	8.4
16	0.265	−0.486	0.728	−0.368	0.169	14.0	11.5	11.0	10.2	7.1	11.4	9.2	8.8	7.1
17	0.152	−0.429	0.643	−0.597	0.152	15.0	16.4	9.8	12.5	7.7	12.5	8.9	10.7	7.7
18	0.182	−0.571	0.642	−0.428	0.214	18.6	14.8	15.0	19.8	8.9	15.1	12.1	15.0	8.9
19	0.273	0.447	0.682	−0.455	−0.161	0.8	11.3	3.2	1.8	0.5	0.6	2.2	1.2	0.5
20	0.152	0.597	0.643	−0.429	−0.152	2.1	16.4	4.5	3.1	0.9	1.5	3.1	2.1	0.9
21	0.307	0.510	0.702	−0.314	−0.234	2.1	10.3	3.8	2.7	1.0	1.6	2.7	2.0	1.0

Table 6.4 MEANINGS OF THE SYMBOLS IN THE SYSTEM NUMBERS

Symbol	Meaning of the symbol
A	Equalization process located entirely at the receiver
B	Equal sharing of the linear equalization process between the transmitter and receiver
1	Linear equalizer (Sections 4.5 and 6.3.3)
2	Pure nonlinear equalizer achieving the accurate equalization of the channel (Section 5.1)
3	Decision-feedback equalizer using zero forcing (Section 5.5)
4	Decision-feedback equalizer using linear and nonlinear equalization of the two factors of the channel response (Section 5.6)
5	Optimum decision-feedback equalizer (Sections 5.8.2 and 6.3.4)

response of the channel have been selected to give a wide range of different signal distortions.

As has been shown in Section 6.3, the theoretical performances of systems 1A, 2A and 5A remain *unchanged* if the equalizer is transferred from the receiver to the transmitter, so that these results hold for both cases.

The tolerance to additive white Gaussian noise, at high signal/noise ratios, of any system using a pure-nonlinear or decision-feedback equalizer at the receiver, may be up to some 2 dB inferior to the value quoted in Table 6.3 due to error-extension effects, and the degradation is often significantly greater in the case of a pure nonlinear equalizer than in the case of a decision-feedback equalizer, due to the increased average number of errors in an error burst. However, as the signal/noise ratio increases, so the degradation becomes smaller, and at extremely high signal/noise ratios it is usually negligible. This reduction in the degradation does not occur when the equalizer is located at the transmitter.

The results in Table 6.3 show that the poorest tolerance to additive white Gaussian noise is likely to be obtained with system 2A, closely followed by system 1A. Thus the use of pure linear or non-linear equalization at the receiver, with no equalizer at the transmitter, will usually (but not always) give a lower tolerance to additive white Gaussian noise than any of the arrangements of combined linear and nonlinear equalization studied here. A decision-feedback equalizer is, of course, taken to be a *combination* of linear and nonlinear equalizers at the receiver. In the case of pure phase distortion (the nearest approach to which is channel 8) system 1A is at least as good as or better than any other system, and when the zeros of the channel z-transform lie inside the unit circle, as for channel 6 or 7, system 2A is at least as good as or better than any other system.

System 1B gains an advantage of up to some 4.5 dB over system 1A. The advantage gained by system 1B over 1A is generally (but not always) greater than that obtained by system 3B over 3A or by system 4B over 4A. System 5B gains no advantage over system 5A, since the linear filter D is here a pure phase equalizer, which gives no improvement in tolerance to noise when shared between the transmitter and receiver (see also, for instance, systems 1A and 1B with channel 8). Systems 1B, 3B and 4B are consistently as good as or better than 1A, 3A and 4A, respectively.

The best tolerance to additive white Gaussian noise is consistently obtained with systems 5A and 5B, with often a considerable advantage over the other systems. It appears therefore that the potentially most cost-effective system is 5A (Section 5.8.2).

REFERENCES

1. Gersho, A. 'Adaptive equalization of highly dispersive channels for data transmission', *Bell Syst. Tech. J.*, **48**, 55–70 (1969)
2. Proakis, J.G. and Miller, J.H. 'An adaptive receiver for digital signaling through channels with intersymbol interference', *IEEE Trans. Inform. Theory*, **IT-15**, 484–497 (1969)
3. Westcott, R.J. 'An experimental adaptively equalized modem for data transmission over the switched telephone network', *Radio Electron. Eng.*, **42**, 499–507 (1972)
4. Harvey, J.D. *Synchronisation of a Synchronous Modem*, SERC report GR/A/1200.7, SERC, Swindon, UK (1980)
5. Clark, A.P. and Hau, S.F. 'Adaptive adjustment of receiver for distorted digital signals', *Proc. IEE*, Pt. F, **131**, 526–536 (1984)
6. Mark, J.W. and Budihardjo, P.S. 'Joint optimization of receive filter and equalizer', *IEEE Trans. Commun.*, **COM-21**, 264–266 (1973)
7. Mark, J.W. and Budihardjo, P.S. 'Performance of jointly optimized prefilter-equalizer receivers', *IEEE Trans. Commun.*, **COM-21**, 941–945 (1973)
8. Forney, G.D. 'Maximum likelihood sequence estimation for digital sequences in the presence of intersymbol interference', *IEEE Trans. Inform. Theory*, **IT-18**, 363–378 (1972)
9. Andersen, I.N. 'Sample whitened matched filters', *IEEE Trans. Inform. Theory*, **IT-19**, 653–660 (1973)
10. Messerschmitt, D.G. 'A geometric theory of intersymbol interference. Part 1: Zero-forcing and decision feedback equalization', *Bell Syst. Tech. J.*, **52**, 1483–1519 (1973)
11. Ungerboeck, G. 'Fractional tap-spacing equalizer and consequences for clock recovery in data modems', *IEEE Trans. Commun.*, **COM-24**, 856–864 (1976)
12. Akashi, F., Tatsui, N., Sato, Y., Koike, S. and Yasushi, M. 'A high performance digital QAM 9600 bit/s modem', *NEC Res. and Develop.*, **45**, 38–48 (1977)
13. Gitlin, R.D. and Weinstein, S.B. 'Fractionally-spaced equalization: An improved digital transversal equalizer', *Bell Syst. Tech. J.*, **60**, 275–296 (1981)
14. Mueller, M.S. 'Least-squares algorithms for adaptive equalizers', *Bell Syst. Tech. J.*, **60**, 1905–1925 (1981)
15. Gitlin, R.D., Meadows, H.C. and Weinstein, S.B. 'The tap-leakage algorithm: An algorithm for the stable operation of a digitally implemented, fractionally spaced adaptive equalizer', *Bell Syst. Tech. J.*, **61**, 1817–1839 (1982)
16. Papoulis, A. *Probability, Random Variables and Stochastic Processes*, McGraw-Hill, New York (1965)

17. Thomas, J.B. *An Introduction to Statistical Communication Theory*, Wiley, New York (1968)
18. Davenport, W.B. *Probability and Random Processes*, McGraw-Hill, New York (1970)
19. Mazo, J.E. 'Optimum timing phase for an infinite equalizer', *Bell Syst. Tech. J.*, **54**, 189–201 (1975)
20. Salz, J. 'On mean-square decision feedback equalization and timing phase', *IEEE Trans. Commun.*, **COM-25**, 1471–1476 (1977)
21. Tufts, D.W. 'Nyquist's problem—the joint optimization of transmitter and receiver in pulse amplitude modulation', *Proc. IEEE*, **53**, 248–259 (1965)
22. Smith, J.W. 'The joint optimization of transmitted signal and receiving filter for data transmission systems', *Bell Syst. Tech. J.*, **44**, 2363–2392 (1965)
23. Aaron, M.R. and Tufts, D.W. 'Intersymbol interference and error probability', *IEEE Trans. Inform. Theory*, **IT-12**, 26–34 (1966)
24. Berger, T. and Tufts, D.W. 'Optimum pulse amplitude modulation. Part I: Transmitter-receiver design and bounds from information theory', *IEEE Trans. Inform. Theory*, **IT-13**, 196–208 (1966)
25. Ericson, T. 'Structure of optimum receiving filters in data-transmission systems', *IEEE Trans. Inform. Theory*, **IT-17**, 352–353 (1971)
26. Hansler, E. 'Some properties of transmission systems with minimum mean-square error', *IEEE Trans. Commun. Technol.*, **COM-19**, 576–579 (1971)
27. Yao, K. 'On the minimum average probability of error expression for binary pulse communication systems with intersymbol interference', *IEEE Trans. Inform. Theory*, **IT-18**, 528–531 (1972)
28. Ericson, T. 'Optimum PAM filters are always bandlimited', *IEEE Trans. Inform. Theory*, **IT-19**, 570–573 (1973)
29. Dwight, H.B. *Tables of Integrals and Other Mathematical Data*, fourth edition, Macmillan, New York (1961)
30. Tomlinson, M. 'New automatic equalizer employing modulo arithmetic', *Electronics Letters*, **7**, 138–139 (1971)
31. Harashima, H. and Miyakawa, H. 'Matched transmission technique for channels with intersymbol interference', *IEEE Trans. Commun.*, **COM-20**, 774–780 (1972)
32. Clark, A.P. and Serinken, M.N. 'Nonlinear equalizers using modulo arithmetic', *Proc. IEE*, **123**, 32–38 (1976)
33. Clark, A.P. and Tint, U.S. 'Linear and non-linear transversal equalizers for baseband channels', *Radio Electron. Eng.*, **45**, 271–283 (1975)

Chapter 7

Developments of the conventional decision-feedback equalizer

7.1 INTRODUCTION

It is shown in Chapter 5 and Table 6.3 that when the channel introduces severe amplitude distortion, a decision-feedback equalizer achieves a much better tolerance to additive white Gaussian noise than a linear equalizer. However, under these conditions, a decision-feedback equalizer sometimes *itself* has tolerance to noise *well below* that where there is no amplitude distortion, and its tolerance to noise is significantly inferior to that obtainable with a maximum-likelihood or near-maximum-likelihood detector operating on the distorted signal[1-61]. This suggests that the performance of the decision-feedback equalizer may perhaps be significantly improved through an appropriate modification to its design. Certainly there is adequate scope for some improvement, as is evidenced by the good performances achieved with a variety of different detection processes[1-61]. In Sections 7.2–7.4 we are not primarily concerned with achieving the *best* available tolerance to noise, such as would be achieved with a maximum-likelihood detector (Section 1.4), but rather with *improving* the performance of a conventional decision-feedback equalizer while still retaining its basic structure. Thus the receiver is constrained to be some combination of linear and nonlinear transversal filters together with a simple threshold-level detector.

Consider first, in a little more detail, the performance of a conventional decision-feedback equalizer. With the transmission of a baseband data-signal having four or more levels, and where the intersymbol interference is not too severe, a conventional decision-feedback equalizer gives a good tolerance to Gaussian noise, which is often no more than some two or three decibels below that obtainable with the corresponding maximum-likelihood (optimum) detector[26,32,48,54,57]. Clearly, the design or adjustment of the equalizer must here be close to that giving the lowest error rate in the detected data-symbols, for *an equalization process*. However, with the transmission of a binary signal and when there is severe amplitude distortion of the transmitted signal, the tolerance to noise may be very considerably below that obtainable with a maximum-likelihood detector[26,32,54]. A recent theoretical investigation into the reason for this large degradation in performance has revealed that, under the particular conditions, the design or adjustment of the conventional decision-feedback equalizer is no longer optimum[53]. In other words, in the presence of additive white Gaussian noise at high signal/noise ratios, the decision-feedback equalizer, that achieves the accurate equalization of the channel, no longer minimizes (even approximately) the error rate in the detected data signal for an equalization process. In order to take account of and hence to correct for this effect, in the particular case where a linear channel has introduced severe amplitude distortion into a binary baseband data-signal, the design of the decision-feedback equalizer for the given channel is now modified, essentially be relaxing the requirement for the exact or even approximate equalization of the channel and hence by using a fundamentally different criterion in the optimization process involved in the equalizer design[53].

Sections 7.3 and 7.4 are concerned with the basic theory of the modified (optimized) equalizer, together with the conditions under which it differs from the conventional equalizer and the simple modification by means of which it is derived from the latter. A suboptimum version of the modified equalizer is also considered. The conventional equalizer is taken to be the ideal decision-feedback equalizer that maximizes the signal/noise ratio in the equalized signal subject to the accurate equalization of the channel. The received signal is assumed to be a binary real-valued baseband signal that has been subjected to severe amplitude distortion. The baseband signal may itself have been fed over the transmission path or else it may have been derived by the linear demodulation of a received suppressed-carrier AM signal. The optimum design of the modified equalizer is derived first and then an approximate expression for its tolerance to additive white Gaussian noise. The latter is used to predict the performance of the modified equalizer over a wide range of channels, and hence to compare its performance with that of a conventional decision-feedback equalizer. The whole procedure is next repeated for an interesting suboptimum design of the modified equalizer. Finally, in Section 7.4, results are presented of computer-simulation tests, with binary signals transmitted over various channels in the presence of additive white Gaussian noise, and the results are compared with the corresponding theoretical predictions. Sections 7.5 and 7.6 are concerned with *simplified* forms of the equalizer considered in Sections 7.3 and 7.4, and Section 7.8 is concerned with more sophisticated near-maximum-likelihood detectors in which the

DEVELOPMENTS OF THE DECISION-FEEDBACK EQUALIZER 389

threshold-level detector is replaced by a more complex detection process.

7.2 CONVENTIONAL EQUALIZER

The model of the data-transmission system when using a conventional decision-feedback equalizer is shown in Fig. 7.1. This is a synchronous serial system in which the information transmitted is carried by the binary data-symbols $\{s_i\}$, where

$$s_i = \pm 1 \tag{7.1}$$

the $\{s_i\}$ being statistically independent and equally likely to have either binary value. Clearly $k=1$, so that the mean-square value of s_i and hence also the average transmitted energy per signal element, at the input to the transmission path, is unity. The impulses at the input to the channel are regularly spaced at intervals of T seconds and form a sequence of binary polar signal-elements. It is assumed that $s_i = 0$ for $i < 0$, so that $s_i \delta(t - iT)$ is the $(i+1)$th transmitted signal-element.

The linear baseband channel (Fig. 7.1) includes the lowpass filter A (transmitter output filter), a linear baseband transmission path and the lowpass filter B (receiver input filter). The transmission path could, of course, itself include a linear modulator at the transmitter, a

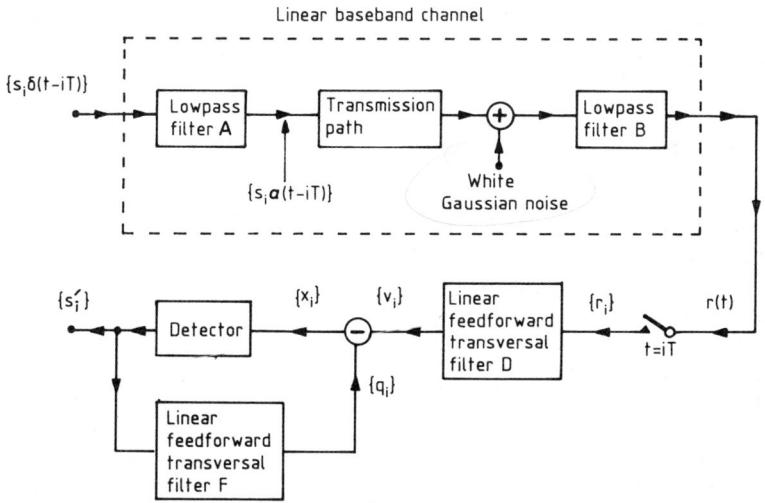

Fig. 7.1 Model of data-transmission system with a conventional decision-feedback equalizer

Control station

bandpass transmission path such as a telephone circuit, and a linear demodulator at the receiver. The resultant channel has an impulse response $y(t)$, and its various properties are as described in Section 1.2. White Gaussian noise is added to the data signal at the output of the transmission path to give the bandlimited Gaussian noise waveform $w(t)$, with zero mean and variance σ^2, at the output of the lowpass filter B. Thus the output waveform from the linear baseband channel is

$$r(t) = \sum_i s_i y(t - iT) + w(t) \tag{7.2}$$

where $r(t)$, $y(t)$ and $w(t)$ have only real values. $y(t-iT)$ is taken to be time invariant over any one transmission, in the sense that its shape is not a function of i, and it is assumed to have a finite duration, at least for practical purposes.

The received waveform $r(t)$ is sampled at the time instants $\{iT\}$ to give the real-valued samples $\{r_i\}$ which are fed to the $(n+1)$-tap linear feedforward transversal filter D, whose corresponding output samples are the $\{v_i\}$. As before (Section 5.8)

$$r_i = \sum_{j=0}^{g} s_{i-j} y_j + w_i \tag{7.3}$$

where the $\{w_i\}$ are statistically independent Gaussian random variables with zero mean and variance σ^2. The sampled impulse-response of the baseband channel is given by the $(g+1)$-component vector (sequence)

$$Y = [y_0 \quad y_1 \quad \ldots \quad y_g] \tag{7.4}$$

and the sampled impulse-response of the linear filter D is given by the $(n+1)$-component vector

$$D = [d_0 \quad d_1 \quad \ldots \quad d_n] \tag{7.5}$$

The resultant sampled impulse-response of the linear baseband channel and filter D (Fig. 7.1) is given by the time-invariant vector (sequence)

$$E = [e_0 \quad e_1 \quad \ldots \quad e_g] \tag{7.6}$$

where the delay in transmission, other than that involved in the time dispersion of the signal, is neglected, so that the component e_0 is taken to have *no* delay. The vector E is given by the *convolution* of Y and D, ignoring the zero-valued components $\{e_i\}$ that precede and follow those in E. Thus $e_0 \neq 0$ and $e_i = 0$ for $i < 0$ and $i > g$. The receiver is assumed to have prior knowledge of Y and E, as well as of the possible values of s_i. Ideally, $n \to \infty$.

The decision-feedback equalizer (Figs. 7.1 and 7.2) is adjusted to

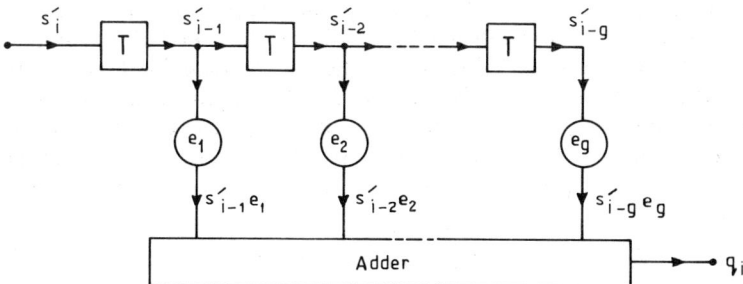

Fig. 7.2 *Linear feedforward transversal filter F*

maximize the signal/noise ratio in the equalized signal subject to the accurate equalization of the channel. As before, the latter is taken to imply the removal of all intersymbol interference from the equalized signal. However, the signal level is *not* now adjusted such that the equalized signal is an *unbiased estimate* of the wanted data symbol (see Section 5.8.1). Instead, the $(n+1)$-tap linear feedforward transversal filter D (Fig. 7.1) is constrained to introduce *no* gain or attenuation into the received signal. The decision-feedback equalizer has the following important properties, which are essentially the same as those described in Sections 5.8.2 and 6.2.4. In the z-transform of the channel and filter D, all zeros (roots) of the z-transform of the channel that lie outside the unit circle in the z plane (ignoring any at infinity) are replaced by their reciprocals, the remaining zeros being left unchanged. Thus all zeros of the z-transform of the channel and filter lie *inside* the unit circle. All zeros of the channel and of the channel and filter, of course, occur in *complex-conjugate* pairs. Furthermore, since the z-transform of the linear filter D represents *pure phase distortion* (Section 3.3), the filter does not change any *amplitude distortion* in the received signal nor does it change the signal or noise level. Hence the filter is an *allpass* network that adjusts the sampled impulse-response of the channel and filter to be *minimum phase* (ignoring the delay of n sampling intervals introduced by the filter). The filter therefore introduces an *orthogonal* transformation into the received signal and so does not change the inner products, lengths, distances or angles between the different possible received signal vectors, corresponding to the different possible received messages (see Section 1.3). In addition, the filter does not change the signal/noise ratio nor any of the noise statistics. Perhaps the most important single function performed by the filter is that it *maximizes* the ratio of the magnitude of the first nonzero component, e_0, of the sampled impulse-response of the channel and filter, to the noise variance at the output of the filter. The various properties just described for the filter D, of course, only hold *exactly* if the filter has an infinite number of taps. However, by using an appropriately large number of taps (which is usually not excessive) the properties of the filter can be made to approach as closely as required to those described.

The signal v_i at the output of the filter D, at time $t = iT$ and when ignoring the delay of n sampling intervals actually introduced by the filter, is

$$v_i = \sum_{j=0}^{g} s_{i-j} e_j + u_i \tag{7.7}$$

where the $\{u_i\}$ have the *same* statistical properties as the $\{w_i\}$, for the reasons just explained. Thus the $\{u_i\}$ are statistically independent Gaussian random variables with zero mean and fixed variance σ^2, and the probability density function of u_i, at a sample value u, is

$$\frac{1}{\sqrt{2\pi\sigma^2}} \exp\left(-\frac{u^2}{2\sigma^2}\right) \tag{7.8}$$

All signals here are *real valued*.

In the conventional arrangement of the decision-feedback equalizer, the linear feedforward transversal filter F is as shown in Fig. 7.2. It has g taps with gains equal to $e_1, e_2, \ldots, e_g$, and operates on the detected data-symbols $\{s'_{i-j}\}$ at the detector output, such that, at time $t = iT$ and just prior to the detection of s_i, the output signal from the filter F is

$$q_i = \sum_{j=1}^{g} s'_{i-j} e_j \tag{7.9}$$

This is subtracted from v_i to give the equalized signal

$$x_i = v_i - q_i \tag{7.10}$$

at the detector input. The symbol q_i here is *not* related to any of the coefficients $q_0, q_1, \ldots, q_f$ in the factor $Y_2(z)$ of the z-transform $Y(z)$ of the channel (see Equations 4.96 and 5.267). With the correct detection of the data-symbols $s_{i-1}, s_{i-2}, \ldots, s_{i-g}$,

$$x_i = s_i e_0 + u_i \tag{7.11}$$

and, regardless of the correct detection of these symbols, s_i is now detected as its possible value s'_i such that $s'_i e_0$ is closest to x_i.

Assume for the moment that Equation 7.11 holds and, as before, let

$$Q(a) = \int_a^\infty \frac{1}{\sqrt{2\pi}} \exp\left(-\tfrac{1}{2} u^2\right) du \tag{7.12}$$

where a is some positive quantity. Then it can be seen from

Expression 7.8 and Equations 7.1 and 7.11 that, with the correct detection of $s_{i-1}, s_{i-2}, \ldots, s_{i-g}$, the probability of error in the detection of s_i from x_i is

$$P_3 = Q\left(\frac{|e_0|}{\sigma}\right) \qquad (7.13)$$

where $|e_0|$ is the absolute value (modulus) of e_0. P_3 is a lower bound to the actual error rate and, at low error rates of below 1 in 10^5, it can be taken as an approximate estimate of this rate.

7.3 MODIFIED EQUALIZER

In the absence of any amplitude distortion or attenuation in the received data-signal, $e_0 = 1$ and $e_i = 0$ for $i = 1, 2, \ldots, g$, any phase distortion in the received signal being removed by the linear filter D in Fig. 7.1 and therefore not affecting the values of $e_0, e_1, \ldots, e_g$. The greater the level of the amplitude distortion in the received signal, the greater the values of $|e_1|, |e_2|, \ldots, |e_g|$ relative to $|e_0|$. When the distortion is severe and involves considerable bandlimiting of the data signal, the latter being transmitted at a rate close to the Nyquist rate[62-64], it is likely that

$$|e_1| > |e_0| \qquad (7.14)$$

This condition is often satisfied in applications such as the very high speed transmission of digital signals over long lengths of a coaxial cable, or where a digital signal is being transmitted at the highest practical rate over any bandlimited channel. The condition is also often observed in data-transmission systems operating at more than 9600 bit/s over the poorer telephone circuits. (The baseband digital signals here may, of course, be complex valued). It will now be shown that, when Equation 7.14 holds (together with the other conditions described in Section 7.2) and when binary data-symbols are transmitted (Equation 7.1), it is possible to achieve an improved tolerance to noise through the appropriate modification of the decision-feedback equalizer[53].

The equalizer is modified by inserting the linear feedforward transversal filter C, as shown in Fig. 7.3, between the filter D and the subtractor circuit in Fig. 7.1. In practice, of course, the filter D would be modified into the equivalent resultant filter, but it is simpler here to consider the two separate filters. The output signal from the filter D, at time $t = iT$, is v_i in Equation 7.7, as before. The output signal from the filter C, at time $t = (i+1)T$, is

$$r'_{i+1} = v_{i+1}c_0 + v_i c_1 \qquad (7.15)$$

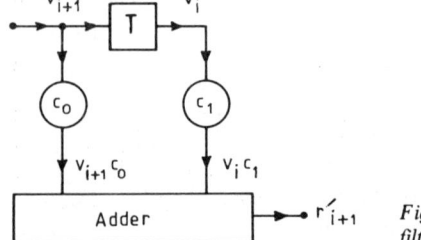

Fig. 7.3 Linear feedforward transversal filter C

and the resultant sampled impulse-response of the channel and filters D and C is given by the $(g+2)$-component row vector

$$Y' = [y'_0 \ y'_1 \ \ldots \ y'_{g+1}] \quad (7.16)$$

whose $(i+1)$th component is

$$y'_i = e_i c_0 + e_{i-1} c_1 \quad (7.17)$$

where, of course, $e_i = 0$ for $i < 0$ and $i > g$. Thus the signal at the output of the filter C, at time $t = (i+1)T$, is

$$r'_{i+1} = s_{i+1} y'_0 + s_i y'_1 + \cdots + s_{i-g} y'_{g+1} + w'_{i+1} \quad (7.18)$$

where

$$w'_{i+1} = u_{i+1} c_0 + u_i c_1 \quad (7.19)$$

Clearly, w'_{i+1} is a Gaussian random variable with zero mean and variance

$$\eta^2 = \sigma^2 (c_0^2 + c_1^2) \quad (7.20)$$

the $\{w'_i\}$ being correlated.

The linear feedforward transversal filter F (Fig. 7.1) has g taps, as before (Fig. 7.2), but now with tap gains equal to $y'_2, y'_3, \ldots, y'_{g+1}$ in place of $e_1, e_2, \ldots, e_g$, respectively. At time $t = (i+1)T$ the detector has just determined s'_{i-1} and the output signal from the filter F is

$$q'_{i+1} = s'_{i-1} y'_2 + s'_{i-2} y'_3 + \cdots + s'_{i-g} y'_{g+1} \quad (7.21)$$

q'_{i+1} is subtracted from r'_{i+1} to give the equalized signal

$$x'_{i+1} = r'_{i+1} - q'_{i+1} \quad (7.22)$$

which is fed to the detector. With the correct detection of $s_{i-1}, s_{i-2}, \ldots, s_{i-g}$,

$$\begin{aligned} x'_{i+1} &= s_{i+1} y'_0 + s_i y'_1 + w'_{i+1} \\ &= s_{i+1} e_0 c_0 + s_i (e_1 c_0 + e_0 c_1) + w'_{i+1} \end{aligned} \quad (7.23)$$

The detected data-symbol s'_i is taken to be the possible value of s_i such that the term

$$s_i(e_1c_0 + e_0c_1) \tag{7.24}$$

in Equation 7.23 is closest to x'_{i+1}. The term $s_{i+1}e_0c_0$ is an intersymbol-interference component in x'_{i+1} and is equally likely to have either value $\pm|e_0c_0|$. It is evident that, for the best tolerance to noise, e_1c_0 and e_0c_1 must both have the *same* sign which, for convenience, is taken to be *positive*. Thus

$$e_1c_0 \geqslant 0, \qquad e_0c_1 \geqslant 0 \tag{7.25}$$

and the detected data-symbol s'_i is taken as 1 or -1 depending upon whether x'_{i+1} is positive or negative, respectively. It can be seen from Equation 7.23 that a necessary condition for the correct operation of the equalizer is that

$$e_1c_0 + e_0c_1 > |e_0c_0| \tag{7.26}$$

which will therefore be assumed.

At high signal/noise ratios, practically all errors in the $\{s'_i\}$ occur when $s_{i+1}e_0c_0$ has the opposite sign to $s_i(e_1c_0 + e_0c_1)$, since under these conditions

$$|s_{i+1}e_0c_0 + s_i(e_1c_0 + e_0c_1)| \tag{7.27}$$

is minimum, and the Gaussian noise component w'_{i+1} needed to produce an error in s'_i now has the smallest magnitude. Thus, the condition under which nearly all errors in the $\{s'_i\}$ occur is that

$$s_{i+1}e_0c_0 = -s_i|e_0c_0| \tag{7.28}$$

because, from Equations 7.25,

$$e_1c_0 + e_0c_1 > 0 \tag{7.29}$$

Hence, most errors occur when

$$x'_{i+1} = s_i(e_1c_0 + e_0c_1 - |e_0c_0|) + w'_{i+1} \tag{7.30}$$

Let
$$dc_0 = e_1c_0 - |e_0c_0| \tag{7.31}$$

bearing in mind that $e_1c_0 \geqslant 0$ (Equations 7.25). Thus, nearly all errors occur when

$$x'_{i+1} = s_i(dc_0 + e_0c_1) + w'_{i+1} \tag{7.32}$$

Since $s_i = \pm 1$ and $dc_0 + e_0c_1 > 0$ (Equations 7.1, 7.26 and 7.31), the condition for an error in the detection of s_i from x'_{i+1} is that w'_{i+1} has a magnitude greater than $dc_0 + e_0c_1$ and the opposite sign to s_i. Furthermore, w'_{i+1} has zero mean and variance η^2 (Equation 7.20), and Equation 7.32 is satisfied with a probability of $\frac{1}{2}$. Thus the probability of error in the detection of s_i, at high signal/noise ratios and given the correct detection of $s_{i-1}, s_{i-2}, \ldots, s_{i-g}$, can be taken to be $\frac{1}{2}Q(a)$, where

$$a = \frac{dc_0 + e_0 c_1}{\eta} = \frac{dc_0 + e_0 c_1}{\sigma \sqrt{(c_0^2 + c_1^2)}} \tag{7.33}$$

from Equation 7.20. An approximate estimate of the actual error rate in the $\{s_i'\}$, at high signal/noise ratios (but, of course, not now assuming the necessity of the correct detection of $s_{i-1}, s_{i-2}, \ldots, s_{i-g}$), can be taken to be $Q(a)$. $Q(a)$ is minimum when a (Equation 7.33) is maximum, which occurs when

$$c_0 = bd, \qquad c_1 = be_0 \tag{7.34}$$

and b is any positive constant. This follows directly from the theory of *matched filters* (Section 1.5)[62-64]. For Equations 7.34 to hold with $c_0 \neq 0$, it is necessary that, in Equation 7.33,

$$dc_0 > 0 \tag{7.35}$$

which means that $e_1 c_0 > |e_0 c_0|$, from Equation 7.31, so that $|e_1| > |e_0|$ as in Equation 7.14. Under these conditions and from Equations 7.31 and 7.34,

$$\frac{c_0^2}{b} = e_1 c_0 - |e_0 c_0| \tag{7.36}$$

so that

$$|c_0| = b(|e_1| - |e_0|) \tag{7.37}$$

and c_0 has the *same sign* as e_1. c_1 is, of course, given by Equations 7.34. Furthermore, from Equation 7.31,

$$|d| = |e_1| - |e_0| \tag{7.38}$$

and d has the *same sign* as e_1. Now, from Equations 7.33–7.38 (where $dc_0 > 0$), the theoretical estimate of the error rate in the $\{s_i'\}$ is

$$P_1 = Q(a) = Q\left(\frac{dc_0 + e_0 c_1}{\sigma \sqrt{[c_0^2 + c_1^2]}}\right)$$
$$= Q\left(\frac{b(|e_1| - |e_0|)^2 + be_0^2}{\sigma \sqrt{[b^2(|e_1| - |e_0|)^2 + b^2 e_0^2]}}\right) = Q\left(\frac{\sqrt{[(|e_1| - |e_0|)^2 + e_0^2]}}{\sigma}\right) \tag{7.39}$$

which can be compared with the corresponding theoretical estimate P_3 (Equation 7.13) for the conventional decision-feedback equalizer.

If $|e_1| \leqslant |e_0|$, then, from Equations 7.25 and 7.31, $dc_0 \leqslant 0$, so that a in Equation 7.33 is maximum when $c_0 = 0$ and $c_1 = be_0$. The arrangement now degenerates into a conventional decision-feedback equalizer and the theoretical estimate of the error rate in the $\{s_i'\}$ becomes

$$Q(a) = Q\left(\frac{dc_0 + e_0c_1}{\sigma\sqrt{c_0^2 + c_1^2}}\right) = Q\left(\frac{be_0^2}{\sigma\sqrt{b^2 e_0^2}}\right) = Q\left(\frac{|e_0|}{\sigma}\right) \quad (7.40)$$

as would be expected. Thus, for any advantage to be gained from the given system it is necessary that Equation 7.14 holds, which will now be assumed.

It follows from Equations 7.13 and 7.39 that, at a given low error probability, the value of σ for the modified equalizer is approximately $\sqrt{[(|e_1| - |e_0|)^2 + e_0^2]}/|e_0|$ times that for the conventional equalizer, which means that, at low error rates, the modified equalizer gains an advantage of about

$$20\log_{10}\left(\frac{\sqrt{[(|e_1| - |e_0|)^2 + e_0^2]}}{|e_0|}\right) = 10\log_{10}\left(\frac{(|e_1| - |e_0|)^2}{e_0^2} + 1\right) \quad (7.41)$$

dB in tolerance to Gaussian noise over the conventional equalizer. Figure 7.4 shows how the theoretical advantage in tolerance to Gaussian noise of the modified equalizer (the *optimum system*) over the conventional equalizer varies with $|e_1|/|e_0|$.

A suboptimum but very interesting arrangement of the modified equalizer is given by

$$c_0 = be_1, \quad c_1 = be_0 \quad (7.42)$$

where b is any positive constant, as before. This ensures that Equations 7.25 and 7.26 are satisfied. From the first part of Equation 7.39, together with Equations 7.31 and 7.42, the theoretical estimate of the error rate in the $\{s_i'\}$ here becomes

$$P_2 = Q\left(\frac{dc_0 + e_0c_1}{\sigma\sqrt{c_0^2 + c_1^2}}\right) = Q\left(\frac{e_1c_0 + e_0c_1 - |e_0c_0|}{\sigma\sqrt{c_0^2 + c_1^2}}\right)$$

$$= Q\left(\frac{e_1^2 + e_0^2 - |e_0 e_1|}{\sigma\sqrt{e_1^2 + e_0^2}}\right) \quad (7.43)$$

so that, at low error rates, this arrangement of the modified equalizer gains an advantage of about

$$20 \log_{10} \left(\frac{e_1^2 + e_0^2 - |e_0 e_1|}{|e_0|\sqrt{e_1^2 + e_0^2}} \right)$$

$$= 10 \log_{10} \left(\frac{(e_1^2 + e_0^2)^2 + e_0^2 e_1^2 - 2|e_0 e_1|(e_1^2 + e_0^2)}{e_0^2(e_1^2 + e_0^2)} \right)$$

$$= 10 \log_{10} \left(\frac{e_1^2 + e_0^2 - 2|e_0 e_1|}{e_0^2} + \frac{e_1^2}{e_1^2 + e_0^2} \right)$$

$$= 10 \log_{10} \left(\frac{(|e_1| - |e_0|)^2}{e_0^2} + 1 - \frac{e_0^2}{e_1^2 + e_0^2} \right) \tag{7.44}$$

dB in tolerance to Gaussian noise over the conventional equalizer, as can be seen from Equations 7.13 and 7.43. The variation of this advantage with $|e_1|/|e_0|$, for the suboptimum system just described, is shown in Fig. 7.4. It is evident from Equations 7.41 and 7.44 that the performance of the suboptimum system (Equations 7.42) is inferior to that of the optimum system (Equations 7.34 and 7.39), but as $|e_1|/|e_0|$ increases so the advantage of the latter over the former rapidly diminishes. It is also interesting to observe that, when $|e_1|/|e_0|$ is less than about 1.55, the performance of the suboptimum system becomes inferior to that of the conventional equalizer, with a maximum degradation of about 3.8 dB (not shown in Fig. 7.4), whereas the optimum system is always at least as good as the equalizer.

A clearer insight into the relationship between the optimum and suboptimum systems is given by the following simple representation of the results obtained. In the suboptimum system, the filter C is matched to the 2-component vector

$$E_2 = [e_0 \quad e_1] \tag{7.45}$$

whereas, in the optimum system, the filter C is matched to the 2-component vector

$$F_2 = [e_0 \quad (e_1 \pm e_0)] \tag{7.46}$$

of the minimum Euclidean length (see Section 1.3). Thus in the optimum system,

$$c_0 = b(e_1 \pm e_0), \qquad c_1 = b e_0 \tag{7.47}$$

where b is a positive constant and the appropriate sign is used in $e_1 \pm e_0$.

A theoretical study has been carried out, along the same lines as above, on two developments of the basic technique described. In the interests of brevity and since the developments are not considered

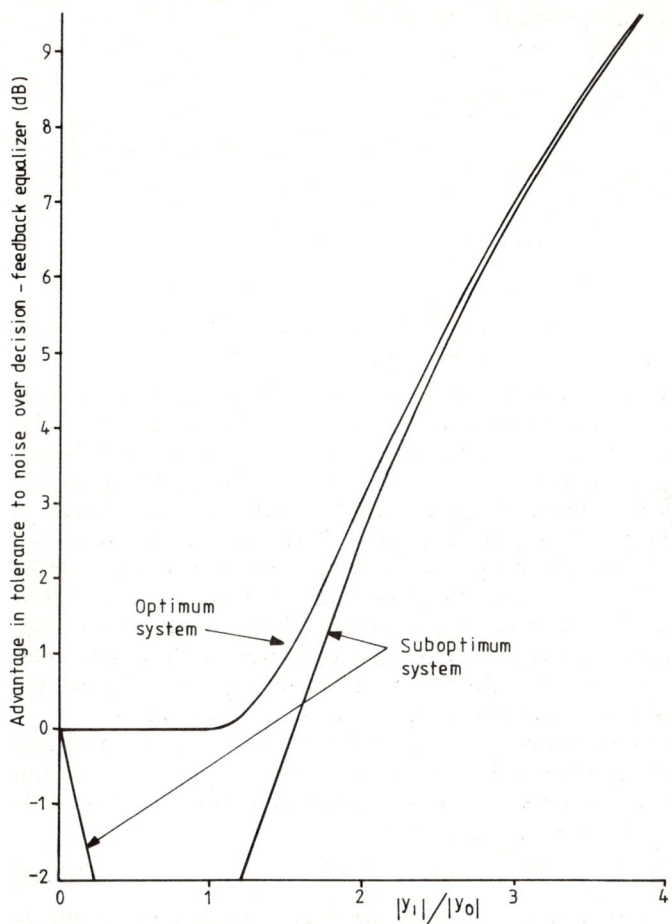

Fig. 7.4 *Theoretically predicted performance of the two modified equalizers at high signal/noise ratios*

further here, only the more important results of this analysis are now presented.

When
$$|e_1| > |e_0| \tag{7.48}$$
and
$$|e_2| > |e_1| + |e_0| \tag{7.49}$$

a further improvement in performance of the optimum system can be achieved, at high signal/noise ratios, by using three taps for the filter C, which is now matched to the two values of the 3-component vector

$$F_3 = [s_i e_0 \quad (s_i e_1 + s_{i+1} e_0) \quad (s_i e_2 + s_{i+1} e_1 + s_{i+2} e_0)] \tag{7.50}$$

of the minimum Euclidean length. s'_i is here determined at time $t = (i+2)T$ from the signal

$$x'_{i+2} = r'_{i+2} - q'_{i+2} \tag{7.51}$$

at the detector input, r'_{i+2} and q'_{i+2} being appropriately modified for the new filter C.

Again, when the filter C has just two taps but 4-level (quaternary) data symbols are used, such that

$$s_i = \pm 1 \text{ or } \pm 3 \tag{7.52}$$

it is necessary that

$$|e_1| > 3|e_0| \tag{7.53}$$

before any advantage in tolerance to noise can be achieved by the optimum arrangement of the modified equalizer over the conventional equalizer. The filter C is now matched to the 2-component vector

$$F_2 = [e_0 \quad (e_1 \pm 3e_0)] \tag{7.54}$$

of the minimum Euclidean length.

Since, in the two cases just described, the amplitude distortion introduced by the channel must be very severe indeed before any useful advantage is gained, it is clear that the most important application is that previously studied where binary data-symbols are used and the filter C has two taps.

7.4 COMPUTER-SIMULATION TESTS

B.A. Hussein has used computer-simulation tests to compare the tolerances to additive Gaussian noise of the optimum and suboptimum arrangements of the modified equalizer with that of a conventional equalizer, for two taps in the filter C, binary signals and the seven different channels shown in Table 7.1[53]. The model of the data-transmission system assumed in the tests is that described in Sections 1.2 and 7.2 and shown in Fig. 7.1.

Each vector E in Table 7.1 has unit length so that no signal gain or attenuation is introduced by any channel. Channel 1 is a well known partial-response channel[64], and channel 2 was obtained with an actual telephone circuit in the transmission path[32]. Whereas the z-transforms of the sampled impulse-responses of channels 2 and 3 have *no* zeros (roots) on the unit circle in the z plane, all zeros of the z-transforms of the channels 1 and 4–7 lie *on* the unit circle[32]. Channels 4 and 6 are unfavourable to nonlinear equalizers[32], whereas channels

Table 7.1 SAMPLED IMPULSE RESPONSE OF THE CHANNEL AND FILTER D, FOR EACH OF THE DIFFERENT CHANNELS

Channel				E			
1	0.408	0.816	0.408				
2	0.548	0.789	0.273	−0.044	0.027		
3	0.321	0.620	0.633	0.322	0.087		
4	0.167	0.500	0.667	0.500	0.167		
5	0.29	0.50	0.58	0.50	0.29		
6	0.085	0.289	0.493	0.577	0.493	0.289	0.085
7	0.19	0.35	0.46	0.50	0.46	0.35	0.19

5 and 7 are particularly unfavourable to maximum-likelihood detectors[65]. The latter channels introduce very severe amplitude distortion. The various channels have been selected to test the different systems over a wide range of values of $|e_1|/|e_0|$, for which Equation 7.14 holds, in order to provide a thorough check for the theoretical results obtained. Furthermore, the tests not only include two practical channels (1 and 2) but also two channels (5 and 7) for which a maximum-likelihood detector is likely to gain only a relatively small advantage in tolerance to additive white Gaussian noise over a conventional decision-feedback equalizer, at the given high levels of amplitude distortion.

The results of the computer-simulation tests, together with the theoretical estimates given by Equations 7.13, 7.39 and 7.43, for the important case where binary data-symbols are transmitted (Equation 7.1), are shown in Figs. 7.5–7.11[53]. The error rate in the detected data symbols $\{s_i'\}$ is here plotted against the signal/noise ratio in dB. The latter is defined as

$$\psi = 10 \log_{10} (1/\sigma^2) \qquad (7.55)$$

where the mean-square value of the data-symbol s_i and hence the average transmitted energy per signal element (at the input to the transmission path) is unity, and σ^2 is the variance of the Gaussian noise components $\{u_i\}$ and $\{w_i\}$, which is numerically equal to the two-sided power spectral density of the white Gaussian noise (at the output of the transmission path). Each individual measurement of the error rate in the $\{s_i'\}$ has been made with the transmission of either 50,000 or 100,000 data-symbols at the given signal/noise ratio, an average of over 3×10^6 data-symbols being used in plotting each curve (Figs. 7.5–7.11)[53]. The 95% confidence limits are up to about ± 0.5 dB. However, for the relative performance of the optimum and suboptimum systems, at all error rates tested, the confidence limits are much better than is suggested by the above figure, through the use of

402 DEVELOPMENTS OF THE DECISION-FEEDBACK EQUALIZER

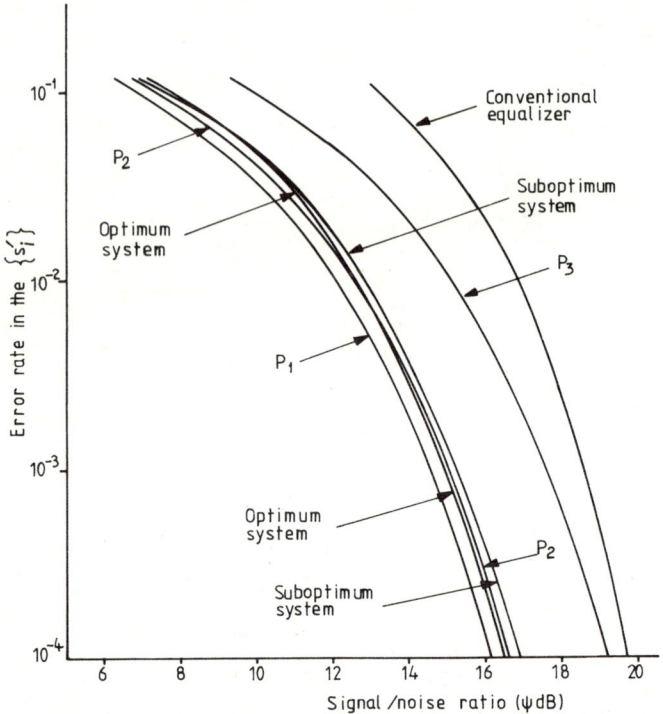

Fig. 7.5 *Performance of equalizers with channel 1*

the same sequence of $\{s_i\}$ and the same sequence of $\{u_i\}$ at any given signal/noise ratio.

Before comparing the theoretical and measured results in Figs. 7.5–7.11 it is necessary first to consider the approximations that have been made in the theoretical estimates P_1, P_2 and P_3 of the error rates in the three systems. The approximate theoretical estimate, P_3 (Equation 7.13), of the probability of error in the conventional equalizer assumes the correct values of the last g detected data-symbols $\{s'_{i-j}\}$. The incorrect detection of s_i means that its inter-symbol interference in the equalized signals x_{i+1}, x_{i+2}, ..., x_{i+g} is likely to cause errors in the detected values of some of the following data symbols (an error-extension effect), so that errors tend to occur in bursts. The actual error rate (when this is less than about 1 in 10) can now be taken to be P_3 multiplied by the average number of errors in an error burst. The latter number typically lies in the range 1 to 100 but may occasionally be a little greater. However, at high signal/noise ratios (giving error rates of less than 1 in 10^5) a change of ten times in

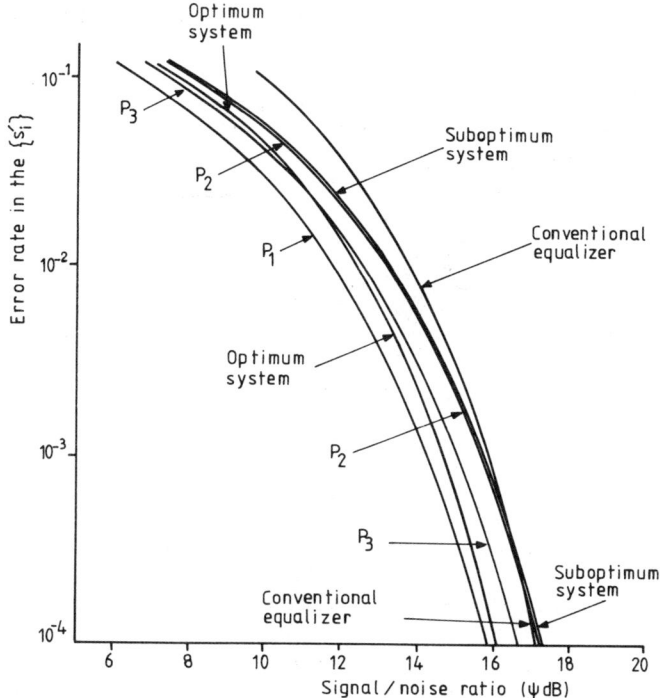

Fig. 7.6 *Performance of equalizers with channel 2*

the error rate, for a given received data signal and a given equalizer, corresponds to a change of less than 1 dB in the signal/noise ratio, so that a discrepancy here in the value of the error rate, of a factor of several times, does not constitute a serious inaccuracy, the error rate being an extremely sensitive measure of the signal/noise ratio[26].

As the error rate with a conventional equalizer becomes smaller so the value of σ (the standard deviation of u_i) that actually causes this error rate becomes closer to the value estimated theoretically by setting P_3 (Equation 7.13) to the given error rate. The theoretically estimated value of σ, for a given error rate, is an upper bound to the actual value, the bound becoming tighter as the error rate becomes smaller or as the number of the larger components of E decreases. The latter holds because the greater the level of the intersymbol interference at the output of the filter D, the greater the average length of the error bursts and hence the greater the degradation in performance relative to the theoretical value.

The estimates P_1 and P_2 (Equations 7.39 and 7.43) are approximate lower bounds to the actual error rates in the optimum and

404 DEVELOPMENTS OF THE DECISION-FEEDBACK EQUALIZER

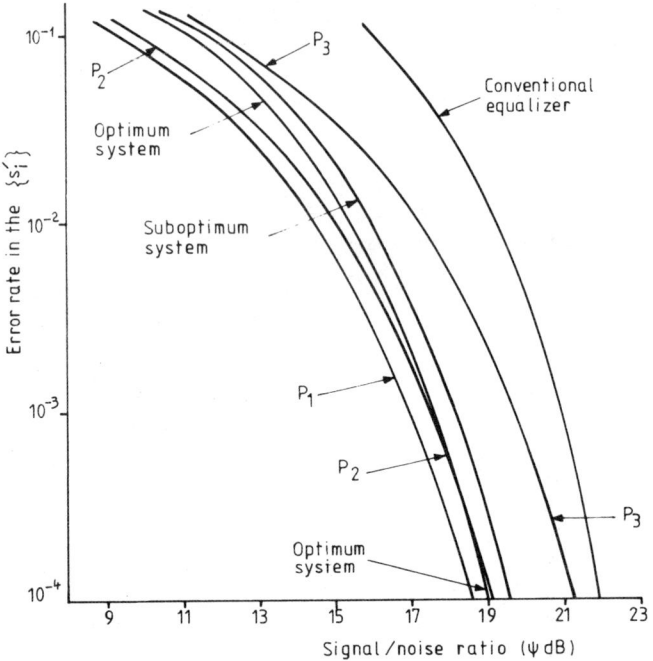

Fig. 7.7 Performance of equalizers with channel 3

suboptimum systems, respectively. Both P_1 and P_2 neglect a factor of $\frac{1}{2}$ and so, in effect, assume an average of two errors in an error burst, which is in general smaller (and sometimes very much smaller) than that experienced here.

It is evident now that the discrepancies between the theoretical and measured results in Figs. 7.5–7.11 are caused mainly by the error-extension effects (error bursts) which are neglected in P_1, P_2 and P_3, the influence of these becoming greater as the error rate increases. For any given channel, the discrepancy is greater with the conventional equalizer than with the optimum or suboptimum system. This is due partly to the factor of $\frac{1}{2}$ that is ignored in each of P_1 and P_2, but is often mainly caused by the greater average length of the error bursts given by the conventional equalizer. The latter is the result of the greater relative magnitudes of the intersymbol interference components removed here (with correct detection) by the transversal filter F and subtractor in Fig. 7.1. The more severe the amplitude distortion introduced by the channel, the greater the average length of the error bursts with any given system, and hence the greater the discrepancy between the theoretical and measured results. At the lower error rates

DEVELOPMENTS OF THE DECISION-FEEDBACK EQUALIZER 405

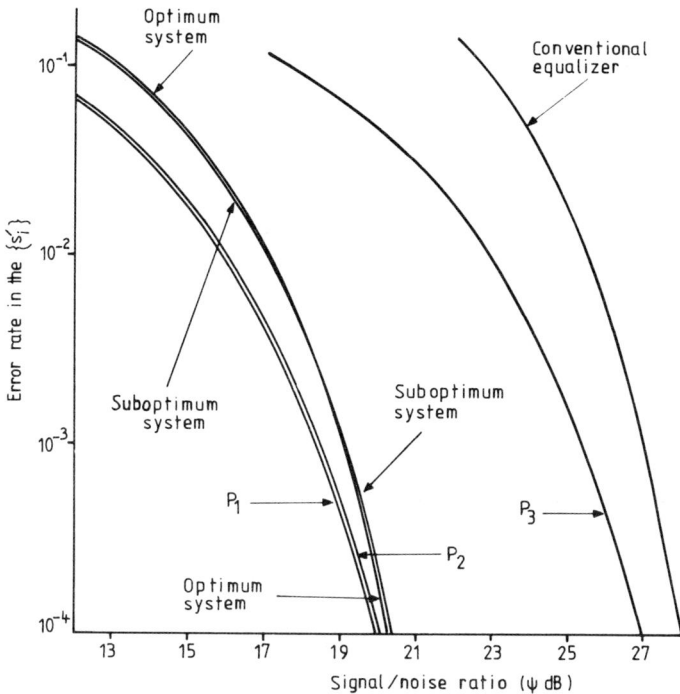

Fig. 7.8 Performance of equalizers with channel 4

tested the discrepancies are not very serious, and at error rates below 1 in 10^5 they should in most cases be well under 1 dB.

At error rates below 1 in 10^3, the relative tolerances to additive white Gaussian noise of the optimum system, suboptimum system and conventional equalizer, as given theoretically by P_1, P_2 and P_3, respectively, are quite close to the relative tolerances to noise as measured by the computer-simulation tests. Thus Fig. 7.4 gives a reasonably accurate comparison of the three systems, at these lower error rates.

Figures 7.5–7.11 show that, at error rates in the range 1 in 10^3 to 1 in 10^4, the optimum system has in every case a better tolerance to noise than both the suboptimum system and the conventional equalizer, and it achieves its best performance relative to the conventional equalizer when operating over channel 6, where it gains a measured advantage of about 9 dB in tolerance to additive white Gaussian noise. With channel 2, the optimum system gains its smallest measured advantage of only about 1 dB over the conventional equalizer and its greatest advantage of about 1 dB over the sub-

406 DEVELOPMENTS OF THE DECISION-FEEDBACK EQUALIZER

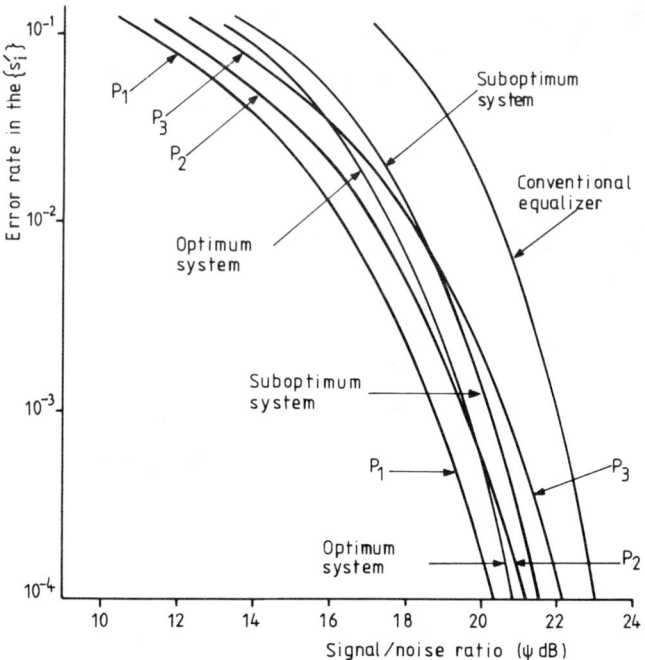

Fig. 7.9   Performance of equalizers with channel 5

optimum system, which, at error rates below 1 in 10^4, now has an inferior performance to that of the conventional equalizer. In the case of channels 4 and 6, the performance of the suboptimum system is very close indeed to that of the optimum system, and in fact becomes very slightly better at the higher error rates. A comparison of the results in Figs. 7.5–7.11 with those in Figs. 5.17–5.23 (Section 5.10), shows that the optimum system has in every case a *better* performance than the corresponding decision-feedback equalizer that operates with a delay in detection of one sampling interval.

Since all zeros (roots) of the z-transform of the sampled impulse-response of each of the channels 1 and 4–7 lie *on* the unit circle in the z plane, which means that, for our purposes, these channels are *minimum phase*, the linear feedforward transversal filter D in Fig. 7.1 is omitted here and the conventional equalizer now becomes a pure nonlinear equalizer (Section 5.1). It can be seen from Figs. 7.4 and 7.5 that in the case of channel 1, which is one of the well known partial-response channels formed by correlative-level coding[64], an improvement of 3 dB in tolerance to noise over that of a conventional system

DEVELOPMENTS OF THE DECISION-FEEDBACK EQUALIZER

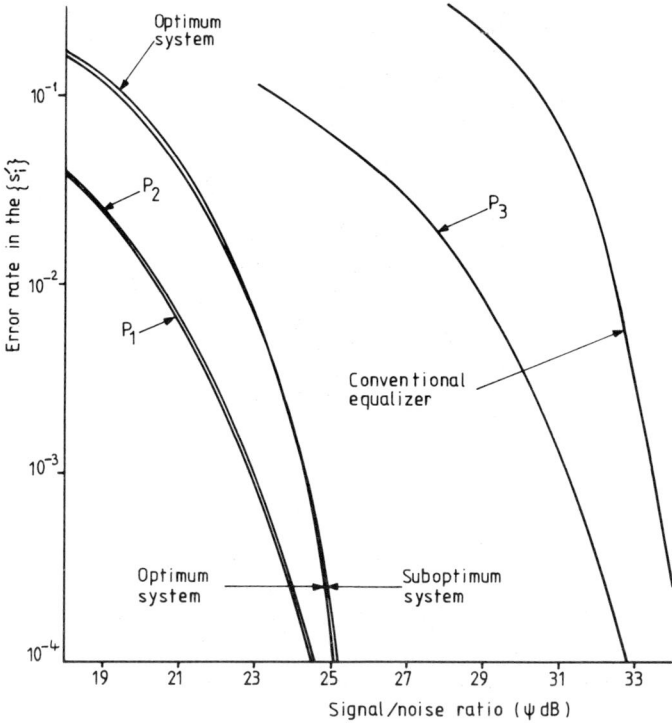

Fig. 7.10 Performance of equalizers with channel 6

can be achieved through the inclusion of the appropriate 2-tap filter C at the input to the subtractor in Fig. 7.1.

It is evident from Equation 7.53 and Table 7.1 that, when 4-level data symbols are transmitted (Equation 7.52), no useful advantage in tolerance to additive white Gaussian noise is likely to be achieved by the optimum system over the conventional equalizer, with any of the channels 1–7 at low error rates, the performance of the suboptimum system being now in every case almost certainly inferior to that of the conventional equalizer.

The essential mechanism used by both the optimum and suboptimum systems is to delay the detection of s_i until the receipt of r'_{i+1} and then to operate on the *two* received samples r'_i and r'_{i+1}, to give the detected value of s_i, thus involving *both* $s_i e_0$ and $s_i e_1$ in the detection of s_i. The conventional equalizer uses only $s_i e_0$, and the equalizer described in Section 5.10 uses only $s_i e_1$. Other techniques have been described whereby the tolerance to noise of an equalizer can be improved by delaying the detection of s_i[26,54], but these all involve a

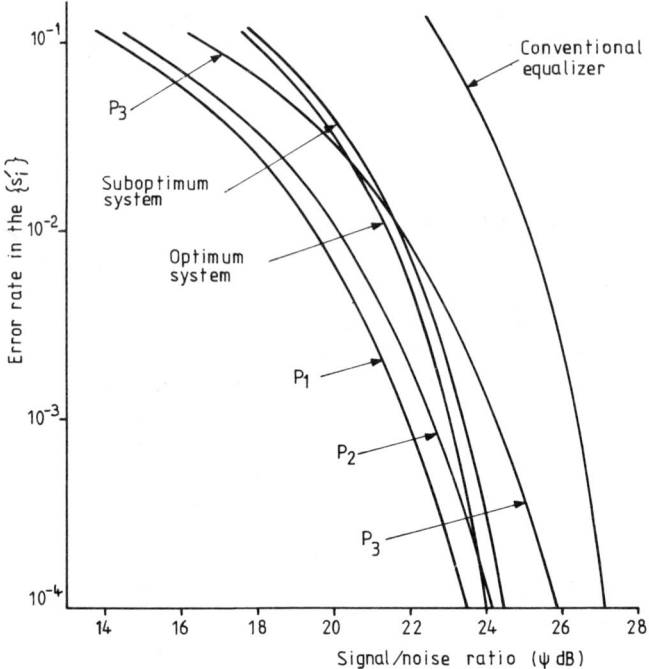

Fig. 7.11 *Performance of equalizers with channel 7*

significant increase in the equipment complexity. They are considered further in Section 7.8. The importance of the techniques described here is that, with a known time-invariant channel and a given filter D in the receiver, the only change required in the equipment is the addition of one tap to the filter D and, of course, the appropriate adjustment of all tap gains.

The results achieved in this investigation can be summarized as follows. When a decision-feedback equalizer is used with a binary signal that has been subjected to severe amplitude distortion, a useful improvement in tolerance to additive white Gaussian noise can be achieved by appropriately modifying the tap gains of the equalizer without changing its basic structure. The equalizer does not now attempt to achieve the accurate equalization of the channel or even to minimize the mean-square error in the equalized signal, but instead deliberately introduces an appreciable level of intersymbol interference into the equalized signal in return for which the noise level here is considerably reduced, giving an overall improvement in tolerance to noise. Unfortunately, the arrangement is not suitable for use with multilevel signals. An important application of the technique appears to be at very high transmission rates, in excess of 1 Mbit/s, where it is required to achieve the highest available transmission rate, with binary signals and using only simple equipment.

7.5 SIMPLIFIED EQUALIZER

7.5.1 Introduction

It is shown in Sections 7.3 and 7.4 that, when a binary polar data signal is transmitted over a linear baseband channel, which introduces severe amplitude distortion into the data signal together with additive white Gaussian noise, the tolerance to noise of a conventional decision-feedback equalizer (Fig. 7.1) can be improved significantly by modifying appropriately the tap gains of the two transversal filters of the equalizer[53]. No changes are made here to the detection (decision) process itself. Two more equalizers are now to be investigated, one of which is a simplified form of the improved equalizer and the other of which is a further development of the simplified equalizer for the case where the received data signal is sampled at *twice* the data-symbol rate[60]. Results are then presented of computer simulation tests over models of various channels, the latter involving three quite different applications, in order to compare the tolerances to additive white Gaussian noise of the simplified equalizers with those or more conventional detectors[60].

The essential feature of the techniques studied here is a *modification* to the *optimum system* studied in Sections 7.3 and 7.4, in which the linear feedforward transversal filter D (Fig. 7.1) is *omitted*, leaving only the 2-tap filter C (that was connected between the filter D and subtractor) thus greatly reducing the complexity of the equalizer. The filter C (Fig. 7.3) is now designed appropriately for the sampled impulse-response of the channel itself, rather than for the sampled impulse-response of the channel and filter D. The modification is suggested by the great improvement in performance sometimes achieved by the optimum system of Section 7.3 over the conventional decision-feedback equalizer.

7.5.2 Model of system

The model of the data-transmission system assumed here is shown in Fig. 7.1, where the conventional decision-feedback equalizer is now replaced by the arrangement shown in Fig. 7.12. The information to be transmitted is carried by the binary data-symbols $\{s_i\}$, where, as in

410 DEVELOPMENTS OF THE DECISION-FEEDBACK EQUALIZER

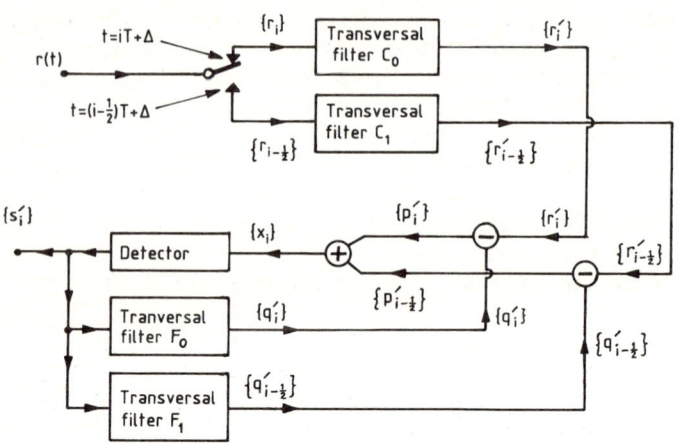

Fig. 7.12 Decision-feedback equalizer and detector

Sections 7.2–7.4,

$$s_i = \pm 1 \tag{7.56}$$

The $\{s_i\}$ are assumed to be statistically independent and equally likely to have either binary value. They are transmitted at $1/T$ symbols/second, in the form of the impulses $\{s_i \delta(t-iT)\}$ at the input to the lowpass filter A. As before, $s_i = 0$ for $i < 0$, so that $s_i \delta(t-iT)$ is the $(i+1)$th transmitted signal-element.

The lowpass filter A has a transfer function $A(f)$ and an impulse response $a(t)$, and it appropriately shapes the transmitted signal-elements $\{s_i a(t-iT)\}$. The transmission path is a linear baseband channel, with transfer function $H(f)$, that could be a cable or else could comprise a modulator, bandpass channel and demodulator, all of which are linear. Stationary white Gaussian noise with zero mean and a two-sided power spectral density of $\tfrac{1}{2}N_0$ is added to the data signal at the output of the transmission path, and the resulting noise waveform at the output of the lowpass filter B (Fig. 7.1) is $w(t)$. The lowpass filter B has the transfer function

$$B(f) = \begin{cases} 1, & -\dfrac{1}{2\tau} < f < \dfrac{1}{2\tau} \\ 0, & \text{elsewhere} \end{cases} \tag{7.57}$$

where

$$\tau = \tfrac{1}{2}T \text{ or } T \tag{7.58}$$

and τ is the sampling interval used at the receiver (Fig. 7.12). The receiver is a development of the conventional decision-feedback equalizer shown in Fig. 7.1 and will be considered in more detail later. Except where otherwise stated, frequency and time are measured in units of Hz and seconds, respectively. For practical purposes, the lowpass filter B (and also, where necessary, the filter A) can be made physically realizable by introducing a sufficient delay into the corresponding impulse response. However, to simplify the analysis, the appropriate non-physically realizable filters A and B may be assumed. It can be shown[64] that the samples of $w(t)$, spaced at regular intervals of τ (Equations 7.57 and 7.58), are statistically independent Gaussian random variables with zero mean and variance

$$\sigma^2 = N_0/2\tau \qquad (7.59)$$

This result follows, in fact, without much further work when the analysis in Section 1.2 is applied to the low pass filter B with transfer function $B(f)$ (Equation 7.57). The lowpass filter A, transmission path and lowpass filter B together form a linear baseband channel with a real-valued impulse response $y(t)$. For practical purposes, $y(t)$ has a finite duration and is also time invariant, such that $y(t-iT)$ is a time-shifted version of $y(t-jT)$, for any integers $i \neq j$. Thus the waveform at the output of the lowpass filter B is

$$r(t) = \sum_i s_i y(t-iT) + w(t) \qquad (7.60)$$

and this is fed to the decision-feedback equalizer and detector (Fig. 7.12). Two different arrangements of the latter, known as systems 1 and 2, will now be described[60].

7.5.3 System 1

The equalizer and detector are here as shown in Fig. 7.1, and the sampling interval is $\frac{1}{2}T$, so that $\tau = \frac{1}{2}T$ in Equations 7.57 and 7.58. The received waveform $r(t)$ is sampled twice per data-symbol s_i, at the time instants $\{iT + \Delta\}$ and $\{(i-\frac{1}{2})T + \Delta\}$, to give the samples $\{r_i\}$ and $\{r_{i-1/2}\}$, where

$$r_i = r(iT + \Delta) \qquad (7.61)$$

and
$$r_{i-1/2} = r((i-\tfrac{1}{2})T + \Delta) \qquad (7.62)$$

i takes on all nonnegative integer values and Δ is a constant given by the delay in the sampling instants needed to compensate for the delay in transmission. Thus two signals, r_i and $r_{i-1/2}$, are received for the evaluation of each detected data-symbol s'_i at the detector output (Fig. 7.1).

Regardless of the impulse-response $y(t)$ of the linear baseband channel, it can be seen that, for appropriate values of Δ and g,

$$r_i = \sum_{j=0}^{g} s_{i-j} y_{0,j} + w_i \quad (7.63)$$

and

$$r_{i-1/2} = \sum_{j=0}^{g} s_{i-j} y_{1,j} + w_{i-1/2} \quad (7.64)$$

where

$$y_{0,j} = y(jT + \Delta) \quad (7.65)$$

$$y_{1,j} = y((j-\tfrac{1}{2})T + \Delta) \quad (7.66)$$

$$w_i = w(iT + \Delta) \quad (7.67)$$

$$w_{i-1/2} = w((i-\tfrac{1}{2})T + \Delta) \quad (7.68)$$

and the noise components $\{w_i\}$ and $\{w_{i-1/2}\}$ are statistically independent Gaussian random variables with zero mean and variance σ^2 (Equation 7.59). For the present it will be assumed that $y_{0,j}$ and $y_{1,j}$ are both zero (at least, approximately) when $j < 0$ and $j > g$. Thus the sampled impulse-responses of the linear baseband channel (Fig. 7.1), at the time instants $iT + \Delta$ and $(i-\tfrac{1}{2})T + \Delta$, are given by the $(g+1)$-component sequences (vectors)

$$Y_0 = [y_{0,0} \quad y_{0,1} \quad \cdots \quad y_{0,g}] \quad (7.69)$$

and

$$Y_1 = [y_{1,0} \quad y_{1,1} \quad \cdots \quad y_{1,g}] \quad (7.70)$$

respectively. The z-transforms of these sampled impulse-responses are

$$Y_0(z) = y_{0,0} + y_{0,1} z^{-1} + \cdots + y_{0,g} z^{-g} \quad (7.71)$$

and

$$Y_1(z) = y_{1,0} + y_{1,1} z^{-1} + \cdots + y_{1,g} z^{-g} \quad (7.72)$$

the respective delays of Δ and $\Delta - \tfrac{1}{2}T$ being, for convenience, neglected here. In each z-transform, z^{-1} corresponds to a delay of T seconds.

The transversal filters C_0 and C_1 (Fig. 7.12) are 2-tap linear feedforward transversal filters, having the tap gains, and hence sampled impulse-responses, given by the 2-component sequences (vectors)

$$C_0 = [c_{0,0} \quad c_{0,1}] \quad (7.73)$$

and

$$C_1 = [c_{1,0} \quad c_{1,1}] \quad (7.74)$$

with z-transforms

$$C_0(z) = c_{0,0} + c_{0,1} z^{-1} \quad (7.75)$$

and

$$C_1(z) = c_{1,0} + c_{1,1} z^{-1} \quad (7.76)$$

respectively.

DEVELOPMENTS OF THE DECISION-FEEDBACK EQUALIZER 413

Consider now the signal r'_{i+1} at the output of the filter C_0, at time $t=(i+1)T+\Delta$, given a very high signal/noise ratio. Suppose first that $|y_{0,1}|>|y_{0,0}|$. It is shown in Section 7.3 that, for the minimum probability of error in the detection of s_i on the receipt of r'_{i+1}, under the assumed conditions,

$$c_{0,1}=y_{0,0} \tag{7.77}$$

and
$$c_{0,0}=y_{0,1}\pm y_{0,0} \tag{7.78}$$

the sign being chosen to minimize $|c_{0,0}|$. Strictly speaking, the given values of $c_{0,1}$ and $c_{0,0}$ may be multiplied by any positive constant, but, for simplicity, this constant has been set to unity. When $|y_{0,1}|\leq|y_{0,0}|$, the tap gains for the minimum error probability are $c_{0,1}$ as before (Equation 7.77) but now

$$c_{0,0}=0 \tag{7.79}$$

Similarly, for the tap gains of C_1, when $|y_{1,1}|>|y_{1,0}|$,

$$c_{1,1}=y_{1,0} \tag{7.80}$$

and
$$c_{1,0}=y_{1,1}\pm y_{1,0} \tag{7.81}$$

the sign being chosen to minimize $|c_{1,0}|$. When $|y_{1,1}|\leq|y_{1,0}|$, $c_{1,1}$ is as before (Equation 7.80) but now

$$c_{1,0}=0 \tag{7.82}$$

Let
$$Y_0(z)C_0(z)=e_{0,0}+e_{0,1}z^{-1}+\cdots+e_{0,g+1}z^{-g-1} \tag{7.83}$$
and
$$Y_1(z)C_1(z)=e_{1,0}+e_{1,1}z^{-1}+\cdots+e_{1,g+1}z^{-g-1} \tag{7.84}$$

where $e_{0,0}=y_{0,0}c_{0,0}$, $e_{0,1}=y_{0,0}c_{0,1}+y_{0,1}c_{0,0}$, and so on. Thus the signal at the output of the filter C_0, at time $t=(i+1)T+\Delta$, is

$$r'_{i+1}=\sum_{j=0}^{g+1} s_{i+1-j}e_{0,j}+w'_{i+1} \tag{7.85}$$

where
$$w'_{i+1}=w_{i+1}c_{0,0}+w_i c_{0,1} \tag{7.86}$$

and the signal at the output of the filter C_1, at time $t=(i+\tfrac{1}{2})T+\Delta$, is

$$r'_{i+1/2}=\sum_{j=0}^{g+1} s_{i+1-j}e_{1,j}+w'_{i+1/2} \tag{7.87}$$

where
$$w'_{i+1/2}=w_{i+1/2}c_{1,0}+w_{i-1/2}c_{1,1} \tag{7.88}$$

The g-tap linear feedforward transversal filter F_0 has tap gains $e_{0,2}$, $e_{0,3}$, ..., $e_{0,g+1}$ and the z-transform

$$F_0(z)=e_{0,2}z^{-1}+e_{0,3}z^{-2}+\cdots+e_{0,g+1}z^{-g} \tag{7.89}$$

and the g-tap linear feedforward transversal filter F_1 has tap gains

$e_{1,2}, e_{1,3}, \ldots, e_{1,g+1}$ and the z-transform

$$F_1(z) = e_{1,2}z^{-1} + e_{1,3}z^{-2} + \cdots + e_{1,g+1}z^{-g} \tag{7.90}$$

Thus, with the correct detection of $s_{i-1}, s_{i-2}, \ldots, s_{i-g}$, the signal at the output of the transversal filter F_0, at time $t = (i+1)T + \Delta$, is

$$q'_{i+1} = s_{i-1}e_{0,2} + s_{i-2}e_{0,3} + \cdots + s_{i-g}e_{0,g+1} \tag{7.91}$$

and the corresponding signal at the input to the adder (Fig. 7.12) is

$$p'_{i+1} = r'_{i+1} - q'_{i+1} = s_{i+1}e_{0,0} + s_i e_{0,1} + w'_{i+1}$$
$$= s_{i+1}y_{0,0}c_{0,0} + s_i(y_{0,0}c_{0,1} + y_{0,1}c_{0,0}) + w_{i+1}c_{0,0} + w_i c_{0,1}$$
$$= c_{0,0}(s_i y_{0,1} + s_{i+1}y_{0,0}) + c_{0,1}s_i y_{0,0} + w_{i+1}c_{0,0} + w_i c_{0,1} \tag{7.92}$$

It may similarly be shown that (under the given conditions)

$$p'_{i+1/2} = r'_{i+1/2} - q'_{i+1/2}$$
$$= c_{1,0}(s_i y_{1,1} + s_{i+1}y_{1,0}) + c_{1,1}s_i y_{1,0}$$
$$+ w_{i+1/2}c_{1,0} + w_{i-1/2}c_{1,1} \tag{7.93}$$

the signals $p'_{i+1/2}$ and $r'_{i+1/2}$ being obtained at time $t = (i+1)T + \Delta$, through the temporary storage of $r'_{i+1/2}$.

It can be seen from Equations 7.56, 7.77–7.82, 7.92 and 7.93 that, when $s_{i-1}, s_{i-2}, \ldots, s_{i-g}$ have been correctly detected and the value of the data-symbol s_{i+1} is such as to minimize the magnitude of the resultant data-signal component in p'_{i+1} or $p'_{i+1/2}$, then

$$p'_{i+1} = s_i(c_{0,0}^2 + c_{0,1}^2) + w_{i+1}c_{0,0} + w_i c_{0,1} \tag{7.94}$$

or $$p'_{i+1/2} = s_i(c_{1,0}^2 + c_{1,1}^2) + w_{i+1/2}c_{1,0} + w_{i-1/2}c_{1,1} \tag{7.95}$$

respectively. These equations hold regardless of whether $|y_{j,1}|$ is greater or less than $|y_{j,0}|$, for $j=0$ or 1, p'_{i+1} or $p'_{i+1/2}$ in fact becoming independent of s_{i+1} when the corresponding $|y_{j,1}| \leq |y_{j,0}|$. In all cases to be studied here, the value of s_{i+1} that minimizes the magnitude of the signal component in p'_{i+1}, for a particular value of s_i (given that $|y_{j,1}| > |y_{j,0}|$ for $j=0$ and 1), is the *same* as that which minimizes the magnitude of the signal component in $p'_{i+1/2}$. To optimize the performance of the system at high signal/noise ratios and with $|y_{j,1}| > |y_{j,0}|$ for $j=0$ and/or 1, the performance must now be optimized for the smallest magnitudes of the signal components in p'_{i+1} and $p'_{i+1/2}$, since the probability of error in s'_i is maximum under these conditions. Under the assumed conditions, the noise components $w_{i-1/2}, w_i, w_{i+1/2}$ and w_{i+1} are statistically independent Gaussian random variables with zero mean and variance σ^2 (Equation 7.59). Thus the noise components in p'_{i+1} and $p'_{i+1/2}$ are statistically independent Gaussian random variables with zero mean and variances $\sigma^2(c_{0,0}^2+c_{0,1}^2)$ and $\sigma^2(c_{1,0}^2+c_{1,1}^2)$, respectively. It may readily be shown from the theory of matched filters (Section 1.5)[62-64] that, when $|y_{j,1}|>|y_{j,0}|$ for $j=0$ and/or 1, and s_{i+1} is such as to minimize the magnitude of the signal component in p'_{i+1} and/or $p'_{i+1/2}$, or else when $|y_{j,1}| \leqslant |y_{j,0}|$ for $j=0$ and 1, regardless of the value of s_{i+1}, then the linear function of p'_{i+1} and $p'_{i+1/2}$ having the maximum signal/noise ratio is just

$$x_{i+1} = p'_{i+1} + p'_{i+1/2} \tag{7.96}$$

or, of course, any constant times the sum of p'_{i+1} and $p'_{i+1/2}$. The signal x_{i+1} is formed in the adder (Fig. 7.12) and fed to the detector. The value of the detected data-symbol s'_i is now taken as 1 or -1, depending upon whether x_{i+1} is positive or negative, respectively. The normal condition in system 1 is that $|y_{j,1}|>|y_{j,0}|$ for $j=0$ and/or 1.

In a practical implementation of the equalizer, the signals r'_{i+1} and $r'_{i+1/2}$ are added, to give

$$r'_{i+1} + r'_{i+1/2} = \sum_{j=0}^{g+1} s_{i+1-j}(e_{0,j} + e_{1,j}) + w'_{i+1} + w'_{i+1/2} \tag{7.97}$$

which is fed to a subtractor. Only a *single* g-tap linear feedforward transfersal filter F is now used, in place of the two filters F_0 and F_1 in Fig. 7.12, and the filter is fed with the detected data-symbols, as before. The filter F has the tap gains

$$(e_{0,2}+e_{1,2}), (e_{0,3}+e_{1,3}), \ldots, (e_{0,g+1}+e_{1,g+1})$$

and the z-transform

$$F(z) = (e_{0,2}+e_{1,2})z^{-1} + (e_{0,3}+e_{1,3})z^{-2} + \cdots + (e_{0,g+1}+e_{1,g+1})z^{-g} \tag{7.98}$$

With the correct detection of $s_{i-1}, s_{i-2}, \ldots, s_{i-g}$, the output signal from this filter, at time $t=(i+1)T+\Delta$, is

$$q'_{i+1} + q'_{i+1/2} = s_{i-1}(e_{0,2}+e_{1,2}) + s_{i-2}(e_{0,3}+e_{1,3}) + \cdots + s_{i-g}(e_{0,g+1}+e_{1,g+1}) \tag{7.99}$$

and this is subtracted from $r'_{i+1} + r'_{i+1/2}$ to give the equalized signal x_{i+1} in Equation 7.96. The modification halves the number of taps in the decision-feedback filters without in any way affecting the performance of the system, as can be seen from Equations 7.85 and 7.87. The resulting arrangement of the receiver is shown in Fig. 7.13.

It is evident from Equations 7.94–7.96 that, when the data-symbols $s_{i-1}, s_{i-2}, \ldots, s_{i-g}$ have been correctly detected and when s_{i+1} is such

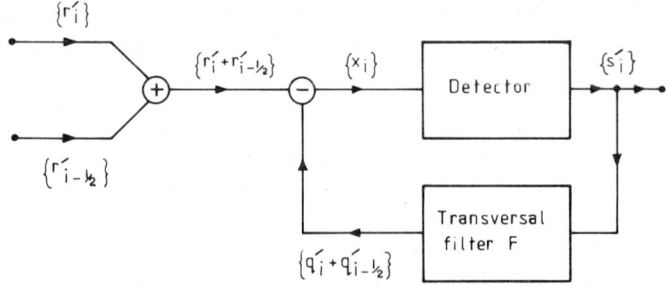

Fig. 7.13 Practical implementation of detector and feedback filter

as to minimize the magnitude of the resultant data-signal component in either p'_{i+1} or $p'_{i+1/2}$, or in both of these, or else where both p'_{i+1} and $p'_{i+1/2}$ are independent of s_{i+1} (in which case, of course, the value of s_{i+1} does not matter), then

$$x_{i+1} = cs_i + u_{i+1} \tag{7.100}$$

where $$c = c_{0,0}^2 + c_{0,1}^2 + c_{1,0}^2 + c_{1,1}^2 \tag{7.101}$$

and $$u_{i+1} = w_{i+1}c_{0,0} + w_i c_{0,1} + w_{i+1/2}c_{1,0} + w_{i-1/2}c_{1,1} \tag{7.102}$$

The $\{c_{j,h}\}$ in Equation 7.101 satisfy Equations 7.77–7.82. When $|y_{j,1}| > |y_{j,0}|$ for $j = 0$ *and* 1, the quantity c in Equation 7.100 becomes

$$c = (|y_{0,1}| - |y_{0,0}|)^2 + y_{0,0}^2 + (|y_{1,1}| - |y_{1,0}|)^2 + y_{1,0}^2 \tag{7.103}$$

Since $w_{i-1/2}$, w_i, $w_{i+1/2}$ and w_{i+1} are statistically independent Gaussian random variables with zero mean and variance σ^2 (Equation 7.59), the noise component u_{i+1} in Equation 7.100 is a Gaussian random variable with zero mean and variance

$$\eta_1^2 = \sigma^2(c_{0,0}^2 + c_{0,1}^2 + c_{1,0}^2 + c_{1,1}^2) \tag{7.104}$$

Thus, with the correct detection of $s_{i-1}, s_{i-2}, \ldots, s_{i-g}$ and the unfavourable value of s_{i+1}, an error occurs in the detection of s_i from x_{i+1} when

$$|u_{i+1}| > c \tag{7.105}$$

and u_{i+1} has the opposite sign to s_i. The probability of this occurring, under the given conditions, is

$$P_4 = \int_c^\infty \frac{1}{\sqrt{2\pi\eta_1^2}} \exp\left(-\frac{u^2}{2\eta_1^2}\right) du = Q\left(\frac{c}{\eta_1}\right)$$
$$= Q\left(\frac{(c_{0,0}^2 + c_{0,1}^2 + c_{1,0}^2 + c_{1,1}^2)^{1/2}}{\sigma}\right) \tag{7.106}$$

There is a probability of $\frac{1}{2}$ that s_{i+1} has the unfavourable binary value (minimizing the magnitude of the signal component in x_{i+1}), and, at high signal/noise ratios, nearly all errors in the $\{s_i'\}$ occur under this condition. When an error occurs in a detected data-symbol, there are likely to be errors in some of the following detected data-symbols, so that errors occur in bursts. Thus the actual error rate in system 1 can be taken to be $\frac{1}{2}bP$, where b is the average number of errors in an error burst. When $b=2$ and there is a high signal/noise ratio, the error rate is, for practical purposes, P_4. Clearly, P_4 can be taken as an approximate (and rather loose) lower bound to the error rate of system 1, the bound becoming tighter as the signal/noise ratio increases and as the intersymbol interference in r_i and $r_{i-1/2}$ is reduced towards a value for which $b=2$.

In a practical implementation of system 1 it may well happen that the first two or three components of both Y_0 and Y_1 are very small. Under these conditions the delay in sampling, which is Δ (Fig. 7.12), is *increased* by the appropriate number of sampling intervals $\{\tau\}$ (where $\tau = \frac{1}{2}T$, Equation 7.58). The components of Y_0 and Y_1 are now moved to the *left* by the corresponding number of places, with an interchange of the two vectors whenever Δ is increased by an odd number of sampling intervals. The adjustment in Δ is such as to maximize c (Equation 7.100). As a result of the operation just described, $y_{0,h}$ and $y_{1,h}$, for the first one or two of the values of h given by $h = -1, -2, \ldots$, become nonzero, but, in all cases studied here,

$$|y_{j,h}| \ll |y_{j,0}| \text{ or } |y_{j,1}| \tag{7.107}$$

for $j = 0, 1$ and $h = -1, -2, \ldots$, and for whichever of $|y_{j,0}|$ and $|y_{j,1}|$ is the greater. The operation just described does not accurately optimize the system (for the given choice of sampling instants), since it takes no account of the *magnitudes* of the $\{y_{j,h}\}$ for $j = 0, 1$ and $h = -1, -2, \ldots$. However, it is probably accurate enough for our purposes. The detection process now operates exactly as previously described except that, in addition to the intersymbol interference from s_{i+1} in x_{i+1}, there is also intersymbol interference from the first one or two of the data-symbols s_{i+2}, s_{i+3}, ... These intersymbol-interference components are treated as noise by the detector, but, since they are all relatively small, their effect on the tolerance to noise is generally not very serious.

7.5.4 System 2

This is a simplified form of system 1 obtained by sampling the received waveform $r(t)$ only once per data symbol, at the time instants $\{iT + \Delta\}$ (Fig. 7.12), so that the signals $\{r_{i-1/2}\}$ are no longer generated. The sampling interval τ is now increased from $\frac{1}{2}T$ to T and the cut-off frequency of the lowpass filter B (Fig. 7.1) is reduced from $1/T$ Hz to $1/2T$ Hz, without otherwise changing the transfer function $B(f)$ (Equation 7.57). Furthermore, the transversal filters C_1 and F_1, together with the adder (Fig. 7.12), are omitted, and

$$x_{i+1} = p'_{i+1} \tag{7.108}$$

The transversal filters C_0 and D_0, together with their output signals, are exactly as described for system 1, and so is the detection of s_i from x_{i+1}. System 2 is a direct development of the *optimum system* described in Section 7.3, in which the filter D is omitted, so that the filter C operates directly on Y (Equation 7.4) instead of on E (Equation 7.6).

The halving of the bandwidth of the lowpass filter B halves the variance of the noise component w_i in the signal r_i, for a given two-sided power spectral density, $\frac{1}{2}N_0$, of the additive white Gaussian noise at the input to the lowpass filter. If the bandwidth of the baseband data signal in $r(t)$ exceeds $1/2T$ Hz, the halving of the filter bandwidth also removes some of the data-signal power at the output of the filter, thus destroying some of the received information, and it also increases the effective time dispersion of the data signal, giving a correspondingly greater value of g.

It is evident from Equations 7.92 and 7.94 that, when $s_{i-1}, s_{i-2}, \ldots, s_{i-g}$ have been correctly detected and when s_{i+1} is such as to minimize the magnitude of the resultant data-signal component in x_{i+1}, for the case where $|y_{0,1}| > |y_{0,0}|$, or else with either value of s_{i+1}, for the case where $|y_{0,1}| \leqslant |y_{0,0}|$, then

$$x_{i+1} = s_i(c_{0,0}^2 + c_{0,1}^2) + w_{i+1}c_{0,0} + w_i c_{0,1} \tag{7.109}$$

where $w_{i+1}c_{0,0} + w_i c_{0,1}$ is a Gaussian random variable with zero mean and variance

$$\eta_2^2 = \sigma^2(c_{0,0}^2 + c_{0,1}^2) \tag{7.110}$$

Equation 7.110 uses the fact that w_i and w_{i+1} are statistically independent Gaussian random variables, with zero mean and variance σ^2. Under the given conditions, the probability of error in the detection of s_i from x_{i+1} is

$$P_5 = Q\left(\frac{c_{0,0}^2 + c_{0,1}^2}{\eta_2}\right) = Q\left(\frac{(c_{0,0}^2 + c_{0,1}^2)^{1/2}}{\sigma}\right) \tag{7.111}$$

$c_{0,0}$ and $c_{0,1}$ being given by Equations 7.77–7.79. The properties of P_5 are similar to those of P_4. Thus P_5 can be taken as an approximate (and rather loose) lower bound to the actual error rate in system 2.

Furthermore, when there are an average of two errors in an error burst and there is a high signal/noise ratio, the error rate is, for practical purposes, P_5.

As in the case of system 1, the detection process should be optimized (at least approximately) for the given choice of sampling instants. Thus the value of Δ (Fig. 7.12) is adjusted in steps of T to maximize $c_{0,0}^2 + c_{0,1}^2$, which may cause the first one or two of the components $y_{0,-1}, y_{0,-2}, \ldots$ to become nonzero, these components now being treated as noise by the detector. The detection process proceeds otherwise exactly as described.

7.6 ASSESSMENT OF SYSTEMS

7.6.1 Computer simulation tests

Three series of tests by computer simulation have been carried out on systems 1 and 2. The first of these, carried out by K. Parama Raj, is a comparison of system 2 with the conventional equalizer, and the second, carried out by M. Slater, is a comparison of system 1 with an optimum detector[60]. The third series of tests, again carried out by M. Slater, is a comparison of both systems 1 and 2 with a conventional linear equalizer, in an application involving the transmission of digital data at 140 Mbit/s over a length of coaxial cable[60].

All tests measure the tolerance to additive white Gaussian noise of the systems concerned, and the conditions assumed in the tests are described in Section 7.5. In particular, every transversal filter in Figs. 7.1 and 7.12 is assumed to be correctly adjusted, and all noise components $\{w_i\}$ and $\{w_{i-1/2}\}$ are statistically independent Gaussian random variables with zero mean and variance σ^2 (Equation 7.59). The average number of data-symbols $\{s_i\}$ transmitted in the plotting of any one curve in Figs. 7.14–7.17, 7.19, 7.21 and 7.22 is over 2×10^6 (sometimes considerably more), giving 95% confidence limits for the results obtained that are generally as good as or better than ± 0.5 dB.

7.6.2 Comparison of system 2 with a conventional decision-feedback equalizer

System 2 and a conventional decision-feedback equalizer are here tested over four different linear baseband channels (Fig. 7.1), which introduce various levels of amplitude distortion but no phase distortion[60]. The absence of phase distortion implies that the sampled impulse-response of the channel is linear phase. It may be assumed that the data signal at the input to the lowpass filter B is bandlimited to the frequency band 0 to $1/2T$ Hz, where T seconds is the sampling

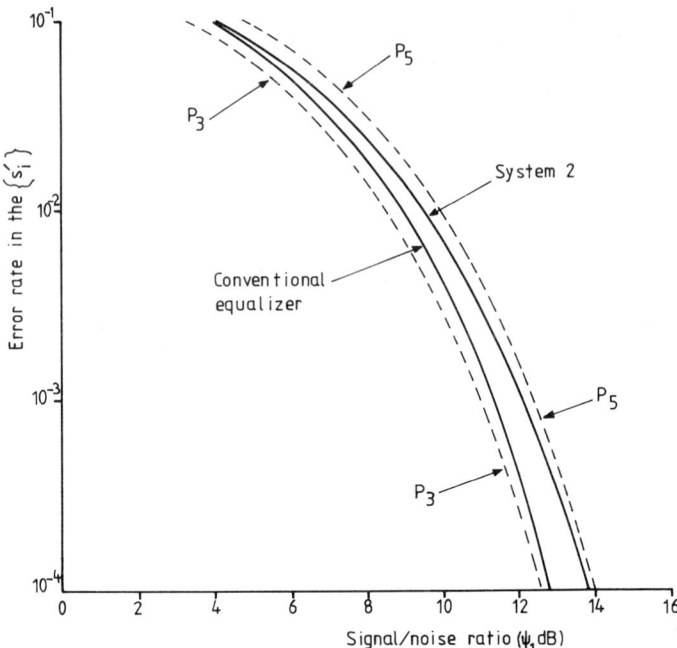

Fig. 7.14 *Performance of equalizers with channel 1*

interval, so that all the useful information in $r(t)$ is contained also in the $\{r_i\}$. However, the results of the tests apply also to the cases where this condition is not satisfied. Table 7.2 shows, for each channel, the sampled impulse-response Y of the channel itself and the sampled impulse-response Y' of the channel and linear feedforward transversal filter D that forms the first part of the conventional decision-feedback equalizer (Fig. 7.1 and Section 7.2).

The results of the tests are shown in Figs. 7.14–7.17, which show also the approximate theoretical estimates, P_3 and P_5, of the error rates in the conventional equalizer (Section 7.2) and system 2, respectively[60]. The signal/noise ratio is taken to be ψ_1 dB, where

$$\psi_1 = 10 \log_{10}(1/\sigma^2) \tag{7.112}$$

the mean-square values of s_i and w_i being 1 and σ^2, respectively. Since we are here concerned only with a *comparison* of two systems, the simple measure of signal/noise ratio given by Equation 7.112 is quite adequate, and avoids making specific assumptions about the lowpass filter A, transmission path or lowpass filter B, other than that the noise components $\{w_i\}$ are statistically independent Gaussian

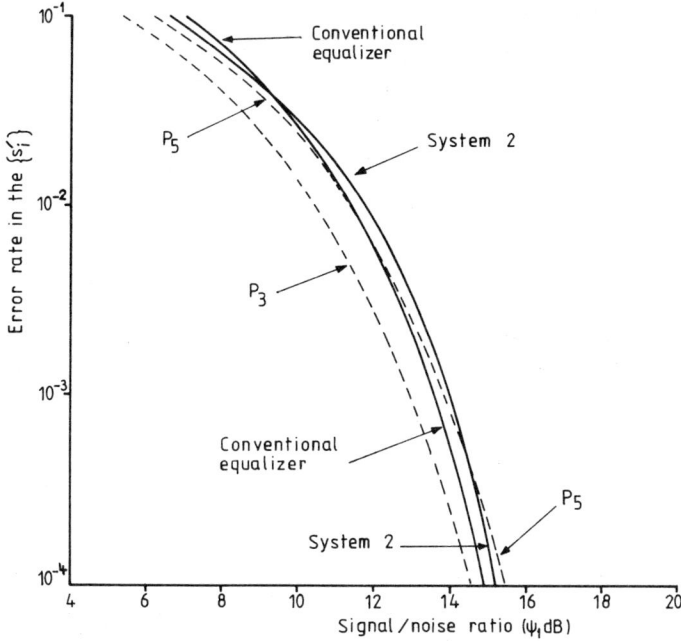

Fig. 7.15 *Performance of equalizers with channel 2*

random variables with zero mean and variance σ^2. It is, of course, assumed here that the *same* relationship between σ^2 and N_0 holds both for the equalizer and system 2, but this may or may not satisfy either Equation 5.7 or Equation 7.59.

The number of taps in the linear feedforward transversal filter D, forming the first part of the conventional equalizer, are approximately 7, 11, 33 and 33 for the channels 1, 2, 3 and 4, respectively. This compares with only 2 taps for the corresponding filter C_0 in system 2, the two systems having the same number of taps in the filter F_0.

Channel 1 (Table 7.2) introduces only small amplitude distortion, channel 2 introduces rather more, and both channels 3 and 4 introduce severe amplitude distortion (see Sections 3.4 and 3.5). It can be seen from Figs. 7.14–7.17 that the performance of system 2, relative to that of the conventional equalizer, improves steadily from channel 1 to 4, being a little inferior over channel 1 and decidedly better over channel 4. With channels 3 and 4, a useful improvement in performance together with a considerable reduction in equipment complexity is achieved through replacing the conventional equalizer by system 2. Clearly, in the presence of the more severe amplitude

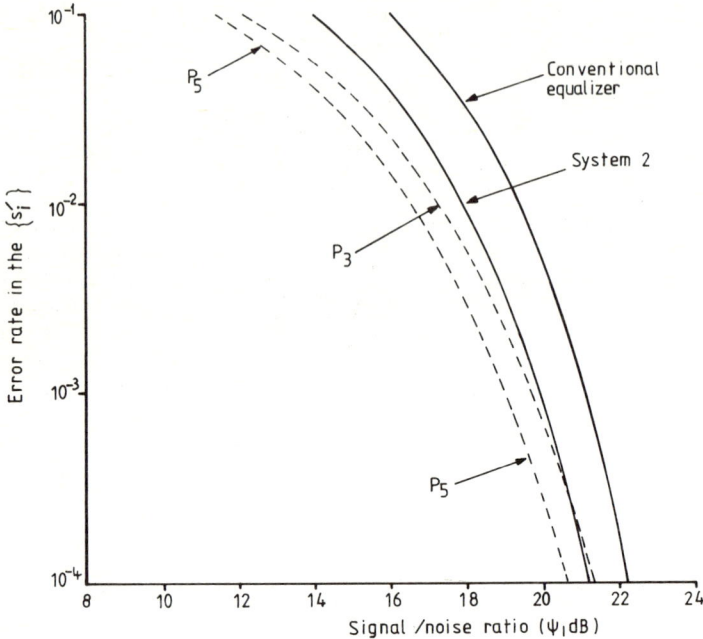

Fig. 7.16 *Performance of equalizers with channel 3*

distortion, system 2 is potentially much more cost-effective than the conventional decision-feedback equalizer.

7.6.3 Comparison of system 1 with an optimum detector

The application considered here is that of very high speed binary data transmission, in short bursts and at high signal/noise ratios, over a strictly band-limited channel, where a useful reduction in the synchronization period at the start of a transmission can be achieved by avoiding the need to adjust the phase of the sampling instants (at the receiver) to any particular value[60]. A required property of the detection process is therefore that it should not be significantly affected by the sampling phase.

It is assumed here that the lowpass filter A (Fig. 7.1) has the 'raised-cosine' transfer function[64]

$$A(f) = \begin{cases} \tfrac{1}{2}T(1+\cos \pi f T), & -\dfrac{1}{T} < f < \dfrac{1}{T} \\ 0, & \text{elsewhere} \end{cases} \quad (7.113)$$

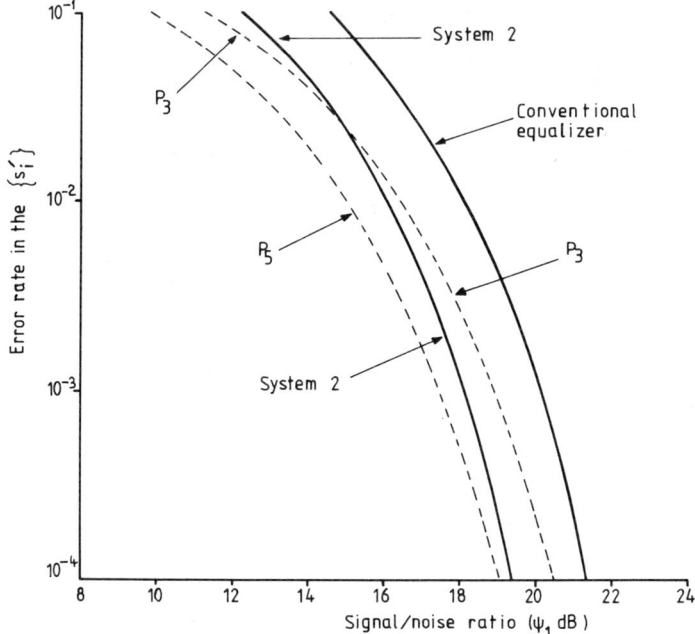

Fig. 7.17 Performance of equalizers with channel 4

Table 7.2 SAMPLED IMPULSE-RESPONSES OF THE DIFFERENT CHANNELS

Channel	Y					Y'				
1	0.236	0.943	0.236			0.880	0.471	0.063		
2	0.348	0.870	0.348			0.696	0.696	0.174		
3	0.167	0.471	0.707	0.471	0.167	0.321	0.620	0.633	0.322	0.087
4	0.070	0.478	0.730	0.478	0.070	0.351	0.708	0.591	0.162	0.014

and impulse-response

$$a(t) = \frac{\sin \pi 2t/T}{\pi 2t/T} + \frac{1}{2}\frac{\sin \pi[(2t/T)+1]}{\pi[(2t/T)+1]} + \frac{1}{2}\frac{\sin \pi[(2t/T)-1]}{\pi[(2t/T)-1]}$$

(7.114)

and that the transmission path introduces no distortion, attenuation or delay. Also $\tau = \frac{1}{2}T$. Thus, from Equations 7.57 and 7.58, $a(t)$ becomes the impulse response of the linear baseband channel in Fig. 7.1 and is shown in Fig. 7.18. $a(t)$ can, of course, be made physically realisable (for practical purposes) by introducing a sufficient delay

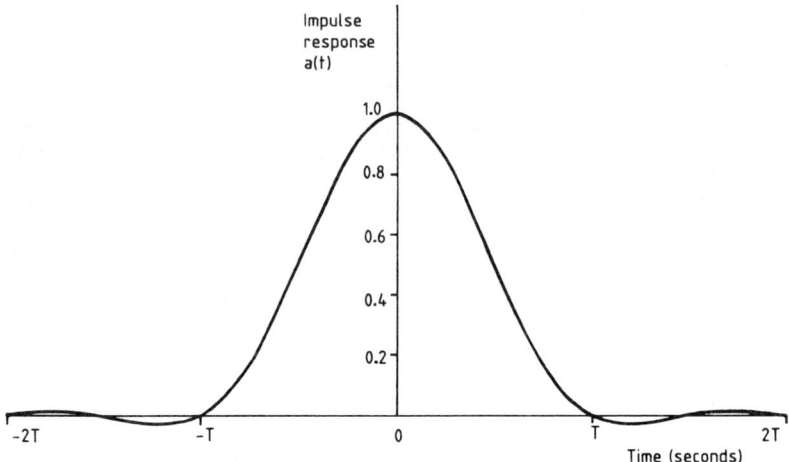

Fig. 7.18 *Impulse response of a linear baseband channel with a raised-cosine transfer function*

into its waveform. System 1 is as shown in Fig. 7.12 and operates as described in Section 7.5.3. All the useful information in the received data-signal is contained in the samples $\{r_i\}$ and $\{r_{i-1/2}\}$. If system 2 were used here, some of the received data-signal power would be lost in the lowpass filter B and there would also be appreciable time dispersion of any individual signal-element at the output of this filter.

Under the assumed conditions, the average transmitted energy per binary signal-element is

$$E_b = \int_{-\infty}^{\infty} |A(f)|^2 \, df \qquad (7.115)$$

which, with some mathematical manipulation, simplifies to

$$E_b = \tfrac{3}{4} T \qquad (7.116)$$

and, from Equation 7.59, the variance of a noise component w_i is

$$\sigma^2 = N_0/T \qquad (7.117)$$

The signal/noise ratio is now taken to be ψ_2 dB, where

$$\psi_2 = 10 \log_{10} (E_b/\tfrac{1}{2} N_0) = 10 \log_{10} (3/2\sigma^2) \qquad (7.118)$$

Computer-simulation tests have been carried out on system 1, with the arrangement just described, for each of four different values of Δ (Fig. 7.12), giving the four values of Y_0 and Y_1 shown in Table 7.3. Of course, Y_0 and Y_1 may now have nonzero components $\{y_{0,h}\}$ and

DEVELOPMENTS OF THE DECISION-FEEDBACK EQUALIZER

Table 7.3 VALUES OF Y_0 AND Y_1 FOR THE FOUR DIFFERENT VALUES OF Δ

Δ	Y_0						
	$y_{0,-2}$	$y_{0,-1}$	$y_{0,0}$	$y_{0,1}$	$y_{0,2}$	$y_{0,3}$	$y_{0,4}$
$-0.500T$	0.0000	0.0000	0.5000	0.5000	0.0000	0.0000	0.0000
$-0.375T$	-0.0022	-0.0125	0.6860	0.3201	0.0072	0.0016	0.0006
$-0.250T$	-0.0037	-0.0243	0.8488	0.1698	0.0081	0.0020	0.0008
$-0.125T$	-0.0031	-0.0246	0.9603	0.0624	0.0046	0.0012	0.0005

Δ	Y_1						
	$y_{1,-2}$	$y_{1,-1}$	$y_{1,0}$	$y_{1,1}$	$y_{1,2}$	$y_{1,3}$	$y_{1,4}$
$-0.500T$	0.0000	0.0000	0.0000	1.0000	0.0000	0.0000	0.0000
$-0.375T$	0.0012	0.0046	0.0624	0.9603	-0.0246	-0.0031	-0.0010
$-0.250T$	0.0020	0.0081	0.1698	0.8488	-0.0243	-0.0037	-0.0012
$-0.125T$	0.0016	0.0072	0.3201	0.6860	-0.0125	-0.0022	-0.0008

$\{y_{1,h}\}$, respectively, for $h < 0$. The results of the tests are shown in Fig. 7.19, where ψ_2 is given by Equation 7.118. The variation of P_4 (Equation 7.106) with ψ_2 is also plotted in Fig. 7.19. Since the value of Δ does not appear to have a great effect on the performance of system 1, only the curves showing the best and poorest performance are shown in each case.

A theoretical upper bound to the performance of the optimum detector over the given channel is obtained by considering the matched-filter detection of an individual received signal-element $s_i a(t - iT)$ under the given conditions but in the *absence* of any other transmitted signal elements. It can be shown[62-64] from Equations 7.116–7.118 that the probability of error in the detection of s_i is now

$$P_6 = Q\left(\sqrt{\frac{E_b}{\tfrac{1}{2}N_0}}\right) = Q\left(\sqrt{\frac{3}{2\sigma^2}}\right) \tag{7.119}$$

The variation of P_6 with ψ_2 (Equation 7.118) is plotted in Fig. 7.19.

It can be seen from Fig. 7.19 that the best performance of system 1 (obtained when $\Delta = -0.375T$) is only some 0.4 dB inferior to that of the ideal matched-filter detector (P_6), whereas the worst performance of system 1 (obtained when $\Delta = -0.5T$) is some 0.8 dB inferior to that of the matched-filter detector. The performance of system 1 is therefore not seriously affected by the phase of the sampling instants at the receiver and is under all conditions within about 1 dB of that of the optimum detector. The performance of the latter is very close to P_6 and, of course, inferior to it.

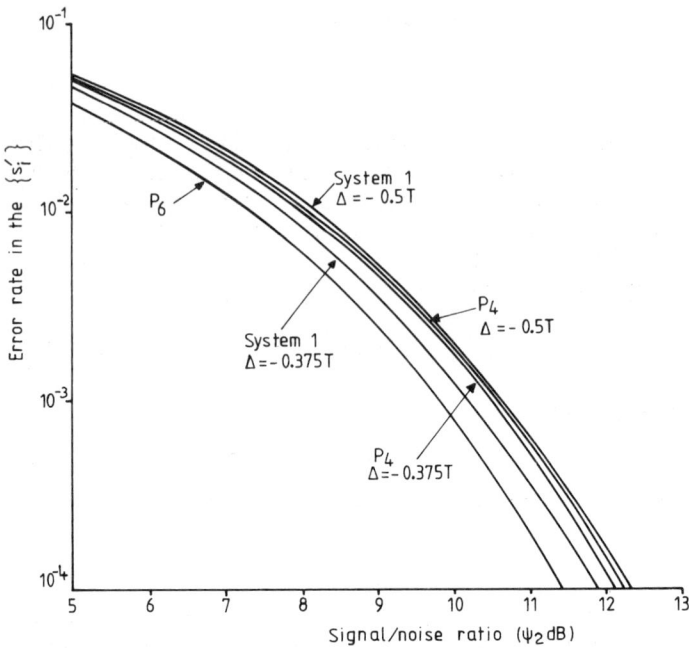

Fig. 7.19 *Performance of equalizers with a channel having a raised-cosine transfer function*

7.6.4 Comparison of systems 1 and 2 with a linear equalizer

The purpose of this series of tests is to study the new equalizers under conditions of extreme time dispersion of the received data signal, when a very large number of taps are required in the transversal filter F of Fig. 7.13. The binary signal elements are here transmitted at 140 Mbit/s over a 1 km length of 4.4 mm coaxial cable. The transmission path formed by this cable is taken to have an attenuation of $5.25\sqrt{F}$ dB at a frequency of F MHz[66,67], and its impulse response is

$$h(t_0) = \frac{k_0}{\sqrt{2\pi t_0^3}} \exp(-k_0^2/2t_0) \qquad (7.120)$$

where $\qquad k_0 = (5.25 \log_e 10)/20\sqrt{2\pi} \qquad (7.121)$

and t_0 is measured in μs[66]. The quantities t, τ and T are measured in seconds, as before, and $T = 7.143 \times 10^{-9}$. The lowpass filter A

(Fig. 7.1) now has the impulse response

$$a(t) = \begin{cases} 1, & 0 < t < \tfrac{1}{2}T \\ 0, & \text{elsewhere} \end{cases} \qquad (7.122)$$

and the lowpass filter B again satisfies Equation 7.57. Ideally, the lowpass filter A should have the same transfer function as the lowpass filter B^{68}, but, since we are here concerned mainly with the receiver design, the more conventional transmitted pulses of short duration (Equation 7.122) are assumed.

The average transmitted energy per binary signal-element is

$$E_b = \int_{-\infty}^{\infty} a^2(t)\,dt = \tfrac{1}{2}T \qquad (7.123)$$

and, from Equation 7.59,

$$N_0 = T\sigma^2 \text{ and } 2T\sigma^2 \qquad (7.124)$$

respectively, for systems 1 and 2. From these results it is easy to evaluate the signal/noise ratio, ψ_3 dB, where

$$\psi_3 = 10\log_{10}(E_b/\tfrac{1}{2}N_0) \qquad (7.125)$$

The impulse response $y(t)$ of the linear baseband channel in Fig. 7.1 has been computed for both systems 1 and 2 (which have, of course, different filters B), and the corresponding curves, which are very similar, are shown in Fig. 7.20. A total of 150 samples of $y(t)$ have been used for each of Y_0 and Y_1 in system 1, and 150 samples for Y_0 in system 2. Since the lowpass filter A and coaxial cable do not form a bandlimited channel, some of the received data-signal power is lost in the lowpass filter B, the loss being however smaller for system 1 than for system 2.

Both of the systems 1 and 2 have been tested by computer simulation for each of several different values of Δ (Fig. 7.12). Any particular value of Δ is now identified by the corresponding value of e, where

$$\Delta = \Delta_0 + eT \qquad (7.126)$$

and Δ_0 is the value of Δ for which $y_{0,1}$ has its maximum value. As before, some of the $\{y_{0,h}\}$ and $\{y_{1,h}\}$ may be nonzero for $h < 0$. The results of the tests are shown in Figs. 7.21 and 7.22 where ψ_3 is given by Equation 7.125. Since the value of Δ does not appear to have a very great effect on the performance of either system 1 or 2, only the curves showing the best and poorest performance are shown in each case. Curves for P_4 (Equation 7.106) and P_5 (Equation 7.111) are also given in Figs. 7.21 and 7.22 for each of the two values of Δ. A most important and perhaps surprising result here is the relatively close agreement

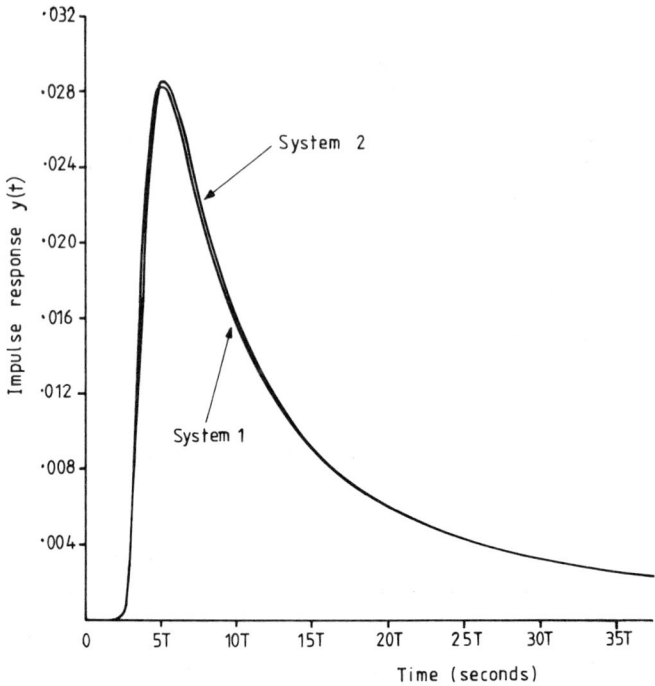

Fig. 7.20 Impulse response of a linear baseband channel formed by a coaxial cable and the equipment filters

between the theoretical and measured performances at low error rates, which shows that, in spite of the extreme time dispersion of each received signal-element, there is *no significant degradation* in the performance of the decision-feedback equalizer due to the occurrence of error bursts. Indeed, the average number of errors in an error burst, at low error rates, appears to be only around 2, for both systems 1 and 2. Much of the difference between the measured and theoretical performances of systems 1 and 2, with $e=0$, is due to the intersymbol interference introduced by $y_{0,-1}$. The components involving the $\{y_{0,h}\}$ and $\{y_{1,h}\}$, for $h = -1, -2, \ldots$, are, of course, ignored (treated as noise) in the detection process and are not taken account of in P_1 and P_2. The reason for the small number of errors in an error burst can be seen intuitively as follows.

Since the incorrect detection of a data symbol results in the doubling of the magnitude of the corresponding signal-element at the detector input (in place of the removal of that element), it follows from Fig. 7.20 that when a data symbol has been incorrectly detected there

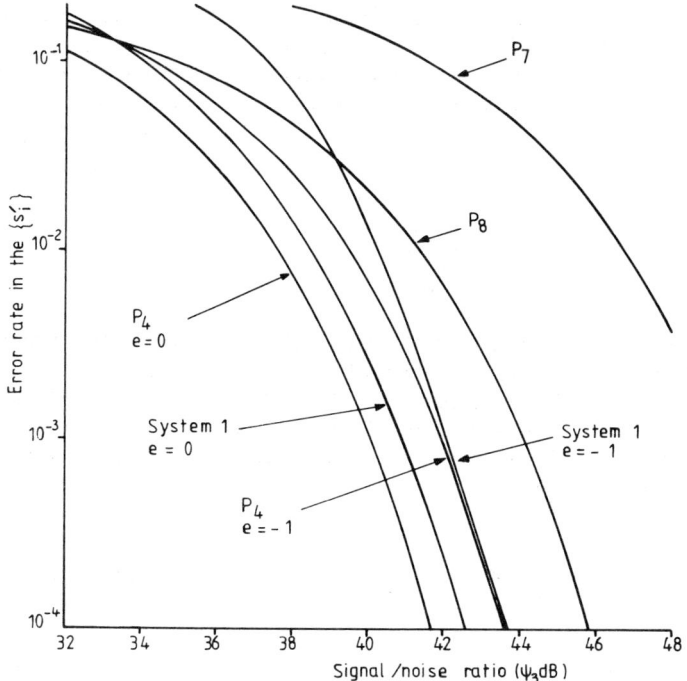

Fig. 7.21 Performance of system 1 and ideal linear equalizers with a coaxial cable as the transmission path

is a high probability that the second data symbol to be incorrectly detected is one of the immediately following symbols, the sign of the second symbol being almost certainly the negative of that of the first. Furthermore, it can be shown that, under the most likely conditions, there is, in fact, a high probability that the second data symbol to be incorrectly detected is the *immediately following* symbol. Clearly, the intersymbol interference at the detector input, resulting from the second wrongly detected symbol, tends to cancel (at least partly) that resulting from the first wrongly detected symbol. The smaller the time interval between the two symbols, the smaller is the resultant intersymbol interference, and so the lower is the probability of further errors, the probability most often being such that these are unlikely to occur. There is therefore a strong tendency for errors to occur in pairs, with just two errors in an error burst.

The performance of a linear equalizer for the given transmitted data signal and transmission path is determined as follows. The lowpass filter B (Fig. 7.1) is replaced by the lowpass filter G, whose

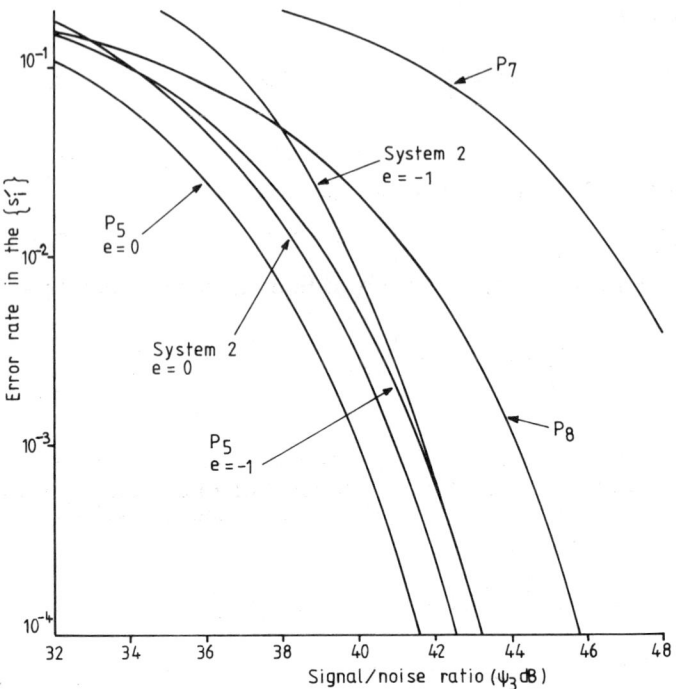

Fig. 7.22 Performance of system 2 and ideal linear equalizers with a coaxial cable as the transmission path

transfer function $G(f)$ is such that the transfer function $L(f)$ of the linear baseband channel, formed by the lowpass filter A, coaxial cable and lowpass filter G, is

$$L(f) = \begin{cases} T, & -\frac{1}{2T} < f < \frac{1}{2T} \\ 0, & \text{elsewhere} \end{cases} \qquad (7.127)$$

The filter G is clearly a linear equalizer for the filter A and coaxial cable. The waveform $r(t)$ at the output of the lowpass filter G is sampled once per data-symbol, at the correct time instants $\{iT\}$, the delay in transmission needed to make the channel physically realizable again being ignored. Thus the signal at the output of the filter G, at time $t = iT$, is[64]

$$r_i = s_i + w_i \qquad (7.128)$$

where w_i is a Gaussian random variable with zero mean, as before, but its variance is

$$\sigma_1^2 = \int_{-\infty}^{\infty} \tfrac{1}{2}N_0|G(f)|^2\,df \tag{7.129}$$

Now
$$L(f) = A(f)H(f)G(f) \tag{7.130}$$

where $A(f)$ is the Fourier transform of $a(t)$ and

$$H(f) = 10^{-(5.25\sqrt{f}/20000)} \tag{7.131}$$

f being measured in Hz, so that (from Equations 7.127 and 7.130)

$$G(f) = \begin{cases} TA^{-1}(f)H^{-1}(f), & -\dfrac{1}{2T} < f < \dfrac{1}{2T} \\ 0, & \text{elsewhere} \end{cases} \tag{7.132}$$

Inserting $G(f)$ in Equation 7.129 gives (at least, in principle) a value for σ_1^2. An approximate solution to an equation of the same type as this one has been derived by D.G.W. Ingram[66], which, when applied to Equation 7.129, gives

$$\sigma_1^2 = \frac{8 \times 10^6 N_0}{\beta^2}\left[(\beta\sqrt{F}-1)\exp(\beta\sqrt{F})+1\right] \tag{7.133}$$

where $F = 70$ and $\beta = 1.2089$, so that

$$\sigma_1 = 1.1098 \times 10^6 \sqrt{N_0} \tag{7.134}$$

The approximation here involves the simplifying assumption that the transmitted rectangular pulses $\{s_i a(t-iT)\}$ (Equation 7.122) are replaced by the corresponding impulses $\{\tfrac{1}{2}Ts_i\delta(t-iT)\}$, which have almost the same magnitude of the spectral density over the passband of the filter G[66].

From Equation 7.128, the probability of error in the detection of s_i, and hence the error rate in the $\{s_i'\}$, is

$$P_7 = Q\left(\frac{1}{\sigma_1}\right) \tag{7.135}$$

The variation of P_7 with ψ_3 (Equation 7.125) is plotted in Figs. 7.21 and 7.22.

A performance analysis has also been carried out for a linear equalizer of the same type as that just considered, with the same information rate of 140 Mbit/s and the same transmission path, but now for a 93.3 Mbaud ternary signal employing a 6B4T code[69], and $s_i = 0$ or ± 1. In this sytem $T = 10.714 \times 10^{-9}$ and, from Equation 7.128, the error rate in the detected (and decoded) binary digits is approximately

$$P_8 = \tfrac{2}{3} Q \left(\frac{0.5}{\sigma_2} \right) \quad (7.136)$$

σ_2^2 is the variance of w_i (Equation 7.128) and satisfies Equation 7.133 (in place of σ_1^2), but with F changed to 46.67, so that

$$\sigma_2 = 0.3922 \times 10^6 \sqrt{N_0} \quad (7.137)$$

The multiplying constant of 2/3 in Equation 7.136 assumes that Gray coding is achieved in the $6B4T$ code, so that the most likely error in s_i' gives only one error in the decoded binary signal. No account is taken in Equation 7.136 of the small redundancy in the $6B4T$ code, and to offset this the additional multiplying constant of 4/3 that should have been applied to the Q function, has been omitted. The $6B4T$ code has been considered here in place of the more conventional $4B3T$ code[67], since it slightly improves the performance of the linear equalizer under the ideal conditions assumed here.

The mean-square value, $\mathscr{E}[s_i^2]$, of a data-symbol s_i in the ternary system is 2/3, and (from Equation 7.122) the average transmitted energy per signal element (or data symbol) is

$$\mathscr{E}[s_i^2] \int_{-\infty}^{\infty} a^2(t)\, dt = \tfrac{1}{3} T \quad (7.138)$$

Since each signal element carries 1.5 binary digits (bits), the average transmitted energy per bit is

$$E_b = \tfrac{2}{9} T \quad (7.139)$$

The signal/noise ratio, ψ_3 dB, is now given by Equations 7.125, 7.137 and 7.139. The variation of P_8 with ψ_3 is plotted in Figs. 7.21 and 7.22.

It is evident from Figs. 7.21 and 7.22 that system 2 has a slightly better performance than system 1. Since it is also a simpler system, it is clearly better suited to the given application. Both systems 1 and 2 gain a considerable advantage in tolerance to additive white Gaussian noise over the linear equalizer that operates on the same binary signal (with error probability P_7), and, at error rates below 1 in 100, they both gain a useful advantage over the ternary system employing a linear equalizer (with error probability P_8). The weakness of systems 1 and 2, in the given application, is that they require a more accurate knowledge of the received signal level than does the corresponding linear equalizer, essentially because of the decision-directed cancellation of intersymbol interference at the detector input. Furthermore, due to the very large number of taps required in the transversal filter F (Fig. 7.13) the latter might be

DEVELOPMENTS OF THE DECISION-FEEDBACK EQUALIZER

better implemented as the corresponding analogue filter. It is nevertheless of great interest to observe that systems 1 and 2 do not suffer from serious error bursts, in spite of the very severe time dispersion of the received signal-elements.

In conclusion, the new techniques are of quite general application with severely band-limited binary signals and should often enable a very good tolerance to additive white Gaussian noise to be achieved with a surprisingly simple receiver. The techniques are particularly well suited to very high speed binary digital-transmission systems, where the sampled impulse-response of the channel has only a few components and where the transmission rate is such as to permit only a small number of operations per received data-symbol.

7.7 EXAMPLE

Problem
The ith sample of a received signal is

$$r_i = s_i + 2s_{i-1} - s_{i-2} + w_i \tag{7.140}$$

where the data-symbols $\{s_i\}$ are statistically independent and equally likely to have either value ± 1, and the noise components $\{w_i\}$ are statistically independent Gaussian random variables with zero mean and variance σ^2.

Derive an expression for the probability of error in the detection of s_i for each of the following three arrangements of decision directed cancellation of intersymbol interference, and compare the relative tolerances to noise of the three systems.

(a) The system operates on r_i such that s_i is detected without intersymbol interference.
(b) The system operates on r_{i+1} such that s_i is detected with intersymbol interference from s_{i+1}.
(c) The system operates on both r_i and r_{i+1} such that s_i is detected with intersymbol interference from s_{i+1}.

Solution
Let s_i' be the detected value of s_i.
(a) The signal $2s_{i-1}' - s_{i-2}'$ is subtracted from r_i to give the equalized signal

$$x_i = r_i - 2s_{i-1}' + s_{i-2}' \tag{7.141}$$

which is fed to the detector.
With the correct detection of s_{i-1} and s_{i-2},

$$x_i = r_i - 2s_{i-1} + s_{i-2} = s_i + w_i \tag{7.142}$$

so that the intersymbol interference in r_i has been eliminated at the detector input and s_i is detected from x_i with no interference from the other data symbols.
When $x_i > 0$, $s_i' = 1$, and when $x_i < 0$, $s_i' = -1$.

The probability density function of w_i at the sample value w is

$$p(w) = \frac{1}{\sqrt{2\pi\sigma^2}} \exp\left(-\frac{w^2}{2\sigma^2}\right) \quad (7.143)$$

With the correct detection of s_{i-1} and s_{i-2}, an error is obtained in s'_i when $|w_i| > 1$ and w_i has the opposite sign to s_i. The probability of this occurring is

$$P_1 = \int_1^\infty p(w)\,dw = Q\left(\frac{1}{\sigma}\right) \quad (7.144)$$

regardless of the sign of s_i. An error in s'_i greatly increases the probability of error in $s'_{i+1}, s'_{i+2}, \ldots$, so that errors tend to occur in bursts. If b is the average number of errors in a burst, the resultant error rate, or average error probability, is bP_1. In the present case $b \simeq 4$ at low error rates, so that the actual error probability is now approximately $4Q(1/\sigma)$. However, for practical purposes, the error probability can be taken to be P_1 in Equation 7.144, when the signal/noise ratio is sufficiently high.

(b) The signal $-s'_{i-1}$ is subtracted from r_{i+1} to give the partially equalized signal

$$x_{i+1} = r_{i+1} + s'_{i-1} \quad (7.145)$$

which is fed to the detector.

With the correct detection of s_{i-1},

$$x_{i+1} = s_{i+1} + 2s_i + w_{i+1} \quad (7.146)$$

so that s_i is detected from x_{i+1} with intersymbol interference from s_{i+1}.
When $x_{i+1} > 0$, $s'_i = 1$, and when $x_{i+1} < 0$, $s'_i = -1$.

At high signal/noise ratios, practically all errors in the $\{s'_i\}$ occur when

$$s_{i+1} = -s_i \quad (7.147)$$

so that

$$x_{i+1} = s_i + w_i \quad (7.148)$$

The probability that $s_{i+1} = -s_i$ is $\frac{1}{2}$. Under these conditions, an error occurs in s'_i when $|w_i| > 1$ and w_i has the opposite sign to s_i, so that the probability of error in the detection of s_i is

$$P_2 = Q\left(\frac{1}{\sigma}\right) \quad (7.149)$$

for either value of s_i. The average number of errors in a burst, b, is now approximately 1, so that the resultant error rate or average error probability, which is effectively $\frac{1}{2}bQ(1/\sigma)$, can be taken to be P_2 in Equation 7.149.

(c) The data-symbol s_i can be detected from *both* r_i and r_{i+1}, using the arrangements of intersymbol interference cancellation just described, such that, with the correct detection of s_{i-1} and s_{i-2},

$$x_i = r_i - 2s_{i-1} + s_{i-2} = s_i + w_i \quad (7.150)$$

$$x_{i+1} = r_{i+1} - s_{i-1} = s_{i+1} + 2s_i + w_{i+1} \quad (7.151)$$

DEVELOPMENTS OF THE DECISION-FEEDBACK EQUALIZER 435

and
$$x_{i+1} + x_i = s_{i+1} + 3s_i + w_{i+1} + w_i \qquad (7.152)$$

When $x_{i+1} + x_i > 0$, $s'_i = 1$, and when $x_{i+1} + x_i < 0$, $s'_i = -1$.
At high signal/noise ratios practically all errors in the $\{s'_i\}$ occur when

$$s_{i+1} = -s_i \qquad (7.153)$$

and now

$$x_{i+1} + x_i = 2s_i + w_{i+1} + w_i \qquad (7.154)$$

For an error in s'_i, the resultant noise component $w_{i+1} + w_i$ must have a magnitude greater than 2 and the opposite sign to s_i. Since w_{i+1} and w_i are statistically independent Gaussian random variables with zero mean and variance σ^2, $w_{i+1} + w_i$ is a Gaussian random variable with zero mean and variance $2\sigma^2$. Thus, when $s_{i+1} = -s_i$ and with the correct detection of s_{i-1} and s_{i-2}, the probability of error in the detection of s_i is

$$P_3 = Q\left(\frac{2}{\sqrt{2}\sigma}\right) = Q\left(\frac{\sqrt{2}}{\sigma}\right) \qquad (7.155)$$

for either value of s_i. The average or actual error probability at high signal/noise ratios, is effectively $\frac{1}{2}bQ(\sqrt{2}/\sigma)$, where the average number of errors in a burst, b, is not large, so that the actual error probability can be taken to be P_3 (Equation 7.155).

At high signal/noise ratios, the arrangement (c) gains an advantage of about 3 dB in tolerance to additive white Gaussian noise over each of the arrangements (a) and (b), the latter being slightly better than the former. A linear equalizer and a matched filter give even better performances here.

7.8 NEAR-MAXIMUM-LIKELIHOOD DETECTORS

7.8.1 Introduction

Throughout this book the actual decision process, that is involved in determining the detected value of a data-symbol s_i, is simple *threshold level detection* that compares the equalized signal with an appropriate threshold. If now, in a decision-feedback equalizer, the detection of a data-symbol s_i is *delayed*, such that, in addition to the original equalized signal containing s_i, several partially equalized signals are available (containing not only s_i but also one or more of the symbols $s_{i+1}, s_{i+2}, \ldots$) then a better tolerance to additive white Gaussian noise can, in general, be achieved than with any of the equalization processes described here[26]. A data symbol is now detected from *several* of the components of the corresponding sampled signal-element (at the output of the linear filter forming the first part of the equalizer) instead of from just *one* component as before, and the further information on s_i provided by the additional components leads to a better detection process.

Consider the data-transmission system of Fig. 7.1 with a conventional decision-feedback equalizer at the receiver. The various properties of this system are as described in Section 7.2. The data-symbols $\{s_i\}$ are statistically independent and equally likely to have any of their given possible values, which are

$$s_i = \pm 1 \tag{7.156}$$

in the case of a 2-level (binary) signal and

$$s_i = \pm \frac{1}{\sqrt{5}} \quad \text{or} \quad \pm \frac{3}{\sqrt{5}} \tag{7.157}$$

in the case of a 4-level (quaternary) signal. In each case s_i has a mean-square value of unity. A finite sequence of m data-symbols, in the form of the impulses $\{s_i \delta(t - iT)\}$ for $i = 0, 1, \ldots, m-1$, is now fed to the linear baseband channel in Fig. 7.1, so that $s_i = 0$ for $i < 0$ and $i > m-1$.

The resultant sampled impulse-response of the channel and filter D in Fig. 7.1 is given by the time-invariant vector (sequence)

$$E = [e_0 \quad e_1 \quad \ldots \quad e_g] \tag{7.158}$$

having the various properties described in Section 7.2. The output signal from the filter D, at time $t = iT$, is

$$v_i = \sum_{j=0}^{g} s_{i-j} e_j + u_i \tag{7.159}$$

where the $\{u_i\}$ are statistically independent Gaussian random variables with zero mean and variance σ^2. As before, all signals are real valued.

The receiver has prior knowledge of E and of the possible values of s_i. At time $t = iT$ it has formed the detected values $s'_{i-1}, s'_{i-2}, \ldots$ of the data-symbols $s_{i-1}, s_{i-2}, \ldots$, respectively, and it forms the signal

$$q_{i,0} = \sum_{j=1}^{g} s'_{i-j} e_j \tag{7.160}$$

at the output of the filter F. It then subtracts $q_{i,0}$ from v_i to give the equalized signal

$$x_{i,0} = v_i - q_{i,0} \tag{7.161}$$

at the detector input. With the correct detection of $s_{i-1}, s_{i-2}, \ldots, s_{i-g}$, this becomes

$$x_{i,0} = s_i e_0 + u_i \tag{7.162}$$

The delay in transmission is, of course, neglected here. In a conventional decision-feedback equalizer the data-symbol s_i is now

DEVELOPMENTS OF THE DECISION-FEEDBACK EQUALIZER 437

detected as its possible value s'_i for which

$$|x_{i,0} - s'_i e_0| \tag{7.163}$$

is *minimum*. The equalizer here achieves the *exact* equalization of the channel, in the sense that *all* intersymbol interference is removed from the equalized signal, but, as in Section 7.2, the latter is *not* an unbiased estimate of s_i, since e_0 is not here, in general, equal to unity. For reasons that will become clear in Section 7.8.2, the symbols used for the equalized signal and the signal at the output of the filter F have been *changed* from those previously used.

7.8.2 System 3

In the arrangement now to be considered the data-transmission system is as shown in Fig. 7.1, except that the threshold-level detector and transversal filter F are replaced by the more complex system to be described here. Suppose first that the detection of s_i is delayed by one sampling interval T, until the receipt of the signal

$$v_{i+1} = \sum_{j=0}^{g} s_{i+1-j} e_j + u_{i+1} \tag{7.164}$$

at the output of the filter D, at time $t = (i+1)T$. The receiver then forms the signal

$$q_{i,1} = \sum_{j=1}^{g-1} s'_{i-j} e_{j+1} \tag{7.165}$$

which is subtracted from v_{i+1}, to give the signal

$$x_{i,1} = v_{i+1} - q_{i,1} \tag{7.166}$$

With the correct detection of $s_{i-1}, s_{i-2}, \ldots, s_{i-g+1}$, this becomes

$$x_{i,1} = s_{i+1} e_0 + s_i e_1 + u_{i+1} \tag{7.167}$$

It is assumed that, in addition to the partially equalized signal $x_{i,1}$, obtained at time $t = (i+1)T$, the receiver has formed the equalized signal $x_{i,0}$ (Equation 7.162) at time $t = iT$, so that *both* $x_{i,0}$ and $x_{i,1}$ are available for the detection of s_i. Furthermore, it can be seen from Equations 7.162 and 7.167 that, with the correct detection of $s_{i-1}, s_{i-2}, \ldots, s_{i-g}$ and for given possible values s'_i and s'_{i+1} of the data-symbols s_i and s_{i+1}, $s'_i e_0$ and $s'_{i+1} e_0 + s'_i e_1$ are the corresponding values of $x_{i,0}$ and $x_{i,1}$, respectively, in the *absence of noise*, so that $s'_i e_0$ and $s'_{i+1} e_0 + s'_i e_1$ are *estimates* of $x_{i,0}$ and $x_{i,1}$. The *smaller* the Euclidean distance between the two-component vectors $[x_{i,0} \ x_{i,1}]$ and $[s'_i e_0 \ (s'_{i+1} e_0 + s'_i e_1)]$ the *better* is the estimate of $[s_i \ s_{i+1}]$ given by $[s'_i \ s'_{i+1}]$. The data-symbol s_i is now detected as its possible value

s_i', for which

$$(x_{i,0} - s_i' e_0)^2 + (x_{i,1} - s_{i+1}' e_0 - s_i' e_1)^2 \qquad (7.168)$$

is minimum over all combinations of possible values of s_i' and s_{i+1}'. Expression 7.168 is, of course, the square of the Euclidean distance between $[x_{i,0} \ x_{i,1}]$ and $[s_i' e_0 \ (s_{i+1}' e_0 + s_i' e_1)]$. Clearly, a detected value s_{i+1}' is determined here for the data-symbol s_{i+1}, but this is discarded, the detected value of s_{i+1} being determined from $x_{i+1,0}$ and $x_{i+1,1}$ at time $t=(i+2)T$, in a manner similar to that just described for s_i.

Whereas the conventional equalizer employs only the first component $s_i e_0$ of a received signal-element in the detection of s_i, the modified system, which is a particular arrangement of system 3, employs the first *two* components $s_i e_0$ and $s_i e_1$. Evidently, if $|e_1| > |e_0|$, a useful improvement in tolerance to noise should be obtained by the latter system. The improvement in performance here relies partly on the fact that the Gaussian noise components u_i and u_{i+1} (in $x_{i,0}$ and $x_{i,1}$) are statistically independent and with the same variance σ^2. It can be seen from Section 1.4, that, given the correct detection of $s_{i-1}, s_{i-2}, \ldots, s_{i-g}$, the detection process just described is the *maximum-likelihood detection* of the vector $[s_i \ s_{i+1}]$ from the vector $[x_{i,0} \ x_{i,1}]$ (and hence from the vector $[v_i \ v_{i+1}]$). Under the assumed conditions, this minimizes the probability of error in the detection of s_i and s_{i+1} from the signals $x_{i,0}$ and $x_{i,1}$. By involving *more* received samples in the detection of s_i and s_{i+1}, a lower error probability can, in general, be achieved.

Suppose next that the detection of s_i is delayed by two sampling intervals. With the correct detection of $s_{i-1}, s_{i-2}, \ldots, s_{i-g+2}$, the partially equalized signal at the detector input, at time $t = (i+2)T$, is

$$x_{i,2} = s_{i+2} e_0 + s_{i+1} e_1 + s_i e_2 + u_{i+2} \qquad (7.169)$$

The three signals $x_{i,0}$, $x_{i,1}$ and $x_{i,2}$ are now used in the detection of s_i, such that its detected value s_i' is taken as its possible value for which

$$(x_{i,0} - s_i' e_0)^2 + (x_{i,1} - s_{i+1}' e_0 - s_i' e_1)^2$$
$$+ (x_{i,2} - s_{i+2}' e_0 - s_{i+1}' e_1 - s_i' e_2)^2 \qquad (7.170)$$

is minimum over all combinations of possible values of s_i', s_{i+1}' and s_{i+2}'. Expression 7.170 is the square of the Euclidean distance between the three-component vectors $[x_{i,0} \ x_{i,1} \ x_{i,2}]$ and

$$[s_i' e_0 \quad (s_{i+1}' e_0 + s_i' e_1) \quad (s_{i+2}' e_0 + s_{i+1}' e_1 + s_i' e_2)]$$

Clearly, with the correct detection of $s_{i-1}, s_{i-2}, \ldots, s_{i-g}$, the detection process just described is the *maximum likelihood* detection of the vector $[s_i \; s_{i+1} \; s_{i+2}]$ from the vector $[x_{i,0} \; x_{i,1} \; x_{i,2}]$. It is evident that the delay in detection can, in principle, be extended to h sampling intervals, such that s_i is detected at time $t = (i+h)T$ from the $h+1$ signals $x_{i,0}, x_{i,1}, \ldots, x_{i,h}$, the detection process being appropriately modified. This arrangement is referred to here as system 3. With a sufficiently large value of h (say, $h > 16$) the process comes close to achieving the best available tolerance to additive white Gaussian noise, for the given delay in detection. Unfortunately, the system now involves either an excessive number of sequential operations per received data-symbol or else becomes unduly complex. Various techniques for simplifying the detection process have therefore been developed and these can often achieve a tolerance to noise within 1 or 2 dB of the optimum available, without the use of complex equipment[9,25-27,54,55,61]. The weakness of these techniques is that they still require quite a large number of sequential operations per received data-symbol and are therefore not well suited to high-transmission rates (over 10,000 bit/s)[26].

Consider now the detection of the *first* data-symbol s_0 (of the sequence of m data-symbols) by an arrangement of system 3 in which the delay in detection of h sampling intervals is increased to $m+g-1$, such that *all* signals $\{v_i\}$ at the output of the filter D, that are dependent on *any* of the m received data-symbols $s_0, s_1, \ldots, s_{m-1}$, are involved in the detection process. Since s_0 is the first of the received sequence of data symbols, *no* decision-directed cancellation of intersymbol interference is carried out here and the detector operates directly on the $m+g$ signals $v_0, v_1, \ldots, v_{m+g-1}$. The detected value s_0' of the data-symbol s_0 is therefore taken as its possible value for which

$$\sum_{i=0}^{m+g-1} \left(v_i - \sum_{j=0}^{g} s_{i-j}' e_j \right)^2 \quad (7.171)$$

is minimum over all combinations of possible values of $s_0', s_1', \ldots, s_{m-1}'$, bearing in mind that $s_i' = s_i = 0$, for $i < 0$ and $i > m-1$. But the detection process just described is the *maximum-likelihood* detection of the *complete sequence* of data symbols

$$[s_0 \quad s_1 \quad \ldots \quad s_{m-1}] \quad (7.172)$$

from the corresponding received sequence of signals

$$[v_0 \quad v_1 \quad \ldots \quad v_{m+g-1}] \quad (7.173)$$

at the output of the filter D (see Section 1.4). Under the assumed conditions, the detection process minimizes the probability of error in the detection of the complete sequence of data symbols, no reduction in error probability being achieved through the use of any of the signals $\{v_i\}$, for $i < 0$ or $i > m+g-1$, in the detection process. Thus *all* data symbols are detected *simultaneously* in a *single* detection process that is *optimum*. Unfortunately, when m is large, the implementation of the *single* optimum detection process cannot be achieved, in practice, by any known techniques. However, essentially the same performance can be achieved by means of appropriate detectors that determine the detected data-symbols *successively* (in sequence) rather than *simultaneously*, and, for our purposes, the most important of these is the *Viterbi-algorithm detector*[8]. Although such detectors are *not* equalizers in the sense assumed here, they will now be considered briefly.

7.8.3 Viterbi-algorithm detector

A practical approach towards the maximum-likelihood detection of a received signal message is the Viterbi-algorithm detector, which operates as follows[8,26,28,32,61]. The data-transmission system is as shown in Fig. 7.1, except that the threshold-level detector and transversal filter F are replaced by a more complex system. It is assumed that l-level data-symbols $\{s_i\}$ are transmitted (Equations 7.156 and 7.157). Just prior to the receipt of the signal v_i from the transversal filter D, the detector holds in store l^g $(h+1)$-component vectors (sequences) $\{Q_{i-1}\}$, where

$$Q_{i-1} = [x_{i-h-1} \quad x_{i-h} \quad \ldots \quad x_{i-1}] \qquad (7.174)$$

The symbol x_j (for any j in the range 0 to $m-1$) may take on any of the l possible values of s_j, and $h > g$. Each vector Q_{i-1} is formed by the last $h+1$ components of the corresponding i-component vector

$$X_{i-1} = [x_0 \quad x_1 \quad \ldots \quad x_{i-1}] \qquad (7.175)$$

which represents a *possible* received sequence of data-symbols $\{s_j\}$. Associated with each vector Q_{i-1} is stored its *cost*

$$c_{i-1} = \sum_{j=0}^{i-1} (v_j - z_j)^2 \qquad (7.176)$$

where

$$v_j = z_j + u'_j \qquad (7.177)$$

$$z_j = \sum_{k=0}^{g} x_{j-k} e_k \qquad (7.178)$$

and $x_j = 0$ for $j < 0$ and $j > m-1$. The symbol z_j is here the value of v_j that would have been received in the absence of noise and for the given vectors E and X_{i-1} (Equations 7.158 and 7.175), and u'_j is the corresponding estimate of u_j (Equation 7.159). The cost c_{i-1} is the square of the Euclidean distance between the vectors (sequences)

$$V_{i-1} = [v_0 \quad v_1 \quad \ldots \quad v_{i-1}] \qquad (7.179)$$

and
$$Z_{i-1} = [z_0 \quad z_1 \quad \ldots \quad z_{i-1}] \qquad (7.180)$$

Since Z_{i-1} is the received sequence (up to the time $t=(i-1)T$) resulting from the transmission of the sequence X_{i-1} (Equation 7.175) in the absence of noise, c_{i-1} is a measure of the probability that this sequence is correct. The *smaller* the value of c_{i-1} the more likely is the corresponding sequence X_{i-1} to be correct (see Section 1.4).

The l^g vectors $\{Q_{i-1}\}$ held in store, at time $t=(i-1)T$, by the Viterbi-algorithm detector, have all l^g different combinations of the possible values of the last g components

$$x_{i-g}, x_{i-g+1}, \ldots, x_{i-1}$$

where each vector Q_{i-1} has the *smallest cost* for its *given combination* of values of the last g components. The vector Q_{i-1} that has the smallest cost of *all* l^g stored vectors forms the last $h+1$-components of the *maximum-likelihood* vector X_{i-1}, which is the possible sequence of values of the data-symbols $\{s_j\}$ most likely to be correct. The detected value s'_{i-h-1} of the data-symbol s_{i-h-1} is now taken as the value of x_{i-h-1} in the vector Q_{i-1} with the smallest cost.

On receipt of the signal v_i at the output of the filter D (Fig. 7.1) at time $t=iT$, each of the l^g stored vectors $\{Q_{i-1}\}$ is used to form l vectors $\{Q_i\}$, in which the first h components are given by the last h components of the original vector Q_{i-1} (that is, by $x_{i-h}, x_{i-h+1}, \ldots, x_{i-1}$ in this vector) and the last component x_i takes on its l possible values (Equations 7.156 and 7.157). Each of the resulting l^{g+1} vectors $\{Q_i\}$ has the cost

$$c_i = c_{i-1} + (v_i - z_i)^2 = c_{i-1} + \left(v_i - \sum_{j=0}^{g} x_{i-j} e_j\right)^2 \qquad (7.181)$$

which is determined using the appropriate stored value of c_{i-1} (Equation 7.176). For every *one* of the l^g possible combinations of values of $x_{i-g+1}, x_{i-g+2}, \ldots, x_i$, there are now l vectors $\{Q_i\}$, which originate from l *different* vectors $\{Q_{i-1}\}$ in the original stored set of l^g vectors, and, for *each* of the l^g combinations of values of $x_{i-g+1}, x_{i-g+2}, \ldots, x_i$ in the vectors $\{Q_i\}$, the detector selects the vector Q_i with the smallest value of c_i. The resulting l^g vectors $\{Q_i\}$ are then stored together with the associated $\{c_i\}$. The vector Q_i with the smallest cost forms the last $h+1$ components of the maximum-likelihood vector X_i, and the value of the first component x_{i-h} of this vector Q_i is now taken as the detected value s'_{i-h} of the data-symbol s_{i-h}. The process continues in this way. It can be seen that, during the processing of the first and last g of the sequence of data-symbols $[s_0 \, s_1 \ldots s_{m-1}]$, the operation of the Viterbi-algorithm detector is modified (simplified) due to the fact that some of the $\{x_j\}$ involved in the detection process are now zero. Furthermore, following the receipt of v_{m+g-1}, the detected values of *all* data symbols remaining to be detected, are given by the values of the corresponding $\{x_j\}$ in the vector Q_{m+g-1} with the smallest cost.

To avoid an unacceptable increase in the values of the costs of the $\{Q_i\}$, over a long message, the smallest of the costs of the l^g stored vectors is each time subtracted from every cost, thus reducing the smallest cost to zero, without, however, changing the differences between the various costs. In the interests of clarity, this process is ignored here.

The Viterbi algorithm *ensures* that the stored vector Q_i with the *smallest* cost forms the last $h+1$ components of the *maximum-likelihood* vector X_i[8,73]. The rigorous derivation of the algorithm is given elsewhere[73], but the basic mechanisms involved in the process are illuminated in the following analysis. As before, on the receipt of v_i, each stored vector Q_{i-1} is used to form l vectors $\{Q_i\}$ having the l different possible values of x_i, the remaining h components being determined by the corresponding h components of Q_{i-1}. Among the l^{g+1} vectors $\{Q_i\}$ so obtained there are now *all* combinations of the possible values of the last $g+1$ components

$$x_{i-g}, x_{i-g+1}, \ldots, x_i$$

giving therefore *all* possible values of the resulting symbol z_i (Equations 7.177 and 7.178).

Consider the general case where z_i has l^{g+1} different possible values. Now, for a *given* value of z_i, the values of $z_{i+1}, z_{i+2}, \ldots$ are *unaffected* by the values of $z_{i-1}, z_{i-2}, \ldots$. It follows that, if c_i is *minimum* for the given value of z_i in Equation 7.181, then, for the given sequence X_i, c_{i-1} must be minimum for the corresponding value of z_{i-1}, since otherwise the value of c_{i-1} could be reduced without affecting z_{i-1} and hence without affecting $(v_i - z_i)^2$ in Equation 7.181, thus reducing c_i. Similarly, c_{i-j} must be *minimum* for the corresponding value of z_{i-j}, for *every* integer j in the range 0 to i. Now, z_{i-j} is uniquely determined by the last $g+1$ components of the corresponding vector Q_{i-j}, so that, for every integer j, the vector Q_{i-j}, forming the corresponding *segment* of $h+1$ components of the *maximum-likelihood* vector X_i, must have the smallest cost for the given set of values of its last $g+1$ components. However, as will now be shown, this condition is by no means sufficient, of itself, to ensure maximum-likelihood detection.

In order to be *sure* of obtaining the maximum-likelihood vector X_i (which is represented by its last $h+1$ components as given by Q_i) the detector must, at time $t = iT$, generate a *complete set* of l^{g+1} vectors

$\{Q_i\}$ having all l^{g+1} different combinations of the values of their last $g+1$ components, and hence forming all l^{g+1} possible values of z_i. These vectors can always be generated, as previously described, from l^g vectors $\{Q_{i-1}\}$ having all l^g different combinations of the values of their last g components, bearing in mind that each vector Q_{i-1} is used to form l vectors $\{Q_i\}$ having the l different values of x_i. Furthermore, the value of x_{i-g-1} affects z_{i-1} but *not* z_i, so that only the last g components of the $\{Q_{i-1}\}$ need, in fact, take on all possible combinations of values. Thus, if *each* of the l^g vectors $\{Q_{i-1}\}$ has the *smallest* possible cost, for the *given* combination of values of x_{i-g}, $x_{i-g+1},\ldots,x_{i-1}$, then it is evident from Equation 7.181 that the one of the resulting l^{g+1} vectors $\{Q_i\}$ with the smallest cost over all these vectors *must* form the last $h+1$ components of the maximum-likelihood vector X_i. Furthermore, from the l^{g+1} vectors $\{Q_i\}$ can always be selected l^g vectors $\{Q_i\}$ having *all l^g* different combinations of the values of their last g components and each vector having the *smallest* possible cost for the given combination. The detector is now ready for the next detection process, and ensures that the required maximum-likelihood vector X_i is *always* obtained. It can be seen that, for every integer j, the vector Q_{i-j}, forming the corresponding *segment* of the *maximum-likelihood* vector X_i, must, in fact, have the smallest cost for the *given* set of values of its last g components, which means, of course, that it *also* has the smallest cost for the given set of values of its last n components, for every $n \geq g$.

The delay in detection of the Viterbi-algorithm detector is h sampling intervals, and this should always be made as large as conveniently possible. If h is made sufficiently large, the resulting sequence of detected data-symbols $[s'_0\ s'_1 \ldots s'_{m-1}]$ is the *same* as that which minimizes Expression 7.171, so that the Viterbi-algorithm detector is now a *maximum-likelihood detector* for the complete message of m data-symbols, and it minimizes the probability of error in the detection of the complete message[8,73].

Ideally, the *optimum* detector for a sequence of m data-symbols is that which minimizes the *total number* of errors in the m detected data-symbols, so that it minimizes the *error rate* in the detected data symbols.[5,50] Unfortunately, this requires a detection process that appears to be considerably more complex than the Viterbi-algorithm detector itself.[50] However, in the presence of additive white Gaussian noise (as in Fig. 7.1), the performances of the two different detectors become asymptotically the *same*, as the signal/noise ratio increases[26]. This is, perhaps, not surprising, since, when there are only one or two errors in the m detected data-symbols, there is no longer any important difference between a detector that minimizes the probability of error in the detected *message* of m data symbols and a detector that minimizes the *number* of errors in the detected data symbols. Our main concern here is with the transmission of digital data signals, where a *low* error rate is normally required, so that, for our purposes, a Viterbi-algorithm detector can be taken to be an *optimum* detector.

The Viterbi-algorithm detector involves the evaluation of l^{g+1} costs and l^g searches through l costs, for each received data-symbol, together with the storage of l^g vectors $\{Q_i\}$ and l^g costs $\{c_i\}$. When l or g become large, the detector becomes unacceptably complex. A technique that has been widely studied, for enabling a Viterbi-algorithm detector to be used with a large number of components in the sampled impulse-response of the channel, is to replace the linear filter D (ahead of the Viterbi-algorithm detector) by a linear filter, which may be adjusted adaptively (Sections 5.7 and 6.1) to reduce the number of components in the sampled impulse-response of the channel and filter to some desired small value[10,12,17,29,33]. The complexity of the Viterbi algorithm detector is now greatly reduced. The disadvantage of the arrangement is that, when the channel introduces appreciable amplitude distortion, the filter inevitably corrects some of this distortion, leading to enhancement of the noise at its output and correlation between neighbouring noise components here, with the consequent degradation in performance[37]. A preferable approach is to modify the detector *itself* in order to reduce its complexity[21,26,28,31]. A weakness of the Viterbi algorithm detector is that a large number of the stored vectors $\{Q_i\}$ (those with the larger costs) play no useful part in the detection process and could therefore well be omitted without affecting its performance[32,48,55,61]. The practical difficulty is in omitting these vectors without also omitting some that are required if maximum-likelihood detection is to be achieved. Various such algorithms have been developed[21-61], and the most widely studied of these will now be described, to illustrate the basic techniques involved[54].

7.8.4 System 4

The detector here is a development of the Viterbi-algorithm detector that uses a process of ranking (placing in order of merit) to select the stored vectors $\{Q_i\}$[54]. No constraints are now placed on the possible values of the $\{x_j\}$ in any of the $\{Q_i\}$, other than that x_j must, of course, take on a possible value of s_j. Furthermore, the number of stored vectors may now be any convenient value k, where typically $4 \leqslant k \leqslant 16$, and k is *not* determined by g or l.

Just prior to the receipt of the signal v_i from the transversal filter D, the detector holds in store k vectors $\{Q_{i-1}\}$ (Equation 7.174) together with their costs $\{c_{i-1}\}$ (Equation 7.176). No two vectors are the same. The detected value s'_{i-h-1} of the data-symbol s_{i-h-1} is now taken as the value of x_{i-h-1} in the vector Q_{i-1} with the smallest cost. Any vector Q_{i-1} in which x_{i-h-1} differs from s'_{i-h-1} is then *discarded*, together with its cost c_{i-1}. For reasonably large values of h and at fairly high signal/noise ratios, the discarding of vectors only occurs very occasionally. On the receipt of v_i, each stored vector Q_{i-1} is replaced by l vectors $\{Q_i\}$, having the same values of $x_{i-h}, x_{i-h+1}, \ldots, x_{i-1}$ as in the original Q_{i-1} and having the l different possible values of x_i, just as in the case of the Viterbi-algorithm detector. The cost c_i of each vector Q_i is now determined according to Equation 7.181. The detector then selects the k vectors $\{Q_i\}$ with the smallest costs, and stores these together with their costs. Finally, the detected value s'_{i-h} of the data-symbol s_{i-h} is taken as the value of x_{i-h} in the vector Q_i with the smallest cost, and all vectors $\{Q_i\}$ in which x_{i-h} differs from s'_{i-h} are discarded, together with their costs. The process continues in this way.

In a practical implementation of the system, the detection of the data-symbol s_{i-h-1} would be carried out *after* the formation of the vectors $\{Q_i\}$. The latter occur in groups of l vectors, each group originating from a different vector Q_{i-1}. Thus, after determining the costs $\{c_i\}$ of the kl vectors $\{Q_i\}$, the vector Q_i with the smallest cost can be identified and the detected value of s_{i-h-1} now taken as the value of x_{i-h-1} in the vector Q_{i-1} from which the given vector Q_i originated. The vectors $\{Q_i\}$ originating from vectors $\{Q_{i-1}\}$ whose values of x_{i-h-1} *differ* from the detected value of s_{i-h-1} are now *discarded*, and the k vectors $\{Q_i\}$ with the smallest costs then selected from the remaining vectors $\{Q_i\}$. The arrangement just described increases the delay in detection from h to $h+1$ sampling intervals. For our purposes, however, the slightly simpler arrangement previously described will be assumed, there being no significant difference in performance between the two systems when appropriately large values of h are used.

The process of discarding any vector Q_i whose component x_{i-h} disagrees with s'_{i-h} has the important property that it prevents the *merging* of the stored vectors, which can be a serious weakness of the corresponding system in which no vectors are discarded[74]. The merging of two or more vectors $\{Q_i\}$ means that these vectors have become the same and have closely similar costs. This effectively reduces the number of stored vectors and hence degrades the performance of the detection process. The discarding of the vectors $\{Q_i\}$ just described ensures that the remaining stored vectors $\{Q_i\}$ are always *all different*, provided only that the stored vectors $\{Q_i\}$ were all different for the *first* detection process. To ensure this, a suitable starting-up procedure must be used at the beginning of a transmission. A convenient technique is to send a *known* sequence of $\{s_i\}$, immediately correct synchronization has been achieved at the receiver, and then to set one of the stored vectors $\{Q_i\}$ to the correct sequence and its associated c_i to zero. The remaining $\{Q_i\}$ may be given any values and their associated $\{c_i\}$ are all set to some very high value. After a few received signals $\{v_i\}$ (at the output of the filter D) all $\{Q_i\}$ will have been derived from the original correct vector and they will all be different. As before, the smallest cost must always be subtracted from every cost, to reduce the smallest cost to zero and so prevent a steady increase in the costs. System 4 involves the evaluation of kl costs and k searches through kl costs, for each received data-symbol, together with the storage of k vectors $\{Q_i\}$ and k costs $\{c_i\}$. For a given number of stored vectors it requires rather more operations per received data-symbol than does the Viterbi-algorithm detector.

The presence of the filter D (Fig. 7.1) ahead of the detector is very important for the satisfactory operation of system 4, since it ensures that the first component e_0 of the sampled impulse-response of the channel and filter is among the larger of the $g+1$ components. If the filter D is omitted, so that the sampled impulse-response E is replaced by the sampled impulse-response Y, then, not only the first component of the sampled impulse-response but sometimes the first two or three components may become very small, and, under these conditions, the performance of system 4 is likely to be seriously degraded[32]. The performance of the Viterbi-algorithm detector, on the other hand, is *unaffected* by the removal of the filter D, assuming, of course, that the detector is appropriately modified to operate with the sampled impulse-response Y. This is because the Viterbi-algorithm detector achieves maximum-likelihood detection for *any* $(g+1)$-component sampled impulse-response of finite energy.

7.8.5 Computer-simulation tests

L.H.C. Lee has used computer-simulation tests to measure the variation of error rate in the $\{s'_i\}$ with signal/noise ratio, for systems 3 and 4, each system being tested over four different channels[54]. The sampled impulse-response E, of the channel and filter D, is shown in Table 7.4, for each of the four channels. Each vector E here has unit length, so that no signal gain or attenuation is introduced by any channel. The channels 1, 2, 3 and 5 here are the *same* as the corresponding channels in Table 7.1, for which reason they are not numbered consecutively. Channel 1 is a well-known partial-response channel[64], and channel 2 was obtained with an actual telephone circuit in the transmission

Table 7.4 SAMPLED IMPULSE-RESPONSE OF THE CHANNEL AND FILTER D, FOR EACH OF THE DIFFERENT CHANNELS

Channel			E		
1	0.408	0.816	0.408		
2	0.548	0.789	0.273	−0.044	0.027
3	0.321	0.620	0.633	0.322	0.087
5	0.29	0.50	0.58	0.50	0.29

path[32]. Whereas the z-transforms of the sampled impulse-responses of channels 2 and 3 have *no* zeros (roots) on the unit circle in the z plane, all zeros of the z-transforms of the channels 1 and 5 lie *on* the unit circle[32]. The sampled impulse-response of channel 5 is the five-component sampled response that most seriously degrades the tolerance to additive white Gaussian noise of a maximum-likelihood detector, with binary signals[65].

Throughout the tests the arrangements of the different systems tested were as described in Sections 7.8.1, 7.8.2 and 7.8.4. Each individual measurement of the error rate in the $\{s'_i\}$ was made with the transmission of between 10,000 and 40,000 data-symbols at the given signal/noise ratio. The latter is expressed as ψ dB, where

$$\psi = 10 \log_{10}\left(\frac{1}{\sigma^2}\right) \tag{7.182}$$

The mean-square value of the data-symbols $\{s_i\}$ is unity here and the mean-square value of the noise-components $\{w_i\}$ and $\{u_i\}$ is σ^2. The results of the tests are shown in Figs. 7.23–7.30[54]. The 95% confidence limits of these results are generally better than about ±0.5 dB, except at the lower error rates where they are somewhat wider. The results are however less accurate than those in Figs. 7.5–7.11, so that some care must be taken in their interpretation. Throughout the Figs. 7.23–7.30, h is the delay in detection, in sampling intervals T, and, for system 4, k is the number of stored vectors $\{Q_i\}$. System 3 with $h = 0$ is the conventional decision-feedback equalizer (Section 7.2).

It can be seen from Figs. 7.23–7.30 that the best of the systems tested is system 4 with $k = 16$ and $h = 15$. This achieves an advantage of up to some 6 dB in tolerance to additive white Gaussian noise over the equalizer, for the given channels tested and with 2-level signals, and an advantage of up to some 5 dB with 4-level signals. When the amplitude distortion in the received signal is not too severe, suitable values of k in system 4 appear to be 4 and 8 for 2-level and 4-level signals, respectively. Larger values of k increase the equipment complexity without, however, giving any very useful improvement in

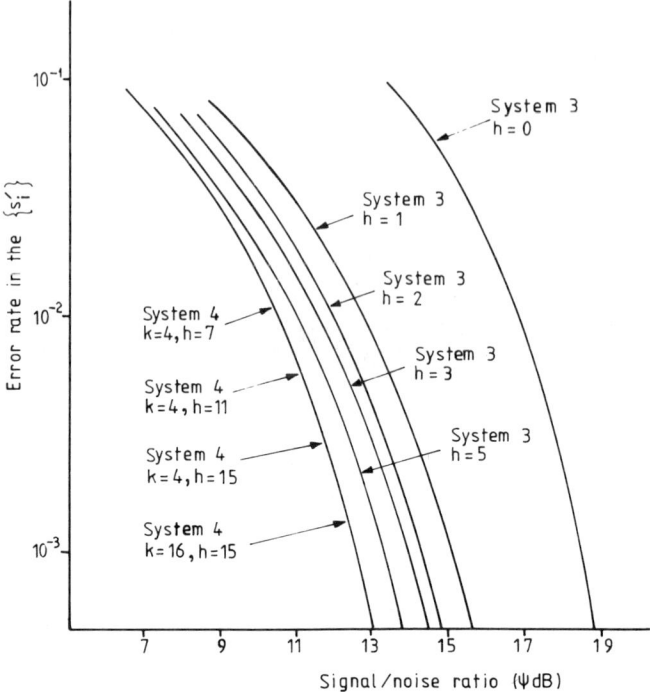

Fig. 7.23 Performance of systems 3 and 4 with 2-level signals transmitted over channel 1

performance over the given channels. Another potentially useful system for 2-level signals is system 3 with $h=1$, which, for the given channels, achieves an advantage of up to some 3 dB over the equalizer. However, with 4-level signals, the system only achieves an advantage of up to around 1 dB. It is evident that a much greater delay in detection must, in general, be used with 4-level signals than with 2-level signals in order to obtain an appreciable improvement in tolerance to noise of system 3 over the equalizer. When 2-level signals are used, system 3 with $h=1$ is not much more complex than the equalizer and, if there is now severe amplitude distortion in the received data-signal, the arrangement of system 3 is likely to be significantly more cost-effective than the equalizer. However, with *2-level* signals, the *optimum system* described in Section 7.3 has a performance at least as good as that of system 3 with $h=1$ and it is, for practical purposes, *no more complex* than the equalizer. Thus for 2-level signals, the *optimum system* is potentially the most cost-effective

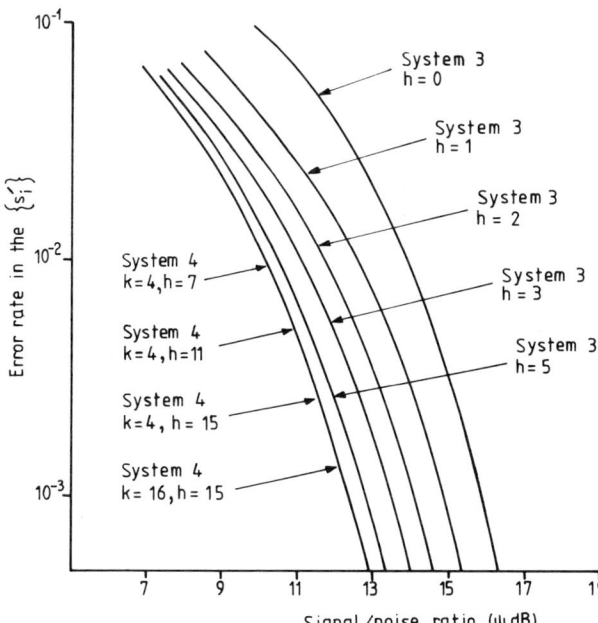

Fig. 7.24 Performance of systems 3 and 4 with 2-level signals transmitted over channel 2

of the three systems just considered. As h increases, so the rate of growth in the complexity of system 3 increases rapidly whereas the rate of improvement in its performance decreases, so that system 3 is unlikely to be cost effective for values of h much greater than 1, particularly with multilevel signals. In the case of system 4, however, the complexity increases only slowly with h, so that a useful improvement in performance can here be achieved through the use of a much larger value of h, without unduly increasing the equipment complexity. In fact, over the channels tested (and for both 2-level and 4-level signals), system 4 with $h = 15$ achieves an advantage of up to some 4 dB, in tolerance to additive white Gaussian noise, over system 3 with $h = 1$, but it is undoubtedly more complex.

For the particular channels tested, where $g \leqslant 4$, and with 2-level signals, a Viterbi-algorithm detector requires no more than 16 stored vectors and it achieves maximum-likelihood detection. Thus, where a good performance is important here, a Viterbi-algorithm detector is likely to be the most suitable arrangement. For channels 2, 3 and 5, and with 4-level signals, a Viterbi-algorithm detector requires 256 stored vectors, so that system 4, with, say, $k = 16$ and $h = 15$, is now

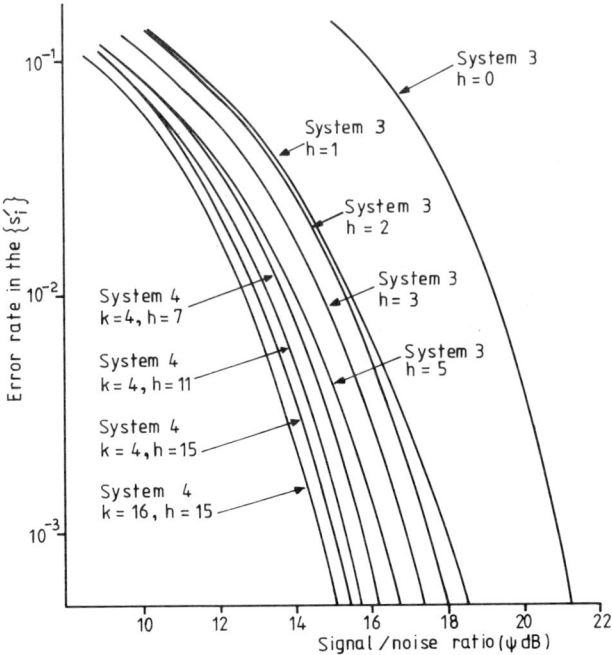

Fig. 7.25 Performance of systems 3 and 4 with 2-level signals transmitted over channel 3

likely to be preferred. In fact, for many practical channels, g can reach 30 or more, which means that an appropriate near-maximum-likelihood detector *must* now be used[6-61].

It is clear that, in the presence of severe amplitude distortion, very useful advantages in tolerance to noise can be achieved by a near-maximum-likelihood detector over the conventional decision-feedback equalizer. It is also evident from Figs. 7.23–7.30 that, to obtain the full potential of a near-maximum-likelihood detector, a relatively large delay in detection must be used. For too small a delay in detection, no matter how good the detection process itself, the best available tolerance to noise is not achieved.

REFERENCES

1. Chang, R.W. and Hancock, J.C. 'On receiver structures for channels having memory', *IEEE Trans. Inform. Theory*, **IT-12**, 463–468 (1966)
2. Gonsalves, R.A. 'Maximum likelihood receiver for digital data transmission', *IEEE Trans. Commun. Technol.*, **COM-16**, 392–398 (1968)

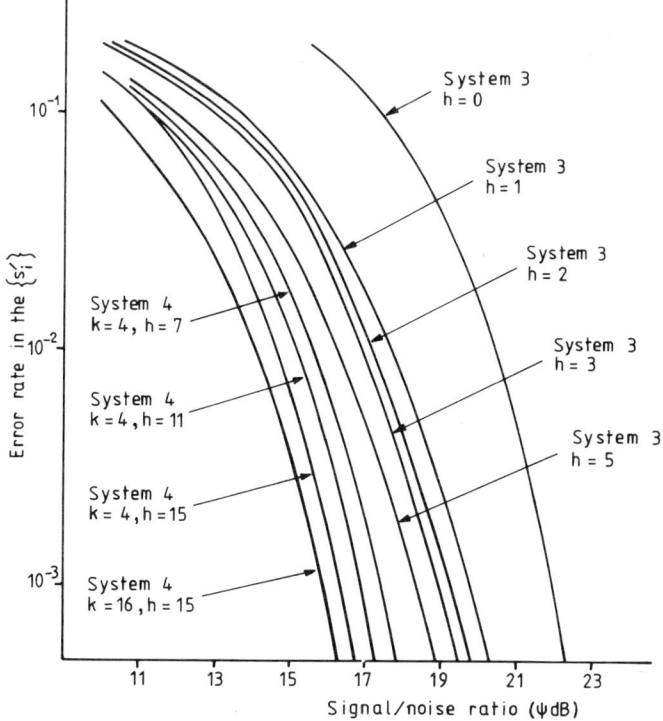

Fig. 7.26 Performance of systems 3 and 4 with 2-level signals transmitted over channel 5

3. Abend, K., Harley, T.J., Fritchman, B.D. and Gumacos, C. 'On optimum receivers for channels having memory', *IEEE Trans. Inform. Theory*, **IT-14**, 819–820 (1968)
4. Bowen, R.R. 'Bayesian decision procedure for interfering digital signals', *IEEE Trans. Inform. Theory*, **IT-15**, 506–507 (1969)
5. Abend, K. and Fritchman, B.D. 'Statistical detection for communication channels with intersymbol interference', *Proc. IEEE*, **58**, 779–785 (1970)
6. Clark, A.P. 'Adaptive detection of distorted digital signals', *Radio Electron. Eng.*, **40**, 107–119 (1970)
7. Clark, A.P. 'A synchronous serial data transmission system using orthogonal groups of binary signal elements', *IEEE Trans. Commun. Technol.*, **COM-19**, 1101–1110 (1971)
8. Forney, G.D. 'Maximum-likelihood sequence estimation for digital sequences in the presence of intersymbol interference', *IEEE Trans. Inform. Theory*, **IT-18**, 363–378 (1972)
9. Clark, A.P. 'Adaptive detection with intersymbol-interference cancellation for distorted digital signals', *IEEE Trans. Commun.*, **COM-20**, 350–361 (1972)
10. Qureshi, S.U.H. and Newhall, E.E. 'An adaptive receiver for data transmission over time-dispersive channels', *IEEE Trans. Inform. Theory*, **IT-19**, 448–457 (1973)
11. Clark, A.P. and Ghani, F. 'Detection processes for orthogonal groups of digital signals', *IEEE Trans. Commun.*, **COM-21**, 907–915 (1973)

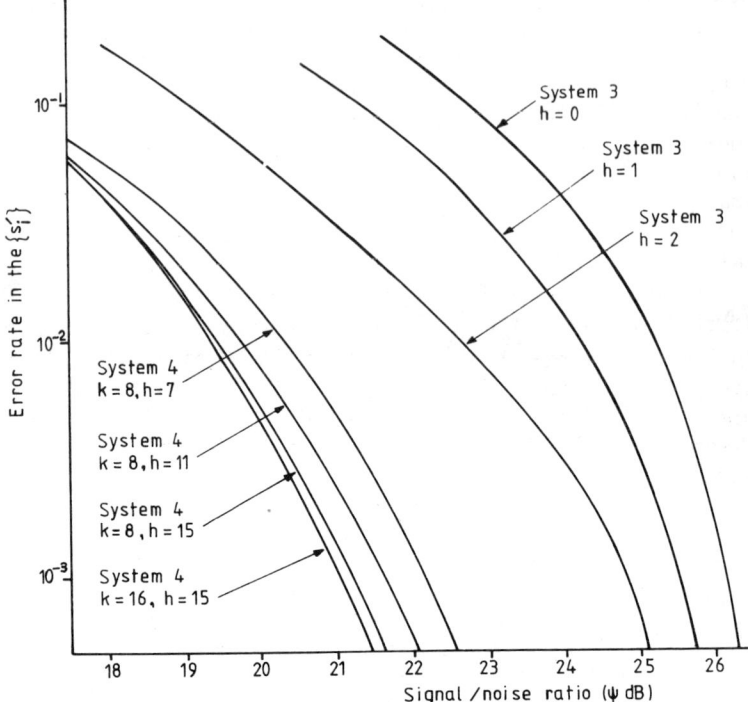

Fig. 7.27 Performance of systems 3 and 4 with 4-level signals transmitted over channel 1

12. Falconer, D.D. and Magee, F.R. 'Adaptive channel memory truncation for maximum likelihood sequence estimation', *Bell Syst. Tech. J.*, **52**, 1541–1562 (1973)
13. Messerschmitt, D.G. 'A geometric theory of intersymbol interference. Part II: Performance of the maximum likelihood detector', *Bell Syst. Tech. J.*, **52**, 1521–1539 (1973)
14. Pulleyblank, R.W. 'A comparison of receivers designed on the basis of minimum mean-square error and probability of error for channels with intersymbol interference and noise', *IEEE Trans. Commun.*, **COM-21**, 1434–1438 (1973)
15. Lucky, R.W. 'A survey of the communication theory literature 1968–1973', *IEEE Trans. Inform. Theory*, **IT-19**, 725–739 (1973)
16. Fredricsson, S.A. 'Optimum transmitting filter in digital PAM systems with a Viterbi detector', *IEEE Trans. Inform. Theory*, **IT-20**, 479–489 (1974)
17. Cantoni, A. and Kwong, K. 'Further results on the Viterbi algorithm equalizer', *IEEE Trans. Inform. Theory*, **IT-20**, 764–767 (1974)
18. Magee, F.R. 'A comparison of compromise Viterbi algorithm and standard equalization techniques over band-limited channels', *IEEE Trans. Commun.*, **COM-23**, 361–367 (1975)
19. Yao, K. and Milstein, L.B. 'On ML (maximum-likelihood) bit detection of binary signals with intersymbol interference in Gaussian noise', *IEEE Trans. Commun.*, **COM-23**, 971–976 (1975)

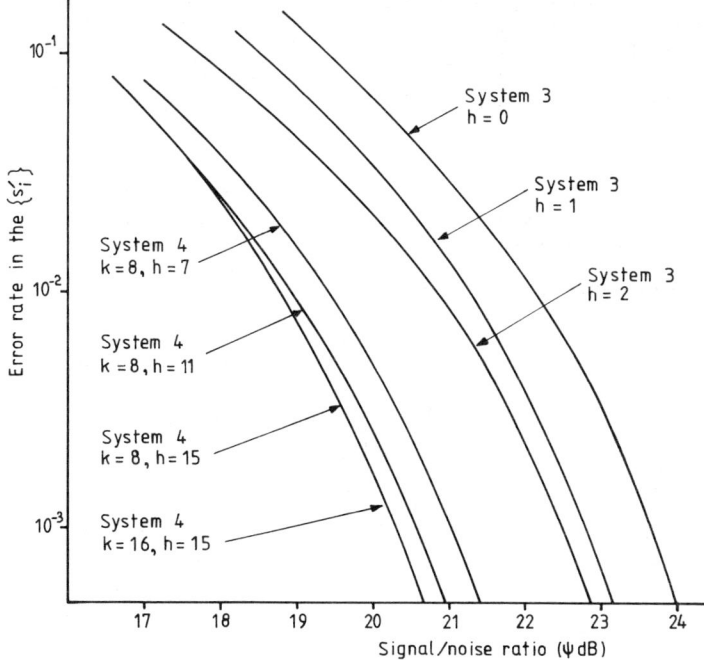

Fig. 7.28 *Performance of systems 3 and 4 with 4-level signals transmitted over channel 2*

20. Fredricsson, S.A. 'Joint optimization of transmitter and receiver filters in digital PAM systems with a Viterbi detector', *IEEE Trans. Inform. Theory*, **IT-22**, 200–210 (1976)
21. Russell, S.P. and Shohara, A. 'Feasibility study of reduced state Viterbi detection', *EASCON 1976 Record*, Washington, USA, 89A-F (1976)
22. Tamburelli, G. 'Decision feedback and feedforward receiver (for rates faster than Nyquist's), *Alta Frequenza*, **45**, 587–594 (1976)
23. Falconer, D.D. and Magee, F.R. 'Evaluation of decision feedback equalization and Viterbi algorithm detection for voiceband data transmission—Part I', *IEEE Trans. Commun.*, **COM-24**, 1130–1138 (1976)
24. Falconer, D.D. and Magee, F.R. 'Evaluation of decision feedback equalization and Viterbi algorithm detection for voiceband data transmission—Part II', *IEEE Trans. Commun.*, **COM-24**, 1238–1245 (1976)
25. Clark, A.P. and Harvey, J.D. 'Detection processes for distorted binary signals', *Radio Electron Eng.*, **46**, 533–542 (1976)
26. Clark, A.P. *Advanced Data-Transmission Systems*, Pentech Press, London (1977)
27. Clark, A.P. and Harvey, J.D. 'Detection of distorted q.a.m. signals', *Electronic Circuits and Systems*, **1**, 103–109 (1977)
28. Clark, A.P., Harvey, J.D. and Driscoll, J.P. 'Improved detection processes for distorted digital signals', *IERE Conference Proceedings*, No. 37, 125–136 (1977)
29. Desblache, A.E. 'Optimal short desired impulse response for maximum likelihood sequence estimation', *IEEE Trans. Commun.*, **COM-25**, 735–738 (1977)

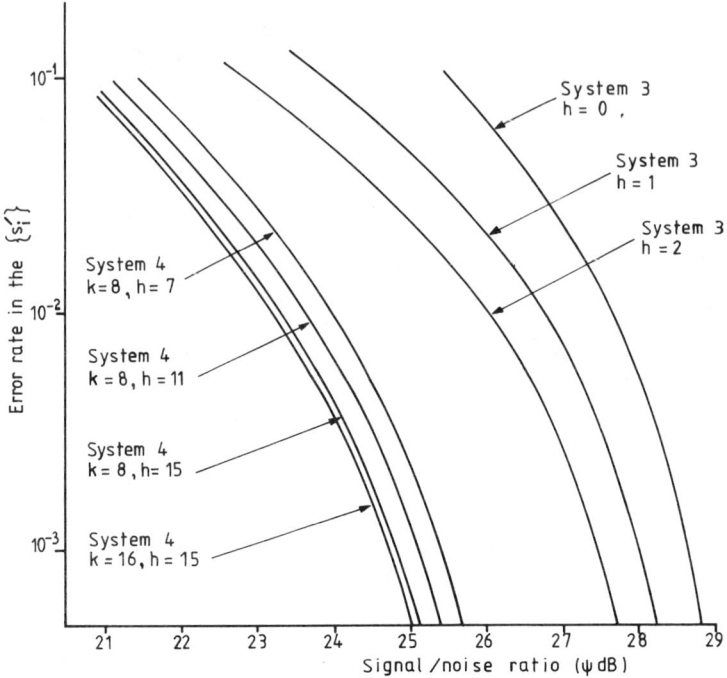

Fig. 7.29 *Performance of systems 3 and 4 with 4-level signals transmitted over channel 3*

30. Lee, W.U. and Hill, F.S. 'A maximum likelihood sequence estimator with decision-feedback equalization', *IEEE Trans. Commun.*, **COM-25**, 971–979 (1977)
31. Foschini, G.J. 'A reduced state variant of maximum-likelihood sequence detection attaining optimum performance for high signal-to-noise ratios', *IEEE Trans. Inform. Theory*, **IT-23**, 605–609 (1977)
32. Clark, A.P., Harvey, J.D. and Driscoll, J.P. 'Near-maximum-likelihood detection processes for distorted digital signals', *Radio Electron. Eng.*, **48**, 301–309 (1978)
33. Beare, C.T. 'The choice of the desired impulse response in combined linear Viterbi algorithm equalizer', *IEEE Trans. Commun.*, **COM-26**, 1301–1307 (1978)
34. Clark, A.P., Kwong, C.P. and Harvey, J.D. 'Detection processes for severely distorted digital signals', *Electronic Circuits and Systems*, **3**, 27–37 (1979)
35. Luvison, A., Sacchi, L. and Tamburelli, G. 'Theory and implementation of adaptive equalizers', *IEEE Int. Conf. on Communications*, Boston, MA, USA, 45.6/1–6 (1979)
36. Dalle Mese, E. and Giuli, D. 'Generalised decision feedback equalisation for digital communication channels', *Electronic Circuits and Systems*, **3**, 145–152 (1979)
37. Kawas-Kaleh, G. 'Double decision feedback equalizer', *Frequenz*, **33**, 146–149 (1979)
38. Shaft, P.D. and Kather, E.C. 'Experimental measurements of Viterbi decoding in burst channels', *IEEE Trans. Commun.*, **COM-27**, 1360–1366 (1979)

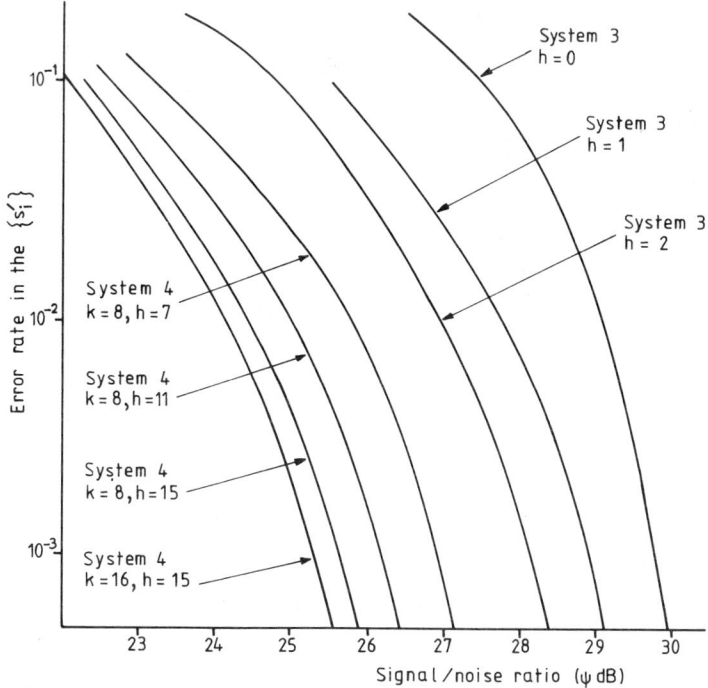

Fig. 7.30 Performance of systems 3 and 4 with 4-level signals transmitted over channel 5

39. Milutinovic, V.M. 'Suboptimum detection procedure based on the weighting of partial decisions', *Electronics Letters*, **16**, 237–238 (1980)
40. Milutinovic, V.M. 'Comparison of three suboptimum detection procedures', *Electronics Letters*, **16**, 681–683 (1980)
41. McLane, P.J. 'A residual intersymbol interference error bound for truncated state Viterbi detectors', *IEEE Trans. Inform. Theory*, **IT-26**, 548–553 (1980)
42. Geist, J.M. and Cain, B. 'Viterbi decoder performance in Gaussian noise and periodic erasure bursts', *IEEE Trans. Commun.*, **COM-28**, 1417–1422 (1980)
43. Harvey, J.D. *Synchronisation of a Synchronous Modem*, SERC Report No. GR/A/1200.7, SERC, Swindon, England (1980)
44. Clark, A.P. 'Equalisation and detection techniques', IEE Coloquium on 'A Review of Modem Techniques', London (1981)
45. Irwin, G.W. 'Predictors of maximum-likelihood performance of an unknown channel', *Proc. IEE*, Pt. F, **128**, 23–27 (1981)
46. Clark, A.P. and McVerry, F. 'Performance of 2400 bit/s serial and parallel modems over an HF channel simulator', *IERE Conference Proceedings*, No. 49, 167–179 (1981)
47. Clark, A.P. and Asghar, S.M. 'Detection of digital signals transmitted over a known time-varying channel', *Proc. IEE*, Pt F, **128**, 167–174 (1981)
48. Clark, A.P. and Fairfield, M.J. 'Detection processes for a 9600 bit/s modem', *Radio Electron. Eng.*, **51**, 455–465 (1981)

49. Gersho, A. and Lim, T.L. 'Adaptive cancellation of intersymbol interference for data transmission', *Bell Syst. Tech. J.*, **60**, 1997–2021 (1981)
50. Hayes, J.F., Cover, T.M. and Riera, J.B. 'Optimal sequence detection and optimal symbol-by-symbol detection: Similar algorithms', *IEEE Trans. Commun.*, **COM-30**, 152–157 (1982)
51. Cheung, R.S.-W. and McLane, P.J. 'Carrier reference error sensitivity of a Viterbi detector for PAM data transmission', *IEEE Trans. Commun.*, **COM-30**, 410–414 (1982)
52. Sano, A. and Yamazaki, J.-I. 'Generalized decision feedback equalizer for approximate maximum likelihood bit detection of digital signals', *Int. J. Electronics*, **52**, 167–176 (1982)
53. Clark, A.P. and Hussein, B.A. 'Non-linear equalizers having an improved tolerance to noise', *Radio Electron. Eng.*, **52**, 145–153 (1982)
54. Clark, A.P., Lee, L.H. and Marshall, R.S. 'Developments of the conventional nonlinear equaliser', *Proc. IEE*, Pt. F, **129**, 85–94 (1982)
55. Clark, A.P., Ip, S.F.A. and Soon, C.W. 'Pseudobinary detection processes for a 9600 bit/s modem', *Proc. IEE*, Pt. F, **129**, 305–314 (1982)
56. Tamburelli, G. 'The parallel decision feedback and feedforward equalizer', *IEEE Trans. Commun.*, **COM-31**, 224–231 (1983)
57. Clark, A.P., Najdi, H.Y. and Fairfield, M.J. 'Data transmission at 19.2 kbit/s over telephone circuits', *Radio Electron. Eng.*, **53**, 157–166 (1983)
58. Clark, A.P. and Najdi, H.Y. 'Detection process of a 9600 bit/s serial modem for HF radio links', *Proc. IEE*, Pt. F, **130**, 368–376 (1983)
59. Clark, A.P. and Najdi, H.Y. 'Performance of a 9600 bit/s serial modem over a model of an HF radio link', *IEE Int. Conf. on Radio Spectrum Conservation Techniques*, Birmingham, England, 151–155 (1983)
60. Clark, A.P., Slater, M. and Parama Raj, K. 'Simple nonlinear equalizers for binary baseband signals', *Proc. IEE*, Pt. F, **130**, 495–505 (1983)
61. Clark, A.P. and Clayden, M. 'Pseudobinary Viterbi detector', *Proc. IEE*, Pt. F, **131**, 208–218 (1984)
62. Wozencraft, J.M. and Jacobs, I.M. *Principles of Communication Engineering*, Wiley, New York (1965)
63. Lathi, B.P. *An Introduction to Random Signals and Communication Theory*, Intertext Books, London (1968)
64. Clark, A.P. *Principles of Digital Data Transmission*, second edition, Pentech Press, London (1983)
65. Magee, F.R. and Proakis, J.G. 'An estimate of the upper bound on error probability for maximum likelihood sequence estimation on channels having a finite-duration pulse response', *IEEE Trans. Inform. Theory*, **IT-19**, 699–702 (1973)
66. Ingram, D.G.W. 'Design of a digital transmission system to meet a specified error rate objective', Lecture Notes 41, Department of Electrical Engineering, Cambridge University (1978)
67. Bylanski, P. and Ingram, D.G.W. *Digital Transmission Systems*, second edition, Peter Peregrinus, London (1978)
68. Belfiore, C.A. and Park, J.H. 'Decision feedback equalization', *Proc. IEEE*, **67**, 1143–1156 (1979)
69. Catchpole, R.J. 'Efficient ternary transmission codes', *Electronics Letters*, **11**, 482–484 (1975)
70. Papoulis, A. *Probability, Random Variables and Stochastic Processes*, McGraw-Hill, New York (1965)
71. Thomas, J.B. *An Introduction to Statistical Communication Theory*, Wiley, New York (1968)
72. Davenport, W.B. *Probability and Random Processes*, McGraw-Hill, New York (1970)

73. Viterbi, A.J. 'Convolutional codes and their performance in communication systems', *IEEE Trans. Commun.*, **COM-19**, 751–772 (1971)
74. Clark, A.P. 'Minimum distance decoding of binary convolutional codes', *Computers and Digital Techniques*, **1**, 190–196 (1978).

Index

Absolute value 21, 22, 90, 111
Adaptive equalizer 7, 193, 210, 324–334, 345, 355
Adder 32, 164
Additive noise 8, 9
Additive white Gaussian noise 9, 13, 17
Advance in time 74–79, 107, 117, 122, 145, 152–154, 175
Allpass filter or network 281–293, 308, 348, 377, 391
Alpha-numeric message 3
Amplitude distortion 7, 50, 54, 96, 98, 115–147, 221, 223, 310
Amplitude modulation effects 7
AM signal 5
Angle between vectors 20, 361
Antipodal signal 157, 158
Aperiodic autocorrelation function 87–91, 96, 97, 102, 103, 124, 136, 148–151, 209
Attenuation 98
 distortion 7, 10
Audio link 6
Autocorrelation function 16
Automatic equalizer 7
Average energy per bit or per signal element 16–18, 160, 232, 358, 368–378, 424, 427, 432
Average power 12, 18
Axes of Euclidean vector space 18, 19

Bandlimited Gaussian noise 13, 17, 18, 52, 159, 161, 216
Bandpass channel 5, 6, 12, 138, 139
Bandpass transmission path 6, 12, 138, 139
Bandwidth 5, 13
Baseband channel 5, 6, 12–18, 50, 158
Baseband signal 5, 12, 13, 50
Bauds 3

Bayes' theorem 30
Biased estimate 210, 283, 284, 309, 392, 436
Binary data-symbol or signal 2, 157, 158, 230
Binary digit 3, 4
Binomial expansion 113, 114, 132, 169, 190
Bit 3, 4, 161
 of information 4
 rate 4–6
Bits per second 4, 7, 11
Business system 7

Cancellation of intersymbol interference 230–239, 310, 311
Carrier link 6, 9
Carrier phase synchronization 5
Channel 1, 5, 7, 139, 158–161, 312, 313, 380, 383, 401
 bandpass 5, 6, 12, 138, 139
 impulse response of 6, 13, 34, 52, 56, 158, 159, 230, 428
 linear baseband 5, 6, 12–18, 50, 158–161
 minimum phase 130, 145, 285, 406
 narrowband .5
 partial response 312, 400, 401
 sampled impulse-response of 13, 34, 52–56, 94, 117, 139, 140, 160, 312, 338, 380, 382, 423, 425
 time invariant 50, 55, 158, 230, 356
 time varying 7, 331–333
 transfer function of 13–16, 56, 70, 113, 115–117
 two factors of z-transform of 131, 179, 180, 257, 285, 347, 377
 two in cascade 125–137, 145–147
 voiceband 6, 7
 z-transform of 55, 56, 106, 107, 131, 162, 231, 285, 339, 347, 412

INDEX 459

Character 3
Circulant matrix 25–27, 60, 68–72, 76, 77, 97, 100–102, 117, 118
Circular convolution 27, 72, 89, 90
Coaxial cable 5, 6, 12, 419, 426–433
Coefficients in z-transform 57, 58, 107–112
Coherent demodulation 5, 6, 9
Column vector 19
Comparison of systems 195–205, 239–253, 297, 299–320, 374, 379–385, 399–408, 419–435, 446–455
Complex conjugate 21, 82, 112, 121, 129, 130, 142
Complex number plane 56, 57, 61, 62, 106
Complex-valued signal 6, 12, 34, 37, 85, 138–147, 215–217, 297–299
Complex vector 21, 22, 34, 37, 63, 87, 120, 124, 217, 298
Components of a vector 19
Conditional probability 30, 31
Conditional probability density 30–32
Conjugate reverse of sequence 142–145
Conjugate transpose of matrix 23, 24, 65, 82
Conjugate transpose of vector 21
Continuous random variable 29
Continuous random vector 29–31
Convergence of adaptive equalizer 326–334
Convergent series 132, 133, 135, 169, 172, 190
Convolution
　circular 27, 72, 89, 90
　linear 27, 59, 60, 70–72, 88–90, 107–111, 120, 121, 133, 140–144
　matrix 26, 27, 59, 60, 68–72, 97, 100–102, 117, 118, 140, 165, 207
Correlation matrix 209
Cost of vector (sequence) 440–446
Cross-correlation 326, 330
Curly bracket 2
Cyclic shift 27, 73–79, 119, 152–155

Data signal 2–6
Data symbols 2–4, 11, 12, 28, 29, 41, 51, 157, 158, 230
　complex-valued 138, 139, 216
　mean-square value of 16, 160, 166, 204–206, 216, 232, 266, 267, 297, 368, 432
　m-level 2–4, 11, 28
　sequence of 58
　statistically independent 2, 4, 157, 166, 206, 216, 230, 266, 267
　statistically orthogonal 2, 16, 166, 206, 218, 266, 267
　uncorrelated 2, 16
　value 28
　zero mean 2, 16, 166, 206, 267
Decision 10, 28, 33, 34, 36, 37
　boundary 35, 36
　region 35, 36
Decision directed cancellation of intersymbol interference 230–239
Decision-feedback equalizer 7, 138, 231–316, 327–334, 345–356, 367, 382–385, 387–435
　achieving exact equalization 230–262, 275–297, 345–356
　achieving linear and nonlinear equalization of the two factors of $Y(z)$ 257–262, 297
　achieving only partial equalization 244–253, 310–320, 393–435
　achieving zero forcing 253–257, 297
　adaptive 7, 327–334, 355
　developments of 387–450
　double sampling 345–356, 411–417
　optimum, subject to exact equalization 275–297
　tap gains of 233, 236–239, 241, 248, 249, 254–261, 265–275, 309, 311, 328, 391, 394
　that minimizes the mean-square error in the equalized signal 262–275, 297–299, 322–334, 351–356
Delay distortion 7, 10
Delay in time 55, 73–79, 98, 106, 115, 122, 123, 145, 152–154
Demodulation 5, 6, 9
Demodulator 1, 5, 6, 12
Descrambling 158, 230
Detector 10, 12, 27–48
　matched-filter 10, 34–42, 47, 48, 210, 217–222, 246, 274
　maximum-likelihood 32–37, 47, 48, 138, 220, 438–444
　near-maximum-likelihood 435–455
　optimum 10, 27–37, 47, 138, 217–222, 438–444

performance of 43–48, 138, 195–205, 239–253, 256, 257, 261, 262, 295–297
suboptimum 42, 48, 435–455
threshold-level 10, 37, 38, 43, 44, 161, 162, 196–205, 234, 235
Viterbi-algorithm 440–444
Diagonal matrix 25, 66, 68, 69, 78, 79
Digit 3, 4
Digital filter 36, 37, 59, 87–89, 101–105
Digital modem 1
Digital signal 1–6
Digitally encoded speech signals 1, 3, 6
Digitally encoded video signals 1, 6
Discrete energy-density spectrum 87–91, 97, 99, 102, 103, 105, 151
Discrete Fourier transform (DFT) 61–93, 98, 110–112, 116–118, 121–124, 130, 131, 140–144, 149, 152–155
  complex conjugate of 82, 86, 87, 110–112, 121, 149
  energy of 92, 360, 361
  inverse 63–68, 110–112, 142, 149
  linearity of 73
  matrix 63
  product of two 71, 72, 110–112, 121, 130, 142, 144, 149, 363
Discrete frequencies 61–65
Discrete random variable 29
Discrete random vector 29–31
Distance between vectors 20–22, 33, 35, 37, 211, 216, 279, 438
Distortion 7, 10, 50, 54, 94–155, 162, 166
  amplitude 7, 50, 54, 96, 98, 115–147, 221, 222, 310
  attenuation 10
  fixed 7
  group delay 7, 10
  linear 7
  mean-square 166
  measure of severity of 137, 138, 148, 150
  nonlinear 7
  peak 166–168, 181, 182, 185
  phase 50, 54, 96, 98–115, 125–130, 133, 134, 139–147, 217–222, 282, 366
  time-varying 7
Divergent series 135, 172, 190
Doppler shift 8
Double sampling equalizer 334–356, 409–417

Double sideband AM signal 5

Eigenvalue 24–26, 69, 70, 96, 98, 118, 140, 271, 272
Eigenvector 24–26, 69, 70
Energy 16, 38, 92
  density spectrum 16, 87–91
  of DFT 92, 360, 361
  of sequence or vector 17, 38, 92, 98, 363
  of signal element 16, 358, 424, 427
  spectral density 16, 99, 102
Equalized signal 161, 164, 171, 175, 177, 197, 234–236, 291
  mean-square error in 166, 196, 207, 213, 216, 217, 269, 298, 343, 344, 351
Equalizer
  achieving exact equalization 163, 168, 191, 192, 230–253, 275–297, 345–356
  adaptive 7, 193, 210, 324–334, 345
  automatic 7
  decision-feedback 7, 138, 231–316, 327–334, 345–356, 367, 382–385, 387–435
  double sampling 334–356, 409–417
  feedback transversal 188–195, 230–253, 368–374
  feedforward transversal 163–188
  fixed 7
  for phase distortion 115, 137, 217–222, 282–293, 377
  linear 7, 12, 115, 143, 156–222, 324–327, 337–345, 429–432
  linear feedback 188–195, 370, 371
  main tap of 179, 289
  maximum delay 179, 289
  maximum phase 179, 289
  minimum delay 171
  minimum mean-square distortion 182, 205–217
  minimum mean-square error 182, 205–217, 262–275, 297–299, 324–334
  minimum peak distortion 167, 168, 181
  minimum phase 171
  nonlinear 230–253, 368–382
  operating on the two factors of $Y(z)$ 176–180, 257–262
  optimum decision-feedback 275–297, 393–409

INDEX 461

pure nonlinear 230–253, 368–374, 433, 434
quantized feedback 232
shared between transmitter and receiver 356–367
simplified 409–435
using modulo-arithmetic 368–382
zero-forcing decision-feedback 253–257
zero-forcing linear 167, 168, 179
Equipment complexity 4, 192, 193, 409, 439, 444
Error
 bursts 8, 240, 244, 245, 264, 310, 312, 402–408, 429
 extension effects 240, 244, 245, 264, 402–408, 429
 mean-square 40, 41, 166, 196, 207, 213, 216, 217, 266–269
 probability 38, 39, 44–46, 196–205, 239–253, 256, 309–320, 360, 393–408, 416–432
 rate 2, 39
Estimate of data signal
 biased 210, 283, 284, 309, 392, 436
 maximum likelihood 36, 37, 46, 47, 437, 438
 minimum mean-square error 209, 269
 unbiased 40, 46, 47, 278, 286
Euclidean distance 20, 33, 35, 211, 279, 438
Euclidean length 20, 98, 105, 196
Euclidean vector space 19, 24, 47, 48, 211, 279
Event 30
Exact equalization 163, 168, 191, 192, 230–262, 275–297, 345–351, 368–382, 389–393
Expected signal energy 16–18, 160, 232, 358, 368–378, 424, 427, 432
Expected noise power 18, 196
Expected value 2

Fast Fourier transform 72
FDM 6
Feedback transversal filter 188–195, 232–238
Feedforward transversal filter 163–188
Fibre optic link 6
Filter 10, 11, 50, 136, 137
 allpass 281–293, 308, 348, 377, 391
 digital 36, 37, 59, 87–89, 101–105

feedback transversal 188–195, 232–238
feedforward transversal 36, 37, 163–188, 195–222
ideal lowpass 11, 410, 411, 418
inverse 102–105, 143, 156–222, 282
linear 10, 11, 36, 37, 50, 136, 137, 143, 156–222, 253, 254
lowpass 11, 50, 336, 337, 389, 410, 411, 418, 427
matched 10, 15, 34–42, 47, 101–105, 122, 137, 143, 217–222, 246, 274
maximum phase 179, 289
minimum phase 171
nonlinear 232–254, 262–265
physically realizable 15, 115, 119, 122, 136, 175, 182, 190, 282
receiver 11, 13–18
transmitter 12–17
transversal 37, 164, 188, 190, 233
Fourier transform 16, 54, 58–64, 74, 102
 inverse 16, 17, 64
Frequency
 modulation effects 7, 8
 offset 8, 9
 spectrum 13–16

Gaussian noise 9, 12, 17, 18, 28, 29, 33, 34, 45, 95, 161, 195, 196, 216
 additive white 9, 12, 17
 samples of 18, 28, 33, 34, 38, 45, 95, 161, 195, 196, 216, 220, 221, 232, 266
 tolerance to 9, 46, 195–205
 waveform 13, 17, 18, 159, 161
Gaussian random process 17, 18
Gaussian random variables
 probability density function of 38, 44, 45, 196, 200, 239
 statistically independent 18, 28, 33, 34, 38, 45, 161, 196, 220, 221, 286, 297, 393, 411
 statistically orthogonal 350
 uncorrelated 18, 33, 38, 161, 196, 216, 220, 221
 variance of 18, 38, 161, 196, 198, 200, 204, 216, 220, 232, 256, 286, 297, 343, 351, 411, 431
Gradient algorithm 324–330
Group delay
 characteristic 7, 10
 distortion 7, 10

462 INDEX

Hermitian matrix 24–26, 140, 217, 298
HF radio link 6, 7, 9, 13
High power amplifier 7
Hyperplane 36
Huffman sequence 112

Identity matrix 23, 66
Imaginary-valued signal 6, 85, 139
Impulse response 6, 13, 34, 52, 56, 158, 159, 230, 428
  complex-valued 6, 138
  real-valued 6, 13, 34, 52, 56
  sampled 13, 34, 52–56, 94, 117, 139, 140, 160, 312, 338, 380, 382, 423, 425
Impulses 12, 34, 51, 53, 56, 61, 99
Impulsive noise 8
Information 1, 2, 4, 11, 28
  rate 2, 4, 11
  unit of 4
Inner product of vectors 20, 21
Integers 28
Interchannel interference 139
Intersection of subspaces (intersection space) 341, 342
Intersymbol interference 4, 10, 42, 54, 95, 161, 166, 207, 233, 242, 244, 311
  decision-directed cancellation of 230–239, 310, 311
  residual 171, 174, 177, 178
Inverse DFT 63–68, 110–112
Inverse filter 102–105, 163–165, 282, 291, 292
Inverse Fourier transform 16, 17
Inverse matrix 22, 23, 63, 67, 79, 82
Inverse sequence 108–112, 142, 143, 149
Inverse transformation 102–105
Inverse z-transform 106–115, 134, 142, 143, 149, 163, 165, 282, 291, 292

Knowledge 7, 28, 37, 157

Laplace transform 57, 58
Length of vector 20–22
  Euclidean 20, 98, 105, 196
  unitary 21, 22
Likelihood function 32
Linear baseband channel 5, 6, 12–18, 50, 158
Linear coherent demodulator 6, 12

Linear combination of DFT's 73
Linear combination of vectors 23, 24, 211, 279–281, 341
Linear convolution 27, 59, 60, 70–72, 88–90, 107–111, 120, 121, 133, 140–144
Linear demodulation 5
Linear distortion 6, 50, 95, 96
Linear equalizer 7, 12, 115, 143, 156–222, 291, 324–327, 337–345, 429–432
  achieving exact equalization 163, 168, 191, 192
  adaptive 7, 324–327, 333, 334, 345
  automatic 7, 210
  double-sampling 334–345
  feedback transversal 188–195, 370, 371
  feedforward transversal 163–188
  fixed 7
  for phase distortion 115, 137, 217–222, 282–293, 377
  for the two factors of $Y(z)$ 168–181
  main tap of 179, 289
  maximum delay 179, 289
  maximum phase 179, 289
  minimum delay 171
  minimum mean-square distortion 215
  minimum mean-square error 182, 205–217, 324–327, 333, 334
  minimum peak distortion 167, 168, 181
  minimum phase 171
  shared between transmitter and receiver 356–368
  tap gains of 163–195, 200, 202, 204, 214, 325, 357–367
  zero forcing 167, 168, 171, 179, 180
Linearly independent vectors 23, 24, 211, 279, 341–343
Loaded audio link 6
Long division 169, 172, 173, 185, 190, 191
Lowpass filter 11, 50, 336, 337, 389, 410, 411, 418, 427

Main tap of equalizer 179, 289
Matched filter 10, 15, 34–42, 47, 101–105, 122, 137, 143, 217–222, 246, 274
  detection 10, 34–42, 47, 48, 137, 210, 217–222, 246, 274
  estimation 36, 47

INDEX   463

Matrix 18–27
  circulant 25–27, 60, 68–72, 76, 77, 97, 100–102, 117, 118
  conjugate transpose of 23, 24, 65, 82
  convolution 26, 27, 59, 60, 68–72, 97, 100–102, 117, 118, 140, 165, 207
  correlation 209
  DFT 63, 65–70
  diagonal 25, 66, 68, 69, 78, 79
  eigenvalue of 24–26, 69, 70, 96, 98, 118, 140, 271, 272
  eigenvector of 24–26, 69, 70
  Hermitian 24–26, 140, 217, 298
  inverse of 22, 23, 63, 67, 79, 82
  inverse DFT 63, 65–68
  orthogonal 23–26, 67, 68, 100–102
  permutation 67, 99, 102
  positive definite 25, 208, 217, 271, 298, 344, 354
  positive semi-definite 25, 272, 344, 354
  real symmetric 24, 25, 67, 68, 118, 208, 269, 344, 354
  square 24, 60, 63
  symmetric 63
  transpose of 23, 24, 81
  unitary 23–26, 82
  with one component 21
Maximum delay network 179, 289
Maximum-likelihood detector 32–37, 47, 48, 138, 220, 438–444
Maximum-likelihood estimate 36, 37, 46, 47, 437, 438
Maximum-likelihood vector 438–443
Maximum phase 179, 289
Maximum signal/noise ratio 40, 275–292, 298, 299
Maximum transmission rate 11
Mean-square distortion 166
Mean-square error 40, 41, 166, 196, 207, 213, 216, 217, 269, 298, 343, 344, 351, 354
Mean-square value 12, 16, 18, 160, 166, 196, 204–206, 216, 266, 267, 232, 297, 368, 432
Merging of vectors 445
Message 3
Microwave link 6
Minimum delay network 171
Minimum error probability 31–34, 37–48, 217–222, 439–444
Minimum phase 130, 145, 171, 282–293, 348, 391, 406

Minimum mean-square error equalizer 182, 205–217, 324–327, 333, 334
Minimum peak-distortion equalizer 167, 168, 181
m-level signal 3, 4, 11, 28
Mobile system 7
Model of channel 11–18
Modem 1, 2
Modulated-carrier signal 5, 6
Modulation
  amplitude 5, 8
  frequency 8
  linear 5
  method 5, 6
  time 8
Modulator 1, 6, 12
Modulo-arithmetic equalizer 368–382
Modulo-m 369
Modulo-n 61, 62, 76
Modulus 21
Multilevel signal 2–4, 34, 37, 138, 139, 436
Multiplication theorem 91, 92
Multiplicative noise 7–9
Multiplier 37, 164, 234

Narrowband channel 5
Near-maximum-likelihood detector 435–455
Noise 7, 17, 18, 28, 33, 34, 38, 45, 95, 161, 195, 196, 220, 221, 230–232
  additive 7–9
  additive white Gaussian 9, 13, 17
  autocorrelation function of 18
  average power of 18
  impulsive 8
  mean-square value of 18, 196
  multiplicative 7–9
  power spectral density of 9, 18, 161
  samples of 18, 28, 29, 33, 34, 38, 45, 95, 161, 196
  tolerance to 195–205, 239–253, 297, 299–320, 374, 379–385, 399–408, 419–435, 446–455
  variance of 18, 38, 161, 196, 198, 200, 204, 216, 220, 232, 256, 286, 297, 343, 411
  vector 27, 28, 35, 38
  waveform 13, 18
  whitening network 336
Nonlinear equalizers 230–253, 368–382
Nonlinear filter 232–254, 262–265, 328, 368–378

INDEX

Nonlinear processor 368–378
Nonlinear transformation 368–378
Nonsingular matrix 22
Nyquist rate 11, 13, 15, 335
Nyquist's vestigial symmetry theorem 14

Optimum decision-feedback equalizer 275–297
Optimum detection process 28–34, 37, 47, 137, 217–222
Optimum estimation 36
Optimum linear equalizer 205–217
Optimum nonlinear equalizer 374–382
Origin of vector space 20, 24, 211
Orthogonal axes 19, 20
Orthogonal matrix 23–26, 67, 68, 100–102
Orthogonal projection 19, 24, 36, 47, 48, 211, 212, 280
Orthogonal signal-elements 17, 101–105, 218, 358
Orthogonal transformation 23–25, 100–105, 107–112, 115, 140, 217–222, 281–293, 308, 348, 377, 391
Orthogonal vectors (sequences) 20, 21, 24, 25, 36, 41, 48, 101–105, 154, 212, 218, 280, 281, 292

Parallel system 3
Partial nonlinear equalization 241–253, 310–320, 393–435
PCM link 6
Peak distortion 166–168, 181, 182, 185
Performance of detectors 43–48, 138, 195–205, 239–253, 256, 257, 261, 262, 295–297, 299–320
Periodic sequence 64, 65
Permutation matrix 67, 99, 102
Phase
 angle 141, 142
 distortion 50, 54, 96, 98–115, 125–130, 133, 134, 139–147, 217–222, 282–293
 equalizer 115, 137, 217–222, 282–293, 377
 quadrature 5, 139
Physically realizable system 15, 108, 114, 115, 117, 119, 122, 132, 136, 142, 175, 182, 190, 282
Pilot carrier 5

PM signal 34, 37
Point in vector space 19, 35
Poles 58, 106–115, 119–130, 134, 135, 143–147, 282–293, 299
 approximation to 113, 114
Polynomial 106–115, 119–125, 131–136, 282–292, 362–366, 369–379
 square root of 131–133, 262, 263
Positive definite matrix 25, 208, 217, 271, 298, 344, 354
Positive semi-definite matrix 25, 272, 344, 354
Power spectral density 9, 18, 161
 inverse Fourier transform of 18
Prior knowledge 7, 28, 37, 157
Probability 28–32
 a posteriori 28, 29, 31
 a priori 28, 32
 conditional 29–31
 density 30, 31, 33, 38, 44, 45, 196, 200, 239, 251
 of error 38, 39, 44–46, 196–205, 239–253, 256, 309–320, 360, 393–408, 416–432
Projection 19, 24, 36, 47, 48, 211, 212, 280
 theorem 211, 212, 280
Pure amplitude distortion 115–130, 133, 135, 139–147
Pure nonlinear equalizer 230–253, 368–374, 433, 434
Pure phase distortion 98–115, 125–130, 133–147, 201, 246, 282–293, 348, 366, 391

QAM signal 5, 34, 37, 138, 139, 215
Q-function 38, 39, 44–46, 392
Quantized-feedback equalizer 232
Quaternary signal 216, 436

Random process 17, 18
Random variables 2, 18, 28, 29, 166, 206, 216, 256, 266, 267
 continuous 29
 discrete 29
 Gaussian 18, 28, 33, 34, 38, 45, 95, 161, 195, 196, 216, 220, 221, 232, 266, 297
 probability density of 30, 31, 33, 38, 44, 45, 196, 200, 239, 251
 sample value of 2, 29, 31

INDEX 465

statistically independent 2, 18, 28, 33, 34, 38, 45, 161, 196, 206, 216, 220, 221, 266, 267
statistically orthogonal 2, 16, 166, 206, 220, 221, 266, 267, 350, 358
uncorrelated 2, 16, 18, 38, 45, 196, 216, 221, 330, 336
variance of 18, 38, 161, 196, 198, 200, 204, 216, 220, 232, 256, 286, 297, 431
Random vector 29–31
sample value of 29, 31
Real orthogonal matrix 23–26, 67, 68, 100–102
Real symmetric matrix 24, 25, 67, 68, 208, 269, 271, 344, 354
Real-valued signal 6, 12, 13, 27, 34, 50, 85, 139, 159
Real vector 19, 34, 38, 47, 48, 59, 63
Received vector 27, 29, 34, 59, 437–441
Receiver 1, 11–13
filter 11–18
synchronization 9, 159
Rectangular or rounded baseband signal 51, 157
Rectangular pulse 427
Residual intersymbol interference 171, 174, 177, 178
Reversal theorem 84–87, 90, 363
Reverse sequence 85–87, 107–112, 118, 120, 121, 124, 134, 142–145, 149, 173, 363
Reverse z-transform 107–112, 120, 121, 124, 134, 142–145, 149, 172, 173, 282, 291, 292, 362–366
Roots of unity 26, 61, 62
Roots of z-transform 57, 58
Row vector 19

Sample value
of signal waveform 10, 57, 59
of random variable 2, 29, 31
of random vector 29
Sampled impulse-response of channel 13, 34, 52, 56, 94, 117, 139, 140, 160, 312, 338, 380, 382, 423, 425
aperiodic autocorrelation function of 96, 97, 102, 103, 148–151, 209
discrete energy density spectrum of 97, 102, 151
DFT of 61–63, 96–98, 110–112, 116–118, 121–124, 130, 131, 140–144, 149

z-transform of 55, 56, 106, 107, 131, 162, 231, 285, 339
Sampler 11–13, 51, 52
Samples 10–13, 18, 50–57, 61
Sampling instant 13
Satellite link 6
Scalar quantity 19, 21, 24
Schwarz inequality 21, 361
Scrambling 158, 230
Sequence 19, 50–56, 58–60
advanced in time 74–79, 107, 117, 122, 145, 152–154
aperiodic autocorrelation function of 87–91, 96, 97, 102, 103, 124, 148–151, 209
complex-valued 34, 37, 63, 87, 120, 124, 139–147, 217, 298
convolution of two 27, 59, 60, 70–72, 88, 107–111, 120, 121, 124, 133
conjugate reverse of 142–145
cyclically shifted 27, 73–79, 119, 152–155
delayed in time 73–79, 106, 122, 145, 152–154
DFT of 61–63, 96–98, 110–112, 116–118, 121–124, 130, 131, 140–144, 149, 152–155
discrete energy density spectrum of 87–91, 97, 99, 102, 103, 105, 151
energy of 38, 92, 98, 363
inverse 108–112, 142, 143, 149
linear transformation of 22, 59, 60, 100–105
orthogonal 20, 21, 24, 25, 36, 41, 48, 101–105, 154, 212, 218
periodic 64, 65
real-valued 27, 34, 47, 48, 50, 58, 59, 82, 94, 124
reverse of 85–87, 107–112, 118, 120, 121, 124, 132–134, 142–145, 149, 173, 364
self orthogonal 154
skew-symmetric 82–84
symmetric 82–84, 117–124
Serial system 3, 41, 50
Series
convergent 132, 133, 135, 169, 172, 190
divergent 135, 172, 190
Shape of signal element 3
Shift theorem 73–79, 87, 89, 90
Signal
AM 5
amplitude 3

466 INDEX

attenuation 98
bandwidth 4
baseband 5, 12, 13, 50
binary 3
binary polar 157, 158
cancellation 230–239
carrier 5
complex-valued 6, 12, 34, 37, 85, 138–147, 215–217, 297–299
delay 55, 98
distorted 94, 95, 98
distortion 7, 10, 50, 54, 94–155
element rate 3, 4
energy 16, 17, 38, 92, 98, 358, 363, 424, 427
equalized 161, 164, 171, 175, 177, 197, 234–236, 291
imaginary-valued 6, 85, 139
mean-square error in 40, 41, 166, 196, 205–217, 269, 298, 343, 344, 351, 354
mean-square value of 12, 16, 18, 160, 166, 196, 204–206, 216, 266, 267, 232, 297, 368, 432
m-level 3, 4, 11, 28
modulated-carrier 5
multilevel 3, 34, 37, 138, 139, 436
orthogonal 17, 20, 21, 24, 25, 36, 41, 48, 101–105, 154, 218, 358
PM 34, 37
QAM 5, 34, 37, 138, 139, 215
real-valued 6, 12, 13, 27, 34, 50, 85, 139, 159
rectangular or rounded 51, 157
sampled 10–13, 18, 50–57
spectrum 13–16
transmitted 2, 5, 12, 16, 17, 50, 51, 157, 158, 356–378, 427
vector 27, 28
waveform 1, 2
Signal element rate 3–7
Signal elements 2–6, 12, 28, 50, 139, 157–159, 230, 231
average energy of 16–18, 160, 232, 358, 368–378, 424, 427, 432
binary polar 157, 158
distorted 94, 95, 98
energy of 16, 358, 424, 427
energy spectral density of 16
orthogonal 17, 101–105, 218, 358
received 51–54, 160
shape of 3, 9
statistically orthogonal 16
uncorrelated 16

waveform of 3
Signal/noise ratio 10, 39, 40, 161, 197, 203–205, 216, 232, 234, 263, 368, 373, 401, 420, 424
Similarity transformation 25, 69
Sinusoidal roll-off 14
Skew-symmetric sequences 82–84
s-plane 57
Square root of a polynomial 131–133, 362, 363
Starting-up procedure 235, 446
Statistically independent random variables 2, 18, 28, 33, 34, 38, 45, 161, 196, 206, 216, 220, 221, 266, 267
Statistically orthogonal random variables 2, 16, 166, 206, 220, 221, 266, 267, 350, 358
Storage element 37, 164, 234
Suboptimum detector 42, 48, 435–455
Subspace 24, 36, 211, 279–281, 341
Sudden carrier-phase changes 8
Sudden signal-level changes 8
Sum of vectors 20, 340, 343, 352
Sum space 341
Suppressed-carrier AM signal 5
Symbol 3, 28, 29
complex-valued 138, 139, 216·
data 2–4, 11, 12, 28, 29, 41, 51, 157, 158, 230
Symmetric sequence 82–84, 117–124
Synchronization
of signal-carrier phase 9
of signal-element timing 9, 159
Synchronous serial system 3, 156, 157

TDM 6
Telephone circuits 5–9
Threshold 10
Threshold-level detector 10, 37, 38, 43, 44, 161, 162, 196–205, 234, 235
Time 2, 12
advance 74–79, 107, 117, 122, 145, 152–154, 175
correlation 88, 326–330
delay 55, 73–79, 98, 106, 115, 122, 123, 145, 152–154
dispersion 4, 10
invariant channel 50, 55, 158, 230, 356
modulation 8

INDEX 467

reversal in 85–87, 107–112, 118–124, 132–134, 142–145, 149, 172, 173, 282, 363–365
shift 73–79, 107
synchronization 9, 159
varying channel 7, 331–333
Tolerance to noise 195–205, 239–253, 297, 299–320, 374, 379–385, 399–408, 419–435, 446–455
Training signal 235, 446
Transfer function 13–16, 56, 70, 98, 113, 115–117, 430, 431
Transient effects 8
Transient interruptions 8
Transformation
  inverse 102–105
  linear 22, 60, 100–105
  orthogonal 23–25, 100–105, 107–112, 115, 140, 217–222, 281–293, 308, 348, 377, 391
  similarity 25, 69
  unitary 140
Transmission path 6–12, 138, 139
Transmission rate 1–7, 11
Transmitted signal 2, 5, 12, 16, 17, 50, 51, 157, 158, 356–378, 427
Transmitter 1, 12
  filter 12–17, 157, 158
Transpose of a matrix 23, 24, 81
Transpose of a vector 19
Transversal filter 37, 163, 164, 188, 190, 233
Twisted pair 5, 6, 12
Two factors of channel z-transform 131, 179, 180, 257, 285, 347, 377

Unbiased estimate 40, 46, 47, 278, 286
Uncorrelated noise samples 18, 33, 38, 161, 196, 216, 220, 221
Unitary distance 22, 37, 216
Unitary length 21, 22, 216
Unitary matrix 23–26, 82, 140
Unitary transformation 140
Unitary vector space 34
Unit circle 56, 57, 106
Unit vector 61
Unloaded audio link 6

Value of vector 29
Variance of noise component 18, 38, 161, 196, 198, 200, 204, 216, 220, 232, 256, 286, 297, 343, 351

Vectors 18–27, 47, 48, 59, 60
  angle between 20, 361
  aperiodic autocorrelation function of 87–91, 96, 97, 102, 103, 124, 148–151, 209
  column 19
  complex 21, 22, 34, 37, 63, 87, 120, 124, 139–147, 217, 298
  conjugate transpose of 21, 22
  components of 19
  continuous random 29
  convolution of 27, 59, 60, 70–72, 88, 107–111, 120, 121, 124, 133
  cost of 440–443
  cyclically shifted 27, 73–79, 119, 152–155
  DFT of 61–63, 96–98, 110–112, 116–118, 121–124, 130, 131, 140–144, 149, 152–155
  discrete energy density spectrum of 87–91, 97, 99, 102, 103, 105, 151
  discrete random 29
  energy of 38, 92, 98, 363
  Euclidean distance between 20, 33, 35, 211, 279, 438
  Euclidean length of 20, 98, 105, 196
  inner product of 20, 21
  linear combination of 23, 24, 211, 279–281, 341
  linear transformation of 22, 59, 60, 100–105
  linearly independent 23, 24, 211, 279, 341–343
  maximum-likelihood 438–443
  merging of 445
  noise 27, 28, 35, 47, 48
  orthogonal 20, 21, 24, 25, 36, 41, 48, 101–105, 154, 212, 218, 280, 281, 292
  real 19–21, 27, 34, 47, 48, 50, 58, 59, 82, 94, 124
  received 27, 29, 34, 59, 437–441
  row 19
  signal 27, 28, 47, 48, 437–446
  squared length of 20, 22
  sum of 20, 340, 343, 352
  transpose of 19
  unit 61
  unitary distance between 22, 34
  unitary length of 21, 22
  value of 29
  zero 20
Vector space 19, 24, 34, 47, 48, 211, 279, 341

Vestigial sideband signal 5
Viterbi algorithm detector 440–444
   reduced-state 444–446
Voiceband channel 5
Voiceband telephone circuit 6

Waveform 1, 3
White Gaussian noise 9, 13, 17
Wideband channel 6, 7
Wiener Kinchine theorem 18
Word 3

Zero forcing linear equalizer 167, 168, 171, 179, 180
Zero forcing decision-feedback equalizer 253–257, 297

Zeros of a z-transform 57, 58, 106–115, 119–130, 134, 135, 143–147, 282–293, 299, 362–366
z plane 56, 57, 106
z-transform 54–59, 63, 64, 104–115, 119–136, 148, 162, 231, 281–296, 369–378
   coefficients in 57, 58, 107–112, 149
   conjugate reverse of 142–145
   inverse of 106–115, 134, 142, 143, 149, 163, 165, 282, 291, 292
   of a channel 55, 56, 106, 107, 131, 162, 231, 285, 339, 347
   product of 59, 104, 107, 111, 120, 122, 124, 125, 133, 282–292
   reverse of 107–112, 120, 121, 124, 134, 142–145, 149, 172, 173, 282, 291, 292, 362–366
   square root of 131–133, 362, 363
   zeros (roots) and poles of 57, 58, 106–115, 119–130, 134, 135, 143–147, 282–293, 299, 362–366

281716

SOUTHEASTERN MASSACHUSETTS UNIVERSITY
TK7872.E7 C57 1985
Equalizers for digital modems

3 2922 00018 753 1

WITHDRAWN

DATE DUE

MAR ~~QED~~ 1987 RETURNED		
OCT 31 1988 RETURNED OCT 28 1991		
NOV 0 9 1993		
DEC 17 1995		
NOV 2 5 1995 MAY 28 1998		
MAY 14 1999		
FEB 13 1999		
FEB 18 1999		
261-2500		Printed in USA